T0310125

LTE-ADVANCED AND NEXT GENERATION WIRELESS NETWORKS

LTE-ADVANCED AND NEXT GENERATION WIRELESS NETWORKS

CHANNEL MODELLING AND PROPAGATION

Editors

Guillaume de la Roche
Mindspeed Technologies, France

Andrés Alayón Glazunov
KTH Royal Institute of Technology, Sweden

Ben Allen
University of Bedfordshire, UK

A John Wiley & Sons, Ltd., Publication

This edition first published 2013
© 2013 John Wiley and Sons Ltd

Registered office
John Wiley & Sons Ltd, The Atrium, Southern Gate, Chichester, West Sussex, PO19 8SQ, United Kingdom

For details of our global editorial offices, for customer services and for information about how to apply for permission to reuse the copyright material in this book please see our website at www.wiley.com.

Library of Congress Cataloging-in-Publication Data

LTE – advanced and next generation wireless networks : channel modelling and propagation / editors, Guillaume de la Roche, Andrés Alayón Glazunov, Ben Allen.
       p. cm.
    Includes bibliographical references and index.
    ISBN 978-1-119-97670-7 (cloth)
    1. Long-Term Evolution (Telecommunications)   I. De la Roche, Guillaume.   II. Glazunov, Andrés Alayón.   III. Allen, Ben (Benjamin Hugh)
    TK5103.48325.L734 2012
    621.39'81–dc23

                                                                        2012015856

A catalogue record for this book is available from the British Library.

ISBN: 9781119976707

Set in 10/12pt Times by Laserwords Private Limited, Chennai, India
Printed and bound in Singapore by Markono Print Media Pte Ltd

Contents

About the Editors   xv

List of Contributors   xvii

Preface   xix

Acknowledgements   xxiii

List of Acronyms   xxv

Part I   BACKGROUND

1   Enabling Technologies for 3GPP LTE-Advanced Networks   3
      Narcis Cardona, Jose F. Monserrat and Jorge Cabrejas

1.1   Introduction   4
1.2   General IMT-Advanced Features and Requirements   5
      1.2.1   Services   5
      1.2.2   Spectrum   5
      1.2.3   Technical Performance   6
1.3   Long Term Evolution Advanced Requirements   11
      1.3.1   Requirements Related with Capacity   13
      1.3.2   System Performance   13
      1.3.3   Deployment   14
1.4   Long Term Evolution Advanced Enabling Technologies   15
      1.4.1   Carrier Aggregation   15
      1.4.2   Advanced MIMO Techniques   19
      1.4.3   Coordinated Multipoint Transmission or Reception   21
      1.4.4   Relaying   23
      1.4.5   Enhancements for Home eNodeBs   26
      1.4.6   Machine-Type Communications   28
      1.4.7   Self-Optimizing Networks (SON)   29
      1.4.8   Improvements to Latency in the Control and User Plane   30
1.5   Summary   33
      References   33

2   Propagation and Channel Modeling Principles   **35**
        Andreas F. Molisch

2.1   Propagation Principles   35
        2.1.1   Free-Space Propagation and Antenna Gain   36
        2.1.2   Reflection and Transmission   36
        2.1.3   Diffraction   37
        2.1.4   Scattering   38
        2.1.5   Waveguiding   39
        2.1.6   Multipath Propagation   40
2.2   Deterministic Channel Descriptions   41
        2.2.1   Time Variant Impulse Response   42
        2.2.2   Directional Description and MIMO Matrix   44
        2.2.3   Polarization   45
        2.2.4   Ultrawideband Description   45
2.3   Stochastic Channel Description   46
        2.3.1   Pathloss and Shadowing   47
        2.3.2   Small-Scale Fading   48
        2.3.3   WSSUS   49
        2.3.4   Extended WSSUS   51
2.4   Channel Modeling Methods   51
        2.4.1   Deterministic Modeling   51
        2.4.2   Modeling Hierarchies   52
        2.4.3   Clustering   53
        2.4.4   Stochastic Modeling   56
        2.4.5   Geometry-Based Stochastic Models   58
        2.4.6   Diffuse Multipath Components   61
        2.4.7   Multi-Link Stochastic Models   61
        References   62

Part II   RADIO CHANNELS

3   Indoor Channels   **67**
        Jianhua Zhang and Guangyi Liu

3.1   Introduction   67
3.2   Indoor Large Scale Fading   69
        3.2.1   Indoor Large Scale Models   69
        3.2.2   Summary of Indoor Large Scale Characteristics   72
        3.2.3   Important Factors for Indoor Propagation   78
3.3   Indoor Small Scale Fading   83
        3.3.1   Geometry-Based Stochastic Channel Model   83
        3.3.2   Statistical Characteristics in Delay Domain   84
        3.3.3   Statistical Parameter in Angular Domain   87
        *3.3.4   Cross-Polarization Discrimination (XPD) for Indoor
                  Scenario*   88

	3.3.5	3-D Modeling for Indoor MIMO Channel	90
	3.3.6	Impact of Elevation Angular Distribution	92
	References		93

4 **Outdoor Channels** **97**
Petros Karadimas

4.1	Introduction	97
4.2	Reference Channel Model	98
4.3	Small Scale Variations	103
	4.3.1	First Order Statistical Characterization	103
	4.3.2	Second Order Statistical Characterization	106
4.4	Path Loss and Large Scale Variations	117
4.5	Summary	119
	Acknowledgements	120
	References	120

5 **Outdoor-Indoor Channel** **123**
Andrés Alayón Glazunov, Zhihua Lai and Jie Zhang

5.1	Introduction	123
5.2	Modelling Principles	124
5.3	Empirical Propagation Models	127
	5.3.1	Path Loss Exponent Model	128
	5.3.2	Path Loss Exponent Model with Mean Building Penetration Loss	128
	5.3.3	Partition-Based Outdoor-to-Indoor Model	130
	5.3.4	Path Loss Exponent Model with Building Penetration Loss	130
	5.3.5	COST 231 Building Penetration Loss Model	131
	5.3.6	Excess Path Loss Building Penetration Models	133
	5.3.7	Extended COST 231 WI Building Penetration at the LOS Condition	134
	5.3.8	WINNER II Outdoor-to-Indoor Path Loss Models	135
5.4	Deterministic Models	137
	5.4.1	FDTD	138
	5.4.2	Ray-Based Methods	138
	5.4.3	Intelligent Ray Launching Algorithm (IRLA)	141
5.5	Hybrid Models	142
	5.5.1	Antenna Radiation Pattern	142
	5.5.2	Calibration	143
	5.5.3	IRLA Case Study: INSA	144
	5.5.4	IRLA Case Study: Xinghai	149
	Acknowledgements	149
	References	149

6   Vehicular Channels   153
        Laura Bernadó, Nicolai Czink, Thomas Zemen, Alexander Paier,
        Fredrik Tufvesson, Christoph Mecklenbräuker and Andreas F. Molisch

6.1   Introduction   153
6.2   Radio Channel Measurements   154
        6.2.1   Channel Sounders   155
        6.2.2   Vehicular Antennas   157
        6.2.3   Vehicular Measurement Campaigns   158
6.3   Vehicular Channel Characterization   160
        6.3.1   Time-Variability of the Channel   160
        6.3.2   Time-Varying Vehicular Channel Parameters   166
        6.3.3   Empirical Results   169
6.4   Channel Models for Vehicular Communications   171
        6.4.1   Channel Modeling Techniques   171
        6.4.2   Geometry-Based Stochastic Channel Modeling   173
        6.4.3   Low-Complexity Geometry-Based Stochastic Channel
                  Model Simulation   177
6.5   New Vehicular Communication Techniques   180
        6.5.1   OFDM Physical (PHY) and Medium Access   180
        6.5.2   Relaying Techniques   181
        6.5.3   Cooperative Coding and Distributed Sensing   182
        6.5.4   Outlook   182
        References   182

7   Multi-User MIMO Channels   187
        Fredrik Tufvesson, Katsuyuki Haneda and Veli-Matti Kolmonen

7.1   Introduction   187
7.2   Multi-User MIMO Measurements   188
        7.2.1   General Information About Measurements   188
        7.2.2   Measurement Techniques   189
        7.2.3   Phase Noise   192
        7.2.4   Measurement Antennas   192
        7.2.5   Measurement Campaigns   193
7.3   Multi-User Channel Characterization   196
7.4   Multi-User Channel Models   200
        7.4.1   Analytical Model   200
        7.4.2   General Cluster Model   202
        7.4.3   Particular Implementation of Cluster Models   206
        References   210

8   Wideband Channels   215
        Vit Sipal, David Edward and Ben Allen

8.1   Large Scale Channel Properties   216
        8.1.1   Path Gain – Range Dependency   217
        8.1.2   Path Gain – Frequency Dependency   217

8.2	Impulse Response of UWB Channel	219
	8.2.1 Impulse Response According to IEEE802.15.4a	220
	8.2.2 Impact of Antenna Impulse Response in Free Space	221
	8.2.3 Manifestation of Antenna Impulse Response in Realistic Indoor Channels	222
	8.2.4 New Channel Model For UWB	223
	8.2.5 UWB Channel Impulse Response – Simplified Model for Practical Use	225
	8.2.6 UWB Channel Impulse Response – Conclusion	225
8.3	Frequency Selective Fading in UWB Channels	226
	8.3.1 Fade Depth Scaling	228
	8.3.2 Probability Distribution Function of Fading	232
8.4	Multiple Antenna Techniques	239
	8.4.1 Wideband Array Descriptors	239
	8.4.2 Antenna Arrays – UWB OFDM Systems	241
8.5	Implications for LTE-A	243
	References	244

9	**Wireless Body Area Network Channels**	**247**
	Rob Edwards, Muhammad Irfan Khattak and Lei Ma	
9.1	Introduction	247
9.2	Wearable Antennas	249
9.3	Analysis of Antennas Close to Human Skin	251
	9.3.1 Complex Permittivity and Equivalent Conductivity of Medium	252
	9.3.2 Properties of Human Body Tissue	253
	9.3.3 Energy Loss in Biological Tissue	256
	9.3.4 Body Effects on the Q Factor and Bandwidth of Wearable Antennas	256
9.4	A Survey of Popular On-Body Propagation Models	259
9.5	Antenna Implants-Possible Future Trends	263
9.6	Summary	265
	References	265

Part III   SIMULATION AND PERFORMANCE

10	**Ray-Tracing Modeling**	**271**
	Yves Lostanlen and Thomas Kürner	
10.1	Introduction	271
10.2	Main Physical Phenomena Involved in Propagation	272
	10.2.1 Basic Terms and Principles	273
	10.2.2 Free Space Propagation	275
	10.2.3 Reflection and Transmission	275
	10.2.4 Diffraction	276
	10.2.5 Scattering	277

10.3   Incorporating the Influence of Vegetation   277
        10.3.1   Modeling Diffraction Over the Tree Canopy   278
        10.3.2   Modeling Tree Shadowing   278
        10.3.3   Modeling Diffuse Scattering from Trees   278
10.4   Ray-Tracing Methods   280
        10.4.1   Modeling of the Environment   280
        10.4.2   Geometric Computation of the Ray Trajectories   281
        10.4.3   Direct Method or Ray-Launching   282
        10.4.4   Image Method Ray-Tracing   283
        10.4.5   Acceleration Techniques   284
        10.4.6   Hybrid Techniques   286
        10.4.7   Determination of the Electromagnetic Field Strength
                  and Space-Time Outputs   287
        10.4.8   Extension to Ultra-Wideband (UWB) Channel Modeling   287
        References   289

11   Finite-Difference Modeling   293
          Guillaume de la Roche

11.1   Introduction   293
11.2   Models for Solving Maxwell's Equations   294
        11.2.1   FDTD   295
        11.2.2   ParFlow   296
11.3   Practical Use of FD Methods   298
        11.3.1   Comparison with Ray Tracing   298
        11.3.2   Complexity Reduction   299
        11.3.3   Calibration   300
        11.3.4   Antenna Pattern Effects   301
        11.3.5   3D Approximation   302
11.4   Results   303
        11.4.1   Path Loss Prediction   303
        11.4.2   Fading Prediction   305
11.5   Perspectives for Finite Difference Models   308
        11.5.1   Extension to 3D Models   308
        11.5.2   Combination with Ray Tracing Models   309
        11.5.3   Application to Wideband Channel Modeling   314
11.6   Summary and Perspectives   314
        Acknowledgements   314
        References   314

12   Propagation Models for Wireless Network Planning   317
          Thomas Kürner and Yves Lostanlen

12.1   Geographic Data for RNP   317
        12.1.1   Terminology   318
        12.1.2   Production Techniques   319
        12.1.3   Specific Details Required for the Propagation Modeling 320

12.1.4	*Raster Multi-Resolution*	321
12.1.5	*Raster-Vector Multi-Resolution*	322
12.2	Categorization of Propagation Models	322
12.2.1	*Site-General Path Loss Models*	323
12.2.2	*Site-Specific Path Loss and Channel Models*	323
12.3	Empirical Models	325
12.3.1	*Lee's Model*	325
12.3.2	*Erceg's Model*	325
12.4	Semi-Empirical Models for Macro Cells	326
12.4.1	*A General Formula for Semi-Empirical Models*	
	for Macro Cells	327
12.4.2	*COST231-Walfisch-Ikegami-Model*	330
12.4.3	*Other Models*	332
12.5	Deterministic Models for Urban Areas	332
12.5.1	*Waveguiding in Urban Areas*	332
12.5.2	*Transitions between Heterogeneous Environments*	333
12.5.3	*Penetration Inside Buildings*	333
12.5.4	*Main Principles of Operational Deterministic Models*	333
12.5.5	*Outdoor-to-Indoor Techniques*	339
12.5.6	*Calibration of Parameters*	339
12.6	Accuracy of Propagation Models for RNP	339
12.6.1	*Measurement Campaign*	340
12.6.2	*Tuning (aka Calibration) Process*	341
12.6.3	*Model Accuracy*	343
12.7	Coverage Probability	344
	References	345
13	**System-Level Simulations with the IMT-Advanced Channel Model**	**349**
	Jan Ellenbeck	
13.1	Introduction	349
13.2	IMT-Advanced Simulation Guidelines	350
13.2.1	*General System-Level Evaluation Methodology*	350
13.2.2	*System-Level Performance Metrics*	352
13.2.3	*Test Environment and Deployment Scenario Configurations*	353
13.2.4	*Antenna Modeling*	356
13.3	The IMT-Advanced Channel Models	357
13.3.1	*Large-Scale Link Properties*	358
13.3.2	*Initialization of Small-Scale Parameters*	363
13.3.3	*Coefficient Generation*	364
13.3.4	*Computationally Efficient Time Evolution of CIRs*	
	and CTFs	365
13.4	Channel Model Calibration	366
13.4.1	*Large-Scale Calibration Metrics*	367
13.4.2	*Small-Scale Calibration Metrics*	368
13.4.3	*CIR and CTF Calibrations*	370

13.5   Link-to-System Modeling for LTE-Advanced   371
        13.5.1   System-Level Simulations vs. Link-Level Simulations   371
        13.5.2   Modeling of MIMO Linear Receiver and Precoder
                   Performance   374
        13.5.3   Effective SINR Values   376
        13.5.4   Block Error Modeling   377
13.6   3GPP LTE-Advanced System-Level Simulator Calibration   379
        13.6.1   Downlink Simulation Assumptions   381
        13.6.2   Uplink Simulation Assumptions   381
        13.6.3   Simulator Calibration Results   382
13.7   Summary and Outlook   385
        References   386

14   Channel Emulators for Emerging Communication Systems   389
        Julian Webber

14.1   Introduction   389
14.2   Emulator Systems   390
14.3   Random Number Generation   391
        14.3.1   Pseudo Random Noise Generator (PRNG)   392
        14.3.2   Gaussian Look-Up-Table   392
        14.3.3   Sum of Uniform (SoU) Distribution   392
        14.3.4   Box-Muller   393
14.4   Fading Generators   394
        14.4.1   Gaussian I.I.D.   395
        14.4.2   Modified Jakes' Model   396
        14.4.3   Zheng Model   396
        14.4.4   Random Walk Model   397
        14.4.5   Ricean K-Factor   398
        14.4.6   Correlation   399
14.5   Channel Convolution   401
14.6   Emulator Development   403
14.7   Example Transceiver Applications for Emerging Systems   403
        14.7.1   MIMO-OFDM   403
        14.7.2   Single Carrier Systems   405
14.8   Summary   407
        References   408

15   MIMO Over-the-Air Testing   411
        Andrés Alayón Glazunov, Veli-Matti Kolmonen and Tommi Laitinen

15.1   Introduction   411
        15.1.1   Problem Statement   412
        15.1.2   General Description of OTA Testing   413
15.2   Channel Modelling Concepts   414
        15.2.1   Geometry-Based Modelling   416
        15.2.2   Correlation-Based Modelling   418
15.3   DUTs and Usage Definition   418

15.4	Figures-of-Merit for OTA	419
15.5	Multi-Probe MIMO OTA Testing Methods	421
	15.5.1 Multi-Probe Systems	421
	15.5.2 Channel Synthesis	422
	15.5.3 Field Synthesis	423
	15.5.4 Two Examples of Field Synthesis Methods	426
	15.5.5 Compensation of Near-Field Effects of Probes and Range Reflections	428
15.6	Other MIMO OTA Testing Methods	429
	15.6.1 Reverberation Chambers	429
	15.6.2 Two-Stage Method	436
15.7	Future Trends	437
	References	437
16	**Cognitive Radio Networks: Sensing, Access, Security**	**443**
	Ghazanfar A. Safdar	
16.1	Introduction	443
16.2	Cognitive Radio: A Definition	443
	16.2.1 Cognitive Radio and Spectrum Management	444
	16.2.2 Cognitive Radio Networks	446
	16.2.3 Cognitive Radio and OSI	447
16.3	Spectrum Sensing in CRNs	448
	16.3.1 False Alarm and Missed Detection	449
	16.3.2 Spectrum Sensing Techniques	450
	16.3.3 Types of Spectrum Sensing	451
16.4	Spectrum Assignment–Medium Access Control in CRNs	452
	16.4.1 Based on Channel Access	452
	16.4.2 Based on Usage of Common Control Channel	453
	16.4.3 CR Medium Access Control Protocols	455
16.5	Security in Cognitive Radio Networks	461
	16.5.1 Security in CRNs: CCC Security Framework	463
	16.5.2 Security in CRNs: CCC Security Framework Steps	466
16.6	Applications of CRNs	468
	16.6.1 Commercial Applications	468
	16.6.2 Military Applications	468
	16.6.3 Public Safety Applications	468
	16.6.4 CRNs and LTE	469
16.7	Summary	470
	Acknowledgements	470
	References	470
17	**Antenna Design for Small Devices**	**473**
	Tim Brown	
17.1	Antenna Fundamentals	474
	17.1.1 Directivity, Efficiency and Gain	475
	17.1.2 Impedance and Reflection Coefficient	476

17.2   Figures of Merit and their Impact on the Propagation Channel   477
        17.2.1   Coupling and S-Parameters   477
        17.2.2   Polarization   479
        17.2.3   Mean Effective Gain   480
        17.2.4   Channel Requirements for MIMO   482
        17.2.5   Branch Power Ratio   482
        17.2.6   Correlation   483
        17.2.7   Multiplexing Efficiency   484
17.3   Challenges in Mobile Terminal Antenna Design   484
17.4   Multiple-Antenna Minaturization Techniques   485
        17.4.1   Folded Antennas   486
        17.4.2   Ferrite Antennas   487
        17.4.3   Neutralization Line   488
        17.4.4   Laptop Antennas   489
17.5   Multiple Antennas with Multiple Bands   489
17.6   Multiple Users and Antenna Effects   491
17.7   Small Cell Antennas   492
17.8   Summary   492
        References   492

18   **Statistical Characterization of Antennas in BANs**   **495**
        Carla Oliveira, Michal Mackowiak and Luis M. Correia
18.1   Motivation   495
18.2   Scenarios   496
18.3   Concepts   498
18.4   Body Coupling: Theoretical Models   500
        18.4.1   Elementary Source Over a Circular Cylinder   500
        18.4.2   Elementary Source Over an Elliptical Cylinder   505
18.5   Body Coupling: Full Wave Simulations   508
        18.5.1   Radiation Pattern Statistics for a Static Body   508
        18.5.2   Radiation Pattern Statistics for a Dynamic Body   511
18.6   Body Coupling: Practical Experiments   513
18.7   Correlation Analysis for BANs   517
        18.7.1   On-Body Communications   517
        18.7.2   Off-Body Communications   520
18.8   Summary   522
        Acknowledgements   523
        References   523

Index   **525**

About the Editors

Guillaume de la Roche is a Wireless System Engineer at Mindspeed Technologies in France. Prior to that he was with the Centre for Wireless Network Design (CWiND), University of Bedfordshire, United Kingdom (2007–2011). Before that he was with Infineon (2001–2002, Germany), Sygmum (2003–2004, France) and CITI Laboratory (2004–2007, France). He was also a visiting researcher at DOCOMO-Labs (2010, USA) and Axis Teknologies (2011, USA). He holds a Dipl-Ing from CPE Lyon, and a MSc and PhD from INSA Lyon. He was the PI of European FP7 project CWNetPlan on radio propagation for combined wireless network planning. He is a co-author of the book *Femtocells: Technologies and Deployment*, Wiley, 2010 and a guest editor of EURASIP JWCN, Special issue on Radio Propagation, Channel Modeling and Wireless Channel Simulation tools for Heterogeneous Networking Evaluation, 2011. He is on the editorial board of European Transactions on Telecommunications. He is also a part time lecturer at Lyon 1 University.

Andrés Alayón Glazunov was born in Havana, Cuba, in 1969. He received the M.Sc. (Engineer-Researcher) degree in physical engineering from the Saint Petersburg State Polytechnic University, Russia and the PhD degree in electrical engineering from Lund University, Lund, Sweden, in 1994 and 2009, respectively. He has held research positions in both the industry and academia. Currently, he holds a Postdoctoral Research Fellowship at the Electromagnetic Engineering Lab, the KTH Royal Institute of Technology, Stockholm, Sweden. From 1996 to 2001, he was a member of the Research Staff at Ericsson Research , Sweden. In 2001, he joined Telia Research, Sweden, as a Senior Research Engineer. From 2003 to 2006 he held a position as a Senior Specialist in Antenna Systems and Propagation at TeliaSonera Sweden. He has actively contributed to international projects such as the European COST Actions 259 and 273, the EVEREST and NEW-COM research projects. He has also been involved in work within the 3GPP and the ITU standardization bodies. His research interests include the combination of statistical signal processing techniques with electromagnetic theory with a focus on antenna-channel interactions, RF propagation channel measurements and simulations and advanced numerical tools for wireless propagation predictions. Dr Alayón Glazunov was awarded a Marie Curie Research Fellowship from the Centre for Wireless Network Design at the University of Bedfordshire, UK, from 2009 to 2010. He is a senior member of the IEEE.

Ben Allen is head of the Centre of Wireless Research at the University of Bedfordshire. He received his PhD from the University of Bristol in 2001, then joined Tait Electronics Ltd, New Zealand, before becoming a Research Fellow and member of academic staff with the Centre for Telecommunications Research, Kings College London, London. Between 2005 and 2010, he worked within the Department of Engineering Science at the University of Oxford. Ben is widely published in the area of wireless systems, including two previous books. He has an established track record of wireless technology innovation that has been built up through collaboration between industry and academia. His research interests include wideband wireless systems, antennas, propagation, waveform design and energy harvesting. Professor Allen is a Chartered Engineer, Fellow of the Institution of Engineering and Technology, Senior Member of the IEEE and a Member of the editorial board of the IET Microwaves, Antennas, and Propagation Journal. He has received several awards for his research.

List of Contributors

Ben Allen, University of Bedfordshire, UK

Laura Bernadó, Forschungszentrum Telekommunikation Wien, Austria

Tim Brown, University of Surrey, UK

Jorge Cabrejas, Universitat Politècnica de València, Spain

Narcis Cardona, Universitat Politècnica de València, Spain

Luis M. Correia, IST/IT – Technical University of Lisbon, Portugal

Nicolai Czink, Forschungszentrum Telekommunikation Wien, Austria

Guillaume de la Roche, Mindspeed Technologies, France

David Edward, University of Oxford, UK

Rob Edwards, Loughborough University, UK

Jan Ellenbeck, Technische Universität München, Germany

Andrés Alayón Glazunov, KTH Royal Institute of Technology, Sweden

Katsuyuki Haneda, Aalto University, Finland

Petros Karadimas, University of Bedfordshire, UK

Muhammad Irfan Khattak, NWFP University of Engineering and Technology, Pakistan

Veli-Matti Kolmonen, Aalto University, Finland

Thomas Kürner, Technische Universität Braunschweig, Germany

Zhihua Lai, Ranplan Wireless Network Design Ltd, UK

Tommi Laitinen, Aalto University, Finland

Guangyi Liu, China Mobile, China

Yves Lostanlen, University of Toronto, Canada

Lei Ma, Loughborough University, UK

Christoph Mecklenbräuker, Vienna University of Technology, Austria

Andreas F. Molisch, University of Southern California, USA

Jose F. Monserrat, Universitat Politècnica de València, Spain

Michal Mackowiak, IST/IT – Technical University of Lisbon, Portugal

Carla Oliveira, IST/IT – Technical University of Lisbon, Portugal

Alexander Paier, Austria

Ghazanfar A. Safdar, University of Bedfordshire, UK

Vit Sipal, University of Oxford, UK

Fredrik Tufvesson, Lund University, Sweden
Julian Webber, Hokkaido University, Japan
Thomas Zemen, Forschungszentrum Telekommunikation Wien, Austria
Jianhua Zhang, Beijing University of Posts and Telecommunications, China
Jie Zhang, University of Sheffield, UK

Preface

In the nineteenth century, scientists, mathematician, engineers and innovators started investigating electromagnetism. The theory that underpins wireless communications was formed by Maxwell. Early demonstrations took place by Hertz, Tesla and others. Marconi demonstrated the first wireless transmission. Since then, the range of applications has expanded at an immense rate, together with the underpinning technology. The rate of development has been incredible and today the level of technical and commercial maturity is very high. This success would not have been possible without understanding radio-wave propagation. This knowledge enables us to design successful systems and networks, together with waveforms, antennal and transceiver architectures. The radio channel is the cornerstone to the operation of any wireless system.

Today, mobile networks support millions of users and applications spanning voice, email, text messages, video and even 3G images. The networks often encompass a range of wireless technologies and frequencies all operational in very diverse environments. Examples are: Bluetooth personal communications that may be outside, indoors or in a vehicle; wireless LAN in buildings, femtocell, microcell and macrocell sites; wireless back-haul; and satellite communications. Examples of emerging wireless technologies include body area networks for medical or sensor applications; ultra wideband for extremely high data rate communications and cognitive radio to support efficient and effective use of unused sections of the electromagnetic spectrum.

Mobile device usage continues to grow with no decrease in traffic flow. Most of the current cellular networks are now in their third generation (3G). Based on Universal Mobile Telecommunication System (UMTS) or Code Division Multiple Access (CDMA), they support data rates of a few megabits per second under low-mobility conditions. During the last few years, the number of cell phones has dramatically increased as wireless phones have become the preferred mode of communication, while landline access has decreased. Moreover, most new wireless devices like smart phones, tablets and laptops include 3G capabilities. That is why new applications are proposed every year and it is now common to use mobile devices not only for voice but also for data, video, and so on.

The direct consequence of this is that the amount of wireless data that cellular networks must support is exploding. For instance, Cisco recently noted in its Visual Networking Index (VNI) Global Mobile Data Forecast that a smart phone generates, on average, 24 times more wireless data than a plain vanilla cell phone. The report also noted that a tablet generates 122 times more wireless data than a feature phone, and a wireless laptop creates 515 times the wireless data traffic of traditional cell phones. Hence in 2009, the International Telecommunication Union – Radiocommunication Sector (ITU-R) organization specified the International Mobile Telecommunication Advanced

(IMT-A) requirements for 4G standards, setting peak speed requirements for 4G service at 100 Mbit/s for high mobility communication (such as from trains and cars) and 1 Gbit/s for low mobility communication (such as pedestrians and stationary users). The main candidate to 4G is the so called Long Term Evolution Advanced which is expected to be released in 2012. Unlike the first Long Term Evolution (LTE) deployments (Rel 8 or Rel 9) which do not fully meet the 4G requirements, LTE-Advanced is supposed to surpass these requirements. That is why LTE-Advanced and beyond networks introduce new technologies and techniques (Multiple antennas, larger bandwidth, OFDMA, and so on) whose aim is to help reach very high capacity even in mobility conditions. 4G and beyond network are not deployed yet, however most of industry and researchers focus on developing new products, algorithms, solutions and applications. Like all wireless networks the performance of 4G and beyond networks depend for a major part on the channel, that is, how the signal propagates between emitters and users. That is why channel modelling and propagation, which is sometimes seen as an old topic, is very important and must have full consideration. Indeed, in order to study the performance of future wireless networks, it is very important to be able to characterize the wireless channel into different scenarios and and to be able to take into account the new situations introduced by future networks such as multiple antennas that can be embedded in high speed cars or worn directly on the body.

Propagation and Channel Models

This book presents an overview of models of how the channel will behave in different scenarios, and how to use these channel models to study the performance of 4G and beyond networks. 4G is imminent, so we believe it is good timing to have a book on channel propagation for these aspects. Moreover, future wireless networks will never stop using larger bandwidth, higher frequencies, more antennas, so this book is not only focused on 4G but on beyond 4G networks as well, where new concepts like cognitive radio or heterogeneous will be ever more important.

This book is divided into three parts as follows:

- This first part includes all the basics necessary to understand the remainder of the book. Therefore the next chapter presents LTE Advanced standard and the new technologies it introduced in order to achieve high data rate and low latency. In particular we will see in this chapter that LTE-Advanced will have to support more antennas, larger bandwidth, more cells and different scenarios compared to traditional cellular networks. Then Chapter 2 will explain the principle of channel modelling and radio propagation as well as the main important concepts and theory.
- The second part of this book details the properties of the radio channel in main scenarios suitable for 4G and beyond wireless networks. First, Chapter 3 discusses the indoor radio channel, which is ever more important when simulating indoor small NodeBs or relays. The following chapter (Chapter 4) focuses on outdoor wireless environments and gives a detailed study of how the spatial and temporal variations occur due to outdoor propagation mechanisms. In LTE-Advanced and beyond cellular networks it is expected that there will be important interactions between indoor and outdoorcells which will lead to interference if the resources are not properly allocated. That is why outdoor to indoor models are also important and will be discussed in Chapter 5. 4G

networks suppose that high mobility users can still expect very high performance,that is why mobility is important to model in LTE-Advanced. Hence, Chapter 6 focuses on vehicular channel models. Moreover, it is also proposed in Long Term Evolution Advanced (LTE-A) and beyond to use more antennas at both emitter and receiver side, and to use larger bandwidth which is referred to as Carrier Aggregation. Hence, Chapter 7 will detail the MIMO channels followed by a description of Wideband channels in Chapter 8. In the future it is also expected that antennas will be deployed directly on or even inside the human body. Hence, Chapter 9 deals with the challenges related to channels for Body Area Networks (BANs).

- After this detailed presentation of the different radio channels for future networks, the last part of this book focuses more on the application of these models from the point of view of performance analysis, simulation, antenna and measurements. One important factor when studying the performance of wireless networks is to use the knowledge on the channel in order to develop accurate models. Hence, Chapter 10 presents the theory and application of ray tracing models which can accurately compute all reflections and diffractions in any given scenarios. Then, Chapter 11 will present an alternative to ray tracing, which is based on FDTD methods, leading to high accuracy. It will also present the challenges that need to be overcome before it can be used for larger and more realistic scenarios. The mainrole for accurate propagation models like ray tracing of Finite-Difference Time-Domain (FDTD) is to be applied to wireless network planning. Hence, Chapter 12 deals with all the wireless network planning, as well as the models, applications and techniques for developing a wireless network planning tool. Simulating the performance of wireless networks requires not only having a good knowledge of the path loss, fading, and so on, but also being able to evaluate the performance of the users in terms of throughput.

That is why Chapter 13 focuses on the use of channel models for performing system level simulations. In more detail it focuses on the IMT-Advanced model which is the model proposed by 3GPP for LTE-Advanced. If software solutions can be a good way to simulate the channel, another alternative is to use channel emulators. Those will be investigated in Chapter 14. For all the channels presented in this book, it is important to consider how to perform measurement and calibratethe models accurately. If most of the chapters present results based on measurements, Chapter 15 focuses on over the air MIMO measurement, which is the most challenging type of measurement and is currently highly regarded by many researchers because multi-user MIMO is a key technology in 4G and beyond networks. Then, Chapter 16 presents different topics related to cognitive radio, which will also play a strong role in future communication systems. If there is one important consideration when studying the performance, it is to take into account the antenna aspects which have a strong interaction with the radio channel. That is why the two last chapters will present the antenna aspects related to future networks. First Chapter 17 will present all the challenges when designing small antennas for a LTE-A system. Finally, Chapter 18 will focus on antennas for BANs and more especially how to perform statistical characterization of antennas in such an environment.

For more information, please visit the companion website – www.wiley.com/go/delaroche_next.

Acknowledgements

As editors of this book, we would first like to express our sincere gratitude to our esteemed and knowledgeable co-authors, without whom this book would not have been accomplished. It is their time and dedication spent on this project that has facilitated the timeliness and high quality of this book. We extend a immensely grateful thank you to all our contributors, from many countries (including Austria, Canada, China, Finland, France, Germany, Pakistan, Portugal, Spain, Sweden, USA and UK) who accepted to share their expertise and contributed to make this book happen – thank you!

We would like to thank Wiley staff and more in particular Anna Smart and Susan Barclay for their help and encouragement during the publication process of this book.

Guillaume de la Roche is very grateful to his family and friends for their support during the time devoted to compiling this book. He also wishes to say thank you to his previous colleagues and more in particular Prof. Jean Marie Gorce for introducing him to the world of radio propagation and Prof Jie Zhang for letting him continue to do research in this area.

Andrés Alayón Glazunov wishes to thank his mother Louise for her encouragement to always pursue his dreams, his children Amanda and Gabriel for being his most precious treasures and his wife Alina for her wonderful love and support. Andrés also wishes to thank his current and former colleagues at KTH Royal Institute of Technology, University of Bedfordshire, Lund University, TeliaSonera/Telia Research and Ericsson Research for the valuable intellectual interactions on wireless propagation and antenna research that have made this project come true

Ben Allen wishes to thank his family, Louisa, Nicholas and Bethany, for their understanding of the dedication and time required for this project. Ben also wishes to thank colleagues at the University of Bedfordshire for making a stimulating and fulfilling work environment that enables works such as this to be possible, and to thank all those who he has collaborated with for making the wireless research community what it is.

List of Acronyms

2D	Two-dimensional
3D	Three-dimensional
3GPP	3rd Generation Partnership Project
3G	Third Generation
4G	Fourth Generation
AAA	Authentication, Authorization and Accounting
ABS	Almost Blank Subframe
ACIR	Adjacent Channel Interference Rejection ratio
ACK	Acknowledgement
ACL	Allowed CSG List
ACLR	Adjacent Channel Leakage Ratio
ACPR	Adjacent Channel Power Ratio
ACS	Adjacent Channel Selectivity
AD	Analog/Digital
ADSL	Asymmetric Digital Subscriber Line
AF	Amplify-and-Forward
AGCH	Access Grant Channel
AH	Authentication Header
AKA	Authentication and Key Agreement
AMC	Adaptive Modulation and Coding
AMPS	Advanced Mobile Phone System
ANN	Artificial Neural Network
ANR	Automatic Neighbor Relation
AOA	Angle-of-Arrival
AOD	Angle-of-Departure
API	Application Programming Interface
APS	Angular Power Spectrum
ARFCN	Absolute Radio Frequency Channel Number
ARQ	Automatic Repeat Request
ASA	Angle Spread of Arrival
ASD	Angle Spread of Departure
AS	Access Stratum
ASE	Area Spectral Efficiency
ASN	Access Service Network

ATM	Asynchronous Transfer Mode
AUC	Authentication Centre
AWGN	Additive White Gaussian Noise
BAN	Body Area Network
BCCH	Broadcast Control Channel
BCH	Broadcast Channel
BCU	Body Central Unit
BE	Best Effort
BF	Beacon Management Frame
BER	Bit Error Rate
BR	Beacon Management Frame
BLER	BLock Error Rate
BP	BandPass
BPSK	Binary Phase-Shift Keying
BPR	Branch Power Ratio
BR	Bit Rate
BS	Base Station
BSC	Base Station Controller
BSIC	Base Station Identity Code
BSS	Blind Source Separation
BTS	Base Transceiver Station
CAC	Call Admission Control
CAM	Cooperative Awareness Message
CAPEX	CAPital EXpenditure
CAZAC	Constant Amplitude Zero Auto-Correlation
CC	Chase Combining
CCCH	Common Control Channel
CCDF	Complementary Cumulative Distribution Function
CCPCH	Common Control Physical Channel
CCTrCH	Coded Composite Transport Channel
CDF	Cumulative Distribution Function
CDM	Code Division Multiplexing
CDMA	Code Division Multiple Access
CGI	Cell Global Identity
CH-SEL	Channel Selection
CH-RES	Channel Reservation
CID	Connection Identifier
CIF	Carrier Indicator Field
CIR	Channel Impulse Response
CN	Core Network
CoC	Component Carrier
CoMP	Coordinated Multipoint transmission or reception
CORDIC	Coordinate Rotational Digital Computer
CP	Cyclic Prefix
CPCH	Common Packet Channel
CPE	Customer Premises Equipment

CPICH	Common Pilot Channel
CPU	Central Processing Unit
CQI	Channel Quality Indicator
CR	Cognitive Radio
CRC	Cyclic Redundance Check
CRN	Cognitive Radio Network
CRS	Channel state information Reference Signal
CSA	Concurrent Spectrum Acces
CS/CB	Coordinated Scheduling and Beamforming
CSG ID	CSG Identity
CSG	Closed Subscriber Group
CSI	Channel State Information
CSI-RS	Channel State Information - Reference Signal
CSMA/CA	Carrier-Sense Multiple Access with Collision Avoidance
CSMA	Carrier-Sense Multiple Access
CTCH	Common Traffic Channel
CTF	Channel Transfer Function
CTS	Clear To Send
CW	Continuous Wave
CWiND	Centre for Wireless Network Design
DAS	Distributed Antenna System
DCCH	Dedicated Control Channel
DCH	Dedicated Channel
DCI	Data Control Indicator
DCS	Digital Communication System
DDH-MAC	Dynamic Decentralized Hybrid MAC
DEM	Digital Elevation Model
DI	Diffuse
DF	Decode-and-Forward
DFP	Dynamic Frequency Planning
DFT	Discrete Fourier Transform
DHM	Digital Height Model
DL	DownLink
DLU	Digital Land Usage
DM RS	Demodulation Reference Signal
DoS	Denial of Service
DoA	Direction of Arrival
DoD	Direction of Departure
DPCCH	Dedicated Physical Control Channel
DPDCH	Dedicated Physical Data Channel
DRX	Discontinuous Reception
DPSS	Discrete Prolate Spheroidal Sequences
DSA	Dynamic Spectrum Access
DS	Delay Spread
DSCH	Downlink Shared Channel
DSD	Doppler Power Spectra Density

DSL	Digital Subscriber Line
DSP	Digital Signal Processor
DFTS-OFDM	DFT Spread-OFDM
DTCH	Dedicated Traffic Channel
DTM	Digital Terrain Model
DUT	Device Under Test
DXF	Drawing Interchange Format
E-SDM	Eigenbeam Space Division Multiplexing
EAB	Extended Access Barring
EAGCH	Enhanced uplink Absolute Grant Channel
EAP	Extensible Authentication Protocol
ECRM	Effective Code Rate Map
EDCH	Enhanced Dedicated Channel
EESM	Exponential Effective SINR Mapping
EHICH	EDCH HARQ Indicator Channel
EIR	Equipment Identity Register
EIRP	Equivalent Isotropically Radiated Power
EM	Electromagnetic Model
EMC	Electromagnetic Compatibility
EMD	IEEE 802.16m Evaluation Methodology Document
EMI	Electromagnetic Interference
EMS	Enhanced Messaging Service
eNB	Evolved NodeB
EPC	Enhanced Packet Core
EPLMN	Equivalent PLMN
ERGCH	Enhanced uplink Relative Grant Channel
ertPS	Extended real time Polling Service
ESP	Encapsulating Security Payload
ETSI	European Telecommunications Standards Institute
EUTRA	Evolved UTRA
EUTRAN	Evolved UTRAN
EVDO	Evolution-Data Optimized
FACCH	Fast Associated Control Channel
FACH	Forward Access Channel
FAP	Femtocell Access Point
FCC	Federal Communications Commission
FCFS	First Come First Served
FCCH	Frequency-Correlation Channel
FCH	Frame Control Header
FCL	Free Channel List
FCS	Frame Check Sequence
FD	Finite Difference
FDD	Frequency Division Duplexing
FDE	Frequency Domain Equalization
FDM	Frequency Division Multiplexing
FDTD	Finite-Difference Time-Domain

FEC	Forward Error Correction
FEM	Finite Element Method
FFRS	Fractional Frequency Reuse Scheme
FFT	Fast Fourier Transform
FGW	Femto Gateway
FIFO	First In First Out
FIR	Finite Impulse Response
FIT	Finite Integration Technique
FMC	Fixed Mobile Convergence
FOM	Figure Of Merit
FPGA	Field Programmable Gate Array
FR-4	Fibreglass Epoxy Resin
FRS	Frequency Reuse Scheme
FSA	Fixed Spectrum Assignment
FTP	File Transfer Protocol
FUSC	Full Usage of Subchannels
GAN	Generic Access Network
GANC	Generic Access Network Controller
GCCC	Global Common Control Channel
GBR	Guaranteed Bit Rate
GERAN	GSM EDGE Radio Access Network
GGSN	Gateway GPRS Support Node
GMSC	Gateway Mobile Switching Center
GO	Geometrical Optics
GPRS	General Packet Radio Service
GPS	Global Positioning System
GPU	Graphics Processing Unit
GSCM	Geometry-based Stochastic Channel Model
GSM	Global System for Mobile communication
GTD	Geometrical Theory of Diffraction
HARQ	Hybrid Automatic Repeat request
HBS	Home Base Station
HCS	Hierarchical Cell Structure
HDFP	Horizontal Dynamic Frequency Planning
HeNB	Home eNodeB
HII	High Interference Indication
HLR	Home Location Register
HNB	Home NodeB
HNBAP	Home NodeB Application Protocol
HNBGW	Home NodeB Gateway
HPLMN	Home PLMN
HR	High Resolution
HRD	Horizontal Reflection and Diffraction
HSCA	Horn Shaped self-Complementary Antenna
HSDPA	High Speed Downlink Packet Access
HSDSCH	High Speed DSCH

HSPA	High Speed Packet Access
HSS	Home Subscriber Server
HSUPA	High Speed Uplink Packet Access
HUA	Home User Agent
I2V	Infrastructure-to-Vehicle
IC	Interference Cancellation
ICI	Intercarrier Interference
I-CI	Inter-Cell Interference
ICNIRP	International Commission on Non-Ionizing Radiation Protection
ICS	IMS Centralised Service
IDFT	Inverse Discrete Fourier Transform
IEEE	Institute of Electrical & Electronics Engineers
IETF	Internet Engineering Task Force
IFFT	Inverse Fast Fourier Transform
IKE	Internet Key Exchange
IKEv2	Internet Key Exchange version 2
ILP	Integer Linear Programming
IMEI	International Mobile Equipment Identity
IMS	IP Multimedia Subsystem
IMSI	International Mobile Subscriber Identity
IMT	International Mobile Telecommunication
IMT-A	International Mobile Telecommunication Advanced
InH	Indoor Hotspot
IO	Interacting Object
IOI	Interference Overload Indication
IP	Internet Protocol
IPsec	Internet Protocol Security
IR	Incremental Redundancy
IRLA	Intelligent Ray Launching
ISB	Incident Shadow Boundary
ISD	Inter-Site Distance
ISI	Intersymbol Interference
ITS	Intelligent Transportation System
ITU	International Telecommunication Union
ITU-R	International Telecommunication Union – Radiocommunication Sector
IWF	IMS Interworking Function
JP	Joint Processing
KPI	Key Performance Indicator
LA	Location Area
LAC	Location Area Code
LAN	Local Area Network
LAI	Location Area Identity
LAU	Location Area Update
LFSR	Linear Feedback Shift Register
LIDAR	Light Detection And Ranging

LIPA	Local IP Access
LLS	Link-Level Simulation
LOS	Line Of Sight
LR	Low Resolution
LSF	Local Scattering Function
LTE	Long Term Evolution
LTE-A	Long Term Evolution-Advanced
LTI	Linear Time-Invariant
LUT	Look Up Table
M2M	Machine-to-Machine
MAC	Medium Access Control
MAP	Media Access Protocol
MaxI	Maximum Insertion
MaxR	Maximum Removal
MBMS	Multicast Broadcast Multimedia Service
MBS	Macrocell Base Station
MBSFN	Multicast-Broadcast Single-Frequency Network
MC	Modulation and Coding
MCS	Modulation and Coding Scheme
MD	Mobile-Discrete
MEG	Mean Effective Gain
MGW	Media Gateway
MIB	Master Information Block
MIC	Mean Instantaneous Capacity
MIESM	Mutual Information Effective SINR Mapping
MIMO	Multiple Input Multiple Output
MinI	Minimum Insertion
MinR	Minimum Removal
MIP	Mixed Integer Program
MM	Mobility Management
MME	Mobility Management Entity
MMIB	Mean Mutual Information per Bit
MMSE	Minimum Mean Square Error
MNC	Mobile Network Code
MNO	Mobile Network Operator
MO	Main Obstacle
MOM	Method Of Moments
MPC	Multipath Component
MR	Measurement Report
MRC	Maximum Ratio Combining
MR-FDPF	Multi Resolution Frequency Domain Parflow
MRTD	Multi Resolution Time Domain
MS	Mobile Station
MSC	Mobile Switching Center
MSISDN	Mobile Subscriber Integrated Services Digital Network Number
MTC	Machine-Type Communications

MU-MIMO	Multi-User MIMO
MUSIC	MUltiple Signal Identification and Classification
NACK	Negative Acknowledgement
NAS	Non Access Stratum
NAV	Network Allocation Vector
NCL	Neighbor Cell List
NDI	New Data Indicator
NGMN	Next Generation Mobile Networks
NIR	Non-Ionisation radiation
NLOS	Non-Line Of Sight
nrtPS	non-real-time Polling Service
NR	Neighbor Relation
NRT	Neighbor Relation Table
NSS	Network Switching Subsystem
NTP	Network Time Protocol
ntp	Network Time Ptotocol
NWG	Network Working Group
OAM	Operations and Maintenance
O2I	Outdoor-to-Indoor
O2V	Outdoor-to-Vehicle
OBU	OnBoard Unit
OC	Optimum Combining
OCC	Orthogonal Cover Code
OCXO	Oven Controlled Oscillator
OFDM	Orthogonal Frequency Division Multiplexing
OFDMA	Orthogonal Frequency Division Multiple Access
OLOS	Obstructed Line Of Sight
OPEX	OPerational EXpenditure
OSA	Opportunistic Spectrum Access
OSI	Open Systems Interconnection
OSS	Operation Support Subsystem
OTA	Over The Air
PAS	Power Azimuth Spectrum
P2MP	Point-to-Multi-Point
P2P	Point-to-Point
PAPR	Peak-to-Average Power Ratio
PC	Power Control
PCCH	Paging Control Channel
PCCPCH	Primary Common Control Physical Channel
PCFICH	Physical Control Format Indicator Channel
PCH	Paging Channel
PCI	Physical Cell Identity
PCPCH	Physical Common Packet Channel
PCPICH	Primary Common Pilot Channel
PDCCH	Physical Downlink Control Channel
PDCP	Packet Data Convergence Protocol

PDF	Probability Density Function
PDP	Power Delay Profile
PDSCH	Physical Downlink Shared Channel
PDU	Packet Data Unit
PES	Power Elevation Spectrum
PF	Proportional Fair
PhD	Doctor of Philosophy
PHICH	Physical Hybrid ARQ Indicator Channel
PHY	Physical
PIC	Parallel Interference Cancellation
PICA	Planar Inverted Cone Antenna
PIFA	Planar Inverted-F Antenna
PKI	Public Key Infrastructure
PLMN ID	PLMN Identity
PLMN	Public Land Mobile Network
PMI	Precoding-Matrix Indicator
PML	Perfect Matched Layer
PN	Pseudorandom Noise
PO	Physical Optics
PoC	Push-to-talk over Cellular
PRACH	Physical Random Access Channel
PRB	Physical Resource Block
PRNG	Pseudo Random Noise Generator
PSC	Primary Scrambling Code
P-SCH	Primary Synchronization Channel
PSSD	Pseudo Spectral Spatial Domain
PSTD	Pseudo Spectral Time Domain
PSTN	Public Switched Telephone Network
PU	Primary User
PUCCH	Physical Uplink Control Channel
PUSC	Partial Usage of Subchannels
PUSCH	Physical Uplink Shared Channel
PWD	Pattern Weighted Difference
QAM	Quadrature Amplitude Modulation
QoS	Quality of Service
QPSK	Quadrature Phase Shift Keying
RAB	Radio Access Bearer
RAC	Routing Area Code
RACH	Random Access Channel
RADIUS	Remote Authentication Dial-In user Services
RAM	Random Access Memory
RAN	Radio Access Network
RANAP	Radio Access Network Application Part
RAT	Radio Access Technology
RB	Resource Block
RBN	Radio Body Network

RC	Reverberation Chamber
RE	Resource Element
RF	Radio Frequency
RFP	Radio Frequency Planning
RI	Random Insertion
RLC	Radio Link Control
RMa	Rural Macrocell
RMS	Root Mean Square
RMSE	Root Mean Square Error
RNC	Radio Network Controller
RNP	Radio Network Planning
RNTP	Relative Narrowband Transmit Power
RPLMN	Registered PLMN
ROM	Read Only Memory
RPT	Radio Planning Tool
RR	Random Removal
RRC	Radio Resource Control
RRPS	Ranplan Radiowave Propagation Simulator
RS	Reference Signal
RSB	Reflection Shadow Boundary
RSRP	Reference Signal Received Power
RSRQ	Reference Signal Received Quality
RSU	RoadSide Unit
RT	Ray Tracing
RTS	Ready To Send
RTL	Register Transfer Level
RRM	Radio Resource Management
RTP	Real Time Transport
rtPS	real-time Polling Service
RV	Redundancy Version
Rx	Receiver
RX	Receiver
SA	Simulated Annealing
SACCH	Slow Associated Control Channel
SAE	System Architecture Evolution
SAEGW	System Architecture Evolution Gateway
SAGE	Space-Alternating Generalized Expectation Maximization
SAIC	Single Antenna Interference Cancellation
SAP	Service Access Point
SAR	Specific Absorption Rate
SBR	Shooting and Bouncing Rays
SCCPCH	Secondary Common Control Physical Channel
SC-FDMA	Single Carrier Frequency Division Multiple Access
SC-FDE	Single Carrier Frequency Domain Equalization
SCH	Synchronization Channel
SCM	Spatial Channel Model

SCME	Spatial Channel Model Extended
SCTP	Stream Control Transmission Protocol
SD	Static-Discrete
SDCCH	Standalone Dedicated Control Channel
SDU	Service Data Unit
SDM	Spatial Division Multiplexing
SFC	Scattered Field Chamber
SFID	Service Flow Identifier
SG	Signalling Gateway
SGSN	Serving GPRS Support Node
SI	State Insertion
SIB	System Information Block
SIC	Successive Interference Cancellation
SIFS	Short Inter Frame Space
SIGTRAN	Signaling Transport
SIM	Subscriber Identity Module
SIMO	Single Input Multiple Output
SINR	Signal to Interference plus Noise Ratio
SIP	Session Initiated Protocol
SIPTO	Selected IP Traffic Offload
SIR	Signal to Interference Ratio
SISO	Single Input Single Output
SLS	System-Level Simulation
SMa	Suburban Macro
SMS	Short Message Service
SNMP	Simple Network Management Protocol
SNR	Signal to Noise Ratio
SOHO	Small Office/Home Office
SON	Self-Organizing Network
SORTD	Spatial Orthogonal-Resource Transmit Diversity
SoU	Sum of Uniform
S-SCH	Secondary Synchronization Channel
SRS	Sounding Reference Signal
SSL	Secure Socket Layer
SU	Secondary User
SU-MIMO	Single-User MIMO
SUI	Stanford University Interim
TA	Tracking Area
TACS	Total Access Communications System
TAI	Tracking Area Identity
TCI	Target Cell Identifier
TDL	Tapped-Delay Line
TAU	Tracking Area Update
TCH	Traffic Channel
TCXO	Temperature Controlled Oscillator
TDD	Time Division Duplexing

TE	Transverse Electric
TIS	Total Isotropic Sensitivity
TDMA	Time Division Multiple Access
TLM	Transmission Line Matrix
TLS	Transport Layer Security
TNL	Transport Network Layer
TP	ThroughPut
TPM	Trusted Platform Module
TS	Tabu Search
TSG	Technical Specification Group
TTG	Transmit/Receive Transition Gap
TTI	Transmission Time Interval
TM	Transverse Magnetic
TRP	Total Radiated Power
TRS	Total Radiated Sensitivity
TV	Television
TX	Transmitter
UARFCN	UTRA Absolute Radio Frequency Channel Number
UDP	User Datagram Protocol
UE	User Equipment
UGS	Unsolicited Grant Service
UICC	Universal Integrated Circuit Card
UK	United Kingdom
UL	UpLink
ULA	Uniform Linear Array
UMA	Unlicensed Mobile Access
UMa	Urban Macrocell
UMi	Urban Microcell
UMTS	Universal Mobile Telecommunication System
URV	Uniform Random Variable
US	Uncorrelated Scattering
USIM	Universal Subscriber Identity Module
UTD	Uniform Theory of Diffraction
UTRA	UMTS Terrestrial Radio Access
UTRAN	UMTS Terrestrial Radio Access Network
UWB	Ultra Wide Band
V2I	Vehicle-to-Infrastructure
V2V	Vehicle-to-Vehicle
VD	Vertical Diffraction
VDFP	Vertical Dynamic Frequency Planning
VHDL	Very High speed integrated circuits hardware Description Language
VLR	Visitor Location Register
VM	Visibility Mask
VoIP	Voice over IP
VPLMN	Visited PLMN
WA	Wearable Antenna

WAVE	Wireless Access in Vehicular Environments
WBAN	Wireless Body area network
WBSN	Wireless Body Sensor Network
WCDMA	Wideband Code Division Multiple Access
WEP	Wired Equivalent Privacy
WG	Working Group
WGNG	White Gaussian Noise Generated
WHO	World Health Organization
WiFi	Wireless Fidelity
WiMAX	Wireless Interoperability for Microwave Access
WINNER	Wireless World Initiative New Radio
WLAN	Wireless Local Area Network
WMTS	Wireless Medical Telemetry Service
WRC	World Radiocommunication Conference
WMAN	Wireless Metropolitan Area Network
WSS	Wide Sense Stationary
WSSUS	Wide Sense Stationary Uncorrelated Scattering
XPR	Cross Polar Ratio
ZF	Zero Forcing

Part One

Background

1

Enabling Technologies for 3GPP LTE-Advanced Networks

Narcis Cardona, Jose F. Monserrat and Jorge Cabrejas
Universitat Politècnica de València, Spain

The specifications of Long Term Evolution (LTE) in 3rd Generation Partnership Project (3GPP) (Release 8) were just finished when work began on the new Long Term Evolution Advanced (LTE-A) standard (Release 9 and beyond). LTE-A meets or exceeds the requirements imposed by International Telecommunication Union (ITU) to Fourth Generation (4G) mobile systems, also called International Mobile Telecommunication Advanced (IMT-A). These requirements were unthinkable a few years ago, but are now a reality. Peak data rates of 1 Gbps with bandwidths of 100 MHz for the downlink, very low latency, more efficient interference management and operational cost reduction are clear examples of why LTE-A is so appealing for operators. Moreover, the quality breakthrough affects not only operators but also end users, who are going to experience standards of quality similar to optical fiber. To reach these levels of capacity and quality, the international scientific community, in particular the 3GPP, are developing different technological enhancements on LTE. The most important technological proposals for LTE-A are: support of wider bandwidth (carrier aggregation), advanced Multiple Input Multiple Output (MIMO) techniques, Coordinated Multipoint transmission or reception (CoMP), relaying, enhancements for Home eNodeB (HeNB) and machine-type communications. To analyze both the context of LTE-A and the new enabling technologies, this chapter is divided as follows:

- Section 1.1: This section introduces LTE-A as an IMT-A technology.
- Section 1.2: This section summarizes the main IMT-A features and its requirements in terms of services, spectrum and performance.
- Section 1.3: This section highlights LTE-A requirements. From a direct comparison with the previous section, it is shown that they are more challenging than those established by the ITU for IMT-A.

LTE-Advanced and Next Generation Wireless Networks: Channel Modelling and Propagation, First Edition.
Edited by Guillaume de la Roche, Andrés Alayón Glazunov and Ben Allen.
© 2013 John Wiley & Sons, Ltd. Published 2013 by John Wiley & Sons, Ltd.

- Section 1.4: This section shows the technological proposals being studied in 3GPP: carrier aggregation, new transmission schemes in both uplink and downlink, CoMP, the use of relays, enhancements for HeNB and machine-type communications and other improvements related to the reduction of latency in the user and control plane.

1.1   Introduction

Along its standardization LTE was designed as an evolution of legacy Third Generation (3G) mobile systems due to the incorporation of a set of technological improvements, such as:

- Dynamic allocation of variable bandwidth.
- New MIMO schemes.
- New transmission schemes with multiple carriers, like Orthogonal Frequency Division Multiplexing (OFDM) in the DownLink (DL) and DFT Spread- OFDM (DFTS-OFDM), also known as Single Carrier Frequency Division Multiple Access (SC-FDMA), in the UpLink (UL).
- Very low latency in the user and control plane.

All these new features represent a qualitative and quantitative leap in system performance that is motivated by different reasons. Market globalization and liberalization and the increasing competence among vendors and operators coming from this new framework has led to the emergence of new technologies. This fact comes together with the popularization of Institute of Electrical & Electronics Engineers (IEEE) 802 technologies within the mobile communications sector. Finally, end users are becoming more discerning and demand new and better services such as Voice over IP (VoIP), video-conference, Push-to-talk over Cellular (PoC), multimedia messaging, multiplayer games, audio and video streaming, content download of ring tones, video clips, virtual private network connections, web browsing, email access, file transfer, and so on. It is precisely this increasing market demand and its enormous economic benefits, together with the new challenges that come with the requirements in higher spectral efficiency and services aggregation that raised the need to allocate new frequency channels to mobile communications systems. That is why the International Telecommunication Union – Radiocommunication Sector (ITU-R) WP 8F started in October 2005 the definition of the future 4G, also known as IMT-A, following the same model of global standardization used with 3G systems. The objective of IMT-A is to specify a set of requirements in terms of transmission capacity and quality of service, in such a way that if a certain technology fulfils all these requirements it is included by the ITU in the IMT-A set of standards. This inclusion firstly endorses technologies and motivates operators to invest in them, but furthermore it allows these standards to make use of the frequency bands specially designated for IMT-A, which entails a great motivation for mobile operators to increase their offered services and transmission capacity. Given this economic outlook, the 3GPP established the LTE standardization activity as an ongoing task to build up a framework for the evolution of the 3GPP radio technologies, concretely Universal Mobile Telecommunication System (UMTS), towards 4G. The 3GPP divided this work into two phases: the former concerns the completion of the first LTE standard (Release 8), whereas the latter intends to adapt LTE to the requirements of 4G through the specification of a new technology called LTE-A (from Release 9 on). Following this plan, the LTE-A Study Item was launched in April 2008 to analyze IMT-A requirements and pose conditions to the new standard:

- LTE-A would be an evolution of LTE. Therefore, backward compatibility between LTE-A and LTE Release 8 must be guaranteed.
- LTE-A requirements should exceed IMT-A ones, following the ITU-R agenda.
- LTE-A would support higher peak data rates, with the aim of fulfilling ITU-R conditions, focusing on low mobility users.

1.2 General IMT-Advanced Features and Requirements

IMT-A systems comprise new capabilities and new services, migrating towards an all-IP network. As happened with the IMT-2000 family of standards, it is expected that IMT-A becomes, through a continuous evolution, the dominant technology designed to support new applications, products and services. Moreover, IMT-A systems must support applications for both low and high speed mobility and for different data rates. The main characteristics of IMT-A systems are:

- Ease of use of applications while maintaining a wide range of services at a reasonable cost.
- Compatibility with International Mobile Telecommunication (IMT) standards and with other fixed networks.
- Interconnection capacity with other radio access systems.
- High quality mobile services.
- Global roaming capabilities.
- Peak data rates of 100 Mbps for high mobility users and 1 Gbps for low mobility users.

Requirements established by ITU-R can be classified in three main categories: services, spectrum and technical aspects. The aim of these requirements is not to limit the performance of the candidate technologies, but to ensure that the IMT-A radio interface technologies fulfil these minimum conditions to become a member of the 4G family of standards.

1.2.1 Services

IMT-A requirements do not specify a set of services but a structure of services that includes service parameters, service classifications and some examples [1]. Capacity requirements are based on the support of a wide range of services comprising basic conversational, advanced and low delay services.

1.2.2 Spectrum

The radio spectrum is a scarce resource that has a considerable economic and social importance. Albeit in the last analysis the governments of every nation must decide on the spectrum allocation, global coordination of the spectrum usage is responsibility of the ITU, which, through spectrum regulation, aims to facilitate the global roaming and to decrease equipment cost by means of global economies of scale. Since 1992, and in the framework of the ITU-R, United Nations have come to quite significant agreements at a global level to designate some specific frequency bands to IMT standards. With the aim of coordinating the global use of spectrum, every three to four years ITU-R holds the World Radiocommunication Conference (WRC), where ITU radio regulations that govern spectrum distribution are adapted.

ITU-R has already started the spectrum regulation tasks concerning IMT-A. The first step was performing an in-depth study of the mobile market forecast and the development of spectrum requirements for the increasing service demand. Reports predicted the total spectrum bandwidth requirements for mobile communication systems in the year 2020 to be 1280 MHz and 1720 MHz for low and high user demand scenarios, respectively. Bearing in mind that the spectrum bandwidth designed by ITU as IMT was much lower than this forecast (693 MHz in Region 1 (Europe, Middle East and Africa, and Russia), 723 MHz in Region 2 (Americas) and 749 MHz in Region 3 (Asia and Oceania)), and given that the time elapsed since the adaptation of the radio regulations until the definitive allocation of a frequency band to operators takes from 5 to 10 years, the WRC-07 that took place in Geneva ended with the identification of new frequency bands for IMT technologies.

Figure 1.1 depicts the current state of the frequency bands reserved for IMT. Despite not fully corresponding to what was targeted, the new spectrum allocated for mobile communications will allow operators to satisfy the initial deployment of technologies towards IMT-A. Furthermore, the increasing demand for mobile services has been progressively recognized with additional spectrum, a trend that is expected to be maintained in future World Radiocommunication Conference (WRC).

1.2.3 Technical Performance

So far, the above requirements may be seen as recommendations. In terms of technical performance requirements, there are ten major indicators [2]. Table 1.1 shows the evaluation method for each technical feature. To calculate cell spectral efficiency, cell edge spectral efficiency and VoIP capacity it is necessary to take into account those signals and channels that consume resources and reduce the capacity of the Physical Downlink Shared Channel (PDSCH) and Physical Uplink Shared Channel (PUSCH). Depending on whether the transmission is uplink or downlink, or the duplexing is Frequency Division

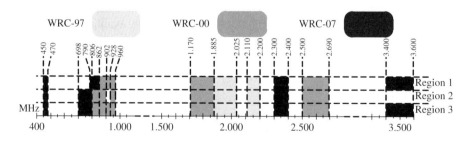

Figure 1.1   Frequency bands allocated to IMT-A.

Table 1.1 Features and their assessment method

Feature	Simulation	Analytical	Inspection
Cell spectral efficiency	×*		
Peak spectral efficiency		×	
Bandwidth			×
Cell edge user spectral efficiency	×*		
Control plane latency		×	
User plane latency		×	
Mobility	×**		
Handover interruption time		×	
Inter-system handover			×
VoIP capacity	×*		

*System-level simulation.
**System and link-level simulation.

Duplexing (FDD) or Time Division Duplexing (TDD), these available resources will be different. For the exact mapping of signals and physical channels, see [3].

Evaluation activities are carried out by analytical means, by inspection or by simulation. The simulation methods are divided into link-level simulations and system-level simulations. The former focuses on defining a simple scenario with one transmitter and one receiver, while system-level simulations aim at studying the global characteristics of a communication system with several mobile users. The IMT-A requirements must be checked for all configurations of the radio access technology, including both the FDD and the TDD mode, for different scenarios and for UL and DL. A technology meets requirements if, for a given configuration, results exceed the established thresholds. For example, LTE-A could meet the requirements on cell edge spectral efficiency with Multi-User MIMO (MU-MIMO) but without using CoMP.

There are four different test environments defined by ITU-R:

- **Indoor**: This is a high user density scenario with high data rate experiences. Cells are considered very small.
- **Microcellular**: This is a dense urban scenario with high traffic and user density. Pedestrian and slow vehicular users are assumed. Distance between base stations is small and antenna mounting is below rooftop.
- **Base Coverage Urban**: Urban scenario with pedestrian and high speed users. Base stations are above rooftop.
- **High Speed**: Macrocellular scenario for high speed vehicles and trains.

Each test environment has one or several deployment scenarios. Table 1.2 summarizes this mapping. The acronyms Indoor Hotspot (InH), Urban Microcell (UMi), Urban Macrocell (UMa) and Rural Macrocell (RMa) are widely used to refer to these scenarios.

The next subsections describe the meaning of the performance indicators included in Table 1.1. All figures are extracted from [2].

Table 1.2 Test deployment scenarios

Test Environment	Indoor	Microcellular	Base Coverage Urban	High Speed
Deployment Scenario	Indoor Hotspot (InH)	Urban Microcell (UMi)	Urban Macrocell (UMa)	Rural Macrocell (RMa)

1.2.3.1 Cell Spectral Efficiency

Cell spectral efficiency is probably the most important parameter that defines the actual capacity of the system. It is defined as the aggregate capacity of all users, that is, the total number of bits that are delivered to the layer 3 in a given time interval, normalized by the channel bandwidth and the number of cells. Cell spectral efficiency has units of b/s/Hz/cell. In a system with N users and M cells, the overall efficiency of the cell is expressed by Equation 1.1.

$$\eta = \frac{\sum_{i=1}^{N} \chi_i}{T \cdot BW \cdot M},$$ (1.1)

where χ_i is the number of bits correctly received by the i-th user, T is the considered time interval and BW the channel bandwidth.

Requirements related to cell spectral efficiency (η) are summarized in Table 1.3 for each environment. These values were derived assuming four transmit antennas and two receive antennas in DL and two transmit antennas and four receive antennas for the UL. It is worth noting that these antenna configurations are not mandatory.

1.2.3.2 Peak Spectral Efficiency

Another parameter to be assessed is the peak spectral efficiency. For its calculation, it is assumed perfect communication and hence coding rate is equal to 1. The minimum values for peak spectral efficiency are shown in Table 1.4.

1.2.3.3 Bandwidth and Scalability

The bandwidth in IMT-A technologies should be scalable to operate in different bands of the spectrum through a single or multiple carriers. Each radio interface technology should be scalable to at least 40 MHz with a recommended maximum bandwidth of 100 MHz.

Table 1.3 Requirements on cell spectral efficiency

	DL Efficiency (b/s/Hz/cell) (4×2)	UL Efficiency (b/s/Hz/cell) (2×4)
Indoor	3	2.25
Microcellular	2.6	1.80
Base Coverage Urban	2.2	1.4
High Speed	1.1	0.7

Table 1.4 Requirements on peak spectral efficiency in downlink and uplink

	Peak spectral efficiency (b/s/Hz)
Uplink (2 × 4)	6.75
Downlink (4 × 4)	15

1.2.3.4 Cell Edge User Spectral Efficiency

Cell edge user spectral efficiency is defined as the 5th percentile of user spectral efficiency, that is, the value over which 95% of users spectral efficiency falls. The spectral efficiency of user i-th (γ_i) is defined as:

$$\gamma_i = \frac{\chi_i}{T_i \cdot BW} \tag{1.2}$$

Requirements established by ITU-R concerning cell edge user spectral efficiency are collected in Table 1.5.

1.2.3.5 Latency in the User and Control Plane

Latency is one of the parameters that most influences the perception of the end user. There are several applications where capacity is not as important as transmission latency, such as VoIP, real-time games, interactive applications, and so on. There are two types of latency: the latency at the user and control plane. The latency at the control plane is the time in the transition between two connection states, for example, from the idle state to the active state.

Figure 1.2 shows the set of connection modes used for the latency analysis and the requirements established by ITU-R. Transition between the dormant and the active state entails that the user is already synchronized.

On the other hand, user plane latency is defined as the time elapsed since the Internet Protocol (IP) packet is available at the base station until this packet is properly received by the IP layer of the end user. ITU-R fixed a maximum latency for this transmission of 10 ms in case of low load of the base station (a single active user) and low size IP packets.

Table 1.5 Cell edge spectral efficiency

	DL Efficiency (b/s/Hz/cell) (4 × 2)	UL Efficiency (b/s/Hz/cell) (2 × 4)
Indoor	0.1	0.07
Microcellular	0.075	0.05
Base Coverage Urban	0.06	0.03
High Speed	0.04	0.015

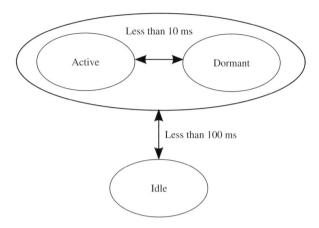

Figure 1.2   Latency in the control plane.

1.2.3.6   Mobility

The proper management of mobility is another feature that ITU-R defined as critical for IMT-A systems. The requirements summarized in Table 1.6 are evaluated for different scenarios and user velocity, both for DL and UL. Spectral efficiency values are assumed for a 4×2 configuration in DL and 2×4 in UL.

1.2.3.7   Handover

Handover time is the time during which the user terminal cannot exchange packets with any base station since it is in the transfer phase from one cell to another. The requirements can be divided into two cases, as shown in Table 1.7: if base stations are transmitting on the same frequency carrier or on different carriers, either in the same band or in different bands.

The handovers described so far are all within the same technology. However, it could be the case that the terminal had to move to another technology. IMT-A systems should be able to ensure that such handovers are also supported.

Table 1.6   Spectral efficiency to assess mobility

	Velocity (km/h)	Efficiency (b/s/Hz/cell)
Indoor	10	1.0
Microcellular	30	0.75
Base Coverage Urban	120	0.55
High Speed	350	0.25

Table 1.7 Handover interruption time

Handover type	Interruption time (ms)
Intra-frequency	27.5
Inter-frequency within a spectrum band	40
Inter-frequency between spectrum bands	50

1.2.3.8 VoIP Capacity

Finally, VoIP Capacity is another parameter to be assessed both for DL and UL, assuming a 12.2 kbps coder and a 50% activity factor. The percentage of users with interruption should be less than 2%, where a user suffers interruption if less than 98% of VoIP packets have been delivered successfully. A package is delivered successfully if transmission delay is less or equal to 50 ms. This delay is the latency in the transmission from the encoder at the transmitter to the decoder at the receiver. The VoIP Capacity is defined as the minimum capacity in both links (DL and UL) normalized by the used bandwidth. Table 1.8 summarizes the IMT-A requirements regarding VoIP Capacity.

1.3 Long Term Evolution Advanced Requirements

The race towards IMT-A was officially started in March 2008, when the ITU-R issued a circular letter asking for the submission of new technology proposals. Previous to this official call, the 3GPP established in 2004 the LTE standardization activity as an ongoing task to build up a framework for the evolution of the 3GPP radio technologies, concretely UMTS, towards 4G. The 3GPP divided this work into two phases: the former concerns the completion of the first LTE standard (Release 8), whereas the latter intends to adapt LTE to the requirements of 4G through the specification of a new technology called LTE-A (Release 9 and beyond). Following this plan, in December 2008 the 3GPP approved the specifications of LTE Release 8 which encompasses the Evolved UTRAN (EUTRAN) and the Enhanced Packet Core (EPC). Otherwise, the LTE-A Study Item was launched in May 2008, being completed in October 2009 according to the ITU-R

Table 1.8 VoIP Capacity

	Minimum VoIP Capacity (Active Users/cell/MHz)
Indoor	50
Microcellular	40
Base Coverage Urban	40
High Speed	30

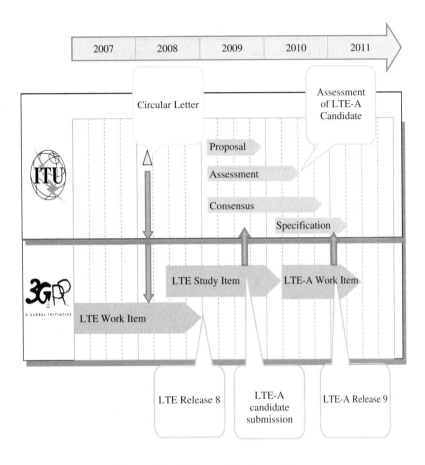

Figure 1.3 LTE specification roadmap.

schedule for the IMT-A process. It was in November 2008 when ITU-R established the technical requirements of radio interface candidates for IMT-A [2]. Figure 1.3 depicts the standardization process of LTE and LTE-A. In September 2009, the 3GPP submitted to the ITU-R the main features of LTE-A candidate for IMT-A. In fact the 3GPP submitted a set of radio interface candidates comprising FDD and TDD modes. The 3GPP also assessed its own candidates, demonstrating that they fulfill IMT-A requirements. Three documents were included in the proposal: a template that describes the main features of the technology [4], a template with a link budget analysis of the four environments [5] and a template with a summary of the self-evaluation of service, spectrum and technical performance requirements [6]. The full self-evaluation can be found in section 16 of [7].

Despite the tight requirements of ITU-R, the 3GPP fixed its own requirements for the development of LTE-A. These requirements are divided into seven items: capacity, system performance, deployment, architecture and migration, radio resource management, complexity, cost and services. Only the first three are further described in this section, since the others are either a consequence of them or the same as for LTE. For more details about LTE-A requirements refer to [8].

1.3.1 Requirements Related with Capacity

1.3.1.1 Peak Data Rate

The minimum requirements for LTE-A, in terms of capacity, are marked by the ITU-R, that is, 100 Mbps for high mobility users and 1 Gbps for low mobility users. Yet this goal is further specified pursuing data rates of 1 Gbps just for the downlink, being 500 Mbps for the uplink.

1.3.1.2 Latency

The expected latency for the control plane in LTE-A must be much smaller than the latency set for LTE Release 8. This latency takes into account the radio access network and core network (excluding the latency of the S1 interface, that is, the interface between the radio access network and the core network) with a lightly loaded system. The transition time from the Camped state (after the allocation of the IP address) to the Active state, including the establishment time of the user plane, but excluding the delay associated with S1 interface, must be less than 50 ms. Latency from the Inactive to Active state must be less than 10 ms. Finally, the standard establishes that the system must support 300 active users in a bandwidth of 5 MHz.

1.3.2 System Performance

1.3.2.1 Peak Spectral Efficiency

As discussed in Section 1.2.3, the peak spectral efficiency is the highest data rate normalized by the bandwidth, considering that the communication is free of errors. With a multi-antenna maximum configuration of up to 8×8, the peak spectral efficiency is 30 b/s/Hz in the downlink, while in the uplink the peak spectral efficiency is 15 b/s/Hz for a maximum configuration of up to 4×4.

1.3.2.2 Mean Spectral Efficiency

Another parameter even more important than peak spectral efficiency, and perhaps more realistic, is the mean spectral efficiency, defined as the sum of users data rates normalized by the bandwidth and the number of cells. In the current definition of LTE-A requirements, the spectral efficiency is calculated using a channel model that is different from the one defined by the ITU-R. The description of this channel model can be found in [9] and is the same as the channel used for the evaluation of LTE. Four different cases were defined including main parameters such as carrier frequency, inter-site distance, bandwidth, penetration losses and speed. Case 1 is used to verify the requirements, which corresponds to the urban macrocellular channel of ITU-R. This channel was used to compare LTE-A to LTE. The results shown in Table 1.9 are a compilation of results of the technical report for LTE Release 8 [8, 10] for LTE-A and [2] for IMT-A. If a resource manager only provides resources to the user with best channel, then that user would enjoy a high spectral efficiency, but others would be penalized thus having a poor spectral efficiency. Therefore, the mean spectral efficiency does not show the reality that

Table 1.9 Mean spectral efficiency requirements

		Antenna Conf.	LTE Rel 8	(LTE-A)	(IMT-A)
Capacity	DL	2 × 2	1.69	2.4	-
(b/s/Hz/cell)		4 × 2	1.87	2.6	2.2
		4 × 4	2.67	3.7	-
	UL	1 × 2	0.74	1.2	-
		2 × 4	-	2.0	1.4

is occurring, as it omits a fair resource allocation. It is in this case when the cell-edge user spectral efficiency metric seems the most appropriate as it ensures that 95% of users are above a certain value of efficiency. Obviously, in a real situation the resource manager offers resources more equitably, reaching an equilibrium.

1.3.2.3   Cell Edge User Spectral Efficiency

The requirements on cell edge user spectral efficiency are summarized in Table 1.10, which shows that these requirements are more stringent than those set by the ITU-R. LTE-A requirements were set considering a gain of 1.4 to 1.6 over LTE Release 8. Moreover, cell edge user spectral efficiency is calculated under the assumption of 10 users uniformly distributed per cell.

1.3.2.4   Others

Requirements of VoIP capacity, mobility, coverage, multicast and broadcast transmission and network synchronization are assumed to be higher than those achieved with LTE Release 8.

1.3.3   Deployment

The deployment of LTE-A is an evolution of LTE deployment since LTE-A could use new frequency bands. One of the most important aspects is that LTE-A was designed to be backward compatible with LTE, so that a LTE user equipment can work in an LTE-A network and vice versa. Still, it is expected that there might be some incompatibility when

Table 1.10 Cell edge user spectral efficiency requirements

		Antenna Conf.	LTE Rel 8	LTE-A	IMT-A
Capacity	DL	2 × 2	0.05	0.07	-
(b/s/Hz/cell)		4 × 2	0.06	0.09	0.06
		4 × 4	0.08	0.12	-
	UL	1 × 2	0.024	0.04	-
		2 × 4	-	0.07	0.03

deploying the network. Another aspect that is being emphasized is the development of femtocellular services, such as remote access of home security, which implies the need of a better indoor deployment of LTE-A.

As discussed in Section 1.2.2, new frequency bands have been reserved for IMT-A. In LTE-A, spectrum can be aggregated up to 100 MHz, either in a contiguous or noncontiguous manner. Moreover, LTE-A and LTE Release 8 must be able to coexist in the same spectrum band.

1.4   Long Term Evolution Advanced Enabling Technologies

To meet the demanding requirements of LTE-A, the 3GPP is in the process of development of certain technological proposals. To this end, 3GPP has focused its attention on different points that required technological innovations: support of wider bandwidth (carrier aggregation), advanced MIMO techniques, relaying, enhancements for HeNB, and so on. Another relevant feature of LTE-A is that latency requirements are more strict than ITU-R ones. The proposed changes to improve LTE-A performance indicators are discussed in the following subsections.

1.4.1   Carrier Aggregation

Carrier aggregation is one of the most important technologies to ensure the success of 4G technologies. This concept involves transmitting data in multiple contiguous or non-contiguous Component Carriers (CCs). Each Component Carrier (CC) takes a maximum bandwidth of 20 MHz to be compatible with LTE Release 8. The maximum number of Resource Blocks (RBs) in a CC is 110 and the bandwidth will be assigned following the same structure as in LTE, that is, 1.4 MHz (6 RB), 3 MHz (15 RB), 5 MHz (25 RB), 10 MHz (50 RB), 15 MHz (75 RB) and 20 MHz (100 RB). However, it is possible that there are certain CCs that are not compatible with LTE.

Carrier aggregation allows both an efficient use of spectrum already deployed and the required support for the resource allocation in new frequency bands. Figure 1.4 shows a schematic of the concept of carrier aggregation in LTE-A. Depending on the capabilities of the mobile terminal, the user may transmit and/or receive from multiple CC, in case

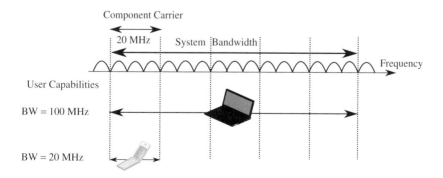

Figure 1.4   Carrier aggregation in LTE-A.

of an LTE-A terminal with a maximum bandwidth of 100 MHz. Meanwhile, an LTE Release 8 equipment may only transmit and/or receive in a single CC, with a maximum bandwidth of 20 MHz. Given the possible bandwidth configurations, the transmitter should support, depending on the capabilities of the mobile terminal, three possible scenarios: aggregation of several contiguous CCs within the same band, aggregation of several non-contiguous CCs within the same band and aggregation of several non-contiguous CCs located in different bands. In [11] various deployment scenarios were considered to meet the requirements of ITU-R. Among them, four scenarios are the most significant, being summarized in Table 1.11 for the FDD and TDD modes. As an example, new frequency bands could be added to existing UMTS bands of 1.8 GHz, 2.1 GHz and 2.6 GHz or new bands could be added at 3.5 GHz.

When working on non-contiguous CCs, it is required null interference between carriers and, therefore, guard bands are added. Otherwise, in cases of aggregation of contiguous CC, there is not such a large guard band, which allows a more efficient use of spectrum. However, in order to be compatible with the LTE frequency raster of 100 kHz and, in turn, preserve the orthogonality of the subcarriers with a spacing of 15 kHz, the distance between carriers must be a multiple of 300 kHz. This can cause certain subcarriers not to be used. Still, these could be used as guard bands.

Based on the three scenarios seen above, there are several architectural alternatives for the transmitter. Figure 1.5 illustrates four of them. As an example, two CCs are frequency multiplexed. Option 1 (first scheme), works on contiguous CCs within the same band, so that two frequency multiplexed signals are mapped to the time domain through an Inverse Fast Fourier Transform (IFFT) and then go through a digital to analog converter. Then the signal is modulated to the corresponding carrier frequency and amplified in power. Option 2 (second diagram), implements the transmission of information conveyed on various CCs contiguously and non-contiguously within the same frequency band. With this aim, it combines both baseband waveforms operating in a first intermediate frequency within the band of the second CC. After this step, the broadband signal is modulated in frequency. Option 3 (third diagram), implements, like option 2, the case of contiguous carrier aggregation within the same frequency band. The difference is that both CCs are modulated to an intermediate frequency before being combined and amplified in power. Option 4 (fourth chart), is the only option that allows for contiguous

Table 1.11   Deployment scenarios of LTE-A

	Deployment scenario	Carrier aggregation	Frequency (GHz)
FDD	Contiguous within the same band	UL: 2 × 20 MHz CC DL: 4 × 20 MHz CC	3.5
	Non-contiguous in different bands	UL/DL: 1 × 10 MHz	1.8
		UL/DL: 1 × 10 MHz	2.1
		UL/DL: 1 × 20 MHz	2.6
TDD	Contiguous within the same band	5 × 20 MHz	2.3
	Non-contiguous in different bands	2 × 20 MHz	1.8
		10 MHz	2.1
		2 × 20 MHz	2.3

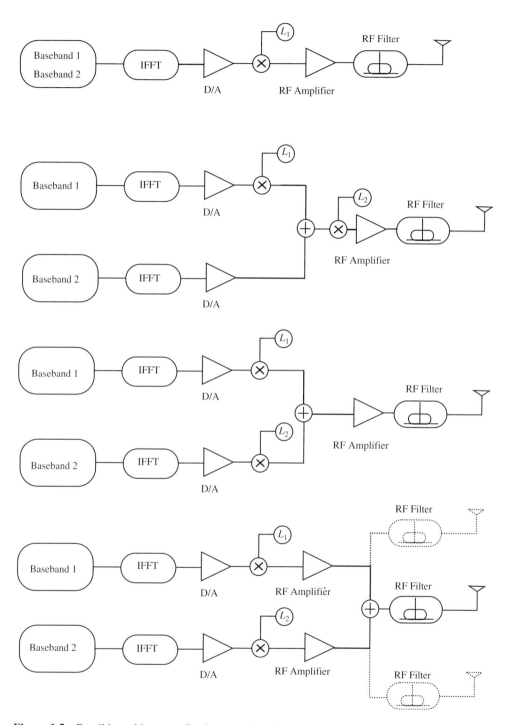

Figure 1.5 Possible architectures for the transmitter in three aggregation scenarios. In order, from top to bottom, are options from 1 to 4.

and non-contiguous aggregation within the same band or in different frequency bands. This scheme employs multiple radio frequency chains and multiple power amplifiers. The combination, therefore, is performed over amplified signals that are transmitted through a single antenna. The choice of the proper architecture depends on the cost, complexity, bandwidth of the amplifier and if the CCs are contiguous or non-contiguous.

In the uplink, one of the problems caused by carrier aggregation is the loss of efficiency in the power amplifier. The cubic metric parameter defined in [12] is useful for assessing the efficiency of power amplifiers. In [13] it is shown that the higher the number of CCs assigned, the higher the cubic metric in case of SC-FDMA. This increase is significant when using two CCs, while from 3 CCs this increment occurs more gradually. Still, the cubic metric is smaller than in the case of Orthogonal Frequency Division Multiple Access (OFDMA). The transmission over multiple carriers is usually restricted to users with good channel conditions, which will reduce the effect of this loss of efficiency. Moreover, cell-edge users will generally not aggregate carriers, so they will not be affected either.

In the physical layer of LTE-A, a single transport block (two in the case of spatial multiplexing) and a single Hybrid Automatic Repeat reQuest (HARQ) entity will be associated with a CC. This allows separate link adaptation to improve the transmission of data from each CC since it will adapt better to the conditions of each CCs. Figure 1.6 shows a description of the basic structure of the physical layer and Medium Access Control (MAC) layer. As shown, LTE-A will support a maximum of five parallel LTE Release 8 processing chains. Of course, the CC may be in different frequency bands.

There are three downlink control channels: the Physical Control Format Indicator Channel (PCFICH), the Physical Downlink Control Channel (PDCCH) and the Physical Hybrid ARQ Indicator Channel (PHICH). The design principle of these channels is, in general, to ensure backward compatibility with LTE Release 8. The signaling control for carrier aggregation is still under study. So far, the following decisions have been taken with respect to these three channels:

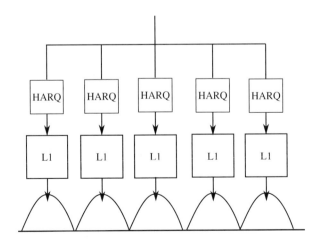

Figure 1.6   Structure of the MAC and PHY layer of LTE-A.

- **PCFICH**. Each CC will have its own information on the size of the control region. Moreover, the same design principles of LTE Release 8 will be followed (modulation, coding and allocation of resource elements).
- **PDCCH**. In this case, there are two ways of allocating resources. On the one hand, the resource manager can allocate resources to the PDSCH and PUSCH in the same CC. The same PDCCH as in LTE Release 8 could be used in each CC as well as the Data Control Indicator (DCI) formats. This allows link adaptation for each CC, so as to improve transmission capacity since each transmission can get adapted to the channel conditions of each CC. The other possibility is that, from a single CC, resources for PDSCH and PUSCH on multiple CCs could be allocated using the Carrier Indicator Field (CIF). This implies that this CC is not compatible with LTE Release 8. However, this option allows for higher scheduling flexibility, being able to balance the load dynamically and reduce interference among CCs.
- **PHICH**. For this channel the same aspects of transmission of LTE Release 8 will be reused, that is, modulation, scrambling code and allocation of resource elements. It will be only transmitted in the CC that was used to transmit scheduling information on the uplink.

Concerning the UL control signaling, HARQ ACK/NACK signaling, scheduling requests and Channel State Information (CSI) have to support up to five DL CCs. A User Equipment (UE) must send a HARQ Acknowledgement (ACK)/Negative Acknowledgement (NACK) for every transport block transmitted in a given CC. Unlike Release 8, in Release 10 if the UE has data to transmit on PUSCH then control signaling (HARQ signaling and CSI) can be time multiplexed with data on the PUSCH.

1.4.2   Advanced MIMO Techniques

In LTE Release 10 spatial multiplexing was extended to support up to four layers in the UL and up to eight layers in the DL. This update improves cell spectral efficiency but also implies changes in the design of the reference signals and in the DL control signaling. The main characteristics of these and other changes are explained below.

1.4.2.1   Uplink Transmission Scheme

In [14] it was specified that spatial multiplexing can be used in the UL with two transport blocks, also called codewords. Each codeword would have its own Modulation and Coding Scheme (MCS). Figure 1.7 shows the structure of the transmitter in the UL. The maximum number of layers supported by LTE-A is four. Depending on the number of layers, modulated symbols associated with each codeword will be mapped using the same philosophy as LTE Release 8. The spatial multiplexing in the UL has the option of using layer shifting in the time domain [14]. If layer shifting is activated, the two transport blocks will have an associated shared HARQ-ACK, New Data Indicator (NDI) and Redundancy Version (RV). Otherwise, each transport block will have associated its own MCS (see Figure 1.7). In the case of layer shifting, when the feedback information from the base station is a NACK without any additional information, the user equipment cannot know the number of transport blocks that have been correctly decoded, if any.

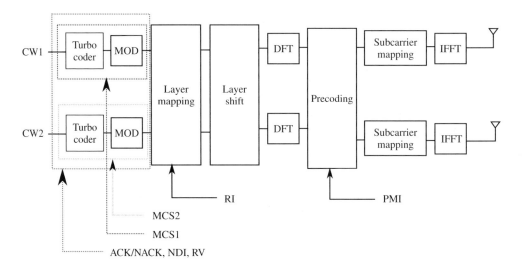

Figure 1.7   Transmitter structure for a ACK/NACK with layer shift.

It is therefore necessary to use any special flag that marks the transport block to be retransmitted so as to avoid redundant information. However, depending on whether both transport blocks have been decoded correctly or incorrectly, two new transport blocks will be transmitted or erroneous ones will be retransmitted. In this last case, the amount of resources allocated to the PHICH could be reduced with the consequent possible increase in data rate.

For both FDD and TDD modes, precoding is performed based on a predefined set of matrices for each CC. To date, these matrices are only defined for two and four antennas with a number of layers less than three [14].

Spatial Orthogonal-Resource Transmit Diversity (SORTD), a kind of UL transmit diversity scheme, is supported in Release 10 for the Physical Uplink Control Channel (PUCCH) with two antenna ports. For four antenna ports, antenna virtualization is used, which consists of managing multiple physical antennas as a single virtual antenna.

Concerning UL reference signals, both Demodulation Reference Signal (DM RSs) and Sounding Reference Signal (SRSs) have been adapted in order to support Single-User MIMO (SU-MIMO). Like DL DM RSs, the UL DM RS are precoded using the same precoding as PUSCH. The generation and mapping of these reference signals can be found in [3]. On the other hand, Release 10 introduced the concept of aperiodic Sounding SRS, which can be dynamically configured through Radio Resource Control (RRC) signaling.

1.4.2.2   Downlink Transmission Scheme

To achieve higher data rates, LTE-A base station supports up to eight antennas. As happened in LTE Release 8, up to two codewords can be transmitted to the same user per CC. Each codeword will have its own encoding and modulation and the HARQ feedback from the user consists of one bit per codeword. So far, the symbols mapping is defined

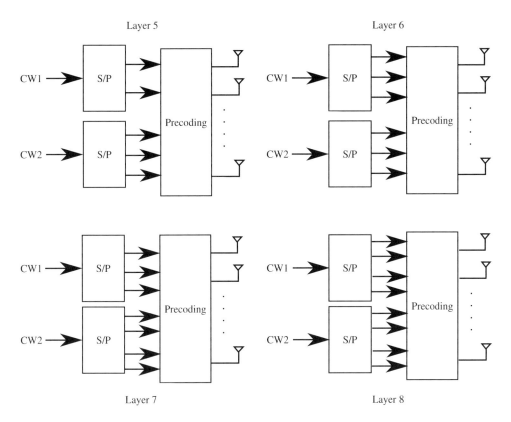

Figure 1.8 Codeword mapping to spatial multiplexing layers.

in [14]. This mapping is the same as in LTE with less than four layers. The mapping of codewords to a number greater than or equal to five layers is depicted in Figure 1.8.

In LTE-A two types of reference signals are used in DL: the DM RS, also known as UE-specific reference signals, and the Channel State Information – Reference Signal (CSI-RSs). Although the CSI-RSs were already introduced in LTE Release 8, their use was limited to single-layer transmission. The DM RS are characterized by being pre-coded in the same way as the PDSCH channel, that is, they are user-specific. They only appear in the resources allocated by the base station to the user and are mutually orthogonal by Code Division Multiplexing (CDM) using Orthogonal Cover Codes (OCCs) in the different layers to avoid the interference between them. On the other hand, the CSI-RSs are cell-specific, that is, all users belonging to the same cell can read these reference signals in order to obtain channel state feedback for up to eight antennas. Mapping of both reference signals can be found in [3].

1.4.3 Coordinated Multipoint Transmission or Reception

The coordinated multipoint transmission/reception is considered by many companies as a clear candidate to improve the system capacity and cell-edge user spectral efficiency, thus fulfilling the requirements of LTE-A. The current LTE Release 8 allows a certain degree of cooperation between base stations in order to reduce interference. However, a big improvement is expected in this technique with LTE-A as compared with LTE Release 8.

1.4.3.1 Coordinated Transmission

In a CoMP system, multiple cells (likely from different base stations) are cooperating in the transmission of data to multiple users. We can distinguish two types of CoMP techniques:

- **Joint Processing (JP) techniques**. Multiple cells transmit the same information to a user. A cell can consist of a set of antenna elements within the base station or outside, geographically separated. This level of cooperation requires user data to be shared between cooperating cells.
- **Coordinated Scheduling and Beamforming (CS/CB) techniques**. The information is only sent from a single cell, but the scheduling and beamforming decisions are made taking into account other cells status so as to coordinate interferences. This technique does not need to share user data among cells.

Cell coverage area is usually understood as the area produced by a set of antenna elements within a base station. A cell is any transmitter that has a physical layer cell identifier, which is detected by the user during the cell search procedure based on the Primary Synchronization Channel (P-SCH) and Secondary Synchronization Channel (S-SCH). Figure 1.9 shows two possible situations of joint processing. In the first case, two base stations transmit the same information to a user (case 1), while in the second case the base station sends the information to different transmission points (cells) in different geographically separated locations (case 2). It might take a third case of coordinated transmission in which two sectors transmit the same information to a user situated in the border of two adjacent cells.

The main limitations of JP is that user data should be coordinated between various transmission points. If the transmission is performed between base stations, then there are

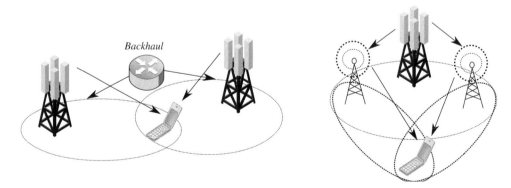

Figure 1.9 Types of joint processing in Coordinated Multipoint transmission or reception (CoMP), case 1 (left) and case 2 (right).

limitations with the latency and backhaul capacity. On the other hand, if coordination is carried out as in case 2, that is, through a distributed antenna system or remote radio heads belonging to the same base station, communication is assumed to be faster. With regard to CS/CB its main limitations are related to the assumption of certain knowledge of the channel quality of cooperating cells. These techniques lack backhaul capacity problems, since cells do not exchange user data. However, latency remains a limitation due to the exchange of channel state information. No major changes are expected in the network architecture to implement this CoMP technique and, therefore, it seems to be the most appropriate technique in scenarios of transmission between base stations.

The LTE-A radio interface must support different feedback mechanisms of cooperating cells. The user dynamically estimates the channel for each one of the transmission points to facilitate the decision on which a set of transmitters will participate in the cooperative transmission. In the case of TDD mode, channel reciprocity could be assumed.

1.4.3.2 Coordinated Reception

OFDM can eliminate the interference within the cell, since the high data rate is divided into parallel flows at lower rates, transmitted over orthogonal subcarriers. However, in a multi-cellular system OFDM cannot remove inter-cell interference. Therefore, coordinated reception has been proposed to reduce this type of interference (especially for cell-edge users), also known as Inter-Cell Interference (I-CI). Coordinated reception is not only expected to minimize interference, but also help improve the average cell efficiency of the cell and the cell-edge efficiency.

As explained before, CoMP is a cooperative technology that coordinates multiple geographically separated cells. The cooperation implies sharing user data, scheduling information and channel quality. Clearly, CoMP reception further affects the implementation rather than the specifications. Figure 1.10 shows an overview of coordinated multi-point reception. There is a base station that serves all the mobile terminals of the cell. The three mobile terminals transmit their data on the same resources simultaneously. The transmitted signal of each user to the base station corresponds to the interference to the neighboring cell shown in the figure with a dotted line. Because users whose information is processed are geographically close together, the interference might be high and there would be a degradation in performance. This interference could be reduced via joint processing. Although in Figure 1.10 it is not showed, a central server is responsible for controlling the behavior of the multiple cells involved in the cooperation. However, it could be physically integrated in any base station.

1.4.4 Relaying

The use of relays was initially envisioned as a tool to increase cell coverage. With the increase of data traffic, operators thought of a new solution that could improve the data rate in certain areas. However, increasing the number of base stations entails a cost that operators cannot afford. This made operators consider the study and development of relays as a means to improve coverage in hotspots, mobility in public transportation vehicles (trains, buses, etc.) and improvement in transmission capacity on the cell edge.

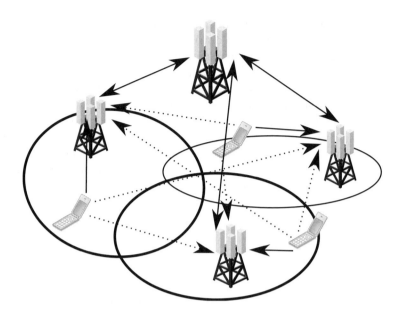

Figure 1.10 CoMP structure in reception.

A general configuration of a relay network can be seen in Figure 1.11. There are three types of link: a base station to user link, a base station to relay link (also known as backhaul) and a relay to user link (also known as access link). It is worth noting that in the picture the relay node is wirelessly connected to the radio access network through the base station or donor cell. Moreover, although the philosophy behind the use of relays is to maintain the compatibility with Release 8 UEs, the physical layer of the access link in the case of relaying may be different from a conventional Base Station (BS) to UE direct link.

There are several classifications of relays depending on the mechanisms for interconnection, functionality and transparency towards LTE Release 8 users:

- **Interconnection**. The interconnection between a relay and a base station may be via cable (fiber, xDSL, etc.), wireless technology (the same LTE) or other technology (microwave). On the other hand, the wireless networking can be divided into in-band or out-band transmission. The difference among them depends on whether you use the same carrier frequency in the base station to relay and the relay to user links. For in-band transmissions, since the relay transmitter causes interference to its own receiver, the base station to relay and relay to user transmissions in the same frequency resources may not be possible unless there was an isolated antenna structure. Similarly, there would be interference when the relay is simultaneously receiving data from the user and transmitting to the base station. One possible solution is illustrated in Figure 1.12. The interference could be avoided by adding gaps where the relay does not transmit to the terminals when data is being received from the base station. During these gaps, the base station transmits data to the relay within the Multicast-Broadcast Single-Frequency Network (MBSFN) subframes. The relay to base station communication can be achieved

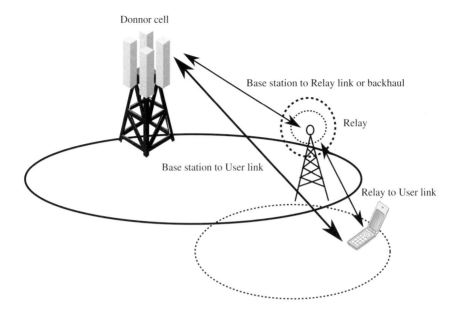

Figure 1.11 Configuration of the relaying network.

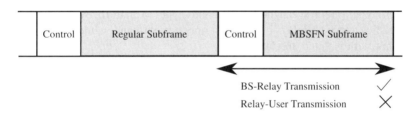

Figure 1.12 Communication Relay-User in regular subframes and Base Station-Relay in MBSFN subframes.

by avoiding the transmission in some subframes of the user to relay link (known as blanking subframes).

- **Functionality**. They can operate as simple repeaters that only amplify and forward the received packets. Despite a repeater increasing the coverage of the cell, its main disadvantage is that it will increase not only the signal level but also interferences, maintaining signal to noise ratio. However, they are often characterized by their low cost and the small delay they introduce in transmission. Other types of relays decode and forward packets. In this case there is an additional delay as compared with repeaters. However, decode and forward relays also enhance the signal to noise ratio. Finally, there are relays whose features are similar to those of the base stations. Such relays manage user connections, schedule traffic, perform HARQ and execute IP layer functionalities. From the point of view of the base station, the relay would be in this case a subordinate in a hierarchical network.

- **Transparency**. The transparency is defined in terms of the characteristics of the link between the user and the relay node. The relay is transparent to the user if the user is not aware of being served by an entity other than a classical base station, that is, the relay. Conversely, if the user is aware of being connected to another entity, relaying would be non-transparent.

Another possible classification differentiates if the relay depends on a donor cell or not. In the former case, the relay has not got a cell identifier but could have a relay identifier. Some of the resource management functions would take place in the base station of the donor cell and the remaining in the relay node. In the latter case, the cells created by the relay would have a unique cell identifier. From the point of view of the user, the access to a relay node would be identical to the access to a base station. In either case, the relay must support LTE Release 8 user equipments.

Based on the above classifications, 3GPP defined two types of relays:

- **Type 1**. The Type 1 relay is perceived by the user as a new LTE base station, providing support to Release 8 terminals. The cells served by the relay have their own cell identifier. Therefore, a relay type 1 is not transparent to the user that will measure the signal level received from the base station and the relay getting connected to the best server. The relay transmits its own synchronization and reference signals. Similarly, the user feeds back information about channel quality and HARQ processes to the relay. This type of relay node increases coverage. However, it implements the same functionalities as a conventional base station and, therefore, its cost could be similar. Moreover, type 1 relays can be divided into type 1a and 1b, which have the same characteristics as type 1 relays but the first operates out-band, while the second operates in-band but with isolated antennas.
- **Type 2**. The Type 2 relay is an in-band node that uses the same cell identifier as the donor cell, allowing the user to move between the base station and the relay node without handovers. Therefore, the type 2 relay would be able to serve users in a transparent manner without requiring the handover when the user moves outside the base station coverage. The CSI-RSs are not sent and, therefore, user equipments do not measure the quality of the relay to user link. A significant cost saving is made by using this type of relay nodes.

As mentioned above, in order to allow for an in-band transmission in the base station to relay link, a time division multiplexing is required. The gap during which the BS transmits data to the relay is less than the subframe duration. To manage this new situation, new physical channels have been defined: R-PDCCH, R-PDSCH and R-PUSCH. The R-PDCCH physical channel is used to allocate resources to the R-PDSCH or R-PUSCH in the backhaul in a dynamic or semi-persistent way. Resource blocks allocated to the R-PDCCH may not be fully utilized. In that case, these free resources could be used by the PDSCH or the R-PDSCH. The processing chain of the R-PDCCH should reuse the Release 8 functionality, but with the possibility of eliminating unnecessary procedures.

1.4.5   Enhancements for Home eNodeBs

Mobile subscribers are increasingly demanding ubiquity connection and higher data rates. 3G Home NodeBs were introduced in the UMTS Terrestrial Radio Access Network

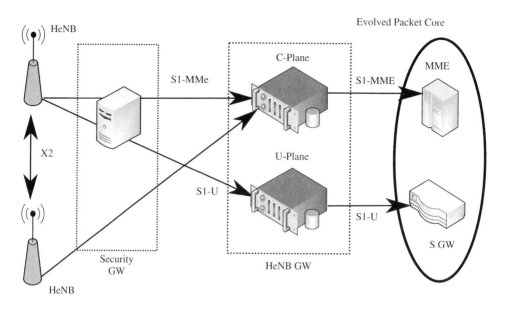

Figure 1.13   HeNB interconnection in the E-UTRAN.

(UTRAN) architecture to provide services and data rates in home environments or small offices. Nevertheless, Home NodeBs introduce some level of lawlessness and because of this the 3GPP began a study item to improve the manageability of 3G femto/pico cells. This activity has continued in the framework of LTE and LTE-A standardization.

In 3GPP LTE terminology, femto-cells are called HeNBs. As shown in Figure 1.13, the EUTRAN architecture includes an optional HeNBs Gateway that routes the packets from different HeNBs to the EPC through the S1. The aim of this gateway is to simplify the functionality of the HeNBs so as to make them cheaper. The specifications also talk about a Security Gateway, which is optional too.

Two types of access control have been defined for LTE HeNBs: closed and open access. In the former, only a Closed Subscriber Group (CSG) can access the HeNB whereas, in the latter, all users can connect to the HeNB. Every cell is identified by CSG ID. A UE subscribed to a CSG would have one or more CSG IDs on which the UE can connect. The UE uses the CSG list along with the CSG ID to select the serving cell. There exists also an hybrid access introduced in Release 9 in which the cell can provide service to all users but still acts as a CSG cell, that is, the subscribed UEs would have higher priority than unsubscribed UEs.

Another remarkable improvement in Release 9 is the inbound handover from a macro eNodeB to a HeNB. Before making a handover decision, the UE monitors the target cell. To optimize this process in a femtocellular scenario, the UE sends the eNodeB a proximity report when camping close to a known HeNB. This allows the system to be prepared in advance to the imminent handover. Besides, to avoid the problems caused by duplicated Physical Cell Identity (PCI), the measurement reports include the Cell Global Identity (CGI) as well as the PCI.

3GPP specifications also allow limiting IP access for HeNB users to a local network, what is referred to as Local IP Access (LIPA). Release 10 extends this concept to limit the access to a corporate network. Moreover, HeNBs must support Selected IP Traffic Offload (SIPTO), which allows internet traffic to flow from the femtocell directly to the internet, bypassing the EPC. Both mechanisms can be enabled/disabled by the mobile operator.

1.4.6 Machine-Type Communications

Machine-to-Machine (M2M) communications, also called Machine-Type Communications (MTC), refer to the type of communication between entities that does not need any human interaction. Some examples of MTC are domotic applications that manage heating and air conditioning systems, alarms, sensors and valves.

There is no consensus about the general network architecture of MTC. 3GPP considers three different scenarios for MTC depicted in Figure 1.14:

- **Scenario A**. The MTC application directly interacts with the UE (MTC device).
- **Scenario B**. The MTC application interacts with the MTC server that is located outside the operator domain.
- **Scenario C**. The MTC application interacts with the MTC server that is located inside the operator domain.

In all cases, the 3GPP radio access network provides transport and communication services, including 3GPP bearer services, IP Multimedia Subsystem (IMS) and Short Message Service (SMS). From Release 10 on, the 3GPP is addressing some specific problems of MTC:

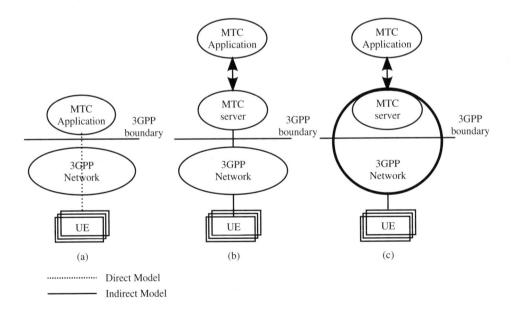

Figure 1.14 MTC scenarios.

- **Signaling congestion and overloading of the core network**. A large number of MTC devices are deployed in a specific area. This can cause, among other things, intolerable delays, packet losses or the service failure. To reduce this problem, load control mechanisms based on different priorities between MTC devices can be implemented. In a similar way, Extended Access Barring (EAB) methods consist of restricting the access to specific UEs. Besides, separate Random Access Channel (RACH) resources could be allocated to M2M.
- **Small data transmission**. MTC does not require high peak data rates, advanced MIMO tecniques, HARQ or sophisticated channel estimations.
- **MTC addressing**. With so many devices, IP addressing can be a problem. As a solution, similar devices can be grouped sharing a common identifier.

1.4.7 Self-Optimizing Networks (SON)

It is apparent that networks are becoming larger in scales and more complex in design. They have grown beyond the limit of manual administration. At the same time, reduction of cost is one of the goals of 4G systems, affecting them anywhere from the cost per transmitted bit until the CAPital EXpenditure (CAPEX) and OPerational EXpenditure (OPEX) reduction. There is a rising pressure from network operators to make networks and systems more manageable, their operations more efficient, and their deployment and maintenance more cost effective.

In the early deployment of 2G networks Operations and Maintenance (OAM) was based on in site operation. Nowadays, with 3G systems, OAM relies on software applications that manage the wireless system in a centralized way. However, the expected complexity of next generation wireless systems is increasing the research community interest towards the design of 4G systems infrastructures by exploiting "cognitive networking" capabilities. Wireless networks cannot remain primitive and the management systems omnipotent. Conversely, a large degree of self-awareness and self-governess must be considered in the new concept of networks and systems. In response, there has been a major push for self-managing networks and systems in the last five years. Although management automaton has been taken into account for decades, never before there has been such a strong concern from both academia and industry, and the need for effective solutions is immediate.

Self-Organizing Network (SON) allows the network to detect changes, make intelligent decisions based upon these inputs, and then implement the appropriate action. The systems must be location and situation-aware, and must take advantage of this information to dynamically configure themselves in a distributed fashion. Applied to resource management, SON allows, for example, making a dynamic and optimum management of radio resources at the border of cells in such a way that there exists an automatic coordination of the radio resource utilization at the cell-edge in order to avoid performance loss or even degradation of service.

In LTE Release 8, SON concepts were associated with initial equipment installation, also known as eNodeB self-configuration. Main procedures included:

- Automatic Inventory.
- Automatic Software Download.

- Automatic Neighbor Relation.
- Automatic PCI Assignment.

The next release of SON (Release 9) provided some procedures covering network optimization. More specifically, the Release 9 standard included these additional use cases:

- Mobility Robustness/Handover Optimization.
- RACH Optimization.
- Load Balancing Optimization.
- Inter-Cell Interference Coordination.

The latest release of SON that appeared in Release 10 provided additional functionalities and methods to manage heterogeneous networks:

- Coverage and Capacity Optimization.
- Enhanced Inter-Cell Interference Coordination.
- Cell Outage Detection and Compensation.
- Self-healing Functions.
- Minimization of Drive Testing.
- Energy Savings.

All SON functionalities are mostly described in [15].

1.4.7.1   Inter-Cell Interference Coordination in LTE

In the downlink, a bitmap known as Relative Narrowband Transmit Power (RNTP) indicator can be exchanged among eNodeBs through the X2 interface. This ON-OFF indicator informs the neighbor cells if the eNodeB intends to transmit on a certain RB over a certain power threshold or not. One bit per RB in the frequency domain is sent. The exact value of the upper limit and the periodicity in the reporting are configurable.

The use of the RNTP indicator allows eNodeBs to choose the proper RBs when scheduling users according to the interference level introduced by their neighbors. The decision making process followed by eNodeBs after receiving RNTP indicators is not standardized, which fosters competence among different implementations.

In the uplink, two messages are exchanged: the Interference Overload Indication (IOI), which indicates the interference level on all RBs, and the High Interference Indication (HII), which informs about the future plans for the uplink transmission. The receiving cells should take this information into account by not scheduling cell-edge users in these RBs.

1.4.8   Improvements to Latency in the Control and User Plane

Although LTE Release 8 technology already meets the requirements of ITU-R in terms of latency (100 ms from camped to connected state and 10 ms from dormant to active state), several mechanisms could be used to reduce latency (about 50 ms from camped to connected state). Improvements related to the transition from camped to connected state are:

- Combined request of RRC connection (User-eNodeB) and Non-Access Stratum (NAS) service (User-Mobility Management Entity (MME)). These two messages are processed in parallel in the eNodeB and the MME, respectively. The motivation for this combined request is that it reduces the amount of information of the radio link specific layers (Packet Data Convergence Protocol (PDCP), Radio Link Control (RLC) and MAC), since the radio interface information control and the NAS signaling are multiplexed together at the RRC protocol. Besides, the request and confirmation process to deliver NAS information is not needed. This improvement is typical of LTE-A, as in 3G systems, LTE Release 8 included, the RRC connection request and the NAS service connection are performed sequentially. Figure 1.15 shows the control plane activation procedure for the FDD mode when using a single message to transmit the RRC and NAS request (in the ideal case without retransmissions). Table 1.12 complements Figure 1.15 with a temporal analysis. The reduction could be of up to 21 ms. It should be noted that in the latency analysis the delay of the interface between the eNodeB and MME (S1-C) was not taken into account, as explicitly stated in the requirements.

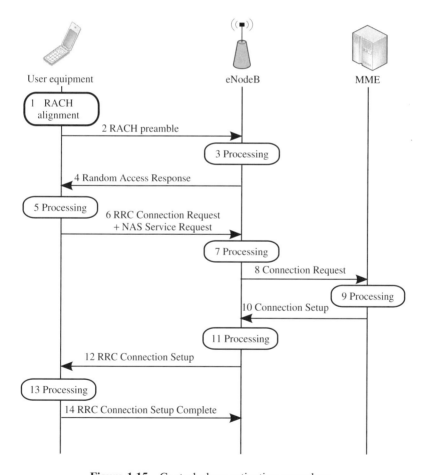

Figure 1.15   Control plane activation procedure.

Table 1.12 Temporal analysis in the control plane

Step	Description	Duration
1	Mean delay due to the RACH scheduling period	2.5 ms
2	RACH preamble	1 ms
3–4	Preamble detection and response transmission (time between the end of the RACH transmission and the reception of the scheduling grant and the temporal alignment) + delay until the nearest downlink subframe	5 ms
5	User processing (scheduling grant decoding, temporal alignment and C-RNTI allocation + L1 message coding of RRC Connection Request) + delay until the nearest uplink subframe	5 ms
6	Message transmission (RRC Connection Request and NAS Service Request)	1 ms
7	Delay of the eNodeB processing (Uu to S1-C)* + delay until the nearest downlink subframe	4 ms
8	Transfer delay S1-C	Ts1c (2–15 ms)
9	Delay of the MME processing	15 ms
10	Transfer delay S1-C	Ts1c (2–15 ms)
11	Delay of the eNodeB processing (S1-C to Uu)* + delay until the nearest downlink subframe	4 ms
12	Message transmission (RRC Security Mode Command and Connection Reconfiguration)	1.5 ms
13–14	User processing	20 ms
	Total delay	59 ms

*Uu: Interface between user and eNodeB.
S1-C: Interface between eNodeB and MME.

- Most of the latency delay is due to the processing delay in the user and in the eNodeB. Therefore, an improvement in processing delays will clearly influence the control plane latency.
- Reduction of the RACH scheduling period.

As for the transition from dormant state to the connected state, the following mechanisms could be used in LTE-A to improve its functioning:

- A smaller PUCCH to reduce the average waiting time of a synchronized user requesting resources in the active state.
- The uplink contest entails that the users transmit data without first having to request resources in the PUCCH and, therefore, reducing the access time of synchronized users in the connected state.

The user plane latency is already below 10 ms in LTE Release 8 with synchronized users. Anyway, the improvements that reduce control plane latency could also be used in the user plane. For example, a shorter RACH scheduling period, a shorter PUCCH and a smaller processing delay would improve user plane latency.

1.5 Summary

This chapter has shown the technological revolution posed by new IMT-Advanced technologies and their evolution. Compared to current systems, IMT-Advanced technologies include a greater bandwidth, new network elements, like relays or femtocells, M2M communications and coordination between transmitters. The channel models that have been used traditionally are no longer valid to analyze all these technological features. Consequently, in the last years there has been significant research activity to develop new channel and propagation models adapted to the new paradigms of wireless communications. The rest of the book addresses this specific issue.

References

1. UIT-R, "Framework for services supported by IMT", UIT, Recommendation M.1822, 2007.
2. UIT-R, "Requirements related to technical performance for IMT-Advanced radio interface(s)", UIT, Report M.2134, 2008.
3. 3GPP, "Physical Channels and Modulation (Release 9)", 3GPP, Technical Specification TR 36.211 v9.1.0.
4. 3GPP, "FDD RIT component of SRIT LTE Release 10 & beyond (LTE-Advanced)", 3GPP, Report RP-090745.
5. 3GPP, "FDD RIT component of SRIT LTE Release 10 & beyond (LTE-Advanced)", 3GPP, Report RP-090746.
6. 3GPP, "FDD RIT component of SRIT LTE Release 10 & beyond (LTE-Advanced)", 3GPP, Report RP-090747.
7. 3GPP, "Feasibility study for Further Advancements for EUTRAs (LTE-Advanced) (Release 9)", 3GPP, Technical Report TR 36.912 v9.0.0.
8. 3GPP, "Requirements for further advancements for Evolved Universal Terrestrial Radio Access (E-UTRA) (LTE-Advanced) (Release 9)", 3GPP, Technical Report TR 36.913 v9.0.0.
9. 3GPP, "Physical layer aspects for evolved Universal Terrestrial Radio Access (UTRA) (Release 7)", 3GPP, Technical Report TR 25.814 v7.1.0.
10. 3GPP, "Feasibility study for evolved Universal Terrestrial Radio Access (UTRA) and Universal Terrestrial Radio Access Network (UTRAN) (Release 9)", 3GPP, Technical Report TR 25.912 v9.0.0.
11. 3GPP, "Further advancements for E-UTRA; LTE-Advanced feasibility studies in RAN WG4 (Release 9)", 3GPP, Technical Report TR 36.815 v9.1.0.
12. 3GPP, "Cubic Metric in 3GPP-LTE", 3GPP, Report R1-060385.
13. 3GPP, "Cubic Metric comparison of OFDMA and Clustered-DFTS-OFDM/NxDFTS-OFDM", 3GPP, Report R1-084469.
14. 3GPP, "Further advancements for E-UTRA physical layer aspects (Release 9)", 3GPP, Technical Report TR 36.814 v9.0.0.
15. 3GPP, "Evolved Universal Terrestrial Radio Access (E-UTRA) and Evolved Universal Terrestrial Radio Access Network (E-UTRAN); Overall description", 3GPP, Technical Report TR 36.300 v9.0.0.

2

Propagation and Channel Modeling Principles

Andreas F. Molisch

University of Southern California, USA

This chapter will explore the basic principles of wireless propagation. It describes the fundamental processes of how signals can get from the transmitter to the receiver and summarizes the description/modeling methods for characterizing the propagation. We will not analyze any specific environments or give quantitative parameterizations of the channel models – those aspects will be treated in the later aspects of the book. Rather, it is the fundamental methodology that is at the center of this chapter. The description partly draws on the presentations [1–5], and the interested reader is referred to those expositions and the references therein for further details.

2.1 Propagation Principles

In almost all wireless communications, analogue passband Radio Frequency (RF) signals are sent from a transmitter to a receiver. At the transmitter, the signal is generated by baseband processing and modulated to a passband, where it is converted to an electromagnetic wave by the transmit antenna. This wave propagates through the channel to the receive antenna, where (part of) it is converted to a (passband) signal that can be further downcoverted and processed by appropriate circuitry. From this description it is clear that both the antennas and the propagation medium have an impact on the characteristics of the signal arriving at the receiver. It is useful to distinguish between the propagation of the electromagnetic waves proper (often characterized as the "propagation channel"), and the concatenation of propagation channel with the Transmitter (TX) and Receiver (RX) antennas (often called the "radio channel"). We assume that the reader is familiar with the basics of antenna theory, including antenna efficiency, directivity, and so on, for more details see [6, 7].

LTE-Advanced and Next Generation Wireless Networks: Channel Modelling and Propagation, First Edition.
Edited by Guillaume de la Roche, Andrés Alayón Glazunov and Ben Allen.
© 2013 John Wiley & Sons, Ltd. Published 2013 by John Wiley & Sons, Ltd.

2.1.1   Free-Space Propagation and Antenna Gain

We start out with the simplest of all possible transmission scenarios: two antennas in free space. Assuming that the transmit antenna is isotropic, it follows from symmetry that the transmitted wave is a spherical wave, and it follows from energy conservation that the power contained on the surface of a sphere of any radius d is equal to the transmit power. This implies, in turn, that the area power density has to decrease linearly with the surface area of the sphere. Taking a receive antenna with effective area A_{rx}, the received signal power is

$$P_{rx}(d) = P_{tx} \frac{1}{4\pi d^2} A_{rx}, \tag{2.1}$$

where $P_{tx/rx}$ is the transmit/receive power. Using the relationship between antenna area and antenna (power) gain G, the receiver power can be written as

$$P_{rx}(d) = P_{tx} G_{rx} \left(\frac{\lambda}{4\pi d} \right)^2. \tag{2.2}$$

The G_{rx} is the antenna power gain in the direction pointing towards the transmitter.

Equation (2.2), which is known as Friis' law, is valid in the far field of the antenna, that is, the TX and RX antennas have to be at least one *Rayleigh distance* apart. The Rayleigh distance (also known as *Fraunhofer distance*) is defined as

$$d_R = \frac{2L_a^2}{\lambda}, \tag{2.3}$$

where L_a is the largest dimension of the antenna.

This receive power can be increased by G_{tx} (antenna gain along the direction pointing towards the receiver) if the transmit antenna is non-isotropic. The receive power is maximized if the antenna gain is maximal along the line connecting TX and RX (in informal language, if the two antennas are "pointing" towards each other). For setting up link budgets, it is advantageous to write Friis' law on a logarithmic scale. Equation (2.2) then reads

$$P_{rx}|_{dBm} = P_{tx}|_{dBm} + G_{tx}|_{dB} + G_{rx}|_{dB} + 20 \log \left(\frac{\lambda}{4\pi d_0} \right) - 20 \log \left(\frac{d}{d_0} \right), \tag{2.4}$$

where $|_{dB}$ means "in units of dB" and d_0 is an (arbitrary) reference distance; it is often set to $d_0 = 1$ m.

Free space propagation usually does not occur in land-mobile radio systems, but it is a useful benchmark for other systems.

2.1.2   Reflection and Transmission

In usual environments, signals are reflected by the ground, buildings, or mountains; or they propagate through building or room walls. Thus, the fundamental process of reflection from, or transmission through, a dielectric layer is of great relevance. We start out with the reflection/transmission at the interface of two dielectric half-spaces, which are governed by Snell's laws. The direction of incidence is characterized by the angle Θ between the wave vector and the normal to the interface. The angle of incident, reflected, and transmitted wave, Θ_e, Θ_r, Θ_r, is determined as

$$\Theta_r = \Theta_e, \tag{2.5}$$

$$\frac{\sin \Theta_t}{\sin \Theta_e} = \frac{\sqrt{\varepsilon_1}}{\sqrt{\varepsilon_2}}, \tag{2.6}$$

where $\varepsilon_{1,2}$ are the dielectric constants of the two materials at each side of the interface.

For the magnitude of the waves, we have to distinguish the cases where the electric field vector is parallel to the interface of the two dielectrics (transversal electric wave, TE wave), and the case where the magnetic field is parallel to the interface, (transversal magnetic field, TM) wave. In the TE case,

$$\rho_{\text{TE}} = \frac{\sqrt{\varepsilon_1} \cos(\Theta_e) - \sqrt{\varepsilon_2} \cos(\Theta_t)}{\sqrt{\varepsilon_1} \cos(\Theta_e) + \sqrt{\varepsilon_2} \cos(\Theta_t)}, \tag{2.7}$$

$$T_{\text{TE}} = \frac{2\sqrt{\varepsilon_1} \cos(\Theta_e)}{\sqrt{\varepsilon_1} \cos(\Theta_e) + \sqrt{\varepsilon_2} \cos(\Theta_t)}, \tag{2.8}$$

and for the TM case

$$\rho_{\text{TM}} = \frac{\sqrt{\varepsilon_2} \cos \Theta_e - \sqrt{\varepsilon_1} \cos(\Theta_t)}{\sqrt{\varepsilon_2} \cos \Theta_e + \sqrt{\varepsilon_1} \cos(\Theta_t)}, \tag{2.9}$$

$$T_{\text{TM}} = \frac{2\sqrt{\varepsilon_1} \cos(\Theta_e)}{\sqrt{\varepsilon_2} \cos \Theta_e + \sqrt{\varepsilon_1} \cos(\Theta_t)}. \tag{2.10}$$

It is noteworthy that for grazing incidence, all reflection coefficients tend to -1, that is, the surface becomes perfectly reflecting.

For a dielectric layer of thickness d_{layer} the overall transmission and reflection coefficients can be computed as

$$T = \frac{T_1 T_2 e^{-j\alpha}}{1 + \rho_1 \rho_2 e^{-2j\alpha}}, \tag{2.11}$$

$$\rho = \frac{\rho_1 + \rho_2 e^{-j2\alpha}}{1 + \rho_1 \rho_2 e^{-2j\alpha}}, \tag{2.12}$$

where T_1 and T_2 are the transmission coefficients at the two interfaces (and similarly for the reflection coefficients $\rho_{1,2}$), and α is the electrical length of the layer

$$\alpha = \frac{2\pi}{\lambda} \sqrt{\varepsilon_2} d_{\text{layer}} \cos(\Theta_t). \tag{2.13}$$

2.1.3 Diffraction

Waves can also "bend" around obstacles, a process known as diffraction. The effectiveness of the process depends on the wavelength: the shorter the wavelength, the more the waves behave like rays that throw sharp "shadows". At cellular frequencies, diffraction can be essential for allowing signals to completely "cover" all points in an area.

A canonical scenario in wireless communications is the diffraction of a wave by an absorbing screen. Let a plane wave propagate along the x-axis, $E = \exp(-jk_0x)$ (where k_0 is the wave number $2\pi/\lambda$ with λ being the wavelength) and impinge on a screen extending at $x = 0$, $-\infty < y < 0$. The electrical field for $x > 0$ is then

$$E_{total} = \exp(-jk_0x)\left(\frac{1}{2} - \frac{\exp(j\pi/4)}{\sqrt{2}} F\left(\nu_F\right)\right), \tag{2.14}$$

where the Fresnel parameter $\nu_F = -2y/\sqrt{\lambda x}$ and the Fresnel integral $F\left(\nu_F\right)$ is defined as

$$F\left(\nu_F\right) = \int_0^{\nu_F} \exp\left(-j\pi\frac{t^2}{2}\right) dt. \tag{2.15}$$

Simple geometrical considerations allow the computation of the Fresnel parameter for arbitrary location of the screen and angle-of-incidence of the wave.

The situation is more complicated when there are multiple screens. As a matter of fact, an exact solution of the multiple-screen problem is only available in a few special cases. However, a number of approximate methods exist [8]. The simplest one is the Bullington method, where an equivalent screen is constructed from the geometry of the actual screens, see Figure 2.1. In this approach, the multiple diffracting screens are replaced by a single, equivalent, screen, whose attenuation can be computed by the equations given above. Note, however, that this method does not correctly reproduce important limiting behaviors, for example, the diffraction loss by many screens of equal height.

2.1.4 Scattering

Specular reflection, as discussed in Section 2.1.2, occurs when a plane wave is incident on a perfectly smooth (and infinitely extended) object. When the surface is rough, radiation is scattered in all directions, see Figure 2.2. For theoretical purposes, it is advantageous

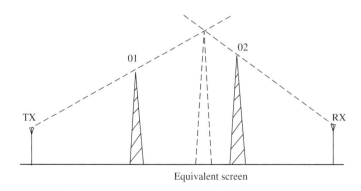

Equivalent screen

Figure 2.1 Equivalent screen after Bullington. Reproduced with permission of John Wiley & Sons, Inc. from [1].

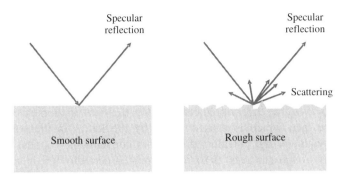

Figure 2.2 Scattering by a rough surface. Reproduced with permission of John Wiley & Sons, Inc. from [1].

to model the surfaces from which scattering occurs as random. However, for modeling purposes it is common to subsume all variations of surfaces that have dimensions smaller than a wavelength into the surface roughness. Thus, for example, a building facade with protruding window sills would be modeled as a surface with random roughness, even though the window sills clearly have a regular structure. The justifications for this approach are rather heuristic: (i) the errors made are smaller than some other error sources in ray tracing predictions, and (ii) there is no better alternative.

There are two main theories of scattering by rough surfaces: the Kirchhoff theory, and the perturbation theory. The Kirchhoff theory assumes that the height variations are so small that different *scattering points* on the surface do not influence each other. Then the surface roughness leads to a reduction of the power of the specularly reflected ray (by the amount that is scattered into other directions). This power reduction can be described by an *effective* reflection coefficient $\rho_{\rm rough}$. In the case of a Gaussian height distribution, this reflection factor becomes

$$\rho_{\rm rough} = \rho_{\rm smooth} \exp\left[-2\left(k_0\sigma_{\rm h}\sin\psi\right)^2\right], \tag{2.16}$$

where $\sigma_{\rm h}$ is the standard deviation of the height distribution, and ψ is the angle of incidence (defined as the angle between the wave vector and the surface). The term $2k_0\sigma_{\rm h}\sin\psi$ is also known as the Rayleigh roughness. Note that for grazing incidence ($\psi \approx 0$), the roughness vanishes, and the reflection becomes specular again.

In contrast to the Kirchhoff theory, the perturbation theory takes into account that some points of a rough surface can "cast a shadow" onto other points. This effect can be quantified by the coherence distance of the surface roughness, that is, how "fast" the surface changes. The theory is often better suited to the physical realities of wireless communications environments, but due to its more complicated nature we will not describe it here in detail; the interested reader is referred to [8].

2.1.5 Waveguiding

Electromagnetic waves can be guided by certain structures, such as dielectric slabs, metallically enclosed cylinders, and so on. Such effects are very important in optical fiber communications, where they lead to propagation of waves that are essentially lossless for many meters or even kilometers. For mobile communications, "approximate" waveguiding exists in structures such as corridors and street canyons. The walls of these structures constrain the propagation of the waves and prevent a "thinning out" such as would occur in free-space propagation. Thus, the received power in such a structure can be larger than what one would obtain in free space – note that this is not a violation of energy conservation.

Conventional waveguide theory predicts a propagation loss that increases exponentially with the distance, and which is caused solely by the losses in the waveguide walls. However, street canyons and corridors differ from conventional (optical or microwave) waveguides by (i) rough surfaces, (ii) periodic or non-periodic openings of the surface (cross streets, doorways, and so on), (iii) the absence of a restricting surface towards the sky (for street canyons). These effects tend to dominate losses, and lead to an attenuation that is governed by a conventional power law.

Propagation prediction can be done either by computing the waveguide modes, or by a geometric-optics approximation. If the waveguide cross section, as well as the objects in it, are much larger than the wavelength, the latter method gives good results.

2.1.6 Multipath Propagation

Propagation for most modern wireless systems is critically impacted by multipath propagation, that is, the fact that the signal can get from the transmitter to the receiver via a multitude of different paths, see Figure 2.3. Consequently, a large number of Multipath Components (MPCs) arrive at the receiver, and add up there. Depending on the relative phases of the MPCs, the addition (superposition) can be constructive or destructive. Generally, MPCs can be characterized by complex amplitude (or equivalently, power and

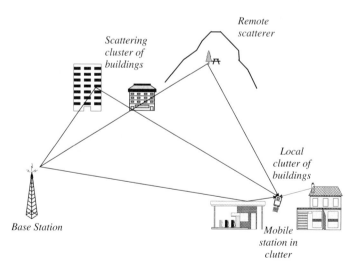

Figure 2.3 Principle of multipath propagation. © 2009 IEEE. Reprinted, with permission, from [3].

phase), delay (time it takes to propagate along a particular path from the TX to the RX), angle of departure from the TX, angle of arrival at the RX, and polarization.

The multipath propagation gives rise to a number of important effects that determine wireless system behavior:

- *fading*: the superposition of the MPCs depends on the relative phases between the MPCs. The phase relationship changes as the TX, the RX, or the Interacting Objects (IOs) (reflectors, scatterers, diffracting objects) move. Thus, the received power changes with the location of those objects, and thus with time. This effect is known as fading, and is generally described statistically, as elaborated in Section 2.3.
- *frequency dispersion*: due to the Doppler effect, movement of TX, RX, or IOs leads to a change in the perceived frequency of a signal. This change depends on the velocity and direction of the movement relative to the direction of the MPCs, and is thus different for different MPCs. Consequently, the Doppler effect leads not only to a shift in the frequency, but to a spreading (dispersion).
- *delay dispersion*: the different MPCs have different "runtimes", that is, the time it takes the signal to propagate from the TX to the RX along a particular path. Thus, if the TX sends out a (delta) pulse, a sequence of pulses arrives at the RX. This impulse response also changes with the position of TX, RX, and IOs. Its statistical description is given in Section 2.3.3.
- *angular dispersion*: if only a Line Of Sight (LOS) connection existed between the TX and RX, then direction of departure and direction of arrival would be uniquely given by the location of TX and RX. Due to the multipath propagation, radiation is leaving the TX in different directions, and similarly, arriving at the RX from different directions. This dispersion is called "angular dispersion".
- *polarization dispersion*: when an MPC undergoes, for example, a reflection process, its polarization state changes. A wave that starts out with purely vertical polarization can end up with a mixed vertical/horizontal polarization by the time it arrives at the RX. The degree of depolarization is different for each MPC, and thus also leads to a polarization dispersion.

Thus, propagation effects generally fall into two categories: (i) multipath propagation, that is, the fact that the signal can propagate via different paths, and that the MPCs interfere with each other at the receiver, and (ii) change of the characteristics of each MPC, which are determined by the propagation process (reflection, transmission, diffraction).

2.2 Deterministic Channel Descriptions

In this section, we will list various methods for deterministic description of the channels, that is, quantifying the impact of wireless propagation for a given time/location of the TX, RX, and IOs. We start out by simple systems where the TX and RX have a single antenna each, so that the system is completely described by the input $x(t)$ and the output $y(t)$ of the channel. We then progress to more complicated, multi-antenna channels, which are highly relevant for Long Term Evolution (LTE), Long Term Evolution Advanced (LTE-A) and other modern communications systems.

2.2.1 Time Variant Impulse Response

In single-antenna systems, input and output of the channel are related by the "time-variant impulse response" $h(t, \tau)$ by

$$y(t) = \int_{-\infty}^{\infty} x(t - \tau)h(t, \tau)d\tau. \tag{2.17}$$

This relationship is analogous to the well-explored input-output relationship in Linear Time-Invariant (LTI) systems; the difference lies in the fact that now the impulse response is time-variant.

An intuitive interpretation is possible if the impulse response changes only slowly with time. Then we can consider the behavior of the system at a particular time t like that of an LTI system. The variable t can thus be viewed as "absolute" time that tells us which impulse response $h(\tau)$ is currently valid. Such a system is also called *quasi-static*. Such an interpretation is meaningful if the timescale on which the channel impulse response changes is much larger than the duration of the impulse response, and furthermore much larger than the transmitted symbol duration. These conditions are fulfilled in almost all practical wireless systems: times over which the channel stay constant are in the order of milliseconds or larger, while duration of impulse responses (and typical symbol durations) are microseconds or smaller.

As the impulse response of a time-variant system, $h(t, \tau)$, depends on two variables, τ and t, we can perform Fourier transformations with respect to either (or both) of them. This results in four different, but equivalent, representations, see Figure 2.4. Fourier-transforming the impulse response with respect to the variable τ results in the

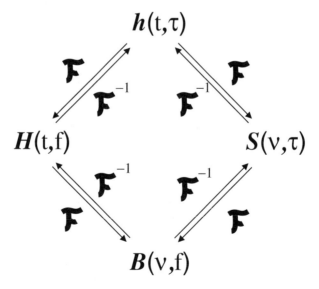

Figure 2.4 Interrelation between the deterministic system functions. Reproduced with permission of John Wiley & Sons, Inc. from [1].

time-variant transfer function $H(f, t)$

$$H(t, f) = \int_{-\infty}^{\infty} h(t, \tau) \exp(-j2\pi f\tau)d\tau. \tag{2.18}$$

The input-output relationship is given by

$$y(t) = \int_{-\infty}^{\infty} X(f)H(t, f) \exp(j2\pi ft)df. \tag{2.19}$$

The interpretation is straightforward for the case of the quasistatic system – the spectrum of the input signal is multiplied with the spectrum of the "currently valid" transfer function, to give the spectrum of the output signal. Thus, for quasi-static channels, the transfer function calculus $Y(f) = H(f)X(f)$ is valid. If we do a Fourier transformation with respect to t we obtain the *delay-Doppler function*, also known as *spreading function* $S(\nu, \tau)$

$$S(\nu, \tau) = \int_{-\infty}^{\infty} h(t, \tau) \exp(-j2\pi \nu t)dt. \tag{2.20}$$

This function describes the spreading of the input signal in the delay/Doppler domains.

The *Doppler-variant transfer function* $B(\nu, f)$ is obtained by transforming $S(\nu, \tau)$ with respect to τ

$$B(\nu, f) = \int_{-\infty}^{\infty} S(\nu, \tau) \exp(-j2\pi f\tau)d\tau. \tag{2.21}$$

The channel impulse response can be related to the representation of the MPCs. Assuming that the propagation channel can be completely represented by summing up the contributions of the different MPCs

$$h(t, \tau) = \sum_{i=1}^{L} a_i(t)\delta(\tau - \tau_i). \tag{2.22}$$

where a_i is the complex amplitude of the i-th MPC; the amplitude of the MPC is the amplitude response of the "radio channel" that includes the antennas, in other words the channel from TX antenna connector to RX antenna connector (we already discussed above the definition of the pure propagation channel, which is defined as not including the antennas). τ_i is the delay associated with it.[1] Note that this impulse response is time-variant – at different absolute times t, a different impulse response $h(\tau)$ characterizes the channel. The actual value of this impulse response is determined by the value of the complex attenuation of the MPCs at time t. Equation (2.22) is valid for systems with large bandwidth, so that we can resolve all the MPCs. Thus, there is no small-scale fading. Furthermore, it is assumed that the only temporal change of the MPCs is a phase shift due to the change of runtime between the TX and RX as the mobile station (MS) (or the interacting objects) move around, which is true for small movement distances.

---

[1] The use of delta functions in the impulse response is a common didactic tool in the derivation of impulse responses for "infinite bandwidth" channels, and also employed here. However, it gives rise to both physical problems (since MPCs in reality are frequency-dependent, see Section 2.1), and mathematical issues in the definition of the power delay profile (squares of delta functions are not defined). For a more exact derivation, see [9].

In a system with finite bandwidth B, the impulse response is completely characterized by the samples taken at the Nyquist rate

$$h(t, \tau) = \sum c_j(t)\text{sinc}(B\tau - j).$$ (2.23)

Note that the samples c_j are impacted by multiple MPCs, and thus their absolute amplitudes are changing with time – in contrast to the a_i.

A further approximation, which is often useful for intuitive interpretations, arises from the fact that we can only distinguish signals that are approximately $\Delta = 1/B$ separated in the delay domain. We can thus divide the delay axis into bins of width Δ, which are called "resolvable delay bins". A discretized impulse response then reads

$$h(t, m) = \sum \tilde{c}_m(t)\delta(\tau - m\Delta),$$ (2.24)

where $\tilde{c}_m$ describes the sum of contributions from different MPCs to the m-th bin.

2.2.2 Directional Description and MIMO Matrix

For multi-antenna systems, the directional characteristics of the MPCs play a major role. Thus, the impulse response (2.22) should be replaced by the *double-directional impulse response* [10], which consists of a sum of contributions from the MPCs

$$h(t, \mathbf{r}_{tx}, \mathbf{r}_{rx}, \tau, \Omega, \Psi) = \sum_{\ell=1}^{L} h_\ell(t, \mathbf{r}_{tx}, \mathbf{r}_{rx}, \tau, \Omega, \Psi)$$ (2.25)

$$= \sum_{\ell=1}^{L} a_\ell \delta(\tau - \tau_\ell)\delta(\Omega - \Omega_\ell)\delta(\Psi - \Psi_\ell),$$

where the locations of transmitter is $\mathbf{r}_{tx}$ and receiver is $\mathbf{r}_{rx}$, the *direction-of-departure (DoD)* Ω, and the *direction-of-arrival (DoA)* Ψ. Just like in the case of (2.22), the phases of the a_i change quickly, while all other parameters, that is, absolute amplitude $|a|$, delay, DoA and DoD, vary slowly with the transmit and receive locations (over many wavelengths). It is noteworthy that in this representation, the MPC amplitudes reflect the complex gain of the propagation channel only, without any consideration of the antennas. The conventional impulse response of the radio channel (including the antennas) can be recovered by weighting the DDIR with the antenna pattern, and then integrating over all angles

$$h(t, \tau) = \int \int h(t, \tau, \Omega, \Psi)\tilde{G}_{tx}(\Omega)\tilde{G}_{RX}(\Psi)d\Omega d\Psi,$$ (2.26)

where $\tilde{G}_{tx}$ and $\tilde{G}_{rx}$ are the complex amplitude antenna patterns (in linear units) for the TX and RX antenna, and we omitted the dependence on the transmitter and receiver location.

For multiple-antenna systems, we are also often interested in the impulse response or channel transfer function of the radio channel (i.e., including the antenna characteristics) from each TX antenna element to each RX antenna element. This is given by the impulse response matrix. We denote the transmit and receive element coordinates as

$\mathbf{r}_{tx}^{(1)}, \mathbf{r}_{tx}^{(2)}, \ldots \mathbf{r}_{tx}^{(N_{tx})}$, and $\mathbf{r}_{rx}^{(1)}, \mathbf{r}_{rx}^{(2)}, \ldots \mathbf{r}_{rx}^{(N_{rx})}$, respectively, so that the impulse response from the j-th transmit to the i-th receive element becomes

$$h_{i,j} = h\left(\mathbf{r}_{tx}^{(j)}, \mathbf{r}_{rx}^{(i)}\right) \tag{2.27}$$

$$= \sum_\ell h_\ell(\mathbf{r}_{tx}^{(1)}, \mathbf{r}_{rx}^{(1)}, \tau, \Omega_\ell, \Psi_\ell) \widetilde{G}_{tx}(\Omega_\ell) \widetilde{G}_{rx}(\Psi_\ell) \delta(\tau - \tau_\ell)$$

$$\exp\left(j\mathbf{k}(\Omega_\ell) \cdot (\mathbf{r}_{tx}^{(j)} - \mathbf{r}_{tx}^{(1)})\right) \exp\left(j\mathbf{k}(\Psi_\ell) \cdot (\mathbf{r}_{rx}^{(i)} - \mathbf{r}_{rx}^{(1)})\right)$$

$\delta(\tau - \tau_\ell)$ where "$\cdot$" denotes the scalar product. For the case that the TX and RX arrays are uniform linear arrays with element spacing d_a we can define a TX *steering vector*, $\alpha_{tx}(\Omega) = \frac{1}{\sqrt{N_t}}[1, \exp(-j2\pi \frac{d_a}{\lambda} \sin(\Omega)), \ldots \exp(-j2\pi(N_{tx} - 1) \frac{d_a}{\lambda} \sin(\Omega))]^T$. The steering vector at the receiver $\alpha_{rx}(\Psi)$ is similarly defined. The impulse response matrix becomes

$$\mathbf{H} = \int \int h(\tau, \Omega, \Psi) \widetilde{G}_{tx}(\Omega) \widetilde{G}_{rx}(\Psi) \alpha_{rx}(\Psi) \alpha_{tx}^\dagger(\Omega) \, d\Psi \, d\Omega. \tag{2.28}$$

2.2.3 Polarization

A further refinement of the propagation model takes the polarization characteristics of channel and antennas into account [11]. Consider a situation where both TX and RX have dual-polarized antennas, that is, antennas that are capable of independently transmitting and receiving orthogonally polarized waves (for the sake of simplicity, we henceforth use vertically and horizontally polarized waves, V and H, though alternative characterizations are possible). We note that the antennas are characterized by two (complex) antenna patterns, $\widetilde{G}^V(\Omega)$ and $\widetilde{G}^H(\Omega)$. The propagation is characterized by four polarisation channels to be considered, $HH, HV, VH,$ and VV, respectively. This gives rise to four impulse responses to be modeled. A generalization of (2.25) reads

$$\mathbf{h}(t, \mathbf{r}_{tx}, \mathbf{r}_{rx}, \tau, \Omega, \Psi) = \sum_{\ell=1}^L \begin{pmatrix} \alpha_\ell^{VV} & \alpha_\ell^{VH} \\ \alpha_\ell^{HV} & \alpha_\ell^{HH} \end{pmatrix} \delta(\tau - \tau_\ell) \delta(\Omega - \Omega_\ell) \delta(\Psi - \Psi_\ell), \tag{2.29}$$

and the generalization of (2.27) thus reads

$$h_{l,m} = h\left(\mathbf{r}_{tx}^{(m)}, \mathbf{r}_{rx}^{(l)}\right) \tag{2.30}$$

$$= \sum_n \begin{pmatrix} \widetilde{G}_{tx}^V(\Omega_n) \\ \widetilde{G}_{tx}^H(\Omega_n) \end{pmatrix}^T \begin{pmatrix} a_n^{VV} & a_n^{HV} \\ a_n^{VH} & a_n^{HH} \end{pmatrix} \begin{pmatrix} \widetilde{G}_{rx}^V(\Psi_n) \\ \widetilde{G}_{rx}^H(\Psi_n) \end{pmatrix}$$

$$\delta(\tau - \tau_n) \exp\left(j\mathbf{k}(\Omega_n) \cdot (\mathbf{r}_{tx}^{(m)} \delta(\tau - \tau_n)\right) \exp\left(j\mathbf{k}(\Psi_n) \cdot (\mathbf{r}_{rx}^{(l)} - \mathbf{r}_{rx}^{(1)})\right).$$

2.2.4 Ultrawideband Description

The channel representation (2.22) is implicitly based on the assumption that each MPC leads to a contribution to the total impulse response in the shape of a delta function, which in turn implies that the propagation processes suffered by an MPC are not

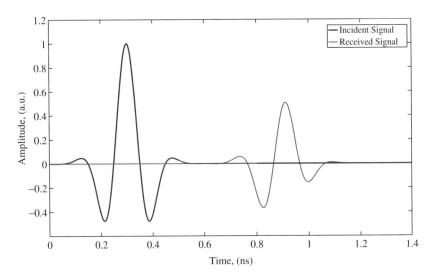

Figure 2.5 UWB pulse diffracted by a semi-infinite screen.

frequency-dependent. As we can see from Section 2.1, this not the case – all the described propagation effects depend in some form on the carrier frequency; for example, diffraction is a more effective process at low frequencies than at high (geometrical optics regime) frequencies. Consequently, when considering very large bandwidths, the impulse response for a specific time becomes

$$h(\tau) = \sum_{i=1}^{N} a_i \chi_i(\tau) \otimes \delta(\tau - \tau_i), \tag{2.31}$$

where $\chi_i(\tau)$ denotes the distortion of the i-th MPC by the frequency selectivity of the interacting objects. One example for a distortion of a short pulse by diffraction is a screen shown in Figure 2.5. Further details on modeling approaches for UWB channels can be found in [12, 13], and references therein.

2.3   Stochastic Channel Description

There are two basic approaches to devise propagation channel models: deterministic and stochastic. A deterministic model aims to correctly predict the channel characteristics (e.g., impulse response), in a specific location, using information about the location of the transmitter and receiver, as well as the surrounding environment. The resulting channel model is thus valid only in this specific location. Such models are commonly used for network planning, that is, estimating how well a *given* system will work in a specific environment. Typical applications are the search for good base station sites in cellular networks and Local Area Networks (LANs).

A stochastic model, on the other hand, does not try to correctly predict each channel realization (channel at a specific location). Rather, it models the *statistical properties* of the channel. This concept can be most easily understood for narrowband channels, where the received field strength is the most important quantity. While a deterministic model would try to predict the correct field strength at, for example, each point in a room, a stochastic model would just specify that the probability density function of the field strength is Rayleigh-distributed. Stochastic models can be made more refined by dividing the environments of interest into different "classes" and using different model parameters in each of those classes. The main purpose of stochastic models is the analysis and design of wireless transceivers and transmission schemes. Clearly, it is not important whether a new equalization algorithm works in one particular square in one particular city, but rather it should be designed in such a way as to work well in the majority of all locations in the prospective deployment area (which could be the whole world).

2.3.1 Pathloss and Shadowing

We first deal with the statistical characterization of the (narrowband) received power. It has been long established that this power varies on different spatial scales. When considering movement within a very short distance, the interference of the MPCs gives rise to small-scale fading, as described in Section 2.1.6, and further elaborated in Section 2.3.2. The average, taken over an area of typically $10\lambda \times 10\lambda$, of the received power is often called the small-scale-averaged (SSA) power. When considering different locations that have the same distance from the TX, we find that MPCs of different strength are able to reach the various locations, or the strength of an MPC is different when reaching different locations – effects that are generally subsumed under the name "shadowing". Consequently, the SSA power shows variations as function of the position x. The probability density function of shadowing is usually modeled as a lognormal distribution, that is, the logarithm of shadowing, S is a Gaussian-distributed real variable. When considering the spatial scale on which significant changes of the shadowing occur, we consider the autocorrelation function along the path of movement: the correlation between S-values for points at x and $x + \triangle x$ can often be approximated by a one-sided exponential function

$$E\{S(x)S(x + \triangle x)\} = \sigma^2 exp(-|\triangle x|/L_S),   \tag{2.32}$$

where L_S is the so-called correlation distance of shadow fading; typical values are on the order of 10 m (in cities) to 500 m (in rural areas).

When averaging the SSA received power over the shadowing process, the resulting large-scale-averaged (LSA) power shows a dependence on the distance between TX and RX. The ratio of the averaged received power $E\{P_{rx}\}$ (where $E\{.\}$ denotes expectation) to the transmit power P_{tx}, also known as path gain G, is usually well described by the following equation

$$G = \frac{E\{P_{rx}\}}{P_{tx}} \sim a \left(\frac{d}{d_0}\right)^{-\gamma},   \tag{2.33}$$

where d_0 is a reference distance, in the same units as d; and a and γ are dimensionless model parameters. It is easy to see that the power gain in free-space propagation is a special case of (2.33), where $a = (\frac{4\pi d_0}{\lambda})^{-2}$ and $\gamma = 2$. In most wireless environments, *the pathloss exponent*, γ, is greater than 2 because, if there are obstructions along the path, power falls off more quickly with distance than predicted by free-space loss alone. Conversely, in some scenarios such as corridors or street canyons with line-of-sight, γ can be less than 2 because of waveguiding effects.

2.3.2 Small-Scale Fading

The interference of the MPCs results in variations of the field strength if the TX, RX, or IO moves a distance in the order of a wavelength or less. Interpreting the sequence of temporal (spatial) samples as a random process, we generally characterize it by the probability density function (pdf), and the autocorrelation function. In the following we will first consider narrowband systems, and then extend the consideration to wideband and MIMO systems.

A variety of models exist for the amplitude pdf. The most common is for the complex amplitude to have a zero-mean circularly symmetric complex Gaussian amplitude distribution. In this case the phase of the resulting field strength has a uniform pdf over $(-\pi, \pi]$; the magnitude r has a Rayleigh pdf [14], that is,

$$f_r(r) = \frac{r}{\sigma^2} \exp\left(-\frac{r^2}{2\sigma^2}\right) \qquad 0 \le r < \infty, \tag{2.34}$$

where σ is the standard deviation of the underlying Gaussian process, and the received power p in Rayleigh fading has a single-sided exponential pdf

$$f_p(p) = \frac{1}{p_m} \exp\left(-\frac{p}{p_m}\right) \qquad 0 \le p < \infty, \tag{2.35}$$

where $p_m = 2\sigma^2$ is the mean power. If one MPC carries significantly more power than the others, the amplitude pdf is Rician

$$f_r(r) = 2r\frac{\exp(-K)}{p_s} \exp\left(-\frac{r^2}{p_s}\right) I_0\left(\sqrt{\frac{4r^2 K}{p_s}}\right) \qquad 0 \le r < \infty, \tag{2.36}$$

where $I_0(x)$ is the modified Bessel function of the first kind and zero order [15], and the Rice factor K (K-factor) describes the ratio of the power carried by the dominant component to that carried in the other MPCs; the sum of the other MPCs is modeled as a zero-mean complex Gaussian process with mean power p_s. The associated distribution of the received power is

$$f_p(p) = \frac{\exp(-K)}{p_s} \exp\left(-\frac{p}{p_s}\right) I_0\left(\sqrt{\frac{4p K}{p_s}}\right) \qquad 0 \le p < \infty. \tag{2.37}$$

Another widely used amplitude distributions the Nakagami distribution

$$f_r(r) = \frac{2}{\Gamma(m)} \left(\frac{m}{\Omega}\right)^m r^{2m-1} \exp\left(-\frac{m}{\Omega}r^2\right) \qquad 0 \le r < \infty, \tag{2.38}$$

for $r \geq 0$ and $m \geq 1/2$; $\Gamma(m)$ is Euler's Gamma function [15]. The parameter Ω is the mean square value $\Omega = E\{r^2\}$, and the parameter m is

$$m = \frac{\Omega^2}{E\left\{\left(r^2 - \Omega\right)^2\right\}}. \tag{2.39}$$

The power follows a Gamma distribution

$$f_p(p) = \frac{m}{\Omega \Gamma(m)} \left(\frac{mp}{\Omega}\right)^{m-1} \exp\left(-\frac{mp}{\Omega}\right) \qquad 0 \leq p < \infty. \tag{2.40}$$

Since the fading results from a change in the relative phase relationships, which occurs continuously as TX, RX, or IOs move, the received amplitude or power changes continuously as well. Furthermore, the rate of change depends on the angles between the directions of the MPC.

We now turn to the temporal evolution of the random process describing the amplitudes. First, the Doppler frequency of a single MPC is

$$v = \frac{v}{\lambda} \cos \alpha, \tag{2.41}$$

where α is the angle between the MPC and the direction of movement and v is the velocity. In a typical wireless environment, there is not just one MPC; there are many. With N MPCs, the received signal is the sum of N MPCs at slightly different frequencies. The result is a random-like variation over time of the envelope of the sum. The set of power+angle of the MPCs is called the Angle-of-Arrival (AoA) Spectrum

$$S_\alpha(\alpha) = \sum_i P_i \delta(\alpha - \alpha_i), \qquad -\pi < \alpha < \pi. \tag{2.42}$$

Using the relationship between the angles and Doppler shifts (2.41), this equation can be transformed to provide the Doppler spectrum.

From the Doppler spectrum, we can derive the second-order statistics of the temporal variation: there exist a Fourier relationship that exists between the Doppler spectrum and the Autocorrelation Function (ACF) of the complex amplitude. In the important case where the Doppler spectrum follows the Clarke-Jakes Doppler Spectrum [14], the Fourier transform is

$$ACF(\triangle t) = J_o(2\pi v_{max} \triangle t) = J_o\left(\frac{2\pi v \triangle t}{\lambda}\right), \tag{2.43}$$

where $\triangle t$ is the time separation between two points on the complex envelope's temporal variation and J_o is the Bessel function zero order, first kind.

2.3.3 WSSUS

A complete stochastic description of the impulse response would require its multidimensional pdf. This is often too complicated, so that [16] suggested a description by the mean (which is usually zero) and autocorrelation function (ACF) only, which is a complete characterization if the fading statistics are complex Gaussian. The ACF is

$$R_h(t, t'; \tau, \tau') = E\left\{h^*(t, \tau)h(t', \tau')\right\}, \tag{2.44}$$

where * denotes complex conjugation. Equivalently, we can define the ACFs of the Fourier transforms of the impulse response

$$R_s(v, v'; \tau, \tau') = E\left\{s^*(v, \tau)s(v', \tau')\right\}, \tag{2.45}$$

$$R_H(t, t'; f, f') = E\left\{H^*(t, f)H(t', f')\right\}, \tag{2.46}$$

$$R_B(v, v'; f, f') = E\left\{B^*(v, f)B(v', f')\right\}. \tag{2.47}$$

A further simplification can be obtained if the random process describing the channel is wide-sense stationary (WSS) and the scatterers are uncorrelated (US). In that case, the ACFs depend only on *two* instead of four variables

$$R_h(t, t + \Delta t, \tau, \tau') = \delta(\tau - \tau')P_h(\Delta t, \tau), \tag{2.48}$$

$$R_H(t, t + \Delta t, f, f + \Delta f) = R_H(\Delta t, \Delta f), \tag{2.49}$$

$$R_s(v, v', \tau, \tau') = \delta(v - v')\delta(\tau - \tau')P_s(v, \tau), \tag{2.50}$$

$$R_B(v, v', f, f + \Delta f) = \delta(v - v')P_B(v, \Delta f), \tag{2.51}$$

where $P_h(\Delta t, \tau)$ is known as the *delay cross power spectral density*, $R_H(\Delta t, \Delta f)$ is the *time-frequency correlation function*, $P_s(v, \tau)$ is the *scattering function*, and $P_B(v, \Delta f)$ is the *Doppler cross-power spectral density* [17].

The function $R_H(0, \Delta f)$ is known as the frequency correlation function; $R_H(\Delta t, 0)$ is the time correlation function. The function $P(\tau) = P_h(0, \tau)$ is usually called the *power delay profile* (PDP), or *delay power density spectrum*, and it describes the expected received power for different delays τ. Assuming ergodicity, the PDP can be computed directly from measured values of the impulse response

$$P(\tau) = E_t\{|h(t, \tau)|^2\}. \tag{2.52}$$

A widely used quantity characterizing delay dispersion is the *rms delay spread*, which is the second central moment of the normalized PDP

$$S_\tau = \sqrt{\frac{\int_{-\infty}^{\infty} P(\tau)\tau^2 d\tau}{\int_{-\infty}^{\infty} P(\tau)d\tau} - \left(\frac{\int_{-\infty}^{\infty} P(\tau)\tau d\tau}{\int_{-\infty}^{\infty} P(\tau)d\tau}\right)^2}. \tag{2.53}$$

Related is the *coherence bandwidth*. More precisely, the coherence bandwidth B_c of level k is defined as the smallest number so that $|\tilde{R}_H(B_c)| < k$, where $\tilde{R}_H(B_c)$ is the normalized frequency correlation function

$$\tilde{R}_H(\Delta f) = \frac{E_t\{H^*(t, f)H(t, f + \Delta f)\}}{E_t\{|H(t, f|^2\}}. \tag{2.54}$$

The levels $k = 0.5, 0.75, 0.9$ have been used in the literature. There is an "uncertainty" relationship between rms delay spread and coherence bandwidth [18]

$$B_k \geq \frac{\arccos(k)}{2\pi} \frac{1}{S_\tau}. \tag{2.55}$$

2.3.4 Extended WSSUS

Analogously to the nondirectional case, we can define condensed descriptions of the directional channel [19, 20]. For example, the angular scattering function $s(\Omega, \tau, \nu)$, which takes into account the DoD Ω (but not the DoA), allows the definition of angular delay power spectrum (ADPS) and the angular power spectrum (APS) as

$$E\{s^*(\Omega, \tau, \nu)s(\Omega', \tau', \nu')\} = P_s(\Omega, \tau, \nu)\delta(\Omega - \Omega')\delta(\tau - \tau')\delta(\nu - \nu'), \tag{2.56}$$

$$ADPS(\Omega, \tau) = \int P_s(\Omega, \tau, \nu)d\nu, \tag{2.57}$$

$$APS(\Omega) = \int APDS(\Omega, \tau)d\tau. \tag{2.58}$$

Note also that an integration of the ADPS over Ω recovers the PDP. Note that we can also define a double-directional delay power spectrum DDDPS, which generalizes the ADPS to take into account the DoA Ψ.

The *azimuthal spread* is defined as the second central moment of the APS if all MPCs are incident in the horizontal plane. In many papers it is defined in a form analogous to (2.53), namely

$$S_\phi = \sqrt{\frac{\int APS(\phi)\phi^2 d\phi}{\int APS(\phi)d\phi} - \left(\frac{\int APS(\phi)\phi d\phi}{\int APS(\phi)d\phi}\right)^2}. \tag{2.59}$$

However, this definition is ambiguous because of the periodicity of the angle. A better definition is [18] (assuming normalized APS)

$$S_\phi = \sqrt{\int \left|\exp(j\phi) - \mu_\phi\right|^2 APS(\phi)d\phi}, \tag{2.60}$$

with

$$\mu_\phi = \int \exp(j\phi)APS(\phi)d\phi. \tag{2.61}$$

2.4 Channel Modeling Methods

2.4.1 Deterministic Modeling

Enabled by faster computers and more efficient algorithms, deterministic field computation methods have developed into one of the most important methods for obtaining channel impulse responses, in particular for site-specific propagation analysis. For small environments (e.g., a few indoor rooms), and for the simulation of the impact of nearby objects on a wireless antenna/transceiver, full electromagnetic solvers have become popular [21]. In particular the FDTD (finite-difference time-domain) method is in widespread use [22]. The use of such method is presented in Chapter 11.

However, for the majority of cases, simplified solutions based on high-frequency approximations are common, where the propagation of waves is approximated by equivalent "rays" that behave similar to optical rays (though diffraction can be taken into account). Ray tracing models will be investigated deeply in Chapter 10. There are two popular variants of high-frequency approximations: "ray tracing" and "ray launching". Ray tracing uses the "imaging method" where reflections at a wall are represented by an "image source", and the ray tracer establishes all image sources of the transmitter that can contribute at a given receiver position. Double reflections lead to an "image of an image", and so on such that the number of images increases exponentially with the number of reflection processes that are taken into account. The method inherently provides the impulse response for one pair of TX/RX locations only.

Ray launching sends off (launches) rays into all directions and follows the rays along their propagation path until they become too weak to be significant [23]. This technique allows the computation of the impulse responses in a large area (for a given TX position), and is thus more efficient for comprehensive site-specific modeling. Ray launching is also capable of dealing with diffraction in a straightforward way; each diffraction edge serves as a secondary source of rays.

Simplification can be achieved – for both ray tracing and ray launching – by considering only a two-dimensional geometry; however, this might lead to exclusion of important propagation paths and thus inaccurate results. A compromise solution is 2.5 dimensional ray tracing, which adds up the contributions in the vertical and the horizontal propagation plane [24].

Ray tracing and ray launching provide the powers, delays, DoAs, and DoDs of the different MPCs. While phases can be computed theoretically, the small-scale fading realizations as computed by ray tracing hardly ever agree with the ones occurring in practice – at best the small-scale amplitude pdfs and autocorrelation functions are correct.

2.4.2 Modeling Hierarchies

Radio propagation depends on topographical and electromagnetic features of the operating environment. To account for variations in these characteristics, stochastic and semi-stochastic models have to have different parameters (and possibly different structures) in different environments. To systematically accommodate a wide variety of environments and transmission situations, some models have adopted a hierarchical structure. For example, the COST 259 Directional Channel Model [4, 25] uses the three-level structure depicted in Figure 2.6 to describe different cellular situations.

At the top level, a first distinction is made by the cell type, namely macro- micro- and picocells, which are defined as the base station (BS) antenna being above rooftop, below rooftop, or indoor, respectively. For each cell type, a number of *radio environments* (REs) have been identified, which represent a whole class of multipath conditions that give rise to similar radio channel characteristics that, in turn, can be related to the surroundings in which the system operates. The features of a RE are defined by a number of *external parameters*, such as the frequency band, the average height of BS and MS antennas, their average distance from each other, and average building heights and separations.

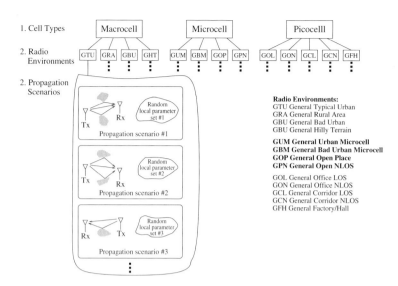

Figure 2.6 Layered structure of the COST259 DCM. © 2006 IEEE. Reprinted, with permission, from [4].

The bottom layer of the hierarchical structure of modeling constructs consists of the *Propagation Scenarios* (PS), which are defined as random realizations of multipath conditions (see Figure 2.6). PSs are not classes in the strict mathematical sense, but represent certain small-scale channel states, each of which exhibits a constant DDDPS, such that large-scale parameters remain constant within a PS.

The propagation conditions of each RE are described by a set of fixed parameters, as well as probability density functions for stochastic parameters and DDDPSs. Since they characterize the propagation conditions of the entire radio environment, they are called *global parameters* (GPs). Propagation scenarios are controlled by the GPs of the associated RE.

Local parameters (LPs) are random realizations of parameters that describe instantaneous channel conditions in a local area. The statistical properties of the LPs are given by the set of GPs. For movements of the MS within a sufficiently small *local area A*, not larger than some tens of wavelengths, the LPs determining the propagation scenario remain approximately constant.

2.4.3 Clustering

Measurement results show that in many REs the MPCs typically arrive in clusters [26–28]. This effect arises, for example, because the MPCs are created by the interaction of the TX signal with objects located in a certain region (IO clusters), like a group of high-rise buildings or mountains, or from waves undergoing similar waveguiding processes in a corridor or street canyon. To be more precise, we need to distinguish between MPC clusters and IO clusters. The former are a group of MPCs that have similar characteristics, while the latter are groups of physical objects that – through interaction with incident waves – can give rise to MPC clusters.

2.4.3.1   Definition of Clusters

Among MPC clusters, we furthermore distinguish two definitions. The most physically motivated is that a cluster is a group of MPCs that show similar large-scale behavior: if an MS moves over a large distance (much larger than a stationarity region), the variations of power, delay, DoAs and DoDs of the MPCs are highly correlated (the threshold for this correlation to be sufficient is an arbitrary parameter to be set for analysis). Clearly, these definitions can only be applied if large-scale measurements (or deterministic simulations), tracking the MS over large distances, are available. Alternatively, we can define an MPC cluster as a group of MPCs with similar (τ, Ω, Ψ), surrounded by "areas" (in the (τ, Ω, Ψ) space) that contain no MPCs.

With either definition, the indices of the MPCs can be grouped into $M \leq L$ *disjoint* classes (or clusters) $C_1, \ldots, C_M$, where each class has $N_m \geq 1$ elements, and

$$\sum_{m=1}^{M} N_m = L, \tag{2.62}$$

so that the double-directional impulse response can be rewritten as

$$h(t, \mathbf{r}_{tx}, \mathbf{r}_{rx}, \tau, \Omega, \Psi) = \sum_{m=1}^{M} \sum_{n \in C_m} h_n(t, \mathbf{r}_{tx}, \mathbf{r}_{rx}, \tau, \Omega, \Psi). \tag{2.63}$$

Visual inspection of DDDPs were long used for clustering, see, for example, [28, 29]. Recent work [30] has established efficient automated clustering.

The concept of clusters is useful because the parameters of a cluster either do not change with time at all, or change only very slowly. To give an example, the PDP of a single cluster is always an exponential function. When the PDP consists of three clusters, we have a total of three exponentials. When the MS moves over large areas, the position of the exponentials relative to each other changes, but the shape of the cluster PDP remains unchanged. In addition, the cluster DDDPS can be written as a product of a cluster PDP and a cluster APS, while it is *not* possible to write the total DDDPS in such a multiplicative way.

2.4.3.2   Visibility Regions

In order to model the appearance and disappearance of clusters, the concept of "Visibility Regions" [4] has found wide acceptance. For each cluster, we define certain physical regions in a coverage area so that if the MS is in such a region, the cluster is active, that is, the MPCs belonging to that cluster contribute to the DDDPS, otherwise they do not. The visibility regions are placed at random in the cell area, with the pdf of the visibility region centers being a parameter of the model. For each cluster, a separate set of visibility regions must be generated. It is furthermore common to define a "transition function" that ensures that when an MS enters a visibility region, the MPC cluster does not activate all of a sudden (which would lead to a discontinuity in power). An alternative to the visibility region is a birth/death process, where clusters are turned on or off at random times.

The cluster of MPCs corresponding to scattering around the MS is always present; the appearance and disappearance of clusters only occurs for "far clusters".

2.4.3.3 Cluster Shape

The cluster shape describes the average relative power of multipath components conditioned on the position in delay and direction at base station and mobile station, and is essentially the DDDPS. It turns out that this cluster shape can often be decomposed: for example, the COST 259 model [25], based on a number of measurements, uses the following simplification

$$P\left(\tau, \theta, \varphi, \theta', \varphi'\right) = P_\tau\left(\tau\right) P_\theta\left(\theta\right) P_\varphi\left(\varphi\right) P_{\theta'}\left(\theta', \tau\right) P_{\varphi'}\left(\varphi', \tau\right), \qquad (2.64)$$

that is, the azimuth and elevation spectra at the MS, $P_{\theta'}$ and $P_{\varphi'}$ depend on the delay, but the other variables are independent of each other.

2.4.3.4 Joint Dispersion Characteristics

Numerous measurement campaigns have found that shadowing, delay dispersion, and angular dispersion are correlated with each other [31, 32]. More specifically, the cluster rms delay spread $\sigma_{\tau,m}$, rms angle spread $\sigma_{\phi,m}$, and shadowing, S_m, of the m-th cluster can be modeled as correlated lognormally distributed random variables according to

$$S_m = 10^{s_{shf}\,X_m/10}, \qquad (2.65)$$

$$\sigma_{\varphi,m} = m_{s\varphi} 10^{s_{s\varphi}\,Y_m/10}, \qquad (2.66)$$

$$\sigma_{\tau,m} = m_{s\tau}\left(\frac{d}{1000}\right)^\varepsilon 10^{s_{s\tau}\,Z_m/10}, \qquad (2.67)$$

where X_m, Y_m, Z_m are random Gaussian variables with zero mean, unit variance, and cross-correlations $\rho_{XY}, \rho_{XZ}, \rho_{YZ}$. The correlations between different clusters are zero, that is, $\rho_{X_m X_n} = \rho_{Y_m Y_n} = \rho_{Z_m Z_n} = \delta_{mn}$, where δ_{mn} is the Kronecker delta. The random variables X_m, Y_m, Z_m have exponential autocorrelation functions with autocorrelation lengths $\{L_S, L_\tau, L_\varphi\}$. The parameters $s_{shf}, s_{s\tau}, s_{s\varphi}$ are standard deviations expressed in dB. The median azimuth spread is $m_{s\varphi}$ while $m_{s\tau}$ is the median delay spread at a distance $d = 1000\,m$. The dimensionless exponent ε determines the distance dependence of the delay spread.

The elevation spread $\sigma_{\theta,m}$ is modeled as uncorrelated with the other cluster spreads also using a lognormal distribution

$$\sigma_{\theta,m} = m_{s\theta} 10^{s_{s\theta}\,W_m/10}, \qquad (2.68)$$

where W_m is a random Gaussian variable with zero mean, unit variance, and exponential autocorrelation function with autocorrelation length L_θ.

2.4.4 Stochastic Modeling

2.4.4.1 Impulse Response and Tapped Delay Line

When modeling the impulse response, many approaches are based on some variant of (2.22) or (2.24), that is,

$$h(t, \tau) = \sum_{i=1}^{L} c_i(t)\delta(\tau - \tau_i), \qquad (2.69)$$

where the $c_i(t)$ are fading. The fading amplitude statistics, together with the temporal autocorrelation function (or equivalently, the Doppler spectrum) of each tap, provide a sufficient characterization of the impulse response within the constraints of the model. Those constraints are: (i) the impulse response does not reflect pathloss and shadowing, which are assumed to be identical for all taps, and have to be superimposed on $h(t, \tau)$, (ii) the number of the MPCs, and the average power carried by each tap, is constant over the time range (area) over which simulations are performed; namely, a "region of stationarity", (iii) the delays of the MPCs are constant over the simulation range. If simulations are to be performed over larger areas, the model has to be combined with, or replaced by, geometrical approaches described in Section 2.4.5.

Tapped delay line models are usually derived from measurements with a specific bandwidth, and/or are intended for the emulation of systems with a particular bandwidth, which in turn determines the spacing between the tap delays τ_i. We note that using models for a bandwidth larger than the one they were designed for is *not* admissible and can cause completely misleading simulation results.

The tap fading statistics are commonly the same as the narrowband fading statistics discussed in Section 2.3.2. If the resolvable delay bin has a duration of at least 10 ns, Rayleigh fading statistics are usually valid within each tap, while for shorter delay bins, Nakagami statistics can occur. In the delay bin corresponding to a line-of-sight path, Rician fading statistics usually apply.

When time variations of the channel are to be included, also the temporal correlation function (or equivalently the Doppler spectrum) needs to be defined. Note that this has to be done for every delay tap, though some models assume a separable scattering function (i.e., the Doppler spectrum is independent of the delay). Simulating a (single) delay tap with a given Doppler spreading is equivalent to that of simulating a flat fading channel [33].

2.4.4.2 Spatial Tapped Delay Line Models

The tapped delay line model can also be extended to the double-directional case. Starting from the formulation (2.25), we can define

$$h(t, \mathbf{r}_{\text{tx}}, \mathbf{r}_{\text{rx}}, \tau, \Omega, \Psi) = \sum_{\ell=1}^{L} c_\ell(t)\delta(\tau - \tau_\ell)\delta(\Omega - \Omega_\ell)\delta(\Psi - \Psi_\ell), \qquad (2.70)$$

with similar validity constraints as for (2.69). We could, again in analogy to (2.69), impose an amplitude pdf and a Doppler spectrum for each tap – though care has to be taken, as the Doppler spectrum has to be consistent with the mobility model and angular spectrum at the moving terminal. In many cases, it is more convenient to use non-fading $c_\ell(t)$, but ensure that multiple taps fall within a resolvable delay bin (while having different DoAs and DoDs). The addition of these taps will then, after appropriate filtering with the bandwidth of interest, automatically provide the correct amplitude pdf and temporal autocorrelation function. This approach is used, for example, in the 3GPP SCM channel model [34] as well as the WINNER/ITU-Advanced channel model [35].

2.4.4.3 Analytical MIMO Models

For many MIMO applications, a stochastic model for the impulse response matrix is desirable. We will in the following only describe models for the narrowband case, which can be employed as a model for each delay tap separately (though the model parameters, such as correlation matrices, might change from tap to tap). We furthermore restrict our attention to the case where the fading for each impulse response is Rayleigh; if a line-of-sight connection exists, the impulse response matrix can be split into a zero-mean stochastic part $\mathbf{H_s}$ and a purely deterministic part $\mathbf{H_d}$

$$\mathbf{H} = \sqrt{\frac{1}{1+K}}\,\mathbf{H_s} + \sqrt{\frac{K}{1+K}}\,\mathbf{H_d}, \tag{2.71}$$

where the matrix $\mathbf{H_d}$ accounts for LOS components, which we neglect in the following, such that $\mathbf{H} = \mathbf{H_s}$. We further define $\mathbf{h} = \mathrm{vec}\{\mathbf{H}\}$ whose correlation matrix

$$\mathbf{R_H} = \mathrm{E}\left\{\mathbf{hh}^H\right\}, \tag{2.72}$$

is known as *full correlation matrix* and describes the spatial MIMO channel statistics. Superscript ^{H} denotes Hermitean transpose. Realizations of MIMO channels can be obtained by

$$\mathbf{H} = \mathrm{unvec}\{\mathbf{h}\}, \qquad \text{with} \quad \mathbf{h} = \mathbf{R_H}^{1/2}\mathbf{g}. \tag{2.73}$$

where $\mathbf{g}$ is a vector with i.i.d. (independent identically distributed) complex Gaussian entries. Since the full correlation matrix requires the specification of $(N_t N_r)^2$ parameters, some simplified models have been defined.

The simplest analytical MIMO model is the i.i.d. model, where $\mathbf{R_H} = \mathbf{I}$, that is, all elements of the MIMO channel matrix $\mathbf{H}$ are uncorrelated and have unit variance [36]. This model is approximately valid in rich scattering environments characterized by independent MPCs whose DoAs and DoDs are uniformly distributed and the spacing of the antenna elements of the transmit and receive arrays is sufficiently large.

The so-called Kronecker model [37] reduces the number of correlation parameters by assuming that spatial TX and RX correlation are separable, which is equivalent to assuming that the correlation matrices can be written as Kronecker product

$$\mathbf{R_H} = \mathbf{R_{tx}} \otimes \mathbf{R_{rx}}, \tag{2.74}$$

with the TX and RX correlation matrices

$$\mathbf{R_{tx}} = \mathrm{E}\{\mathbf{H}^H\mathbf{H}\}, \qquad \mathbf{R_{rx}} = \mathrm{E}\{\mathbf{HH}^H\}, \tag{2.75}$$

respectively. This reduces the number of model parameters from $(N_t N_r)^2$ to $N_t^2 + N_r{}^2$. Channel realizations can then be computed from

$$\mathbf{h} = (\mathbf{R}_{tx} \otimes \mathbf{R}_{rx})^{1/2} \mathbf{g} \quad \Longleftrightarrow \quad \mathbf{H} = \mathbf{R}_{rx}^{1/2} \mathbf{G} \mathbf{R}_{tx}^{1/2}, \tag{2.76}$$

with $\mathbf{G}$ an i.i.d. unit-variance MIMO channel matrix. Note that the Kronecker model is not able to reproduce the coupling between DoDs and DoAs, which is a physical feature of single-interaction processes. Nonetheless, the model (2.76) has been successfully used for the theoretical analysis of MIMO systems and for MIMO channel simulation.

To lift the limitation to separable DoA-DoD spectra, the Weichselberger model [36, 37] first performs an eigendecomposition of the TX and RX correlation matrices,

$$\mathbf{R}_{tx} = \mathbf{U}_{tx} \Lambda_{tx} \mathbf{U}_{tx}^H, \qquad \mathbf{R}_{rx} = \mathbf{U}_{rx} \Lambda_{rx} \mathbf{U}_{rx}^H, \tag{2.77}$$

where $\mathbf{U}_{tx}$ and $\mathbf{U}_{rx}$ are unitary matrices whose columns are the eigenvectors of $\mathbf{R}_{tx}$ and $\mathbf{R}_{tx}$, respectively, and Λ_{tx} and Λ_{tx} are diagonal matrices with the corresponding eigenvalues. The channel matrices $\mathbf{H}$ are then constructed as

$$\mathbf{H} = \mathbf{U}_{rx} (\boldsymbol{\Phi} \odot \mathbf{G}) \mathbf{U}_{tx}^T, \tag{2.78}$$

where $\mathbf{G}$ is again an i.i.d. MIMO matrix, $\odot$ denotes element-wise multiplication. $\boldsymbol{\Phi}$ is a matrix whose elements determine the average power coupling between the TX and RX eigenmodes. The Kronecker model can be seen as a special case of the Weichselberger model.

A somewhat related model is the "virtual channel representation" [40]

$$\mathbf{H} = \mathbf{F}_n (\boldsymbol{\Omega} \odot \mathbf{G}) \mathbf{F}_m^H. \tag{2.79}$$

Here, the matrices $\mathbf{F}_m$ and $\mathbf{F}_n$ contain the steering vectors for N_t virtual TX and N_r virtual RX scatterers, and in the case of uniform linear arrays are simply Complex Fourier Transform (beam) matrices; the model can also be seen as a special case of the Weichselberger model.

2.4.5 Geometry-Based Stochastic Models

Geometry-based stochastic channel models (GSCM) choose the scatterer (IOs) locations in a random way, according to a certain probability distribution, and then obtain the actual channel impulse response by a simplified ray tracing procedure that typically involves no more than two interactions on the way from TX to RX [41–45], see Figure 2.7. GSCMs have a number of important advantages over purely stochastic modeling methods [46]:

- close connection to physical reality, so that important parameters can often be determined via simple geometrical considerations;
- many effects are implicitly reproduced: small-scale fading is created by the superposition of waves from individual IOs; angle and delay drifts caused by MS movement are implicitly included;
- TX, RX and IO movement as well as shadowing and the disappearance of propagation paths (e.g., due to blocking by obstacles) can be easily implemented; as a matter of fact visibility regions and similar concepts are a natural extension of GSCM concepts.

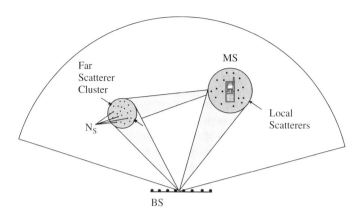

Figure 2.7 Principle of the GSCM (BS = base station, MS = mobile station). Reproduced with permission of John Wiley & Sons, Inc. from [46].

We next discuss the implementation of GSCM when only single-interaction processes occur, often called single-bounce scattering. We then move to the description of multiple-interaction processes. We note that generally GSCMs serve to provide normalized versions of the (double-directional) impulse response; pathloss (and possibly shadowing) are super-imposed on these.

2.4.5.1 Single-Bounce Scattering

The single-bounce scattering assumption makes ray tracing extremely simple: apart from the LOS, all paths consist of two subpaths connecting the scatterer to the TX and RX, respectively. These subpaths characterize the DoD, DoA, and propagation time (which in turn determines the overall attenuation, usually according to a power law). The scatterer interaction itself can be taken into account via an additional random phase shift and/or attenuation.

Different versions of the GSCM differ mainly in the proposed scatterer distributions. The simplest GSCM is obtained by assuming that the scatterers are spatially uniformly distributed; an alternative approach suggests placing the scatterers randomly around the MS [43, 45]. A unique mapping exists between the distribution of the IOs and the ADPS.

To make the density or strength of the scatterers depend on distance, two implementations are possible. In the "classical" approach, the probability density function of the scatterers is adjusted such that scatterers occur less likely in the region where the IO pdf takes on small values. Alternatively, the "non-uniform scattering cross section" method places scatterers with uniform density in the considered area, but down-weights their (power) contributions according to the IO pdf [46].

2.4.5.2 Multiple-Bounce Scattering

In the single-bounce scattering model, the position of an IO completely determines DoD, DoA, and delay. However, many environments, multiple scattering processes occur. If the directional channel properties need to be reproduced only for *one* link end (i.e., multiple

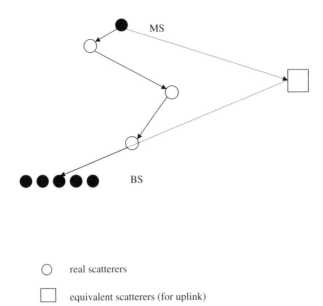

Figure 2.8 Equivalent scatterers for multiple antennas at BS only. Reproduced with permission of John Wiley & Sons, Inc. from [46].

antennas only at the TX or RX), multiple-bounce scattering can be incorporated into a GSCM via the concept of *equivalent scatterers*. These are virtual single-bounce scatterers whose position is chosen such that they mimic multiple bounce contributions in terms of their delay and DoA (see Figure 2.8).

In a MIMO system, the equivalent scatterer concept fails since the angular channel characteristics are reproduced correctly only for one link end. As a remedy, [47] suggested the use of double scattering where the coupling between the scatterers around the BS and those around the MS is established by means of a so-called illumination function (essentially a DoD spectrum relative to that scatterer). Another approach to incorporate multiple-bounce scattering into GSCM models is the twin-cluster concept pursued within COST 273 [48], see Figure 2.9. A twin cluster is composed of two representations of an identical IO cluster with respect to the TX and RX. The centers of the twin clusters are

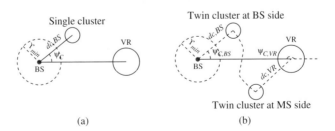

Figure 2.9 Clusters for single interaction processes (a) and twin clusters for the description of multiple-interaction processes (b). From [49]. Reproduced with permission of Springer.

chosen such that the correct angular spectra are seen. An extra cluster-link distance and -delay are introduced to compensate the delay mismatch in the twin cluster. The ratio of twin clusters to the total number of clusters is determined by a selection factor [48].

2.4.6　Diffuse Multipath Components

Numerous measurements showed that the channel impulse response consists of several well-concentrated strong paths (specular MPCs) and a huge number of weak paths (dense or diffuse MPCs). Most of the previously discussed models are based on a superposition of a finite number of (specular) MPCs. The accuracy of this approach can be controlled – within certain limits – via the number of propagation paths. However, in order to keep the number of model parameters reasonably small, it is often preferable to model a relatively small number of MPCs according to (2.25), and include an additional component into the model to describe non-specular contributions, termed dense or diffuse MPCs [50]. The diffuse components are usually described by their DDDPS, which can often be decomposed into a (usually single-exponential) temporal decay, and a uniform APS at TX and RX. Alternatively, DMCs have also been modeled as consisting of a sum of diffuse clusters (with non-uniform APSs), associated with discrete MPC clusters.

2.4.7　Multi-Link Stochastic Models

As multi-user MIMO, Cooperative Multipoint (CoMP), relays, and ad-hoc networks gain importance, the correct modeling of multiple simultaneous links becomes more and more important. It is generally accepted that small-scale fading of different links is uncorrelated, but large-scale parameters including shadowing, angular spreads, and so on, can be correlated.

Two main modeling approaches exist to describe these correlations. The first is based on a purely stochastic approach, and describes the correlation coefficient between two parameters. Consider, for example, the shadowing of the links from one cellular MS to

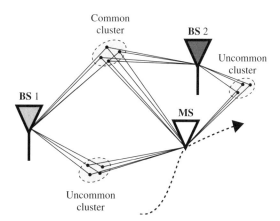

Figure 2.10　Common and uncommon clusters for multi-link simulations. From [49]. Reproduced with permission of Springer.

two BSs. The shadowing are lognormal random variables; a positive correlation coefficient between them indicates that if the link to one BS is in a shadowing dip, there is a high probability that the other link is suffering the same fate.

An alternative is based on a physical interpretation of the propagation processes. Of particular importance are propagation processes or IOs that are common for two or more links, that is, common scatterers (or clusters of scatterers). In order to quantify the amount of energy that propagates via the same scatterers in different links, a measure called the significance of common scatterers was introduced in [51]. Figure 2.10 shows an example of a multi-BS scenario, where links between the MS and BS 1 and BS 2 have one uncommon cluster per link, and one common cluster that contributes to both links.

References

1. A. F. Molisch, *Wireless Communications*, 2nd ed. IEEE Press - Wiley, 2011.
2. A. F. Molisch and F. Tufvesson, *Handbook of Signal Processing for Wireless Communications*. CRC, 2004, ch. 4.7 Multipath propagation models for broadband wireless systems.
3. A. F. Molisch, L. J. Greenstein, and M. Shafi, "Propagation issues for cognitive radio," *Proceedings of the IEEE*, vol. 97, pp. 787–804, 2009.
4. A. F. Molisch, H. Asplund, R. Heddergott, M. Steinbauer, and T. Zwick, "The COST259 directional channel model - i. overview and methodology," *IEEE Trans. Wireless Comm.*, vol. 5, pp. 3421–3433, 2006.
5. A. F. Molisch and H. Hofstetter, *Mobile Broadband Multimedia Networks*, L. Correia, Ed. Academic Press, 2006, ch. 4.7 The COST 273 channel model.
6. C. Balanis, *Antenna theory*. Wiley, 2010.
7. W. L. Stutzman and G. A. Thiele, *Antenna Theory and Design*. New York: John Wiley & Sons, Inc., 1981, iSBN: 0-471-04458-X.
8. R. Vaughan and J. B. Andersen, *Channels, Propagation and Antennas for Mobile Communications*. IEE Publishing, 2003.
9. L. Greenstein, M. Shafi, and A. F. Molisch, "Propagation effects in cognitive radio," in *Cognitive Radio Principles*, E. Bilglieri, A. Goldsmith, L. Greenstein, and H. V. Poor, Eds. Cambridge University Press, 2012.
10. M. Steinbauer, A. F. Molisch, and E. Bonek, "The double-directional radio channel," *IEEE Antennas and Propagation Magazine*, vol. 43, pp. 51–63, August 2001.
11. M. Shafi, M. Zhang, A. L. Moustakas, P. J. Smith, A. F. Molisch, F. Tufvesson, and S. H. Simon, "Polarized MIMO channels in 3-D: models, measurements and mutual information," *IEEE Journal on Selected Areas in Communications*, vol. 24, pp. 514–527, March 2006.
12. A. F. Molisch, "Ultrawideband propagation channels-theory, measurement, and modeling," *IEEE Transactions on Vehicular Technology*, vol. 54, pp. 1528–1545, Sept. 2005.
13. A. F. Molisch, "Ultrawideband propagation channels," *Proceedings of the IEEE*, vol. 97, pp. 353–371, 2009.
14. W. C. Jakes, *Microwave Mobile Communications*. Piscataway, NJ: IEEE Press, 1974.
15. M. Abramowitz and I. A. Stegun, *Handbook of Mathematical Functions*. Dover, 1965.
16. P. Bello, "Characterization of Randomly Time-Variant Linear Channels," *IEEE Trans. Comm.*, vol. 11, pp. 360–393, 1963.
17. R. Kattenbach, *Characterization of Time-Variant Indoor Radio Channels by Means of Their System- and Correlation Functions (in German)*. PhD thesis at University GhK Kassel, published by Shaker-Verlag, Aachen, 1997.
18. B. H. Fleury, "An uncertainty relation for WSS processes and its application to WSSUS systems," *IEEE Trans. Comm*, vol. 44, pp. 1632–1634, 1996.
19. R. Kattenbach, "Statistical modeling of small-scale fading in directional radio channels," *IEEE J. Selected Areas Comm*, vol. 20, pp. 584–592, 2002.
20. B. H. Fleury, "First- and second-order characterization of direction dispersion and space selectivity in the radio channel," *IEEE Trans. Information Theory*, vol. 46, pp. 2027–2044, 2000.

21. A. Valcarce, G. de la Roche, L. Nagy, J.-F. Wagen, and J.-M. Gorce, "A new trend in propagation prediction," *Vehicular Technology Magazine, IEEE*, vol. 6, no. 2, pp. 73–81, June 2011.

22. K. S. Kunz and R. J. Luebbers, *The Finite Difference Time Domain Method for Electromagnetics*. CRC Press, 1993.

23. G. Durgin, N. Patwari, and T. Rappaport, "An advanced 3d ray launching method for wireless propagation prediction," in *Vehicular Technology Conference, 1997, IEEE 47th*, vol. 2, May 1997, pp. 785–789 vol.2.

24. J. Li, J.-F. Wagen, and E. Lachat, "Propagation over rooftop and in the horizontal plane for small and micro-cell coverage predictions," in *Vehicular Technology Conference, 1997, IEEE 47th*, vol. 2, May 1997, pp. 1123–1127.

25. H. Asplund, A. A. Glazunov, A. F. Molisch, K. I. Pedersen, and M. Steinbauer, "The COST259 directional channel model II - macrocells," *IEEE Trans. Wireless Comm.*, vol. 5, pp. 3434–3450, 2006.

26. U. Martin, "Spatio-temporal radio channel characteristics in urban macrocells," *IEE Proc. Radar, Sonar and Navigation*, vol. 145, no. 1, pp. 42–49, 1998.

27. T. Zwick, C. Fischer, D. Didascalou, and W. Wiesbeck, "A stochastic spatial channel model based on wave-propagation modeling," *IEEE Journal on Selected Areas in Communications*, vol. SAC-18, no. 1, pp. 6–15, Jan 2000.

28. M. Toeltsch, J. Laurila, K. Kalliola, A. F. Molisch, P. Vainikainen, and E. Bonek, "Statistical characterization of urban spatial radio channels," *IEEE J. Selected Areas Comm.*, vol. 20, pp. 539–549, 2002.

29. L. Vuokko, P. Vainikainen, and J. Takada, "Clusterization of measured direction-of-arrival data in an urban macrocellular environment," *Proc. 14th IEEE PIMRC* pp. 1222–1226, 2003.

30. N. Czink, P. Cera, J. Salo, E. Bonek, J.-P. Nuutinen, and J. Ylitalo, "Improving clustering performance using multipath component distance," *Electronics Letters*, vol. 42, no. 1, pp. 33–35, Jan. 2006.

31. L. J. Greenstein, V. Erceg, Y. S. Yeh, and M. V. Clark, "A new path-gain/delay-spread propagation model for digital cellular channels," *IEEE Transactions on Vehicular Technology*, vol. 46, no. 2, pp. 477–485, May 1997.

32. A. Algans, K. Pedersen, and P. Mogensen, "Experimental analysis of the joint statistical properties of azimuth spread, delay spread, and shadow fading," *Selected Areas in Communications, IEEE Journal on*, vol. 20, no. 3, pp. 523–531, April 2002.

33. M. Paetzold, *Mobile Fading Channels*. Wiley, 2002.

34. TSG RAN WG4, "Deployment aspects," 3rd Generation Partnership Project (3GPP), Tech. Rep. 3G TR 25.493V2.0.0, 2000.

35. International Telecommunications Union, "Imt-advanced channel models," Tech. Rep., 2010.

36. G. J. Foschini and M. J. Gans, "On limits of wireless communications in fading environments when using multiple antennas," *Wireless Personal Comm.*, vol. 6, pp. 311–335, 1998.

37. J. Kermoal, L. Schumacher, K. Pedersen, P. Mogensen, and F. Frederiksen, "A Stochastic MIMO Radio Channel Model with Experimental Validation," *IEEE J. Sel. Areas Comm.*, vol. 20, no. 6, pp. 1211–1226, Aug. 2002.

38. W. Weichselberger, M. Herdin, H. Ozcelik, and E. Bonek, "A stochastic mimo channel model with joint correlation of both link ends," *Wireless Communications, IEEE Transactions on*, vol. 5, no. 1, pp. 90 – 100, Jan. 2006.

39. W. Weichselberger, "Spatial Structure of Multiple Antenna Radio Channels," Ph.D. dissertation, Institut für Nachrichtentechnik und Hochfrequenztechnik, Vienna University of Technology, Vienna, Austria, Dec. 2003, downloadable from http://www.nt.tuwien.ac.at/mobile.

40. A. Sayeed, "Deconstructing Multiantenna Fading Channels," *IEEE Trans. on Signal Proc.*, vol. 50, no. 10, pp. 2563–2579, Oct. 2002.

41. P. Petrus, J. Reed, and T. Rappaport, "Geometrical-based Statistical Macrocell Channel Model for Mobile Environments," *IEEE Trans. Comm.*, vol. 50, no. 3, pp. 495–502, Mar. 2002.

42. J. C. Liberti and T. Rappaport, "A Geometrically Based Model for Line-Of-Sight Multipath Radio Channels," in *Proc. IEEE Vehicular Technology Conf.*, Apr.May 1996, pp. 844–848.

43. J. Blanz and P. Jung, "A Flexibly Configurable Spatial Model for Mobile Radio Channels," *IEEE Trans. Comm.*, vol. 46, no. 3, pp. 367–371, Mar. 1998.

44. O. Norklit and J. Andersen, "Diffuse Channel Model and Experimental Results for Array Antennas in Mobile Environments," *IEEE Trans. on Antennas and Propagation*, vol. 46, no. 6, pp. 834–843, June 1998.

45. J. Fuhl, A. F. Molisch, and E. Bonek, "Unified Channel Model for Mobile Radio Systems with Smart Antennas," *IEEE Proc. – Radar, Sonar and Navigation: Special Issue on Antenna Array Processing Techniques*, vol. 145, no. 1, pp. 32–41, Feb. 1998.

46. A. F. Molisch, A. Kuchar, J. Laurila, K. Hugl, and R. Schmalenberger, "Geometry-based Directional Model for Mobile Radio Channels – Principles and Implementation," *European Trans. Telecomm.*, vol. 14, pp. 351–359, 2003.

47. A. F. Molisch, "A Generic Model for the MIMO Wireless Propagation Channels in Macro- and Micro-cells," *IEEE Trans. on Signal Proc.*, vol. 52, no. 1, pp. 61–71, Jan. 2004.

48. A. F. Molisch and H. Hofstetter, "The COST273 Channel Model," in *COST 273 Final Report*, L. Correia, Ed. Springer, 2006.

49. R. Verdone and A. Zanella, Eds., *Pervasive Mobile and Ambient Wireless Communications*. Springer, 2012.

50. A. Richter, C. Schneider, M. Landmann, and R. Thomä, "Parameter Estimation Results of Specular and Dense Multipath Components in Micro-Cell Scenarios," in *Proc. WPMC'04*, Padova, Italy, Sept. 2004.

51. V.-M. Kolmonen, P. Almers, J. Salmi, J. Koivunen, K. Haneda, A. Richter, F. Tufvesson, A. Molisch, and P. Vainikainen, "A dynamic dual-link wideband mimo channel sounder for 5.3ghz," *Instrumentation and Measurement, IEEE Transactions on*, vol. 59, no. 4, pp. 873–883, April 2010.

Part Two

Radio Channels

3

Indoor Channels

Jianhua Zhang[1] and Guangyi Liu[2]

[1] *Beijing University of Posts and Telecommunications, China*
[2] *China Mobile, China*

3.1 Introduction

According to the statistics and prediction of some institutions, the mobile radio communication services will increase dramatically in the next ten years. Especially, the requirement
of the services which happened indoor will be the main increasing part, which will be
much more than 50 percent of the total service. Although there are various technologies
for the indoor coverage, like wireless local area network (WLAN) and femtocell, all
of them are still facing challenges to satisfy the increasing requirements. Therefore,
developing new technologies and upgrading the existing systems for indoor services are
urgent issues not only at present but also in the future. The study on indoor propagation
characteristics, as an indispensable and fundamental step, is of great importance. So the
indoor propagation characteristics will be discussed in detail in this chapter.

Indoor propagation characteristics have been widely studied over the world since the
1990s. Lots of field measurements were carried out in various indoor environments,
such as offices, factories, houses, and so on. Meanwhile, standardization organizations and research institutions have also defined several indoor channel models
for different applicant purposes, as summarized in Table 3.1. International mobile
telecommunication–advanced (IMT-A) system, also known as the 4th generation mobile
telecommunication system, defines one essential indoor scenario called indoor hotspot
for the technology evaluation and system simulation in international telecommunication
union–radio sector (ITU-R) M.2135 [1]. The 3rd generation partnership project
(3GPP) TR 36.814 have defined the channel model for femtocell simulation based
on the ITU-R M.2135 indoor channel model [2], and IEEE 802.11n WLAN has
developed corresponding models for four different indoor scenarios [3]. Wireless
world initiative new radio (WINNER) as a consortium of 41 partners has published
WINNER-Phase II channel model (WIM2) in 2007 [4]. The indoor scenarios in WIM2

LTE-Advanced and Next Generation Wireless Networks: Channel Modelling and Propagation, First Edition.
Edited by Guillaume de la Roche, Andrés Alayón Glazunov and Ben Allen.
© 2013 John Wiley & Sons, Ltd. Published 2013 by John Wiley & Sons, Ltd.

Table 3.1 Indoor channel models of research institutions and standardization organizations

Item	Frequency band	Scenario	Small-scale modeling	Spatial modeling
ITU-R M.2135	2 GHz ~ 6 GHz	indoor hotspot	GBSM	2-D
3GPP TR 36.814	2 GHz ~ 6 GHz	femtocell	GBSM	2-D
IEEE 802.11n	2 GHz and 5 GHz	residential small office typical office large space	TDL	2-D
WIM2	2 GHz ~ 6 GHz	indoor hotspot indoor office	GBSM	2-D
COST 231	150 MHz~2 GHz	dense, open, large, corridor	–	2-D*

Note: GBSM stands for geometry based stochastic model. TDL stands for time delay line. 2-D stands for 2-dimension. * indicates that the spatial modeling is studied in COST 259 and 273.

are categorized into indoor hotspot and indoor office. European cooperation in the field of information science and technology research (COST) has many initiatives dedicated to channel modeling. COST 207, 231, 259, 273 and 2100 channel models are developed for different objectives, which are general reference channel models to be widely used [5].

Various scenarios have been defined by standardization organizations. Indoor hotspot and indoor office scenarios are paid much attention, because the demand for mobile communications service in these two scenarios is supposed to be large in the future. Indoor office environment is referred to as several small rooms with desks and chairs. Few people are moving in an indoor office. However, indoor hotspot environment is defined as a relatively large indoor space with lots of people moving, such as a shopping mall, factory, train station and airport. As the difference on the propagation environment, the propagation characteristics in the two scenarios will definitely be different. Lots of measurement results in these two scenarios will be summarized and analyzed in this chapter.

Indoor propagation, which is more versatile than outdoor propagation, is impacted by several factors. Nowadays, the layout of the building is of utmost diversity, which brings great difficulties into indoor scenario categorization and definition. Walls and floors also introduce attenuations to indoor propagation besides the traditional factors like frequency. Even the population density and the shape of the human body affect the indoor propagation to a certain extent. All these factors impact on the indoor propagation jointly, which makes it more difficult to exactly describe the indoor propagation characteristics. Therefore, those important factors impacting on the indoor propagation will also be focused on in this chapter.

The main part of the chapter will be divided into two sections: Section 3.2 is large scale fading, and Section 3.3 is small scale fading. In Section 3.2, the modeling of path loss (PL) and shadow fading (SF) will be discussed. Then the statistical characteristics in delay domain and angular domain will be investigated in Sections 3.3.2 and 3.3.3.

3.2   Indoor Large Scale Fading

Large scale fading is the average loss of the received signal power on a large separation distance between the transmitter (TX) and the receiver (RX), which is extremely important for wireless network planning and optimization. Generally, large scale fading is divided into two parts, one is PL and the other is SF. PL is the average change of the signal power loss over the distance. SF indicates the slow fluctuation around the average loss, which is caused by the obstruction of scatterers.

In this section, we will mainly focus on the indoor PL in order to give the readers a general insight into the recent indoor PL modeling study. Besides, the study on SF in indoor environments will also be presented. In the following, some widely used empirical indoor PL models will be introduced first. Then the indoor large scale models which are presented by research groups and standardization organizations will be compared from the view of the IMT- A system requirements. Finally, several impact factors on indoor large scale propagation will be studied based on the results of indoor field channel measurements.

3.2.1   Indoor Large Scale Models

Various indoor PL models are proposed by lots of research institutes and standard organizations. However, most of those models can be categorized into a few PL model forms. Those typical PL model forms will be introduced in the following.

3.2.1.1   Free-Space (FS) Model

Free-space model, which is extracted from the free-space propagation principal of electromagnetic waves, is a basic PL model, not only for indoor propagation, but also for outdoor propagation. The electromagnetic wave propagation principal in free space can be described by the Friis transmission equation

$$\frac{P_r}{P_t} = G_r G_t \left(\frac{\lambda}{4\pi d} \right)^2, \tag{3.1}$$

where P_r is the received power, and P_t is the transmitted power. G_t and G_r represent the transmitting and receiving antenna gains, respectively. d is the distance between the TX and the RX whereas λ is the wave length of the electromagnetic wave. Transforming the equation into a formulation in logarithmic scale, and utilizing the definition of PL, free-space PL model can be written as

$$PL_{FS} = P_t - P_r + G_t + G_r \approx 32.4 + 20 \lg(d) + 20\log(f_c), \tag{3.2}$$

where PL_{FS} indicates the PL value in unit dB, f_c and d are in unit MHz and Km, or GHz and m, respectively. For ease of description, the models to be introduced in the following will omit the frequency item.

Almost all the other indoor PL models are originated from the FS model, which only have some slight changes according to different practical environments. Some popular evolutionary PL models will be introduced in the following.

3.2.1.2 Single-Slope (SS) Model

The SS model is also known as the log-distance model, which can be expressed as

$$PL_{SS} = PL_0 + 10n \lg(d), \tag{3.3}$$

where n indicates the PL exponent. PL_0 is the PL value at a reference distance which is usually selected as 1 m. The model expresses that PL increases with the distance at a fixed rate, which is mostly used for modeling PL in certain simple environments with similar propagation mechanisms at different spots, such as a huge lobby or a spacious hall. Due to this, it is certainly widely used for the PL modeling in indoor environments.

3.2.1.3 Multi-Slope Model

$$PL_{MS} = PL_s + 10n_s \lg(d), \quad d_{s-1} \le d \le d_s, s \in \{1, 2, \ldots\}, \tag{3.4}$$

where PL_s and n_s denote the reference PL value and PL exponent corresponding to the sth distance range of (d_{s-1}, d_s). The multi-slope model can be either continuous or discontinuous at the breakpoints, which is decided by the real environments. Considering a compromise of accuracy and simplicity of the model, the multi-slope model is usually simplified to dual-slope model in which there are only two different PL exponents.

3.2.1.4 Linear-Distance (LD) Model

Devasirvatham proposed the LD model to modeling the indoor PL base on the measurements in two buildings of different structures at 850 MHz, 1.7 GHz, and 4.0 GHz [6]. The LD model can be written as

$$PL_{LD} = PL_{FS} + \alpha d, \tag{3.5}$$

where α is an attenuation constant in the unit of dB/m. The LD model includes two parts. One is the free-space PL, the other is the additional attenuation which is increasing linearly with the distance between the TX and the RX.

3.2.1.5 Attenuation Factor (AF) Model

$$PL = PL_{base} + A_F, \tag{3.6}$$

where PL_{base} can be PL_{FS}, PL_{SS} and PL_{LD} [8, 9]. A_F is the attenuation factor, which indicates the additional attenuation caused by obstacles like walls and floors. If only obstacles in one floor are considered, A_F can be expressed as

$$A_F = \sum_{m=1}^{M} K_m A_m, \tag{3.7}$$

where K_m is the number of the mth obstacle and A_m is the attenuation of the mth obstacle. Multi-wall-and-floor (MWF) model is a practical example of the AF model, which is proposed by Mathias Lott in 2001 [9]. In MWF model, PL_{base} is PL_{SS}. And the attenuation factor A_F consists of the additional attenuations of walls and floors, which can be expressed as

$$A_F = \sum_{p=1}^{P} \sum_{k=1}^{K_p^w} A_{pk}^w + \sum_{q=1}^{Q} \sum_{k=1}^{K_q^f} A_{qk}^f, \tag{3.8}$$

where

A_{pk}^w : attenuation caused by the kth wall of type p
A_{qk}^f : attenuation caused by the kth floor of type q
P : number of wall types
Q : number of floor types
K_p^w : number of walls of type p
K_q^f : number of walls of type q.

The parameters of the model can be extracted by ray tracing technique or field channel measurements. It is apparent that the AF model is more accurate, but more complex than the log-distance models on the prediction of indoor propagation. However, it is difficult to precisely achieve all the attenuation factors whose accuracy has a strong impact on the AF model.

SF which indicates the fluctuation of PL is commonly modeled by a log-normal distribution. Generally, the modeling of SF is based on the modeling of PL. The completed large scale model can be expressed as

$$L = PL + X_\sigma, \tag{3.9}$$

where PL is the chosen PL model. X_σ which denotes the SF is a Gaussian random variable with zero mean and σ^2 variance. The parameter σ can be calculated statistically by the measured PL value subtracting the estimated PL value of the model used. So the σ of the SF will be different if different PL models are chosen to model the same measured PL data set. The variance of SF depends on the propagation environment. The relationship between SF and the distance between TX and RX is investigated in [10]. When RX moves away from TX, SF is found to increase. Moreover, the SF spatial autocorrelation properties between different locations are evaluated in [10] and [11]. The correlation of S between two points separated by distance Δx is defined as [12]

$$< S(x), S(x + \Delta x) > = \sigma^2 \exp\left(-\frac{|\Delta x|}{X_c}\right), \tag{3.10}$$

where X_c is correlation distance of shadow fading. σ is standard deviation. Typical value of correlation distance of shadow fading ranges from 10 m to 500 m. In [10], the effective correlation distance is given by:

$$X_c = -\frac{D}{\ln(\varepsilon_D)}, \tag{3.11}$$

where ε_D is correlation coefficient of two points at distance D. From Equation 3.11, the shadow fading spatial autocorrelation properties between different locations can be evaluated. As is deduced, the correlation of SF decreases with the distance, that is, it is highly correlated at a shorter distance.

3.2.2   Summary of Indoor Large Scale Characteristics

Because of the importance of indoor scenarios in communication systems and the complexity of indoor propagation, extensive research has been performed. Many standardization organizations, research institutes, and colleges are dedicated to investigate the indoor propagation characteristics. Some indoor large scale models have been standardized by ITU-R, 3GPP and IEEE 802.11n. The indoor large scale models proposed by WINNER and COST 231 are also widely used for link or system level simulation and network planning and optimization.

3.2.2.1   ITU-R M.2135 Model

The ITU-R M.2135 [1] report provides the guidelines to evaluate the proposed IMT-A radio interface technologies (RITs) or sets of RITs (SRITs) for a number of test environments and deployment scenarios for evaluation. Based on the field measurement by Beijing University of Posts and Telecommunications (BUPT), China proposes the indoor hotspot simulation scenario and models, and has been accepted by ITU-R M.2135 [13]. The field measurement environment is shown in Figure 3.1, which is a large space with 120 m long and 45 m width. The measurement is carried out at the carrier frequency of 2.35 and 5.25 GHz with 100 MHz bandwidth taking both line-of-sight (LOS) and non-line-of-sight (NLOS) propagation conditions into account, as Grid A and Grid B illustrated in Figure 3.1, respectively. Combined with the results of other proposals, the final PL models of indoor hotspot scenario have been defined in ITU-R M.2135, which is displayed in Table 3.2. In addition, the delay and spatial characteristics obtained in the measurement are also included in ITU-R M.2135 [13, 14]. Besides the main contribution to indoor hotspot models, lots of study achievements on other scenarios are also included as references in ITU-R M.2135, such as outdoor environment [15–17] and outdoor-to-indoor environment [18].

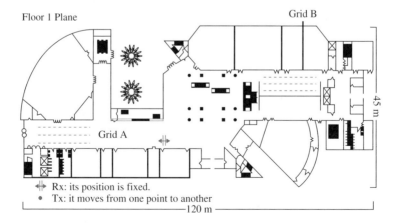

Figure 3.1   Measurement environment for indoor hotspot [13].

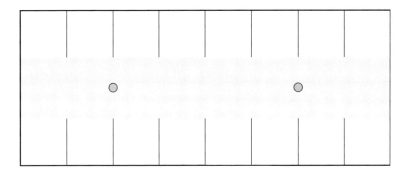

Figure 3.2 Layout of simulation scenario for femtocell in 3GPP TR 36.814 [2].

Table 3.2 The indoor PL models defined in ITU-R M.2135

Indoor hotspot	PL (dB) f_c in GHz, d in m.	SF (dB)	Distance range
LOS	$PL = 16.9\lg(d)+32.8+20\lg(f_c)$	$\sigma = 3$	3 m $< d <$100 m
NLOS	$PL = 43.3\lg(d)+11.5+20\lg(f_c)$	$\sigma = 4$	10 m $< d <$150 m

3.2.2.2 3GPP TR 36.814

The first femtocell standard has been officially published by 3GPP, paving the way for standardized femtocells to be produced in large volumes. Simulation scenario is defined and models are provided in 3GPP TR 36.814 [2]. The simulation scenario consists of a single floor of a building shown in Figure 3.2 which is the same as the ITU-R M.2135 indoor hotspot sketch. The height of the floor is 6 m. The floor contains 16 rooms of 15 m $\times$ 15 m and a long hall of 120 m $\times$ 20 m. Two sites are placed in the middle of the hall at 30 m and 90 m with respect to the left side of the building [2]. The simulation PL models are given in Table 3.2.

3.2.2.3 IEEE 802.11n

In IEEE 802.11n [3], six PL models marked A $\sim$ F are defined for different indoor scenarios. The indoor scenarios include residential, small office, typical office and large space where LOS and NLOS cases are considered. Model mapping to a particular scenario is based on the root-mean-square (rms) delay spread of the scenario, which is presented in [3]. The framework of PL models A $\sim$ F are based on the dual-slope model which is a special case of the multi-slope PL model (Equation 3.4). Their parameters are summarized in Table 3.3 and illustrated in Figure 3.3.

3.2.2.4 WIM2 Model

WIM2 interim model has been published in the deliverable D1.1.1. The final WIM2 model in D1.1.2 [4] has been available since 2007. The PL models and channel parameters are

Table 3.3 The parameters of PL models (A-F) in IEEE 802.11n

New Model	d_{BP} (m)	Slope before d_{BP}	Slope after d_{BP}	SF before d_{BP}, LOS (dB)	SF after $d_{BP}, NLOS$ (dB)	Delay spread (ns)
A	5	2	3.5	3	4	0
B	5	2	3.5	3	4	15
C	5	2	3.5	3	5	30
D	10	2	3.5	3	5	50
E	20	2	3.5	3	6	100
F	30	2	3.5	3	6	150

Note: Model A is not used for system performance comparison.

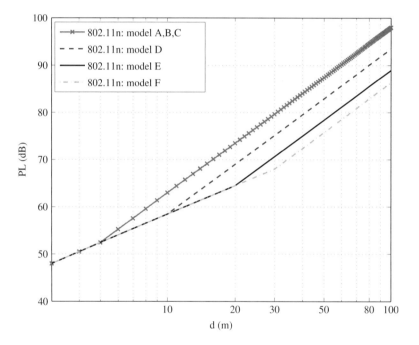

Figure 3.3   PL models in IEEE 802.11n.

mainly based on 2 GHz and 5 GHz measurements. However, the frequency bands are
extended from 2 to 6 GHz. The indoor environments here are mainly divided into two
scenarios: indoor office and indoor hotspot. For indoor office, the layout is shown in
Figure 3.4 and different conditions (LOS and NLOS) have been discussed. In Figure 3.4,
when the base stations (access points) are assumed to be in corridor, LOS case is corridor-
to-corridor. For NLOS case, basic PL model just considers rooms adjacent to the corridor
where the access point is placed. As to rooms farther away from the corridor (several
walls will be penetrated), losses caused by walls must be applied for the walls parallel to
the corridors. Moreover, the floor to floor case is also modeled. All the floors are assumed

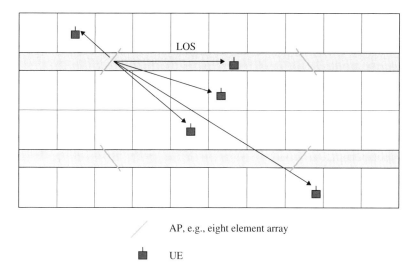

AP, e.g., eight element array

UE

Figure 3.4 Layout of the indoor scenario in WIM2 [4].

Table 3.4 Model parameters in WINNER D1.1.2

Scenarios	PL (dB)	SF (dB)	Notation
indoor office			
LOS	$PL = 18.7\lg(d)+46.8+20\lg(f_c)$	$\sigma = 3$	$3\,\text{m} < d <100$
NLOS	$PL = 36.8\lg(d)+43.8+20\lg(f_c)$	$\sigma = 4$	$3\,\text{m} < d <100$
wall-penetration	light wall: $A_{wall} = 5(n_w - 1)$	–	PL_{base} should
	n_w is the number of walls		be considered
	heavy wall: $A_{wall} = 12(n_w - 1)$	–	
	n_w is the number of walls		
floor penetration	$A_{floor} = 17(n_f - 1)$	–	PL_{base} should
	n_f is the number of floor		be considered
indoor hotspot			
LOS	$PL = 13.9\lg(d)+64.4+20\lg(f_c)$	$\sigma = 3$	$5\,\text{m} < d <100$
NLOS	$PL = 37.8\lg(d)+36.5+23\lg(f_c)$	$\sigma = 4$	$5\,\text{m} < d <100$

identical. The floor loss is modeled as constant for the same distance between floors, but increases with the number of floors linearly.

While indoor hotspot is characterized by larger open space whose typical dimension could range from 20 m × 20 m up to more than 100 m in length and width and up to 20 m in height. In such scenarios, both LOS and NLOS propagation conditions could exist. The PL models both for indoor office and indoor hotspot are summarized in Table 3.4.

3.2.2.5 COST 231 Hata

COST 231 Hata [5] is the extension of Hata model and can be used for 900 MHz and 1800 MHz. Indoor PL models in COST 231 Hata contains three forms of PL models:

the SS model (Equation 3.3), the LD model (Equation 3.5), and the multi-wall model (MWM). MWM is similar to the MWF Model (Equation 3.8), but the wall loss and floor loss are defined in different ways. Wall-penetration loss is defined in MWM as:

$$A^{wall} = \sum_{i=1}^{I} k_{wi} L_{wi}, \qquad (3.12)$$

where k_{wi} is the number of penetrated walls of type i, but here two types of wall are differentiated. L_{wi} is loss of wall type i, I is the number of wall types.

As for floor-penetration loss,

$$A_{floor} = k_f^{\left[\frac{k_f+2}{k_f+1}-b\right]} L_f, \qquad (3.13)$$

where k_f is the number of penetrated floors, L_f loss between adjacent floors, b is an empirical parameter.

From Equation 3.13, it can be found that the total floor loss is a non-linear function of the number of floors. An empirical factor b is inducted to depict the non-linear relation. Moreover, the wall types are discussed in the MWM. The MWM is also used for different indoor scenarios. The model parameters for different scenarios are listed in Table 3.5. The indoor long term fading follows the log-normal distribution with $\sigma = 2.7 \sim 5.3 \, dB$.

3.2.2.6 Comparison of Indoor Channel Models

In order to learn the difference between PL models from different standardization organizations and research groups mentioned above, some comparisons are made in Table 3.6, Figures 3.5 and 3.6.

In Figure 3.5, the path loss predicted by IEEE 802.11n B and C models at 2 GHz is 10 dB better than that in dense scenario, and 9.8 dB worse than that in corridor scenario for COST 231 Hata at 1.8 GHz, when $d = 10$ m. Besides, as depicted in Figure 3.6, in NLOS condition, when d is larger than 50 m, PL in WIM2 and ITU-R M.2135 are approximately the same and at least 5.8 dB worse than that in IEEE 802.11n model B $\sim$ C at 5.25 GHz. Moreover, PL in WIM2 is 11.6 dB worse than that in ITU-R M.2135 for LOS case, when d is from 3 m to 100 m.

Table 3.5 Model coefficients for different PL models at 1800 MHz [5]

Environment	SS model		LD model		MWM		
	PL_0(dB)	n	α (dB/m)	L_{w1} (dB)	L_{w2} (dB)	L_f (dB)	b
Dense							
one floor	33.3	4.0	0.62	3.4	6.9	18.3	0.46
two floor	21.9	5.2	–	–	–	–	–
multi floor	44.9	5.4	2.8	–	–	–	–
Open	42.7	1.9	0.22	3.4	6.9	18.3	0.46
Large	37.5	2.0	–	3.4	6.9	18.3	0.46
Corridor	39.2	1.4	–	3.4	6.9	18.3	0.46

L_{w1} is the penetration loss for light wall like plasterboard.
L_{w2} is the penetration loss for heavy wall like a load-bearing wall.

Table 3.6 Comparison of indoor PL models proposed by institutes

Institutes	Frequency	Indoor environment	Factors considered
ITU-R M.2135	2 GHz ∼ 6 GHz	Indoor hotspot: LOS and NLOS condition	distance, frequency
3GPP TR 36.814	2 GHz ∼ 6 GHz	Femtocell	distance, frequency
IEEE 802.11n	2 GHz and 5 GHz	4 scenarios residential, small office, typical office and large space	distance
WIM2	2 GHz ∼ 6 GHz	Indoor hotspot and indoor office different condition LOS and NLOS, and so on	distance, wall, floor and frequency
COST 231 Hata	0.9 GHz and 1.8 GHz	4 scenarios: dense, open, large	distance, wall, floor and frequency

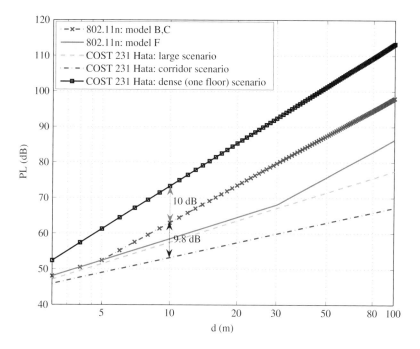

Figure 3.5 Comparison of indoor PL models in IEEE 802.11n and COST 231 Hata.

The candidate deployment frequency bands of long term evolution-advanced (LTE-A) proposed by World Radio Conference in 2007 (WRC07) including 450 MHz–470 MHz, 698 MHz–862 MHz, 790 MHz–862 MHz, 2.3 GHz–2.4 GHz, 3.4 GHz–4.2 GHz, 4.4 GHz–4.99 GHz, and so on. It can be seen that, in addition to 2.3 GHz-2.4 GHz deployed in traditional cellular systems, new bands are either high or low. In order to develop channel models for LTE-A communication systems, channel models should cover a wide range of frequency bands. As high frequency band for LTE-A, the ITU-R models (such as ITU-R M.2135) and WIM2 can be referred; while for low band, it can turn to

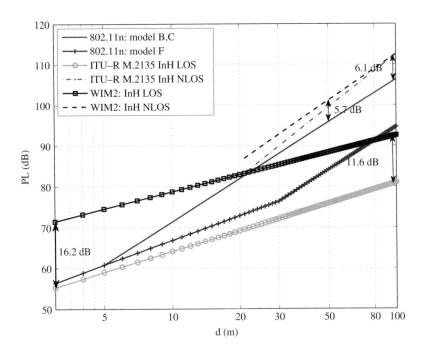

Figure 3.6   Comparison of indoor PL models in WIM2, ITU-R M.2135 and COST 231 Hata.

COST 231 Hata model. Since every standardized channel model includes a set of channel models, whichever standardized channel model you adopt, a specific model for actual deployment depends on concrete conditions such as scenarios, frequencies, and so on.

3.2.3   Important Factors for Indoor Propagation

Though these channel models and channel parameters mentioned above are widely used, they are just representative of a global view of some communication system but are not intended to relate to any specific implementation. What is more, these models mainly target the evaluation candidate radio technology, a choice of channel models and channel parameters has to be taken into consideration to balance the complexity of evaluation methodology. More practical propagation mechanism and effects should be added to channel models or channel parameters, especially for specific employment.

There are many individuals, companies, colleges and institutes which are dedicated to channel model. In their models or parameters, more factors are taken into consideration, such as antenna height and frequency, complex obstacles loss such as wall or floor, and so on. Their achievements not only validate the existing PL models by these organizations mentioned above, but also enrich PL models to make models predict propagation characteristics more accurately, especially for some employments with a specific frequency or environment. This research can be mainly divided into the following aspects.

3.2.3.1   Categories of Buildings

Many measurements have been conducted by companies, individuals, universities and institutes. It can be demonstrated by these measurement results that the types of building

Table 3.7 PL model parameters for different types of buildings

Building	Frequency (MHz)	n	σ (dB)
Literature [7]			
Retail stores	914	2.2	8.7
Grocery store	914	1.8	5.2
Literature [19]			
Office, hard partition	1500	3.0	7.0
Office, soft partition	1900	2.6	14.1
Textile/chemical [factory]	1300	2.0	3.0
Textile/chemical [factory]	4000	2.11	7

have an impact on propagation property. So, the channel models and channel parameters may be different for various types of buildings.

In order to model propagation characteristics precisely, many researchers classify the types of buildings mainly into the following categories [7, 19]: traditional office building, factory buildings and grocery stores, retail stores, residential homes in suburban areas, and residential homes in urban areas, and so on. The difference of structures, materials and internal layout may exist among them. For example, there are a few hard partitions, but a large amount of metal goods in some factory buildings and grocery stores. But, as for retail stores, traditional office buildings and residential homes, many walls made of plaster and metal lathe exist. These differences of structures will bring about the diversity in propagation characteristics. Some measurements concentrating on these differences are studied in [7] and [19], as shown in Table 3.7.

3.2.3.2 Carrier Frequency

Many literatures show that the propagation property depends on carrier frequency [10, 20, 21]. Generally, the radio signals with higher frequency tend to suffer from more loss introduced by distance and obstacles for its poor abilities of diffraction and reflection. So carrier frequency has a very important effect on propagation characteristics. However, the exact relationship between PL and carrier frequency is still unknown.

In ITU-R M.2135, WIM2 and Walfisch-Bertoni model, a frequency factor $(20 \cdot \lg(f))$ is introduced to model the relationship between PL and carrier frequency [22]. Some measurement results for both high-tier and low-tier scenarios in outdoor environment are in line with it [23]. Taking the signal strength for instance, the signal strength at 151 MHz is 6 dB better than that at 433 MHz, 9 dB better than that at 902 MHz, and 19 dB better than that at 2400 MHz at the same distance point. The difference between measurement results and the calculation values by using $20 \lg(f_c)$ is $2 \sim 6$ dB. Other frequency factor parameters are also proposed in the COST 231 Hata and Walfisch-Ikegami model [22], where the coefficients of the frequency term are adjusted to 26 and $26 \sim 29$, instead of 20.

Besides, the carrier frequency impacts the PL model differently for LOS, obstructed-LOS (OLOS) and NLOS case [10, 24]. In LOS case, the intercept of PL model changes with frequency, while the PL exponent is very close and changes mildly for different frequency. Just as [10] shown, the exponent is 1.74 in 5 GHz and 1.64 at 2 GHz in LOS

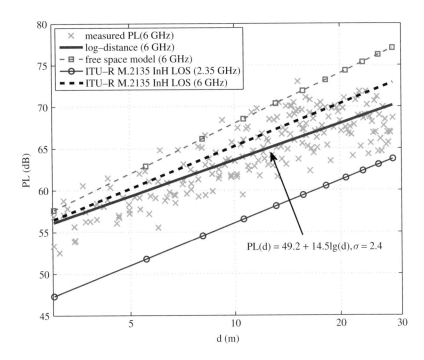

Figure 3.7 The comparison of path loss (LOS) between 2.35 GHz and 6 GHz.

condition, while the PL value at 2.4 GHz is 7.9 dB less than that at 5.25 GHz. The same phenomenon are reported in [24], the best fit PL exponents for 5 GHz band and the 2.4 GHz band over a 1 ~ 10 m range are both 1.9 in the LOS case. The measurements done by BUPT show similar results as shown in Figure 3.7. But in the OLOS or NLOS case, the PL exponent increases with frequency. In [24], PL exponents in the NLOS case extracted from all measurements over a 3 ~ 14 m range are 3.7 and 4.6 for 2.4 GHz band and the 5 GHz, respectively. In the case that both TX and RX antennas are 1.5 m in height [25], the PL exponents measured at 1 GHz, 5.5 GHz, 10 GHz and 18 GHz are 1.27, 1.45, 1.26 and 1.76. In [26] and [27], the similar results are found. Moreover, SF also depends on carrier frequency [26]. The σ increases with frequency as SF of different frequencies shown Table 3.8.

Table 3.8 Large scale parameters of different frequencies [26]

Frequency	LOS/OLOS	PL exponent	PL_0(dB)	SF (dB)
2.4 GHz	LOS	1.86	41.5	1.6
	OLOS	3.33	37.7	3.6
4.75 GHz	LOS	1.98	47.2	2.0
	OLOS	3.75	41.9	4.1
11.5 GHz	LOS	1.94	55.9	2.3
	OLOS	4.46	48.7	5.0

3.2.3.3   Multi-Floors and Multi-Walls

When radio wave propagates through walls or floors, it will be attenuated. So the additional loss introduced by penetrated walls and floors should be taken into consideration. The models for penetration loss can be divided into AF model and LD model. While for attenuation factor PL model proposed, penetration loss is modeled as constant. The additional loss is added to basic PL model. As literatures [5] and [28] suggests attenuation factor PL model provides a more reliable, less error-prediction. So the following discussion is based on the AF model.

As WIM2 proposes, the loss may be 5 dB for a light wall, and 12 dB for a heavy wall [4]. And the propagation loss through one floor is 17 dB. In COST 231 Hata, penetration loss is 18.3 dB for one floor, 3.4 dB for a light wall and 6.9 dB for a heavy wall. Actually, the penetration loss of walls and floors are not constant values. Different attributes of obstacles such as thickness and material and frequency will cause different penetration loss [4, 29–31]. Some measurement results are listed in Tables 3.9 and 3.10. In Table 3.10, the results of the indoor hotspot scenario measured are provided by BUPT.

Many measurement results in [7, 19, 30, 32] show that the loss between floors or walls does not increase linearly in dB with an increasing number of floors or walls. The attenuation caused by the first traversed floor or wall is greater than the incremental attenuation caused by each additional floor or wall. In [19], typical values of attenuation for one floor is 15 dB and an additional 6 to 10 dB per floor from two to four floors. As for five or more floors, PL will increase by only a few dB for each additional floor. As stated in [7, 19, 30], and [32], the propagation mechanism is different for the first floor and other floors. In the first floor, the received signal mainly comes through the floors. Instead, the signal in the higher floors is mainly composed of diffracted paths.

What is more, the penetration loss would also change with frequency. The signal with higher frequency tends to be more attenuated. [31] reports the penetration loss over the frequency range of 900 MHz to 18 GHz for typical indoor walls. The results demonstrate that the penetration loss of reinforced concrete wall increases with frequency. Besides, the relation between penetration loss and frequency depends on actual transmission condition to a certain extent, such as types of walls or floors. Just as COST 231 Hata multi-wall model [5], a difference of 1.5 dB for the light wall loss and a difference of 3.5 dB for the floor loss between 900 MHz and 1800 MHz are reported. More measurement results are shown in Table 3.11.

Table 3.9   Penetration loss in open literatures

Material Type	Loss (dB)	Frequency
All metal wall [29]	26	815 MHz
Aluminum siding [29]	20.4	815 MHz
Foil insulation [29]	3.9	815 MHz
Internal walls (in flat) [28]	6–8	2.4 GHz
Walls separating flats [28]	12–15	2.4 GHz
Floor penetration [28]	15	2.4 GHZ
A 35 cm concrete wall [31]	22	1 to 4 GHz
A 12 cm concrete wall [31]	12–15	1 to 4 GHz

Table 3.10 Floor Attenuation Factor (*FAF*)

Scenarios	Total *FAF* (dB)	One floor *FAF* (dB)	Frequency (MHz)
Office building [30]			
Through 1 floor	12.9	12.9	914
Through 2 floor	18.7	5.8	914
Through 3 floor	18.7	5.7	914
Through 4 floor	27	2.6	914
High-rise building [30]			
Through 1 floor	26.2	26.2	1900
Through 2 floor	33.4	7.2	1900
Through 3 floor	35.2	1.8	1900
Through 4 floor	38.4	3.2	1900
Through 5 floor	46.4	8	1900
Indoor hotspot*			
Through 1 floor	23.8	23.8	2350
Through 2 floor	30.0	6.2	2350
Through 3 floor	38.3	8.3	2350
Through 4 floor	46.9	8.6	2350
Indoor hotspot*			
Through 1 floor	21.7	21.7	4900
Through 2 floor	29.4	6.7	4900

*The results are from the measurements in BUPT.

Table 3.11 Floor attenuation factor for different frequency [30]

Building	Penetration loss (dB)	
	at 915 MHz	at 1900 MHz
San Ramon		
One floor	29.1	35.4
Two floor	36.6	35.6
Three floor	39.6	35.2
SF PacBell		
One floor	13.2	26.2
Two floor	18.1	33.4
Three floor	24.0	35.2
Four floor	27.0	38.4
Five floor	27.1	46.4

There are many other factors which influence the PL, such as antenna height, antenna pattern, people, mobility of terminal, and so on. But for indoor scenario, some factors can be negligible. Since the antenna height and the speed of the mobile station are limited for indoor deployment. Anyhow the simple model might be severely inaccurate in different scenarios and frequencies. In order to improve the accuracy, more site-specific information about the environment and the propagation characteristic in a specific site is required. In order to determine reasonable design guidelines and propagation parameters for indoor systems, more factors should be taken into account depending on the practical deployment.

3.3 Indoor Small Scale Fading

3.3.1 Geometry-Based Stochastic Channel Model

Multiple input and multiple output (MIMO) is one of the key technologies for IMT-A system and beyond. So most of the small scale models are MIMO channel models. They can be divided into two major categories: the correlation based models and the GBSM. The GBSM modeling methodology has been adopted by ITU-R M.2135 for the evaluation of IMT-A systems, due to its higher accuracy in recreating real propagation environment [1]. Considering a single down link case for a IMT-A GBSM channel model, if a wideband MIMO system with an S element base station (BS) array and a U element mobile stations (MS) array, the channel impulse response (CIR) at time t and delay τ is modeled as

$$H(\tau, t) = \sqrt{\frac{K}{K+1}} H_0(t)\delta(\tau) + \sqrt{\frac{1}{K+1}} \sum_{n=1}^{N} H_n(t)\delta(\tau - \tau_n), \qquad (3.14)$$

where K is the Rician K-factor on a linear scale, $H_0(t)$ is the channel coefficient matrix corresponding to the LOS ray, $H_n(t), n = 1, 2, \ldots, N$ is the nth NLOS channel coefficient component and $\delta(t)$ is the Dirac delta function. The elements of the $U \times S$ matrix $H_n(t) = (h_{u,s,n}(t))$ are given by

$$h_{u,s,n}(t) = \begin{bmatrix} F_{rx,u,V}(\varphi_{LOS}) \\ F_{rx,u,H}(\varphi_{LOS}) \end{bmatrix}^T \begin{bmatrix} \exp(j\Phi_{LOS}^{VV}) & 0 \\ 0 & \exp(j\Phi_{LOS}^{HH}) \end{bmatrix} \begin{bmatrix} F_{tx,s,V}(\phi_{LOS}) \\ F_{tx,s,H}(\phi_{LOS}) \end{bmatrix}$$
$$\cdot \exp\left(jd_s 2\pi\lambda_0^{-1}\sin(\phi_{LOS})\right) \exp\left(jd_u 2\pi\lambda_0^{-1}\sin(\varphi_{LOS})\right) \exp(j2\pi\upsilon_{LOS}t)$$
$$(3.15)$$

for $n = 0$, and

$$h_{u,s,n}(t) = \sqrt{P_n}$$
$$\cdot \sum_{m=1}^{M} \begin{bmatrix} F_{rx,u,V}(\varphi_{n,m}) \\ F_{rx,u,H}(\varphi_{n,m}) \end{bmatrix}^T \begin{bmatrix} \exp(j\Phi_{n,m}^{VV}) & \sqrt{\kappa^{-1}}\exp(j\Phi_{n,m}^{VH}) \\ \sqrt{\kappa^{-1}}\exp(j\Phi_{n,m}^{HV}) & \exp(j\Phi_{n,m}^{HH}) \end{bmatrix} \begin{bmatrix} F_{tx,s,V}(\phi_{n,m}) \\ F_{tx,s,H}(\phi_{n,m}) \end{bmatrix}$$
$$\cdot \exp\left(jd_s 2\pi\lambda_0^{-1}\sin(\phi_{n,m})\right) \exp\left(jd_u 2\pi\lambda_0^{-1}\sin(\varphi_{n,m})\right) \exp(j2\pi\upsilon_{n,m}t)$$
$$(3.16)$$

for $n = 1, 2, \ldots, N$. $F_{rx,u,V}$ and $F_{rx,u,H}$ are the field patterns of the uth receiving antenna element for vertical and horizontal polarizations, respectively. Similarly F_{tx} is the transmitting antenna pattern. P_n is the power resulting from the cluster. M is the number of subpaths. $\{\Phi_{n,m}^{vv}, \Phi_{n,m}^{vh}, \Phi_{n,m}^{hv}, \Phi_{n,m}^{hh}\}$ are the random initial phases for each ray m of each cluster n and for four different polarization combinations (vv, vh, hv, hh). $\{\Phi_{LOS}^{vv}, \Phi_{LOS}^{hh}\}$ are the initial random phases for the LOS ray. κ is the inverse of cross polarization power ratio. d_s and d_u are the uniform distances (m) between TX elements and RX

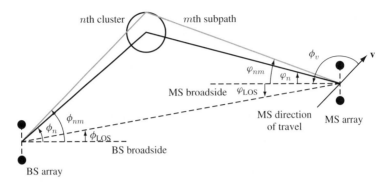

Figure 3.8 Angles in GBSM channel model.

elements respectively, and λ_0 is the wavelength of the carrier frequency. $v_{n,m}$ is the doppler frequency shift for each ray m of each cluster n. ϕ_{LOS} is the angle of departure (AOD) for the LOS ray with respect to the BS broadside, φ_{LOS} is the angle of arrival (AOA) for the LOS ray with respect to the MS broadside, is the AOD for the mth subpath of the nth cluster with respect to the BS broadside, while $\varphi_{n,m}$ is the AOA with respect to the MS broadside. The angles above are illustrated in Figure 3.8.

Table 3.12 displays a part of important parameters used in IMT-A system simulation. The parameters are all calculated by statistical method and estimation based on the data from a large quantity of field channel measurements.

3.3.2 Statistical Characteristics in Delay Domain

Table 3.13 shows the statistical results of the delay parameters in indoor hotspot scenario, measured in the building as shown in Figure 3.1. The measurement campaign was

Table 3.12 Key parameters of indoor hotspot in ITU-R M.2135 [1]

Items		Indoor hotspot	
		LOS	NLOS
Delay spread(DS)	μ	−7.70	−7.41
$\log_{10}(s)$	σ	0.18	0.14
AOD spread(ASD)	μ	1.60	1.62
$\log_{10}$(degrees)	σ	0.18	0.25
AOA spread(ASA)	μ	1.62	1.77
$\log_{10}$(degrees)	σ	0.22	0.16
XPR(dB)	μ	11	20
Correlation distance(m)	DS	8	5
	ASD	7	3
	ASA	8	3
	SF	10	6
	K	4	–
Delay distribution	–	Exponential	Exponential
AOD and AOA distribution	–	Laplacian	Laplacian

Table 3.13 Delay parameters in indoor hotspot

Literature	Scenario	Center frequency (MHz)	LOS/NLOS	$\bar{\tau}_{mean}$ (ns)	$\bar{\tau}_{rms}$ (ns)
*	Indoor hotspot	2350	LOS	26	19
			NLOS	61	42
[34]	Indoor hotspot	3705	LOS	387	50.9
			NLOS	416.2	80
[33]	Indoor hotspot	3705	LOS	224.3	47.79
			NLOS	303.8	64.66
[35]	Indoor office	2440	LOS	30.37	420.39
			NLOS	42.90	270.20

*the results are from the measurements in BUPT.

conducted with uniform panel antenna array and omnidirectional antenna array at the TX and the RX, respectively. It is obvious from Table 3.13 that the delay parameters of NLOS condition are larger than those of LOS condition. The measurement campaigns presented in [33] and [34] are performed at 3.7 GHz in a big hall which is a departure platform. Though the center frequency is different, the similar phenomenon is observed. However, the opposite phenomenon occurred in other indoor scenarios like office and laboratory, and so on [35]. So it can be concluded that the delay statistical characteristics are strongly related to the indoor environments.

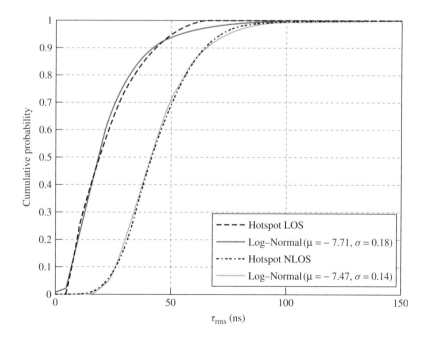

Figure 3.9 Probability distribution of the rms delay spread in indoor hotspot.

Table 3.14 Delay parameters of multi-floors in indoor hotspot

Center frequency (MHz)	Floor	$\bar{\tau}_{mean}$ (ns)	$\bar{\tau}_{rms}$ (ns)	$\bar{\tau}_{max}$ (ns)
580	2 Floor	88	60	320
	3 Floor	216	84	500
	4 Floor	248	80	500
	5 Floor	228	84	500
2350	1 Floor	22	17	110
	2 Floor	86	79	375
	3 Floor	226	109	540
	4 Floor	224	81	480
	5 Floor	206	75	455
4900	1 Floor	25	24	150
	2 Floor	67	62	335
	3 Floor	154	109	455
	4 Floor	193	93	490
	5 Floor	174	63	320

The root-mean-square (rms) delay spread is the most important delay parameter whose probability distribution is extensively investigated. As for indoor hotspot scenario, lognormal distribution fits the measured results of the rms delay spread both in LOS and NLOS condition. One example is given in Figure 3.9.

The radio propagation in indoor environment is complex. Table 3.14 displays the delay parameters of different frequencies and floors in indoor hotspot scenario. The measurement campaign is also carried out in the same building. The TX with the vertical dipole antenna is fixed in the first floor, and the receiver moves on different floors.

3.3.2.1 Dependence of rms Delay Spread on Frequency

Referring to Table 3.14, taking $\bar{\tau}_{rms}$ for an instance, no stable relationship with center frequency is shown. Similarly, the frequency dependence on rms delay spread is also not clearly observed in [36]. However, the decreasing trend in LOS condition and the increasing trend in the NLOS condition are presented in [37]. The measurement campaigns in [36] and [37] are both conducted in a single floor, which is different from the measurement settings in BUPT. Besides, the frequency band measured in [37] is $2.4 \sim 24$ GHz whereas that in [36] is 450 MHz ~ 5.8 GHz. Therefore, considered in the same floor, it can be preliminarily concluded that the rms delay spread is increasing with frequency in the high frequency band but does not apparently show frequency dependence in the low frequency band. And more detailed measurement should be carried out to confirm this conclusion.

3.3.2.2 Impact of Multi-Floor on rms Delay Spread

Referring to Table 3.14, no matter what center frequency is, the rms delay spread increases from the first floor to the third floor, and then decreases from the third floor to the fifth floor. With the floor increasing, the multi-paths caused by the reflection and the diffraction increases correspondingly so that the rms delay spread becomes larger. However, the power of the radio wave is attenuated to a large extent for much higher floors so that the multipaths with large delay and small power are drowned in the noise.

3.3.3 Statistical Parameter in Angular Domain

The presence of reflecting objects and scatterers creates an environment that dissipates the signal energy in amplitude, phase, and time. These effects result in multiple versions of the transmitting signal that arrives at the receiving antenna. This distortion in angular domain that causes the space selective fading will have an adverse influence on the performance of the MIMO system. Power angular spectrum (PAS) and angular spread (AS) are important MIMO channel parameters for a wireless system design.

3.3.3.1 PAS for Indoor Scenario

A MIMO field measurement campaign is performed at center frequency of 2.35 GHz with 100 MHz bandwidth at BUPT, Beijing, China. The measurement environment is described in detail in [38]. Based on the measured CIRs, the space alternating generalized expectation-maximization (SAGE) algorithm is utilized to extract a set of the channel parameters [39]:

$$\Theta_l = \{A_l, \tau_l, \phi_{1,l}, \theta_{1,l}, \phi_{2,l}, \theta_{2,l}, f_{d,l}\}, \tag{3.17}$$

where Θ_l is the parameter vector characterizing the lth path. A_l, τ_l, $\phi_{1,l}$, $\theta_{1,l}$, $\phi_{2,l}$, $\theta_{2,l}$, and $f_{d,l}$ are, respectively, its polarization matrix, propagation delay, azimuth angle of departure, elevation angle of departure, azimuth angle of arrival, elevation angle of arrival, and doppler frequency. In order to guarantee the accuracy of estimation, up to 50 paths are extracted.

And PAS can be calculated from the estimated channel parameters. In ITU-R M.2135, the PAS distributions of the azimuth AOD and the azimuth AOA are both close to Laplacian distribution [1]. In WIM2, the Wrapped Gaussian distribution is applied to fit the PAS distributions of the azimuth AOD and AOA in each cluster in indoor scenario [4]. The differences between Laplacian distribution and Wrapped Gaussian distribution can be recognized in Figure 3.10. Compared with Wrapped Gaussian distribution, Laplacian distribution has a sharper peak. That means the power of signal in angular domain fitted by Laplacian distribution is more centralized than fitted by Wrapped Gaussian distribution. The Laplacian distribution fits the measured PAS in an indoor hotspot well as shown in Figures 3.11 and 3.12.

3.3.3.2 AS for Indoor Scenario

In order to compare different multipath channels and to develop some general design guidelines for wireless systems, AS, which can be determined from PAS, is used to quantify the spatial dispersive property. It is important to note that the rms AS is defined as the square root of the second central moment of PAS. To avoid the ambiguous effect due to the circular wrapping of the angles, the circular angle spread (CAS) is calculated. CAS is constant regardless of the value of the angle shift as described in 3GPP spatial channel model (SCM) specification [40]. The obtained cumulative distribution functions (CDFs) of azimuth rms spreads for LOS and NLOS in the field measurement can be best fitted by log-normal distribution. And both in ITU-R M.2135 and WIM2, the log-normal distribution is applied to fit the CAS observations for both NLOS and LOS conditions. The mean value and standard deviation of the fitted log-normal distribution for rms AS of AOA and AOD in indoor hotspot scenario of ITU-R M.2135 are given in Table 3.12.

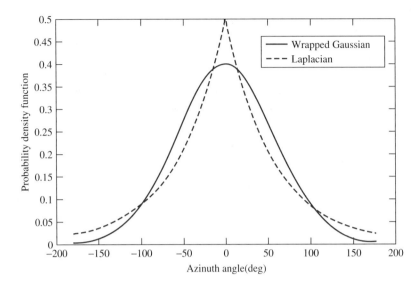

Figure 3.10   Comparison of azimuth PAS models in ITU-R M.2135 and WIM2.

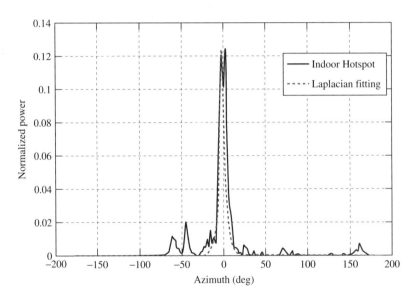

Figure 3.11   PAS of TX azimuth and Laplacian fitting in indoor hotspot at 2.35 GHz.

3.3.4   Cross-Polarization Discrimination (XPD) for Indoor Scenario

Cross-polarized systems are of interest since they are able to double the antenna numbers for half the spacing needs of co-polarized antennas. Meanwhile, data multiplexing and diversity gains can also be achieved in the polarization domain. XPD is a critical factor for

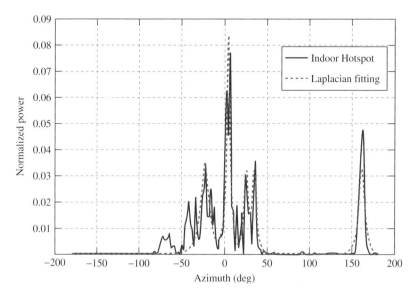

Figure 3.12   PAS of RX azimuth and multi-cluster Laplacian fitting in indoor hotspot at 2.35 GHz.

the cross-polarized antenna and defined as the ratio of the co-polarized antenna received signal power to the cross-polarized received power.

$$XPD = 10log_{10}\left(\left|\left(\frac{\alpha_{V,V} + \alpha_{H,H}}{\alpha_{H,V} + \alpha_{V,H}}\right)\right|^2\right) (dB) \qquad (3.18)$$

where α_{p_1,p_2} are the entries of polarization matrix A, which can be estimated by SAGE algorithm.

There is a general consensus in the literatures that the XPD in dB has a nonzero-mean Gaussian distribution. Hence, we can write XPD $\sim (N, \sigma^2)$. Depending on the environment and the existence of a LOS component, the mean values of XPD measured in the literatures vary from 0 to 18 dB, with the standard deviations typically in the order of 3–8 dB.

3.3.4.1   Dependence of XPD on Distance

In [41], it is found that in a corridor environment with LOS, the decay with distance can be modeled as an exponential decay $d^{-\gamma}$, although the traditional law also gives satisfactory performance. Decay exponents are different for all polarizations as shown in Table 3.15. The XPD is large (around 15 dB), and increases with distance. And in [42], the mean values of XPD for LOS and OLOS are 8.3 and 2.8 dB, respectively. In LOS condition, the received signal is dominated by the LOS component, which is not depolarized. Conversely, in OLOS condition, reception is primarily due to reflection, diffraction and scattering, which depolarizes the transmitted signal. In fact, [43] even reported negative XPD in some OLOS channels.

Table 3.15 Decay exponents for all polarizations in corridor environment [41]

	P_{VV}	P_{HH}	P_{VH}	P_{HV}
γ (dB)	1.07	1.20	1.49	1.48

3.3.4.2 Dependence of the XPD on Azimuth and Elevation

In [44], measurements were performed with vertically and horizontally polarized antennas of different antenna patterns, and a definite dependence of XPD on antenna pattern was found. However, it is difficult to extract the dependence of the XPD on azimuth and elevation from these results. Most references suggest that the azimuth spread is independent of the polarization. [45] suggest that the XPD has been shown to have a weak negative correlation with the azimuth spread. In other words, a larger azimuth spread at the base leads to a lower XPD. Again, this result is intuitive, as a larger azimuth spread indicates stronger scattering. However, the ITU-R M.2135 and the WIM2 find for indoor environments that the mean (in dB) is independent of the azimuth spread.

3.3.4.3 Dependence of XPD on Delay Spread

Results in [42] did not find a dependence between delay spread and polarization in indoor environments. While in [46], it was found that the co-polarized and cross-polarized components had different decay time constants. Analyzing cluster decay constants in a microcell scenario, they showed that the VV component decayed with 8.9 dB, while the co-polarized component decayed with 11.8 dB. This indicates that the XPD increases with increasing delay. However, it is found that the mean value (in dB) is independent of the delay spread for indoor environments in ITU-R M.2135 and the WIM2.

3.3.4.4 Dependence of XPD on Penetration

In [47] the study found that building penetration encountered by indoor coverage has a minor influence on the XPD perceived at the BS. Thus the diversity potential remains fairly unchanged, when applying power control to compensate for the penetration loss.

3.3.5 3-D Modeling for Indoor MIMO Channel

The 3GPP only developed a cross-polarized 2-D channel model for MIMO systems without considering the elevation spectrum. However, the assumption of 2-D propagation breaks down when in some propagation environments the elevation angle distribution is significant. The estimation of ergodic capacity assuming a 2-D channel coefficient alone can lead to erroneous results. Therefore, it is necessary to modify the above 2-D model to incorporate such effects.

Several papers have proposed to extend the existing 2-D models to 3-D ones. For example, in [48] the SCM model is extended to 3-D by adding a 3-D channel component to the original 2-D channel model. The impact of the elevation angle spread on the MIMO

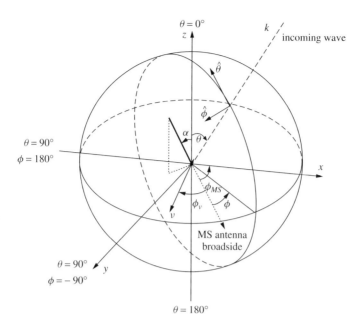

Figure 3.13   Spherical coordinate system for 3-D model [48].

channel capacity is studied based on this 3-D model in [49]. An algorithm is proposed in [50] to incorporate 3-D antenna radiation patterns in SCM. In [51], a 3-D antenna array model is proposed to describe the antenna orientation in different coordinates and is used in the WIM2 [4].

With the consideration of the low antenna heights and local scattering interactions in the vicinity of both the transmit and receive arrays and the existing 2-D model captures many of the effects of indoor propagation, in [48] the authors extend the traditional 2-D model to cases where the mobile antennas are located at 3-D indoor environments. A simplified sketch of the model is given in Figure 3.13.

In a 3-D channel model, the response vector $\vec{F}$ in its θ and ϕ components is expressed as

$$\vec{F}_{MS}(\vec{k}) = \begin{bmatrix} F_\theta(\vec{k}) \\ F_\phi(\vec{k}) \end{bmatrix} = \begin{bmatrix} \cos\alpha\,\sin\theta + \sin\alpha\,\cos\theta\,\sin\phi \\ \sin\alpha\,\cos\phi \end{bmatrix} e^{i\vec{k}\vec{r}} \qquad (3.19)$$

where F_θ, F_ϕ are the θ and ϕ polarized responses of the antenna at direction $\vec{k}$. The vector $\vec{k}$ is defined in terms of θ and ϕ

$$\vec{k} = \frac{2\pi}{\lambda} \begin{bmatrix} \sin\theta\,\cos\phi, \sin\theta\,\sin\phi, \cos\theta \end{bmatrix} \qquad (3.20)$$

The fading channel coefficient $h_{su}^{3D}(t)$ for this component of the propagation between BS S antenna and MS antenna U is defined as

$$h_{su}^{3D}(t) = \sqrt{\frac{1}{M} \sum_{i=1}^{M} \left(\vec{F}_{s,BS}^T \left(\vec{k}_{i,BS} \right) H_i^{3D} \vec{F}_{u,MS} \left(\vec{k}_{i,MS} \right) e^{-i\vec{k}_{i,MS}\vec{v}t} \right)} \qquad (3.21)$$

The $\vec{k}_{i,MS}$ are independently chosen for the 2-D and 3-D channel coefficients and the matrix H_i^{3D} for the i_{th} wave component is given by

$$H_i^{3D} = \begin{bmatrix} z_i^{v\theta} & z_i^{v\phi} \\ z_i^{h\theta} & z_i^{h\phi} \end{bmatrix} \qquad (3.22)$$

The z_i terms in Equation 3.22 are the random coefficients of the i_{th} wave component of the sum for each of the V and H channels and their respective components in the θ and ϕ polarizations. The antenna responses $\vec{F}$ for the BS are the same for both 2-D and 3-D models. At the MS, the antenna responses are different due to the 3-D character of the radiation. Assuming that the composite channel consists of two independent terms, namely, the 2-D and 3-D channel coefficients, scaled by their relative powers. The 2-D and 3-D channel coefficients are used to model a composite channel coefficient. Thus, the composite channel coefficient between antennas S and U can be written as

$$h_{su}(t) = \sqrt{\frac{1}{1+g}}h_{su}^{2D}(t) + \sqrt{\frac{g}{1+g}}h_{su}^{3D}(t) \qquad (3.23)$$

where g is the ratio of powers of the 3-D to 2-D components of the channel. The relative strength of the 3-D radiation at the mobile to that of the already existing 2-D radiation depends on several factors, including the distance of the mobile from openings. There are very few measurements of g. In the case of indoor mobiles, which are far from open spaces, one can assume that $g = inf$ and keep only the 3-D components of the channel. For indoor channels close to a window, a reasonable value for g is $g = -4\,dB$.

3.3.6 Impact of Elevation Angular Distribution

In [49], the authors discuss the sensitivity of angular distribution to a variety of different azimuth and elevation power distributions and other system parameters (Figure 3.14). For discussing how sensitive ergodic capacity is to the choice of elevation AOA distributions in the 3-D component, the authors simulate four cases: elevation angle is spherical uniform, power is concentrated close to the horizontal plane of the scattering sphere, elevation angle is uniform and power concentrated in the top and bottom of the scattering sphere. The 3-D elevation AOA distributions are shown in Figure 3.6.

According to Equation 3.23, the ergodic capacity of the pure 3-D propagation environment ($g = 1$) almost doubles the corresponding value for the 2-D case ($g = 0$). For the large angle spread case, the 3-D component still leads to the ergodic capacity increase of about 30 percent. These results are intuitive, as the 3-D distribution has a wider angular spread, and thus leads to a better decorrelation of the signals at the different antenna elements.

Moreover, the largest capacity increase is for the case when the incoming powers are concentrated on the top and bottom of the scattering sphere, followed by the uniform, spherical uniform and the case when power concentrates in the horizontal plane of the scattering sphere, respectively. All these results show that the largest impact of the 3-D

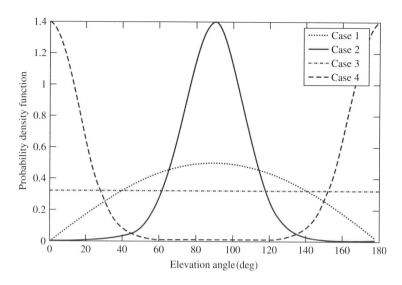

Figure 3.14 Elevation AOA distributions for the 3-D component [49].

component is for cases when the incoming power is coming from the top (or bottom) of the scattering sphere. Furthermore, ignoring the 3-D component greatly underestimates the ergodic capacity.

In [52] the distribution of elevation AOA is not a good fit to a Laplacian function, it is well represented by a Gaussian function.

References

1. "Guidelines for evaluation of radio interface technologies for imt- advanced," ITU-R M.2135, Tech. Rep., 2008.

2. "Further advancements for e-utra (physical layer aspects)," 3GPP TR 36.814 v9.0.0, Tech. Rep., March 2010.

3. V. Erceg, L. Schumacher, and P. Kyritsi, "Tgn channel models," IEEE P802.11Wireless LANs, Tech. Rep., 2004 May.

4. "Winner ii channel models," IST-WINNER II Deliverable 1.1.2, Tech. Rep., 2008.

5. "Digital mobile radio towards future generation systems, final report," COST Action 231, European Communities, EUR 18957, Tech. Rep., 1999.

6. D. M. J. Devasirvatham, C. Banerjee, M. J. Krain, and D. A. Rappaport, "Multi-frequency radiowave propagation measurements in the portable radio environment," in *Proc. IEEE International Communications ICC '90, Including Supercomm Technical Sessions. SUPERCOMM/ICC '90. Conf. Record. Conf*, 1990, pp. 1334–1340.

7. S. Y. Seidel and T. S. Rappaport, "914mhz path loss prediction models for indoor wireless communications in multifloored buildings," *IEEE Transactions on Antennas and Propagation*, vol. 40, no. 2, pp. 207–217, 1992.

8. R. R. Skidmore, T. S. Rappaport, and A. L. Abbott, "Interactive coverage region and system design simulation for wireless communication systems in multifloored indoor environments: smt plus," in *Proc. 5th IEEE Int Universal Personal Communications Record. Conf*, vol. 2, 1996, pp. 646–650.

9. M. Lott and I. Forkel, "A multi-wall-and-floor model for indoor radio propagation," in *Proc. VTC 2001 Spring Vehicular Technology Conf. IEEE VTS 53rd*, vol. 1, 2001, pp. 464–468.

10. D. Laselva, X. Zhao, J. Meinila, T. Jamsa, J.-P. Nuutinen, P. Kyosti, and L. Hentila, "Empirical models and parameters for rural and indoor wideband radio channels at 2.45 and 5.25ghz," in *Proc. IEEE 16th Int. Symp. Personal, Indoor and Mobile Radio Communications PIMRC 2005*, vol. 1, 2005, pp. 654–658.

11. A. F. Molisch, L. J. Greenstein, and M. Shafi, "Propagation issues for cognitive radio," *Proceedings of the IEEE*, vol. 97, no. 5, pp. 787–804, 2009.

12. D. Giancristofaro, "Correlation model for shadow fading in mobile radio channels," *Electronics Letters*, vol. 32, no. 11, pp. 958–959, 1996.

13. "Proposed new test environments and channel models of preliminary draft new report m.[imt.eval]," ITU-R WP 8F 1252, Tech. Rep., 16 May 2007.

14. J. Zhang, X. Gao, P. Zhang, and X. Yin, "Propagation characteristics of wideband mimo channel in hotspot areas at 5.25ghz," in *Proc. IEEE 18th Int. Symp. Personal, Indoor and Mobile Radio Communications PIMRC 2007*, 2007, pp. 1–5.

15. W. Dong, J. Zhang, X. Gao, P. Zhang, and Y. Wu, "Cluster identification and properties of outdoor wideband mimo channel," in *Proc. VTC-2007 Fall Vehicular Technology Conf. 2007 IEEE 66th*, 2007, pp. 829–833.

16. X. Gao, J. Zhang, G. Liu, D. Xu, P. Zhang, Y. Lu, and W. Dong, "Large-scale characteristics of 5.25ghz based on wideband mimo channel measurements," *IEEE Antennas and Wireless Propagation Letters*, vol. 6, pp. 263–266, 2007.

17. J. Zhang, D. Dong, Y. Liang, X. Nie, X. Gao, Y. Zhang, C. Huang, and G. Liu, "Propagation characteristics of wideband mimo channel in urban micro- and macrocells," in *Proc. IEEE 19th Int. Symp. Personal, Indoor and Mobile Radio Communications PIMRC 2008*, 2008, pp. 1–6.

18. Y. Lu, J. Zhang, X. Gao, P. Zhang, and Y. Wu, "Outdoor-indoor propagation characteristics of peer-to-peer system at 5.25ghz," in *Proc. IEEE 66th Vehicular Technology Conf. 2007 VTC-2007 Fall*, 2007, pp. 869–873.

19. J. B. Andersen, T. S. Rappaport, and S. Yoshida, "Propagation measurements and models for wireless communications channels," *IEEE Communications Magazine*, vol. 33, no. 1, pp. 42–49, 1995.

20. X. Zhang, T. W. Burress, K. B. Albers, and W. B. Kuhn, "Propagation comparisons at vhf and uhf frequencies," in *Proc. IEEE Radio and Wireless Symp. RWS '09*, 2009, pp. 244–247.

21. J. Walfisch and H. L. Bertoni, "A theoretical model of uhf propagation in urban environments," *IEEE Transactions on Antennas and Propagation*, vol. 36, no. 12, pp. 1788–1796, 1988.

22. M. Hata, "Empirical formula for propagation loss in land mobile radio services," *IEEE Transactions on Vehicular Technology*, vol. 29, no. 3, pp. 317–325, 1980.

23. Y. Oda, R. Tsuchihashi, K. Tsunekawa, and M. Hata, "Measured path loss and multipath propagation characteristics in uhf and microwave frequency bands for urban mobile communications," in *Proc. VTC 2001 Spring Vehicular Technology Conf. IEEE VTS 53rd*, vol. 1, 2001, pp. 337–341.

24. D. Heung and C. A. Prettie, "Path loss comparison between the 5ghz unii band (802.11a) and the 2.4ghz ism band (802.11b)," *Intel Labs Comporation*, 2002.

25. G. Santella and E. Restuccia, "Analysis of frequency domain wide-band measurements of the indoor radio channel at 1, 5.5, 10 and 18ghz," in *Proc. Communications: The Key to Global Prosperity Global Telecommunications Conf. GLOBECOM '96*, vol. 2, 1996, pp. 1162–1166.

26. G. J. M. Janssen, P. A. Stigter, and R. Prasad, "Wideband indoor channel measurements and ber analysis of frequency selective multipath channels at 2.4, 4.75, and 11.5ghz," *IEEE Transactions on Communications*, vol. 44, no. 10, pp. 1272–1288, 1996.

27. D. M. J. Devasirvatham, C. Banerjee, R. R. Murray, and D. A. Rappaport, "Four-frequency radiowave propagation measurements of the indoor environment in a large metropolitan commercial building," in *Proc. Countdown to the New Millennium. Featuring a Mini-Theme: Personal Communications Services Global Telecommunications Conf. GLOBECOM '91*, 1991, pp. 1282–1286.

28. T. Chrysikos, G. Georgopoulos, and S. Kotsopoulos, "Attenuation over distance for indoor propagation topologies at 2.4ghz," in *Proc. IEEE Symp. Computers and Communications (ISCC)*, 2011, pp. 329–334.

29. D. Cox, R. Murray, and A. Norris, "Measurement of 800mhz radio transmission into building with metallic walls," *Bell Systems Technical Journal*, vol. 62, no. 9, pp. 2695–2717, 1983.

30. T. S. Rappaport, *Wireless Communications: Principles and Practice*. PrenticeHall, New Jersey, 1996.

31. Y. P. Zhang and Y. Hwang, "Measurements of the characteristics of indoor penetration loss," in *Proc. IEEE 44th Vehicular Technology Conf*, 1994, pp. 1741–1744.

32. W. Honcharenko, H. L. Bertoni, J. L. Dailing, J. Qian, and H. D. Yee, "Mechanisms governing uhf propagation on single floors in modern office buildings," *IEEE Transactions on Vehicular Technology*, vol. 41, no. 4, pp. 496–504, 1992.

33. M.-D. Kim, H. K. Kwon, B. S. Park, J. J. Park, and H. K. Chung, "Wideband mimo channel measurements in indoor hotspot scenario at 3.705ghz," in *Proc. 4th Int Signal Processing and Communication Systems (ICSPCS) Conf*, 2010, pp. 1–5.

34. M.-D. Kim, H. K. Kwon, B. S. Park, J. J. Park, and H. K. Chung, "Multipath channel parameters based on indoor hotspot channel measurements at 3.7ghz," in *Proc. 13th Int Advanced Communication Technology (ICACT) Conf*, 2011, pp. 579–583.

35. H. MacLeod, C. Loadman, and Z. Chen, "Experimental studies of the 2.4-ghz ism wireless indoor channel," in *Proc. 3rd Annual Communication Networks and Services Research Conf*, 2005, pp. 63–68.

36. A. Affandi, G. El Zein, and J. Citerne, "Investigation on frequency dependence of indoor radio propagation parameters," in *Proc. VTC 1999 - Fall Vehicular Technology Conf. IEEE VTS 50th*, vol. 4, 1999, pp. 1988–1992.

37. D. Lu and D. Rutledge, "Investigation of indoor radio channels from 2.4ghz to 24ghz," in *Proc. IEEE Antennas and Propagation Society Int. Symp*, 2003, pp. 134–137.

38. X. Nie, J. Zhang, Y. Zhang, G. Liu, and Z. Liu, "An experimental investigation of wideband mimo channel based on indoor hotspot nlos measurements at 2.35ghz," in *Proc. IEEE Global Telecommunications Conf. IEEE GLOBECOM 2008*, 2008, pp. 1–5.

39. B. H. Fleury, P. Jourdan, and A. Stucki, "High-resolution channel parameter estimation for mimo applications using the sage algorithm," in *Proc. Networking Broadband Communications Access, Transmission 2002 Int. Zurich Seminar*, 2002, pp. 30–31.

40. "Spatial channel model for multiple input multiple output (mimo) simulations(rel.6)," 3GPP, TR 25.996, 2003.

41. P. Kyritsi, D. C. Cox, R. A. Valenzuela, and P. W. Wolniansky, "Effect of antenna polarization on the capacity of a multiple element system in an indoor environment," *IEEE Journal on Selected Areas in Communications*, vol. 20, no. 6, pp. 1227–1239, 2002.

42. T. S. Rappaport and D. A. Hawbaker, "Wide-band microwave propagation parameters using circular and linear polarized antennas for indoor wireless channels," *IEEE Transactions on Communications*, vol. 40, no. 2, pp. 240–245, 1992.

43. D. Cox, R. Murray, H. Arnold, A. Norris, and M. Wazowicz, "Cross-polarization coupling measured for 800mhz radio transmission in and around houses and large buildings," *IEEE Transactions on Antennas and Propagation*, vol. 34, no. 1, pp. 83–87, 1986.

44. P. Soma, D. S. Baum, V. Erceg, R. Krishnamoorthy, and A. J. Paulraj, "Analysis and modeling of multiple-input multiple-output (mimo) radio channel based on outdoor measurements conducted at 2.5ghz for fixed bwa applications," in *Proc. IEEE Int. Conf. Communications ICC 2002*, vol. 1, 2002, pp. 272–276.

45. M. Nilsson, B. Lindmark, M. Ahlberg, M. Larsson, and C. Beckman, "Measurements of the spatio-temporal polarization characteristics of a radio channel at 1800mhz," in *Proc. IEEE 49th Vehicular Technology Conf*, vol. 1, 1999, pp. 386–391.

46. A. Kara and H. L. Bertoni, "Blockage/shadowing and polarization measurements at 2.45ghz for interference evaluation between bluetooth and ieee 802.11 wlan," pp. 376–379, 2001, Antennas and Propagation Society International Symposium, 2001. IEEE.

47. P. C. F. Eggers, I. Z. Kovacs, and K. Olesen, "Penetration effects on xpd with gsm 1800 handset antennas, relevant for bs polarization diversity for indoor coverage," in *Proc. 48th IEEE Vehicular Technology Conf. VTC 98*, vol. 3, 1998, pp. 1959–1963.

48. M. Shafi, M. Zhang, A. L. Moustakas, P. J. Smith, A. F. Molisch, F. Tufvesson, and S. H. Simon, "Polarized mimo channels in 3-d: models, measurements and mutual information," *IEEE Journal on Selected Areas in Communications*, vol. 24, no. 3, pp. 514–527, 2006.

49. M. Shafi, M. Zhang, P. J. Smith, A. L. Moustakas, and A. F. Molisch, "The impact of elevation angle on mimo capacity," in *Proc. IEEE Int. Conf. Communications ICC '06*, vol. 9, 2006, pp. 4155–4160.

50. H. Kanj, P. Lusina, S. M. Ali, and F. Kohandani, "A 3d-to-2d transform algorithm for incorporating 3d antenna radiation patterns in scm," *IEEE Antennas and Wireless Propagation Letters*, vol. 8, pp. 815–818, 2009.

51. M. Narandzic, M. Kaske, C. Schneider, M. Milojevic, M. Landmann, G. Sommerkorn, and R. S. Thoma, "3d-antenna array model for ist-winner channel simulations," in *Proc. VTC2007-Spring Vehicular Technology Conf. IEEE 65th*, 2007, pp. 319–323.

52. Y. Zhang, A. K. Brown, W. Q. Malik, and D. J. Edwards, "High resolution 3-d angle of arrival determination for indoor uwb multipath propagation," *IEEE Transactions on Wireless Communications*, vol. 7, no. 8, pp. 3047–3055, 2008.

4

Outdoor Channels

Petros Karadimas
University of Bedfordshire, UK

4.1 Introduction

In outdoor wireless environments, the received signal is subject to spatial and temporal variations due to several inherent propagation mechanisms. According to the nature of the environment, the transmitted signal bandwidth and the overall distance that electromagnetic waves travel from the transmitting antenna to the receiving antenna, propagation can be confined within resolved clusters [1]. A resolved cluster is uniquely determined by its joint directional and delay characteristics [1]. The number of resolved clusters increases when the transmitted signal bandwidth increases or/and the scatterers[1] are more densely distributed (e.g., in an urban environment compared to a suburban). A generic modeling approach characterizing spatial and temporal variations of the received signal forms the subject of this chapter. The latter is carried out by considering the outdoor propagation channel to be jointly frequency selective, time selective and space selective at both the transmitter and receiver sides. Several common channel categories encountered in the literature such as time varying mobile and mobile-to-mobile channels are seen as special cases of the model adopted here.

In future wireless networks, the requirements for increased capacity and data rates will be met through combating propagation effects and transforming them into mechanism providing benefits. The latter is achieved through diversity combining schemes such as frequency diversity, time diversity and antenna diversity. The performance of diversity systems is strongly related to the propagation effects of the wireless channel [2]. Thus, knowledge and modeling of propagation is necessary during the selection of appropriate diversity schemes to meet the customers', requirements and to maintain the complexity and cost of the system to an acceptable level.

---

[1] The term scatterer refers to any object interacting with propagating multipath components. Accordingly, the term scattering incorporates any other propagation mechanism such as reflection and diffraction.

LTE-Advanced and Next Generation Wireless Networks: Channel Modelling and Propagation, First Edition.
Edited by Guillaume de la Roche, Andrés Alayón Glazunov and Ben Allen.
© 2013 John Wiley & Sons, Ltd. Published 2013 by John Wiley & Sons, Ltd.

This chapter is organized as follows. Section 4.2 presents the reference outdoor channel model. In Section 4.3, small scale variations are characterized from the reference model. Distance power losses and large scale variations are characterized in Section 4.4. Finally, Section 4.5 presents a summary of this chapter with a synopsis of the main results and outcomes.

4.2 Reference Channel Model

We consider an outdoor propagation environment as in Figure 4.1. A number, L, of multi-path components depart from the transmitting antenna and arrive at the receiving antenna having been scattered by the environment's scatterers. By using two local coordinate systems (one at the transmitting and one at the receiving antenna), the vectors $\mathbf{r_T}$ and $\mathbf{r_R}$ determine the location in space of the transmitting and receiving antennas, respectively, as shown in Figure 4.1. The received signal arises by extending the directional representation in [3] to that of double-directional representation as in [4]. Thus, if $x(t)$ is the transmitted signal in complex baseband form, the complex baseband form of the received signal $y(t, r_T, r_R)$ will be[2]

$$y(t, \mathbf{r_T}, \mathbf{r_R}) = \sum_{l=1}^{L} a_l \exp\left[j2\pi\left(\Omega_{\mathbf{T,l}} \cdot \mathbf{r_T}\right)/\lambda_0\right] \exp\left[j2\pi\left(\Omega_{\mathbf{R,l}} \cdot \mathbf{r_R}\right)/\lambda_0\right]$$
$$\times \exp\left(j2\pi v_{d,l}t\right) x\left(t - \tau_l\right) \tag{4.1}$$

where t is time, and λ_0 the wavelength. The remaining parameters of the lth multipath component are as follows: a_l is the complex amplitude, τ_l is the delay, $v_{d,l}$ is the Doppler frequency, $\Omega_{\mathbf{T,l}}$ is the unit vector for the direction-of-departure (DOD) and $\Omega_{\mathbf{R,l}}$ is the unit vector for the direction-of-arrival (DOA).

The reference model in 4.1 does not take into account dual antenna polarization, with such an extension being straightforward. In dual polarization channel modeling, four similar expressions to 4.1 are required, accounting for all possible combinations of co-polarized and cross-polarized transmitting-receiving antenna pairs. Thus, for the sake

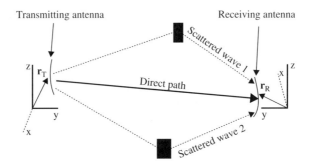

Figure 4.1 Multipath propagation in an outdoor environment.

---

[2] In general $L, a_l, \tau_l, v_{d,l}, \Omega_{\mathbf{T,l}}$ and $\Omega_{\mathbf{R,l}}$ are time dependent parameters.

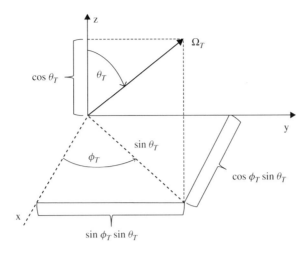

Figure 4.2 Unit vector for the DOD with respect to the azimuth and elevation AOD. The unit vector for the DOA arises by replacing subscript T with R.

of simplicity, we consider single antenna polarization which captures all the important channel characteristics presented here.

The unit vectors $\Omega_{\mathbf{T,l}}$ and $\Omega_{\mathbf{R,l}}$ are related to the azimuth and elevation angle-of-departure (AOD) and angle-of-arrival (AOA) of the lth multipath component, respectively, through the following forms

$$\Omega_{\mathbf{T,l}} = \mathbf{x}\cos\varphi_{T,l}\sin\theta_{T,l} + \mathbf{y}\sin\varphi_{T,l}\sin\theta_{T,l} + \mathbf{z}\cos\theta_{T,l}$$

$$\Omega_{\mathbf{R,l}} = \mathbf{x}\cos\varphi_{R,l}\sin\theta_{R,l} + \mathbf{y}\sin\varphi_{R,l}\sin\theta_{R,l} + \mathbf{z}\cos\theta_{R,l} \tag{4.2}$$

where $\mathbf{x}$, $\mathbf{y}$, $\mathbf{z}$ are the unit vectors in the x, y, z directions, respectively, and $\phi_T, \phi_R \in [-\pi, \pi]$, $\theta_T, \theta_R \in [0, \pi]$, are the azimuth and elevation AOD, AOA, respectively. Figure 4.2 shows the azimuth and elevation AOD together with the unit vector for the DOD. The Doppler frequency, $v_{d,l}$, results from the interaction of the lth multipath

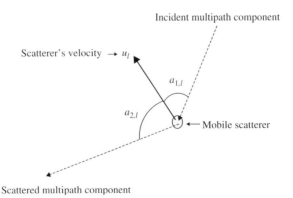

Figure 4.3 Interaction of a multipath component with a mobile scatterer.

component with a mobile scatterer, as illustrated in Figure 4.3. This is derived by adding two Doppler frequency contributions, that is one contributed by the arrival to the mobile scatterer (that is, $(u_l/\lambda_0)\cos a_{1,l}$) and another contributed by the departure from the mobile scatterer (that is, $(u_l/\lambda_0)\cos a_{2,l}$). Thus [5, 6].

$$v_{d,l} = \left(u_l/\lambda_0\right)\left(\cos a_{1,l} + \cos a_{2,l}\right) \tag{4.3}$$

with u_l the amplitude of the scatterer velocity, $a_{1,l}$ the AOA and $a_{2,l}$ the AOD with respect to the direction of the scatterer velocity.

The rationale behind Equation 4.1 is that the received signal consists of arbitrarily weighted and delayed replicas of the transmitted signal, having the form $a_l x(t - \tau_l)$. The complex amplitude depends on the complex field of the lth multipath component and also on the field patterns of the transmitting and receiving antennas. The complex exponentials in Equation 4.1 characterize the phase variations of the received signal causing small scale fading (that is, variations within distances comparable to the carrier wavelength). Particularly, $\exp\left[j2\pi\left(\Omega_{T,l}\cdot\mathbf{r}_T\right)/\lambda_0\right]$ and $\exp\left[j2\pi\left(\Omega_{R,l}\cdot\mathbf{r}_R\right)/\lambda_0\right]$ characterize phase variations due to space variations at the transmitter and receiver sides, respectively. Such space variations occur when the transmitter and/or the receiver move in the wireless environment. The term $\exp\left(j2\pi v_{d,l}t\right)$ characterizes phase variations due to time variations of the propagation environment. For outdoor environments, time variations arise from mobile scatterers such as vehicular traffic, wind-blown trees and vegetation and pedestrian mobility. In this chapter, spatial and temporal variations are considered separately following the rationale in [2]. However, in classical textbooks such as [7] and [8] spatial and temporal variations are connected through substitutions of the form $\mathbf{r}_T = \mathbf{u}_T \cdot t$ and $\mathbf{r}_R = \mathbf{u}_R \cdot t$, where $\mathbf{u}_T$ and $\mathbf{u}_R$ are the velocity vectors of the mobile transmitter and receiver, respectively. Thus, as demonstrated in [2], maintaining separate variability with respect to both space and time is more general, as the plain temporal variability is a special case of the space-time variability arising after the aforementioned substitutions. Moreover, space-time representation is advantageous when multi-element antennas are employed at the transmitter and/or the receiver sides with antenna elements occupying different positions in space.

The way that the channel model is presented in Equation 4.1 allows us to identify several channel categories commonly encountered in the international literature. More specifically, if the delayed replicas of the transmitted signal are independent of their time delay, τ_l, that is, $x\left(t - \tau_i\right) \approx x(t)$, then the channel is characterized as frequency non-selective or narrowband [7]. Otherwise, the channel pertains to the category of frequency-selective or wideband [7]. If both the transmitter and receiver are in motion, the channel is characterized as mobile-to-mobile, as reported in [9–11]. If the transmitter is static and the receiver is in motion or vice versa, the channel is characterized as mobile, as reported in [12–17]. Mobile channels can be further characterized by considering which part transmits and which receives. Thus, if the receiver is in motion together with the base station transmitting, the wireless link is called the downlink [12–15], whereas in the reverse case with a static base station receiver, the wireless link is called the uplink [16, 17]. In both mobile-to-mobile and mobile channels a static propagation environment is often assumed, that is, the mobile scatterers' contribution is assumed to be negligible compared to the static scatterers' contribution. However, recent studies in mobile and mobile-to-mobile channels have considered the impact of mobile scatterers [18, 19]. As a final channel category, the so called fixed wireless channel can be also found in the literature, for example [20–22]. For such channels, both the transmitter and receiver are static resulting in no spatial variations and temporal variations attributed only to scatterers' mobility.

The input-output relation in Equation 4.1 can be written in an equivalent integral form as follows[3]

$$
y(\mathbf{r_T}, \mathbf{r_R}, t) = \iiiint \exp[j2\pi(\Omega_\mathbf{T} \cdot \mathbf{r_T})/\lambda_0] \exp[j2\pi(\Omega_\mathbf{R} \cdot \mathbf{r_R})/\lambda_0]
$$
$$
\times \exp(j2\pi v_d t) x(t - \tau) h(\Omega_\mathbf{T}, \Omega_\mathbf{R}, \tau, v_d) d\Omega_\mathbf{T} d\Omega_\mathbf{R} d\tau dv_d \tag{4.4}
$$

where $h(\Omega_\mathbf{T}, \Omega_\mathbf{R}, \tau, v_d)$ is the double-directional-delay-Doppler variant channel response defined as

$$
h(\Omega_\mathbf{T}, \Omega_\mathbf{R}, \tau, v_d) = \sum_{l=1}^{L} a_l \delta\left(\Omega_\mathbf{T} - \Omega_{\mathbf{T},l}\right) \delta\left(\Omega_\mathbf{R} - \Omega_{\mathbf{R},l}\right)
$$
$$
\times \delta\left(\tau - \tau_l\right) \delta\left(v_d - v_{d,l}\right) \tag{4.5}
$$

with $\delta(.)$ being the Dirac delta function. Equation 4.4 can be written in a more compact form as follows

$$
y(\mathbf{r_T}, \mathbf{r_R}, t) = \int H(\mathbf{r_T}, \mathbf{r_R}, \tau, t) x(t - \tau) d\tau \tag{4.6}
$$

where $H(\mathbf{r_T}, \mathbf{r_R}, \tau, t)$ is the space-delay-time variant channel response obtained by the triple inverse Fourier transform of $h(\Omega_\mathbf{T}, \Omega_\mathbf{R}, \tau, v_d)$ with respect to the two direction vectors $\Omega_\mathbf{T}, \Omega_\mathbf{R}$ and the Doppler frequency v_d. Thus,

$$
H(\mathbf{r_T}, \mathbf{r_R}, \tau, t) = \iiint h(\Omega_\mathbf{T}, \Omega_\mathbf{R}, \tau, v_d) \exp[j2\pi(\Omega_\mathbf{T} \cdot \mathbf{r_T})/\lambda_0]
$$
$$
\times \exp[j2\pi(\Omega_\mathbf{R} \cdot \mathbf{r_R})/\lambda_0] \exp(j2\pi v_d t) d\Omega_\mathbf{T} d\Omega_\mathbf{R} dv_d \tag{4.7}
$$

By further defining $x(t)$ with respect to its Fourier transform, $X(f)$, that is, $x(t) = \int X(f) \exp(j2\pi ft) df$, Equation 4.6 can be written as

$$
y(\mathbf{r_T}, \mathbf{r_R}, t) = \int g(\mathbf{r_T}, \mathbf{r_R}, f, t) X(f) \exp(j2\pi ft) df \tag{4.8}
$$

where $g(\mathbf{r_T}, \mathbf{r_R}, f, t)$ is the space-frequency-time variant channel response obtained by the Fourier transform of $H(\mathbf{r_T}, \mathbf{r_R}, \tau, t)$ with respect to the time delay τ. Thus,

$$
g(\mathbf{r_T}, \mathbf{r_R}, f, t) = \int H(\mathbf{r_T}, \mathbf{r_R}, \tau, t) \exp(2\pi f\tau) d\tau = \iiiint h(\Omega_\mathbf{T}, \Omega_\mathbf{R}, \tau, v_d)
$$
$$
\times \exp[j2\pi(\Omega_\mathbf{T} \cdot \mathbf{r_T})/\lambda_0] \exp[j2\pi(\Omega_\mathbf{R} \cdot \mathbf{r_R})/\lambda_0]
$$
$$
\times \exp(j2\pi v_d t) \exp(-j2\pi f\tau) d\Omega_\mathbf{T} d\Omega_\mathbf{R} d\tau dv_d \tag{4.9}
$$

---

[3] It is implied that the integral limits when not given cover the entire range of integration of the integrated variables. The same practice applies for the remainder of this text.

Substituting Equation 4.5 into Equation 4.9 and after some algebraic manipulations we obtain the equivalent form for Equation 4.9 based on a summation

$$g(\mathbf{r_T}, \mathbf{r_R}, f, t) = \sum_{l=1}^{L} a_l \exp[j2\pi(\Omega_{\mathbf{T},l} \cdot \mathbf{r_T})/\lambda_0] \exp[j2\pi(\Omega_{\mathbf{R},l} \cdot \mathbf{r_R})/\lambda_0]$$
$$\times \exp(j2\pi v_{d,l}t)\exp(-j2\pi f \tau_l)$$

(4.10)

All channel representations in Equations 4.5, 4.7 and 4.9 are equivalent representing channel variations in different domains being connected through Fourier transform relations. Particularly, as can be seen from Equation 4.9, $g(\mathbf{r_T}, \mathbf{r_R}, f, t)$ and $h(\Omega_{\mathbf{T}}, \Omega_{\mathbf{R}}, \tau, v_d)$ constitute a complete Fourier transform pair with respect to all four dependencies. The received signal variations represented by Equation 4.4 or Equation 4.6 or Equation 4.8 occur within a local area, that is, variations in space comparable to the size of the wavelength of the carrier. Those variations are characterized as small scale variations around the mean signal level or small scale fading. A detailed analysis and description of the underlying physics of local area modeling, which constitutes the foundations of small scale variations, can be viewed in [2]. Apart from small scale variations, wireless channels experience large scale variations or large scale fading, which is the result of variations in the average received local power (averaged over about ten wavelengths, see [1]) for a given transmitter-receiver distance. Large scale variations arise from variations in space in the order of hundreds of wavelengths and are attributed to shadowing or shadow fading by large objects occupying the path from the transmitter to the receiver. Thus, large scale variations cause different received signal power levels when moving distances of hundreds of wavelengths for a given transmitter-receiver distance. Both small scale and large scale variations exhibit random behavior, thus statistical tools and analyses are employed for their characterization. The wireless channel exhibits a deterministic variation arising from variations of the average received power with respect to the distance between the transmitting and receiving antennas. The underlying effect is called the path loss and results in a monotonic decrease of average received power with respect to distance. Figure 4.4 depicts the variations of the received signal level in dB scale as a function of distance with small and large scale variations overlaid on the path loss.

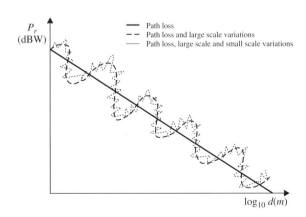

Figure 4.4 Path loss, large scale and small scale variations.

4.3   Small Scale Variations

The time varying random nature of the model parameters in Equation 4.10, namely, L, a_l, τ_l, $v_{d,l}$, $\Omega_{\mathbf{T},l}$ and $\Omega_{\mathbf{R},l}$ makes the channel response described by Equations 4.9 and 4.10 a random space-frequency-time varying complex process, that is a complex stochastic process.[4] The latter means that the channel response is an ensemble of several space-frequency-time realizations having the form given by Equations 4.9 and 4.10. Thus, statistical tools should be employed for characterizing the random nature of the outdoor radio channel response. The two universally accepted types of characterization are the first order and the second order statistical characterization. First order characterization arises when only one sample in Equations 4.9 and 4.10 with respect to space, frequency and time is used to characterize channel behavior. Second order characterization uses two samples in Equations 4.9 and 4.10 with respect to space and/or frequency and/or time. One common assumption essential for both types of characterization is that of wide sense ergodicity (WSE) [2, 7]. WSE means that all possible expected values of Equations 4.9 and 4.10 with respect to one or two space-frequency-time samples can arise by only one space-frequency-time realization without considering the whole ensemble of realizations. If the latter holds for all expected values with respect to any number of space-frequency-time samples (and not for only one or two), the process will be ergodic in the strict sense.

4.3.1   First Order Statistical Characterization

First order characterization considers only one sample of the channel response with respect to space, frequency and time, thus Equation 4.10 can be treated as complex random variable of the form

$$g = \sum_{l=1}^{L} |a_l| \exp(j\Phi_l) = G \exp(j\Phi) \tag{4.11}$$

where $|a_l|$ is the amplitude of a_l (i.e., $a_l = |a_l| \exp(j\phi_l)$) and Φ_l sums all the phase terms in Equation 4.9 (including that of a_l). First order statistical characterization accounts for statistically modeling of the amplitude G and phase Φ of g. We focus on the amplitude distribution as being the most essential for the majority of applications in wireless communications because the phase terms are modeled as uniformly distributed random variables in $[-\pi, \pi]$. A simple, yet physical, way to model the amplitude term is to group the terms $|a_l| \exp(j\Phi_l)$ in equation 4.11 into specular and diffuse components [2, 23, 24]. Specular components are a small number of strong multipath components and diffuse components are a large number of weak multipath components. Propagation mechanisms such as line-of-sight (LOS) propagation and reflections from large smooth surfaces create specular components. Diffuse components are inherent to almost every outdoor propagation channel and are attributed to different propagation mechanisms such as scattering from rough surfaces, multiple interactions with scatterers and waveguiding through street canyons.

---

[4] Space-frequency-time variations are clearly evident from Equations 4.9 and 4.10 and also implied by Equations 4.5 and 4.7 through their respective Fourier interrelations.

Grouping the multipath components into specular and diffuse, results in the following representation of Equation 4.11 [2, 23, 24]

$$g = \sum_{l=1}^{L_s} |b_l| \exp[j\Phi_{s,l}] + \sum_{l=1}^{L_d} |c_l| \exp[j\Phi_{d,l}] = B \exp(j\Phi_s) + C \exp(j\Phi_d) \quad (4.12)$$

where L_s is the number of specular components (relatively small) and L_d is the number of diffuse components (large enough, i.e., $L_s << L_d$). The pairs $b_l, \Phi_{s,l}$, and $c_l, \Phi_{d,l}$, represent the amplitude and phase terms of the specular and diffuse components, respectively. In turn, B, Φ_s and C, Φ_d represent the total amplitude and phase of the specular and diffuse parts, respectively.

The above grouping seems arbitrary, however, some general principles apply. The rationale is to consider a small number of strong specular components and incorporate the remaining multipath components into the diffuse part of Equation 4.12. Besides, it has been demonstrated that an arbitrary increase of the number of specular components results in the power arising by them to resemble the form of diffuse power [2, 23]. Even a small number of six non-dominant random multipath components can create an almost purely diffuse signal [25]. Each diffuse component carries a small amount of multipath power and the total number of diffuse components gives rise to the so-called diffuse power which is much larger than the power of an individual diffuse component. A condition of the following form characterizes diffuse components and total diffuse power

$$\max\left(|c_l|^2\right) << \sum_{l=1}^{L_d} |c_l|^2 \quad (4.13)$$

Thus, we can group all the multipath components into the diffuse part of Equation 4.12 and then exclude those components that do not satisfy condition Equation 4.13. The process ends when condition Equation 4.13 is satisfied by all the remaining components, whereas the excluded components will constitute the specular part. In the most general case treated in [24], amplitudes b_l in Equation 4.12 are dependent positive random variables. The phases $\Phi_{s,l}$ are independent of b_l and also independent of each other and uniformly distributed in $[-\pi, \pi]$. Thus, the specular part will not obey the central limit theorem (CLT) being a complex non-Gaussian random variable. In the diffuse part, as the number of multipath components is large enough, the CLT applies making the diffuse part a complex Gaussian random variable with Rayleigh distributed amplitude C. In this generic case, the amplitude probability density function (PDF) $p_g(z)$ of g will be [24]

$$p_g(z) = z \exp(-z^2/2) \sum_{n=0}^{\infty} \frac{h_n(1/4)}{(-4)^n n!} L_n(z^2), z \geq 0 \quad (4.14)$$

where $L_n(.)$ is the Laguerre polynomial of order n [26]. A truncation formula of the infinite sum in Equation 4.14 is also presented in [24]. The coefficients $h_n(.)$ are defined as

$$h_n(z) = \int_{A_{\min}}^{A_{\max}} \int_0^{\infty} \exp(-za^2) a^{2n+1} u J_0(au) M(u) du da \quad (4.15)$$

with $J_0(.)$ the Bessel function of first kind and zero order [26], and $[A_{min}, A_{max}]$ defines the interval over which the PDF of B, $p_b(z)$ (the amplitude of the specular part in Equation 4.12) is non-zero. $M(.)$ is the characteristic function of the PDF of B defined as [27]

$$M(u) = E\left[\prod_{l=1}^{L_s} J_0(b_j u)\right] = \int\int \cdots \int \prod_{l=1}^{L_s} J_0(b_l u) p_{b_1,b_2,\ldots,b_{Ls}}(z_1, z_2, \ldots, z_{Ls})$$

$$dz_1 dz_2 \ldots dz_{Ls} \tag{4.16}$$

where $E[\cdot]$ is the expectation operator and $p_{b_1,b_2,\ldots,b_{Ls}}(z_1, z_2, \ldots, z_{Ls})$ is the joint PDF of the amplitudes of the specular components.

The generic modeling approach of Equation 4.12 leading to the amplitude PDF of Equation 4.14 generates several amplitude PDFs encountered in the literature as special cases [2, 23, 27]. If there is no specular part, that is, $L_s = 0$ in Equation 4.12, only the diffuse part of Equation 4.12 exists leading to a Rayleigh distributed amplitude PDF for g. Thus [2]

$$p_g(z) = p_{Rayleigh}(z) = z \exp[-z^2/(2\sigma_0)^2]/\sigma_0^2, z \geq 0 \tag{4.17}$$

where σ_0^2 is the variance of the real and imaginary parts of $C \exp(j\Phi_d)$ as according to the CLT, both parts of $C \exp(j\Phi_d)$ are independent Gaussian distributed random variables with zero mean and equal variances [2]. The quantity σ_0^2 determines also the mean power of the diffuse part as $E[G^2] = E[C^2] = 2\sigma_0^2$. The Rayleigh distribution is appropriate when multipath propagation occurs in dense scattering environments, such as urban environments where no LOS path exists between the transmitting and receiving antennas [1]. The existence of one specular component, that is, $L_s = 1$ in equation 4.12, with constant deterministic amplitude b_1 together with the diffuse part of equation 4.12 leads to a Rice distributed amplitude PDF for g, thus [2]

$$p_g(z) = p_{Rice}(z) = z \exp[-(z^2 + b_1^2)/(2\sigma_0)^2] I_0(zb_1/\sigma_0^2)/\sigma_0^2, z \geq 0 \tag{4.18}$$

where $I_0(.)$ is the modified Bessel function of zero order [26]. An important variable of the Rice distribution is the Rician K factor which is the power ratio of the specular to the diffuse part, that is, $K = b_1^2/(2\sigma_0)^2$. The Rice distribution is suitable in environments where the LOS path is not obstructed by any of the environment's obstacles [8]. Figures 4.5 and 4.6 present the Rayleigh and Rice PDFs, respectively, for several values of their related parameters.

Apart from the Rayleigh and Rice PDFs, the literature contains several other PDFs based on the general model in Equation 4.12. More specifically, [27] contains the PDF for an arbitrary number of random components with statistically dependent amplitudes and no diffuse part ($L_d = 0$). In [2] and [23] the following PDFs are presented:

1. The one wave PDF, that is, the PDF that results when only one specular component with a constant amplitude exists ($L_d = 0$ and $L_s = 1$).

2. The two wave PDF, that is, the PDF that results when only two specular components with a constant amplitude exist ($L_d = 0$ and $L_s = 2$).

3. The three wave PDF, that is, the PDF that results when only three specular components with a constant amplitude exist ($L_d = 0$ and $L_s = 3$).
4. The two wave and diffuse power PDF, that is, the PDF that results when two specular components with a constant amplitude together with a diffuse part exist ($L_d >> 0$ and $L_s = 2$).

4.3.2 Second Order Statistical Characterization

Second order characterization considers two samples of the channel response with respect to space, frequency and time. Of particular importance is the multi-dimensional autocorrelation function (ACF), $R(.)$, defined by taking the following expectation formula of the

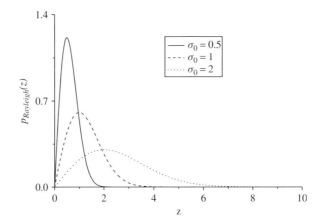

Figure 4.5 Rayleigh PDF.

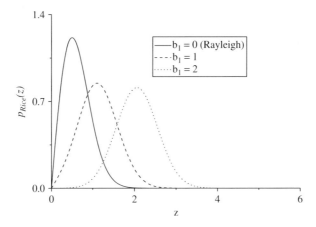

Figure 4.6 Rice PDF ($\sigma_0 = 0.5$).

response given by Equation 4.10. Thus[5]

$$R(\mathbf{r_{T1}}; \mathbf{r_{T2}}, \mathbf{r_{R1}}; \mathbf{r_{R2}}, f_1; f_2, t_1; t_2) = E[g^*(\mathbf{r_{T1}}, \mathbf{r_{R1}}, f_1, t_1)g(\mathbf{r_{T2}}, \mathbf{r_{R2}}, f_2, t_2)] \quad (4.19)$$

A common assumption is to consider the channel response as a wide sense station-ary (WSS) stochastic process with respect to all four dependencies. That assumption makes the ACF in Equation 4.19 depend on the difference among the samples, that is, $R(\mathbf{r_{T1}}; \mathbf{r_{T2}}, \mathbf{r_{R1}}; \mathbf{r_{R2}}, f_1; f_2, t_1; t_2) = R(\Delta\mathbf{r_T}, \Delta\mathbf{r_R}, \Delta f, \Delta t)$, where, for example, $\Delta\mathbf{r_T} = \mathbf{r_{T2}} - \mathbf{r_{T1}}$, and similar definitions hold for all other dependencies. In such cases, [2]

$$R(\Delta\mathbf{r_T}, \Delta\mathbf{r_R}, \Delta f, \Delta t) = \sum_{l=1}^{L} |a_l|^2 \exp\left[j2\pi\left(\Omega_{T,l} \cdot \Delta\mathbf{r_T}\right)/\lambda_0\right]$$
$$\times \exp\left[j2\pi\left(\Omega_{R,l} \cdot \Delta\mathbf{r_R}\right)/\lambda_0\right]\exp\left(j2\pi v_{d,l}\Delta t\right)\exp\left(-j2\pi\tau_l\Delta f\right) \quad (4.20)$$

The ACF in Equation 4.20 characterizes the channel selectivity in terms of all dependen-cies, that is, selectivity in space, frequency and time.

WSS implies that an expectation formula of the response in Equation 4.5 similar to that in Equation 4.19, will result in an expression of the form [3]

$$E\left[h^*\left(\Omega_{T1}, \Omega_{R1}, \tau_1, v_{d1}\right) h\left(\Omega_{T2}, \Omega_{R2}, \tau_2, v_{d2}\right)\right] = P\left(\Omega_{T1}, \Omega_{R1}, \tau_1, v_{d1}\right)$$
$$\times \delta\left(\Omega_{T2} - \Omega_{T1}\right)\delta\left(\Omega_{R2} - \Omega_{R1}\right)\delta\left(\tau_2 - \tau_1\right)\delta\left(v_{d2} - v_{d1}\right) \quad (4.21)$$

where the function $P(\Omega_T, \Omega_R, \tau, v_d)$ is the power spectral density (PSD) describing the manner in which the power is distributed with respect to all four spectral domains. The formal definition of the PSD is [2]

$$P\left(\Omega_T, \Omega_R, \tau, v_d\right) = \sum_{l=1}^{L} |a_l|^2 \delta\left(\Omega_T - \Omega_{T,l}\right)\delta\left(\Omega_R - \Omega_{R,l}\right)\delta\left(\tau - \tau_l\right)\delta\left(v_d - v_{d,l}\right)$$

$$(4.22)$$

The PSD in Equation 4.22 characterizes the channel dispersion in terms of all dependen-cies, that is, dispersion in direction, delay and Doppler frequency.

The formula in Equation 4.21 simply states that the spectral components of $h(\Omega_T, \Omega_R, \tau, v_d)$ are uncorrelated, thus the terminology of uncorrelated scattering (US) arises and accordingly, the well-known terminology of WSSUS channel models is readily adopted. This terminology is traced back to 1963 in the seminal paper of Bello [28] where WSS channel models with respect to frequency and time were analyzed. WSS with respect to space at the receiver's side was adopted in [2] and [3] and in this chapter, WSS with respect to space at the transmitter's side is further adopted. Theoretical and empirical investigations arguing over the assumption of WSSUS can be viewed in [29] and [30]. A final important property of the adopted modeling approach arises by a simple inspection of Equations 4.20 and 4.22, from which we can see that the ACF and PSD are Fourier transform pairs. That property is widely known in the literature as the Wiener-Khintchine theorem [2, 7]. Thus, we adopt the notation,

---

[5] Indices 1 and 2 denote two different samples with respect to space, frequency and time.

$R(\Delta\mathbf{r_T}, \Delta\mathbf{r_R}, \Delta f, \Delta t) \rightleftharpoons P(\Omega_\mathbf{T}, \Omega_\mathbf{R}, \tau, v_d)$, where $\rightleftharpoons$ indicates the Fourier transform pair, and we get

$$P(\Omega_\mathbf{T}, \Omega_\mathbf{R}, \tau, v_d) = \int\int\int\int R(\Delta\mathbf{r_T}, \Delta\mathbf{r_R}, \Delta f, \Delta t) \exp[-j2\pi(\Omega_\mathbf{T} \cdot \Delta\mathbf{r_T})/\lambda_0]$$

$$\times \exp[-j2\pi(\Omega_\mathbf{R} \cdot \Delta\mathbf{r_R})/\lambda_0] \exp(-j2\pi v_d \Delta t) \qquad (4.23)$$

$$\times \exp(j2\pi\tau\Delta f) d\Delta\mathbf{r_T} d\Delta\mathbf{r_R} d\Delta t d\Delta f$$

Equations 4.20, 4.22 and 4.23 offer a complete second order statistical characterization from which several other channel metrics can arise by removing one or more of the other dependencies. In the general case of multi-clustered propagation [1], the dependencies in Equation 4.22 cannot be decomposed as each specific cluster will be uniquely characterized by its delays τ_l, Doppler frequencies $v_{d,l}$, DOD $\Omega_{\mathbf{T},l}$ and DOA $\Omega_{\mathbf{R},l}$. For example, the spatial AFC, $R_{ST}(\Delta\mathbf{r_T})$, at the transmitter's side arises from equation 4.20 by equating all other dependencies to zero, that is,

$$R_{ST}(\Delta\mathbf{r_T}) = R(\Delta\mathbf{r_T}, 0, 0, 0) \qquad (4.24)$$

In turn, $R_{ST}(\Delta\mathbf{r_T})$ is a Fourier transform pair with the directional PSD at the transmitter's side, $P_{DT}(\Omega_\mathbf{T})$, that is, $R_{ST}(\Delta\mathbf{r_T}) \rightleftharpoons P_{DT}(\Omega_\mathbf{T})$, defined as

$$P_{DT}(\Omega_\mathbf{T}) = \int\int\int P(\Omega_\mathbf{T}, \Omega_\mathbf{R}, \tau, v_d) d\Omega_\mathbf{R} d\tau dv_d \qquad (4.25)$$

In a similar way, they have defined the remaining Fourier transform pairs, namely:

1. Spatial AFC, $R_{SR}(\Delta\mathbf{r_R})$ – directional PSD, $P_{DR}(\Omega_\mathbf{R})$, at the receiver's side, that is, $R_{SR}(\Delta\mathbf{r_R}) \rightleftharpoons P_{DR}(\Omega_\mathbf{R})$.
2. Frequency ACF, $R_F(\Delta f)$ – time delay PSD or power delay profile (PDP) as it is widely known, $P_T(\tau)$, that is, $R_F(\Delta f) \rightleftharpoons P_T(\tau)$.
3. Temporal ACF, $R_T(\Delta\mathbf{t})$ – frequency PSD or Doppler spectrum, $P_{vd}(v_d)$, that is, $R_T(\Delta\mathbf{t}) \rightleftharpoons P_{vd}(v_d)$.

A widely accepted and validated PDP for outdoor environments is the one sided exponential profile [31–33] which can be further generalized to a multi-clustered exponential profile [1]. Other PDPs encountered in the literature include the Gaussian, triangular, rectangular and double spike [34–36] profiles. These PDPs are mostly of mathematical value being suitable for specific scenarios and not complying with the physical strength of an exponential or a multi-clustered exponential profile [1]. For example, a double spike PDP is appropriate in a two-clustered scenario where both clusters have a very dominant multipath component with insignificant power dispersion in terms of the dominant component. A rectangular PDP assumes equal powers with respect to all time delays, while the Gaussian and triangular profiles predict power concentration into a specific centralized time delay (for example, caused by a dominant cluster) and even power dissipation with respect to higher and lower delays. Two important quantities characterizing a PDP is the mean delay, μ_T, and the delay spread, σ_T, defined as [1]

$$\mu_T = \frac{\int \tau P_T(\tau) d\tau}{P} \qquad (4.26)$$

$$\sigma_T = \sqrt{\frac{\int (\tau - \mu_T)^2 P_T(\tau) d\tau}{P}} \tag{4.27}$$

The exponential PDP is defined as

$$P_T(\tau) = P \exp(-\tau/\sigma_T)/\sigma_T \tag{4.28}$$

where $P = E\left[|h(\Omega_\mathbf{T}, \Omega_\mathbf{R}, \tau, v_d)|^2\right] = \sum_{l=1}^{L} |a_l|^2$ is the mean power of the channel response.

For the frequency PSD induced by a time varying environment (that is, mobile scatterers), both theoretical and empirical studies revealed a very sharp peak at zero Doppler frequency with a rapid even decay as the absolute value of Doppler frequency increases [5, 22, 37, 38].

For the directional PSDs at the transmitter and the receiver sides, it is of paramount importance for practical applications in outdoor environments to relate them to their corresponding power angular distribution or power angular spectrum (PAS), $P_{T(\Phi,\Theta)}(\phi_T, \theta_T)$, $P_{R(\Phi,\Theta)}(\phi_R, \theta_R)$, as it is widely known. Such relation can be found in [3] resulting in

$$P_{T(\Phi,\Theta)}(\phi_T, \theta_T) = \sin\theta_T P_{DT}(\Omega_\mathbf{T})$$
$$P_{R(\Phi,\Theta)}(\phi_R, \theta_R) = \sin\theta_R P_{DR}(\Omega_\mathbf{R}) \tag{4.29}$$

The DOD, DOA vectors, $\Omega_\mathbf{T}$, $\Omega_\mathbf{R}$, should be replaced by equivalent formulas with Equation 4.2 relating them with the azimuth and elevation AOD, AOA as follows

$$\Omega_\mathbf{T} = \mathbf{x}\cos\phi_T \sin\theta_T + \mathbf{y}\sin\phi_T \sin\theta_T + \mathbf{z}\cos\theta_T$$
$$\Omega_\mathbf{R} = \mathbf{x}\cos\phi_R \sin\theta_R + \mathbf{y}\sin\phi_R \sin\theta_R + \mathbf{z}\cos\theta_R \tag{4.30}$$

From Equation 4.29, the azimuth and elevation PAS can be derived by integrating with respect to one dependency. For example, for the azimuth PAS at the transmitter's side we have

$$P_{T(\Phi)}(\Phi_T) = \int P_{T(\Phi,\Theta)}(\phi_T, \theta_T) d\theta_T \tag{4.31}$$

The PAS is a physically intuitive quantity and it has two dependencies, whereas the directional PSD has three (see Equation 4.30). Multipath power arising from diffuse and specular components is better represented by a PAS. More specifically, diffuse power is represented by continuous intervals, whereas specular power is represented by discrete impulses [2]. In Figure 4.7, an illustrative azimuth PAS at the transmitter's side is shown arising from diffuse and specular components.

If Equation 4.29 is normalized with respect to the mean power P, it is referred to as an AOD and AOA PDF, with investigations in the literature predicting various PDFs for the AOD and AOA. Additionally, reciprocity predicts a similar behaviour for the AOD and AOA PDFs. In a mobile receiver-static transmitter scenario (downlink case), the AOD and AOA PDFs will be the same with the AOA and AOD PDFs, if the roles of receiver and transmitter are interchanged and the frequency remains the same. In a mobile-to-mobile scenario, the AOD and AOA PDFs will be the same with the AOA and AOD PDFs, if the location of the transmitter becomes that of the receiver and vice versa. Generally, in dense scattering outdoor environments, multipath power propagates in three dimensions

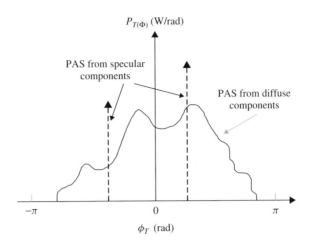

Figure 4.7 Azimuth PAS from diffuse and specular components.

(3-D) [39–41]. The azimuth and elevation AOD and AOA PDFs cannot be decomposed in generalized multi-clustered scenarios (i.e., they are statistically dependent) [42]. The first 3-D scattering model can be found in the classical paper of Aulin [15], in which a uniform 3-D scattering scenario truncated to a maximum symmetrical elevation angle is considered. The model in [43] presents a completely uniform 3-D scenario (3-D isotropic scattering) and as such, it is a special case of the model in [15]. Modified sinusoidal functions for the elevation AOA PDF that do not have the sharp discontinuities evident in [15] and hence, being closer to reality were presented in [44] and [45]. A Gaussian PDF was used for the elevation AOA in [46], which means that arrival in the elevation plane is evenly distributed around a mean elevation AOA. A uniform distribution was considered for the azimuth AOA in all the models reported in [44–46]. In [47], uniform and Gaussian PDFs were used for the distribution of the incoming power with respect to the azimuth and elevation AOA (i.e., the distribution of $P_{DR}(\Omega_R)$ in Equation 4.29 after replacing Ω_R with Equation 4.30), respectively, representing dense scattering outdoor scenarios with symmetric contributions in the elevation plane. In reference [48], the respective azimuth and elevation AOA PDFs were both Gaussian representing directional scattering at both the azimuth and elevation plane and in [49], these distributions were the uniform and a generalized double exponential representing non-symmetric contributions of scatterers in the elevation plane. In [50] and [51], new discontinuous PDFs for the azimuth AOA were considered, while the PDF for the elevation AOA was that presented in [15]. In [52] and [53], two bivariate joint PDFs for the azimuth and elevation AOA were used, namely, the Fisher-Bingham PDF [52] and the von Mises-Fisher PDF [53] as appropriate models for statistically dependent azimuth and elevation AOAs. 2-D propagation or from-the-horizon propagation as referred to in [2], occurs when the receiver and/or the transmitter are far away from the surrounding scatterers. Accordingly, several PDFs for the azimuth AOD, AOA can be found in the case of 2-D scattering. Thus, one can find the uniform PDF [12], the power of cosine PDF [54], the restricted uniform PDF [55], the Gaussian PDF [56], the Laplacian PDF [16] and the von Mises PDF [57]. The uniform PDF represents 2-D isotropic scattering outdoor scenarios, whereas the remaining PDFs predict 2-D directional anisotropic scattering. Their appropriateness depends on the nature of the specific propagation environment.

Similarly with Equations 4.26 and 4.27, the mean direction and direction spread at the transmitter's side are defined as follows

$$\mu_{\Omega T} = \int \Omega_{\mathbf{T}} P_{DT}(\Omega_T) d\Omega_T = \iint \Omega_{\mathbf{T}} P_{T(\Phi,\Theta)}(\phi_T, \theta_T) d\phi_T d\theta_T \tag{4.32}$$

$$\sigma_{\Omega T} = \sqrt{\left|\Omega_T - \mu_{\Omega T}\right|^2 P_{DT}(\Omega_T) d\Omega_T} = \sqrt{\int \int \left|\Omega_T - \mu_{\Omega T}\right|^2 P_{T(\Phi,\Theta)}(\phi_T, \theta_T) d\phi_T d\theta_T} \tag{4.33}$$

where in both Equations 4.32 and 4.33, the formula in Equation 4.30 should be used for $\Omega_{\mathbf{T}}$. Similar formulas with Equations 4.32 and 4.33 can be defined at the receiver's side.

We now consider an application of the above second order statistical description by studying a mobile wireless channel (mobile receiver) in a static environment and without frequency selectivity. More specifically, we consider the mobile scenario presented in [51], where a mobile receiver moves in the positive y axis with velocity v_r being subject to 3-D multipath scattering. We transform the spatial variations into temporal variations by defining in Equation 4.10 $\mathbf{r_R} = \mathbf{y} \cdot v_r t$. The remaining other dependencies are zero and Equation 4.10 will describe a time varying wireless channel. The new form for Equation 4.10 will be $g_m(t) = g(0, \mathbf{y} \cdot v_r t, 0, 0)$, in which using Equation 4.2 and after some algebraic manipulations we have

$$g_m(t) = \sum_{l=1}^{L} a_l \exp(j2\pi v_r t \sin\phi_{R,l} \sin\theta_{R,l}/\lambda_0) \tag{4.34}$$

Accordingly, the temporal ACF in Equation 4.20 will arise after substituting $\Delta\mathbf{r_R} = \mathbf{y}.v_r\Delta t$ with the remaining dependencies being equal to zero, that is, $R_m(\Delta t) = R(0, \mathbf{y}.v_r\Delta t, 0, 0)$. Thus

$$R_m(\Delta t) = \sum_{l=1}^{L} \left|a_l\right|^2 \exp(j2\pi v_r \Delta t \sin\phi_{R,l} \sin\theta_{R,l}/\lambda_0) \tag{4.35}$$

To further simplify the analysis, we define an alternative equivalent azimuth and elevation AOA as $\alpha = \pi/2 - \phi_R$ and $\beta = \pi/2 - \theta_R$ such that these angles have the receiver's direction of motion as a reference, that is, the positive y axis. α counts from the value $-\pi$ in the negative y axis returning to the same point in the clockwise direction and β is zero on the x-y plane, $\pi/2$ on the positive z axis and $-\pi/2$ on the negative z axis. Figure 4.8 shows the geometry of this mobile channel scenario together with the definitions of the alternative azimuth and elevation AOA α and β, respectively. Using the alternative azimuth and elevation AOA, equation 4.35 is rewritten as

$$R_m(\Delta t) = \sum_{l=1}^{L} \left|a_l\right|^2 \exp(j2\pi v_r \Delta t \cos\alpha_l \cos\beta_l/\lambda_0) \tag{4.36}$$

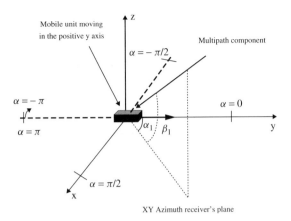

Figure 4.8 Geometry of the mobile channel scenario.

Now, assuming that the propagating multipath power consists of one LOS (i.e., specular) component combined with diffuse power, the amplitude PDF of $g_m(t)$ will be Rician as depicted in Section III.a, which is repeated here for the sake of convenience. Thus

$$p_{gm}(z) = z \exp[-(z^2 + \rho^2)/(2\sigma_0)^2] I_0(z\rho/\sigma_0^2)/\sigma_0^2 \qquad (4.37)$$

where ρ is the amplitude of the LOS component (replaces b_1 in Equation 4.18 and $2\sigma_0^2$ is the mean power of the diffuse part. Thus, the temporal ACF in Equation 4.36 can be written in the following compact form [51]

$$R_m(\Delta t) = R_d(\Delta t) + \rho^2 \exp(j2\pi v_r \Delta t \cos\alpha_0 \cos\beta_0/\lambda_0) \qquad (4.38)$$

where α_0 and β_0 define the deterministic azimuth and elevation AOA, respectively, of the LOS component and $R_d(\Delta t)$ represents the contribution of the diffuse part defined as [51]

$$R_d(\Delta t) = 2\sigma_0^2 \iint \exp(j2\pi v_r \Delta t \cos\beta \cos\alpha \Delta t/\lambda_0) P_{\alpha,\beta}(\alpha, \beta) d\alpha d\beta \qquad (4.39)$$

where $P_{\alpha,\beta}(\alpha, \beta)$ constitutes the joint PDF of α and β. As a result, the product $2\sigma_0^2 P_{\alpha,\beta}(\alpha, \beta)$ constitutes the PAS of the diffuse part distributed over α and β. In order to proceed, we consider $P_{\alpha,\beta}(\alpha, \beta) = P_\alpha(\alpha) P_\beta(\beta)$ (that is, statistical independence between α and β) and define the PDFs for the azimuth and elevation AOA $P_\alpha(\alpha)$ and $P_\beta(\beta)$, respectively as follows [51]

$$P_\alpha(\alpha) = \begin{cases} \sigma_1^2/(2\sigma_0)^2, & \alpha_1 \le \alpha \le \pi - \alpha_2 \\ \sigma_2^2/(2\sigma_0)^2, & -\pi + \alpha_4 \le \alpha \le -\alpha_3 \\ 0, & \text{otherwise} \end{cases} \qquad (4.40)$$

$$P_\beta(\beta) = \begin{cases} \dfrac{\cos\beta}{2\sin b_m}, & |\beta| \le b_m \le \dfrac{\pi}{2} \\ 0, & \text{otherwise} \end{cases} \qquad (4.41)$$

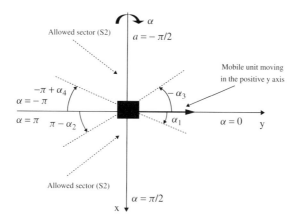

Figure 4.9 Sectors of arrival in the azimuth plane.

where $0 \le \alpha_i \le \pi/2$, $(i = 1, 2, 3, 4)$, restrict $P_\alpha(\alpha)$ in two azimuth sectors, that is, S1: $\alpha_1 \le \alpha \le \pi - \alpha_2$ and S2: $-\pi + \alpha_4 \le \alpha \le \alpha_3$, as illustrated in Figure 4.9 and $0 \le b_m \le \pi/2$ is the maximum elevation AOA in each allowed sector [15]. The parameters σ_1 and σ_2 are related to $2\sigma_0^2$ in such a way that integrating Equation 4.40 in the allowed angular range, the result equals to one. Thus,

$$2\sigma_0^2 = \sigma_1^2(\pi - \alpha_1 - \alpha_2) + \sigma_2^2(\pi - \alpha_3 - \alpha_4) \tag{4.42}$$

Substituting Equations 4.40 and 4.41 in Equation 4.39, we take the temporal ACF as follows [51]

$$R_d(\Delta t) = J_1(\Delta t) + J_2(\Delta t) \tag{4.43}$$

where

$$J_1(\Delta t) = \frac{\sigma_1^2}{\sin b_m} \int_0^{b_m} \int_{\alpha_1}^{\pi - \alpha_2} \cos \beta \exp(j2\pi v_r \Delta t \cos \alpha \cos \beta / \lambda_0) d\alpha d\beta \tag{4.44}$$

and

$$J_2(\Delta t) = J_1(\Delta t)|_{\sigma_1 \to \sigma_2, \alpha_1 \to \alpha_3, \alpha_2 \to \alpha_4} \tag{4.45}$$

By Fourier transforming equation 4.38, we find the Doppler spectrum of the mobile channel, $S_m(f_m)$, as follows

$$S_m(f_m) = S_d(f_m) + \rho^2 \delta(f_m - f_\rho) \tag{4.46}$$

where $f_m = v_r \cos \alpha \cos \beta / \lambda_0$ is the Doppler frequency shift parameter, $f_\rho = v_r \cos \alpha_0 \cos \beta_0 / \lambda_0$ is the deterministic Doppler frequency shift of the LOS component, and $S_d(f_m)$, the Doppler spectrum of the diffuse part defined as

$$S_d(f_m) = F[J_1(\Delta t)] + F[J_2(\Delta t)] \tag{4.47}$$

In Equation 4.47, $F[.]$ is the Fourier transform operator transforming the field of Δt into that of f_m. The Fourier transforms in equation 4.47 have been derived in [51] as

$$
F[J_1(\Delta t)] = \begin{cases}
\dfrac{\sigma_1^2}{2f_{\max}\sin b_m}\left[\dfrac{\pi}{2} - \arcsin\dfrac{2\cos^2 b_m - 1 - (f_m/f_{\max})^2}{1 - (f_m/f_{\max})^2}\right], \\
\quad -f_{\max}\cos\alpha_2\cos b_m \le f_m \le f_{\max}\cos\alpha_1\cos b_m \\[2mm]
\dfrac{\sigma_1^2}{2f_{\max}\sin b_m}\left[\dfrac{\pi}{2} - \arcsin\dfrac{f_m^2(\sin^2\alpha_1 + 1) - f_{\max}^2\cos^2\alpha_1}{(f_{\max}^2 - f_m^2)\cos^2\alpha_1}\right], \\
\quad f_{\max}\cos\alpha_1\cos b_m \le f_m \le f_{\max}\cos\alpha_1 \\[2mm]
\dfrac{\sigma_1^2}{2f_{\max}\sin b_m}\left[\dfrac{\pi}{2} - \arcsin\dfrac{f_m^2(\sin^2\alpha_2 + 1) - f_{\max}^2\cos^2\alpha_2}{(f_{\max}^2 - f_m^2)\cos^2\alpha_2}\right], \\
\quad -f_{\max}\cos\alpha_2 \le f_m \le -f_{\max}\cos\alpha_2\cos b_m \\[2mm]
0, \text{ otherwise}
\end{cases}
\tag{4.48}
$$

and

$$
F[J_2(\Delta t)] = F[J_1(\Delta t)]|_{\sigma_1\to\sigma_2,\alpha_1\to\alpha_3,\alpha_2\to\alpha_4}
\tag{4.49}
$$

with $f_{\max} = v_r/\lambda_0$ being the maximum Doppler frequency shift. Figure 4.10 presents the Doppler spectrum in Equation 4.46 for the following parameter set: $b_m = \pi/5, \alpha_1 = \pi/6, \alpha_2 = \pi/9, \alpha_3 = \pi/3, \alpha_4 = \pi/3.5, f_{\max} = 91 Hz, f_\rho = 0.25, f_{\max}, \rho = 1$ and $\sigma_1 = \sigma_2 = 0.5$. The delta impulse represents the contribution of the LOS component and the remainder represents the diffuse components. As shown in Figure 4.10, the Doppler spectrum becomes discontinuous and asymmetric, with these properties being strongly dependent on the parameters of the azimuth and elevation AOA PDFs.

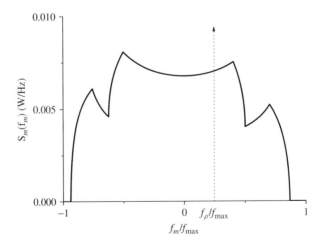

Figure 4.10 Doppler spectrum of a mobile channel ($b_m = \pi/5, \alpha_1 = \pi/6, \alpha_2 = \pi/9, \alpha_3 = \pi/3, \alpha_4 = \pi/3.5, f_{max} = 91 Hz, f_p = 0.25 f_{max}, \rho = 1$ and $\sigma_1 = \sigma_2 = 0.5$).

Two more metrics being of paramount importance when dealing with second order statistical characterization are the average level crossing rate (LCR), or simply the LCR and average fade duration (AFD). We first consider the LCR, where, $N(z)$, is the average number of crossings per second that $g_m(t)$ crosses a specified signal level, z, with positive slope. The formal definition of the LCR is [7]

$$N(z) = \int_0^\infty y' p_{g_m, g'_m}(z, y') dy' \tag{4.50}$$

with $p_{g_m, g'_m}(z, y')$ being the joint PDF of $g_m(t)$ with its time derivative $g'_m(t)$ at the same time instant. For a Rician channel model, the LCR will be [7]

$$N(z) = \frac{z\sqrt{2d_1}}{\pi^{3/2}\psi_0} \exp\left(-\frac{z^2 + \rho^2}{2\psi_0}\right) \int_0^{\pi/2} \{\exp[-(d_2\rho\sin\theta)^2] +$$
$$\sqrt{\pi} d_2\rho\sin\theta \cdot erf(d_2\rho\sin\theta)\} \cosh\left(\frac{z\rho\cos\theta}{\psi_0}\right) d\theta \tag{4.51}$$

where $erf(.)$ is the error function defined in [7]. The following parameters should be further determined in Equation 4.51, namely, $\psi_0 = R_d(0)/2$, $\phi_{01} = Im[R'_d(0)]/2$ and $\psi_{02} = R_d''(0)/2$, where the primes denote derivatives with respect to the time difference Δt and $Im[.]$ refers to the imaginary part of the bracketed term. These parameters will be [51]

$$\psi_0 = \sigma_0^2 = \frac{\sigma_1^2(\pi - \alpha_1 - \alpha_2) + \sigma_2^2(\pi - \alpha_3 - \alpha_4)}{2} \tag{4.52}$$

$$\phi_{01} = \frac{\pi f_{\max}}{2}\left(\cos b_m + \frac{b_m}{\sin b_m}\right)\left[\sigma_1^2\left(\sin\alpha_2 - \sin\alpha_1\right) + \sigma_2^2\left(\sin\alpha_4 - \sin\alpha_3\right)\right] \tag{4.53}$$

$$\phi_{01} = \frac{\pi^2 f_{\max}^2\left[\cos(2b_m) + 5\right]}{12}\{\sigma_1^2\left(\sin(2\alpha_1) + \sin(2\alpha_2) + 2(\alpha_1 + \alpha_2 - \pi)\right)$$
$$+ \sigma_2^2\left(\sin(2\alpha_3) + \sin(2\alpha_4) + 2(\alpha_3 + \alpha_4 - \pi)\right)\} \tag{4.54}$$

Moreover, d_1 and d_2 in Equation 4.51 will be defined as [51]

$$d_1 = -\psi_{02} - (\phi_{01}^2/\psi_0) \tag{4.55}$$

$$d_2 = \left[2\pi f_\rho - (\phi_{01}/\psi_0)\right]/\sqrt{2d_1} \tag{4.56}$$

From Equations 4.53, 4.54 and 4.55, we can see that the LCR in Equation 4.51 is proportional to $f_{\max}$.

We now consider the AFD, where, $T(z)$, determines the mean value of the time intervals that $g_m(t)$ remains below a specified signal level z. The following definition is employed for the AFD [7]

$$T(z) = \frac{F_{gm}(z)}{N(z)} \tag{4.57}$$

with $F_{gm}(z)$ the cumulative distribution function (CDF) of $g_m(t)$ defined as

$$F_{gm}(z) = \int_0^z p_{gm}(x) dx \tag{4.58}$$

From Equation 4.51, we can see that the AFD in Equation 4.57 is inversely proportional to $f_{\max}$.

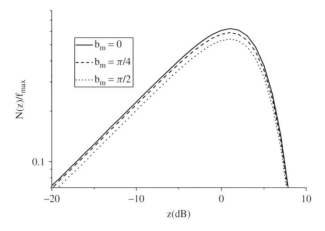

Figure 4.11   LCR with the maximum elevation AOA as parameter ($\alpha_1 = \pi/6, \alpha_2 = \pi/9, \alpha_3 = \pi/3, \alpha_4 = \pi/3.5, f_{max} = 91$ Hz, $f_\rho = 0.25 f_{max}, \rho = 1$ and $\sigma_1 = \sigma_2 = 0.5$).

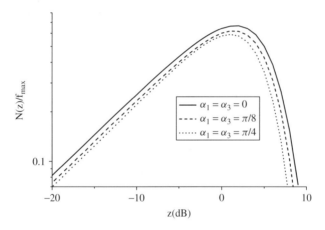

Figure 4.12   LCR with the width of sectors S1 and S2 as parameters ($b_m = \pi/5, \alpha_2 = \pi/9, \alpha_4 = \pi/3.5, f_{max} = 91$ Hz, $f_\rho = 0.25 f_{max}, \rho = 1$ and $\sigma_1 = \sigma_2 = 0.5$).

In order to determine the impact of diffuse scattering on the time variability of the wireless channel, we plot the normalized LCR $N(z)/f_{max}$ as a function of b_m and α_1, α_3, in Figures 4.11 and 4.12. The remaining parameters are the same as those in Figure 4.9 (apart from b_m in Figure 4.11 and α_1, α_3 in Figure 4.12). It is clear from Figure 4.11 that with increased elevation AOA (b_m increases) the LCR decreases because the multipath propagation reduces its influence as being projected to the receiver's azimuth plane. Thus, fluctuations occur less frequently. From Figure 4.12, with decreased diffuse multipath power (the width of sectors S1 and S2 decrease when α_1 and α_3 increase, thus, the diffuse multipath power decreases, see Figure 4.9 and Equation 4.42) the LCR decreases and thus fluctuations occur less frequently, because the diffuse part reduces its impact.

4.4   Path Loss and Large Scale Variations

Superimposed on the small scale variations are the large scale variations which arise when spatial changes at the transmitter and/or the receiver sides exceed distances of hundreds of wavelengths. A full description of variations in outdoor propagation environments also considers a deterministic variation attributed to the path loss effect, hence making the received power monotonically decrease with distance (see Figure 4.4). A joint description in dB scale of path loss with respect to distance and large scale variations is given by [58]

$$PL(d) = PL(d_0) + 10 n log(d/d_0) + X_s \qquad (4.59)$$

where $PL(d)$ and $PL(d_0)$ is the power loss in dB at distances d and d_0, respectively, between the transmitting and receiving antennas, n is the path loss exponent and X_s is a zero mean Gaussian process accounting for the large scale variations (i.e., the shadowing effect). The reference distance, d_0, is typically chosen to be 1m [58] and must be in the far field of the transmitting antenna [59]. The formal definition of path loss in dB is given by the following equation

$$PL(d) = 10 log[P_T/P_R(d)] \qquad (4.60)$$

with $P_R(d)$ being the received power at distance d and P_T the transmitted power (that is, the power the transmitted antenna is fed by). In the case of free space propagation, the received power is given by the Friis Equation [59]

$$P_R(d) = P_T G_R G_T \lambda_0^2/(4\pi d)^2 \qquad (4.61)$$

where G_R and G_t are the gains of the receiving and transmitting antennas, respectively. In fact, from Friis equation the simplest path loss model is derived, that is, the free space path loss model, which predicts $n = 2$ in Equation 4.59. Substituting Equation 4.61 in equation 4.60, we take the deterministic free space path loss, $PL_{fs}(d)$, as follows

$$PL_{fs}(d) = PL_{fs}(d_0) + 20 log(d/d_0) + log[(4\pi)^2/(\lambda_0^2 G_R G_T)] \qquad (4.62)$$

The free space path loss model has very limited applicability in outdoor propagation environments as it requires a purely unobstructed LOS path without any other component (specular or diffuse). Thus, more sophisticated path loss models are required that take into account several other parameters such as antenna heights, propagation environment type, building heights, terrain type and so on [60, 61]. Reference [61] contains a comparative study of the most important path loss models used in the planning of modern long term evolution (LTE) networks.
Of particular importance are:

1. The COST 231 Hata urban model.
2. The COST 231 Walfish-Ikegami non-LOS model.
3. The COST 231 Walfish-Ikegami street canyon LOS model.

The first is considered appropriate for suburban and urban macrocell environments and the latter two for microcell non-LOS and LOS environments [33]. The COST 231 Hata urban model predicts the path loss in dB as follows [33, 60]

$$PL_{Hata}(d) = \left[44.9 - 6.55log(h_T)\right]log(d/1000) + 45.5$$
$$+ (35.46 - 1.1h_R)log(f_0) - 13.82log(h_R) + 0.7h_R + C \qquad (4.63)$$

where h_T and h_R are the transmitting and receiving antenna height in meters, respectively, f_0 is the carrier frequency in MHz, $d \in [1000\,\text{m}, 20000\,\text{m}]$ is the distance among the transmitting and receiving antennas in meters and C is a constant ($C = 0\,\text{dB}$ for urban environments).

Considering the parameter set: transmitting antenna height $= 12.5\,\text{m}$, receiving antenna height $= 1.5\,\text{m}$, building height $= 12\,\text{m}$, building-to-building distance $= 50\,\text{m}$, street width $= 25\,\text{m}$, orientation for all paths $= 30$ and selection of metropolitan center, the COST 231 Walfish-Ikegami non-LOS model predicts the path loss as follows [33, 60]

$$PL_{WI,NLOS}(d) = -55.9 + 38log(d) + [24.5 + (f_0/616.67)]log(f_0) \qquad (4.64)$$

Finally, the COST 231 Walfish-Ikegami street canyon LOS model predicts the path loss in dB as follows [33, 60]

$$PL_{WI,LOS}(d) = -35.4 + 26log(d) + 20log(f_0) \qquad (4.65)$$

For both models in Equations 4.64 and 4.65, $d \in [20\,\text{m}, 5000\,\text{m}]$ is in meters and f_0 is in MHz.

A complete model description requires a statistical characterization of both path loss and large scale variability stemming from Equation 4.59. As was mentioned in Equation 4.59, X_s is a zero mean Gaussian process. Thus, we infer that the received power in dB units is a random variable for large scale variations following a zero mean Gaussian PDF. If we consider a normal scale of power units (for example, Watts), then the received power PDF in large scale will be log-normal [7]. Consequently, the envelope PDF will also be log-normal [8]. The latter effect constitutes the log-normal shadowing for the large scale variations, as referred to in references [62–64]. It has also been demonstrated that the gamma PDF can substitute the lognormal for describing the large scale variability of wireless channels having the advantage of providing analytical envelope PDFs in mixture models accounting for both small and large scale variations [65]. The log-normal PDF is described by the following form [7]

$$p_L(z) = \exp[-(lnz - m)^2/(2s^2)]/(\sqrt{2\pi}sz) \qquad (4.66)$$

where the parameters s and m characterize the inherent large scale variations of the wireless channel.

A complete small scale-large scale statistical description requires a new stochastic process, Z, arising by mixing two random stochastic processes, that is, a process, X, and a process, Y, accounting for small scale and large scale variations, respectively. There are two methods of mixing them, namely, Z arises by either taking the mean value of X

to follow the log-normal PDF of Y or by multiplying X and Y. Both methods are shown to be equivalent [8]. If we consider X following a Rice PDF (see Equation 4.37) and applying the multiplicative method, the composite Rice-log-normal PDF, also known as extended Suzuki PDF, will be [7]

$$p_{RL}(z) = \frac{z}{2\sqrt{2\pi}\sigma_0^2 s} \int_0^\infty \exp\left[-\frac{z^2 u + \rho^2}{2\sigma_0^2}\right] I_0\left(\frac{2z\rho\sqrt{u}}{2\sigma_0^2}\right) \exp\left[-\frac{(lnu + 2m)^2}{8s^2}\right] du \tag{4.67}$$

Finally, if we consider the mobile wireless channel of the previous section (see Figure 4.8), the LCR of the Rice-log-normal composite process will be defined as [7]

$$N_{RL}(z) = \int_0^\infty \frac{F(z,y)}{y^2} \exp\left[-\frac{(z/y)^2 + \rho^2}{2\sigma_0^2}\right] \exp\left[-\frac{(lny - m)^2}{2s^2}\right] \int_0^{\pi/2} \cosh\left(\frac{z\rho\cos u}{\sigma_0^2 y}\right)$$

$$\times \left[\exp\left[-\left(\frac{d_2\rho\sin u}{F(z,y)}\right)^2\right] + \frac{\sqrt{\pi}d_2\rho\sin u}{F(z,y)} erf\left[\frac{d_2\rho\sin u}{F(z,y)}\right]\right] du\, dy \tag{4.68}$$

where $F(z,y) = \sqrt{1 + \frac{q}{d_1}\left(\frac{zs}{y}\right)^2}$ and q a parameter characterizing the time variability of the log-normal process [7].

4.5 Summary

A reference model which considered four dependencies in the channel response, namely, spatial at the transmitter and receiver sides, frequency and temporal, has been the basis for a complete characterization of outdoor propagation channels. Other channel responses can arise by Fourier transform relations with respect to one or more of those dependencies. The impact of antenna patterns and dual antenna polarization was not taken into account. A complete first order statistical characterization results from the reference model, with the Rayleigh and Rice processes representing specific propagation scenarios. In turn, second order statistical characterization results by considering correlation functions of the channel responses. With the assumption of WSS with respect to all four dependencies, it is possible to derive the channel PSD which constitutes a Fourier transform pair with the channel ACF. The ACF models the channel selectivity, that is, selectivity in space, frequency and time, whereas the PSD accounts for the channel dispersion, that is, dispersion in direction, delay and Doppler frequency. The opposite of selectivity is coherence, described by the coherence parameters of coherence distance, coherence bandwidth and coherence time. The coherence parameters are inversely proportional to the second centered moments of the single PSDs (e.g., see Equation 4.27 for the second centered moment of the time delay PSD). Finally, the stochastic large scale variations due to shadowing, together with deterministic path loss completes the set of variations that the outdoor propagation channel experiences. It is possible though to derive composite stochastic processes accounting for both small scale and large scale variations.

Acknowledgements

The author acknowledges the support of the FP7 MC-IPLAN project, as material of this chapter was created during the author's employment on that project. The help and support of the editor Prof. B. Allen during the preparation of this chapter is also acknowledged. Finally, the author wants to acknowledge the help of his colleagues Mr. M. Nabeel Asghar and Dr. G. Epiphaniou in preparing the presentation format of this chapter.

References

1. A. Molisch, *Wireless Communications*. Wiley-IEEE Press, 2005.
2. G. Durgin, *Space-time Wireless Channels*. Prentice Hall, 2003.
3. B. Fleury, "First-and second-order characterization of direction dispersion and space selectivity in the radio channel," *Information Theory, IEEE Transactions on*, vol. 46, no. 6, pp. 2027–2044, Sep. 2000.
4. M. Steinbauer, A. Molisch, and E. Bonek, "The double-directional radio channel," *Antennas and Propagation Magazine, IEEE*, vol. 43, no. 4, pp. 51–63, 2001.
5. R. Torres, B. Cobo, D. Mavares, F. Medina, S. Loredo, and M. Engels, "Measurement and statistical analysis of the temporal variations of a fixed wireless link at 3.5ghz," *Wireless Personal Communications*, vol. 37, no. 1, pp. 41–59, 2006.
6. P. Karadimas, "Statistical investigation of scattered multipath power and received signal in time varying wireless channels," Ph.D. dissertation, 2006. [Online]. Available: http://nemertes.lis.upatras.gr/jspui/bitstream/10889/2702/6/\Nimertis_Karadimas%28ele%29.pdf
7. M. Pätzold, *Mobile Fading Channels*. John Wiley & Sons, 2002.
8. G. Stuber, *Principles of mobile communication*. Norwell, MA: Kluwer Academic Publishers, 2001.
9. A. Akki and F. Haber, "A statistical model of mobile-to-mobile land communication channel," *Vehicular Technology, IEEE Transactions on*, vol. 35, no. 1, pp. 2–7, Feb. 1986.
10. F. Vatalaro and A. Forcella, "Doppler spectrum in mobile-to-mobile communications in the presence of three-dimensional multipath scattering," *Vehicular Technology, IEEE Transactions on*, vol. 46, no. 1, pp. 213–219, Feb. 1997.
11. C. Patel, G. Stuber, and T. Pratt, "Simulation of rayleigh-faded mobile-to-mobile communication channels," *Communications, IEEE Transactions on*, vol. 53, no. 11, pp. 1876–1884, Nov. 2005.
12. R. H. Clarke, "A statistical theory of mobile-radio reception," *Bell Syst. Tech. J.*, vol. 47, pp. 957–1000, Jul./Aug. 1968.
13. W. Braun and U. Dersch, "A physical mobile radio channel model," *Vehicular Technology, IEEE Transactions on*, vol. 40, no. 2, pp. 472–482, May 1991.
14. M. Gans, "A power-spectral theory of propagation in the mobile-radio environment," *Vehicular Technology, IEEE Transactions on*, vol. 21, no. 1, pp. 27–38, Feb. 1972.
15. T. Aulin, "A modified model for the fading signal at a mobile radio channel," *Vehicular Technology, IEEE Transactions on*, vol. 28, no. 3, pp. 182–203, Aug. 1979.
16. K. Pedersen, P. Mogensen, and B. Fleury, "Power azimuth spectrum in outdoor environments," *Electronics Letters*, vol. 33, no. 18, pp. 1583–1584, Aug. 1997.
17. A. Turkmani and J. Parsons, "Characterisation of mobile radio signals: base station crosscorrelation," *Communications, Speech and Vision, IEE Proceedings I*, vol. 138, no. 6, pp. 557–565, Dec. 1991.
18. T. Feng and T. Field, "Statistical analysis of mobile radio reception: an extension of clarke's model," *Communications, IEEE Transactions on*, vol. 56, no. 12, pp. 2007–2012, Dec. 2008.
19. J. Karedal, F. Tufvesson, N. Czink, A. Paier, C. Dumard, T. Zemen, C. Mecklenbrauker, and A. Molisch, "A geometry-based stochastic mimo model for vehicle-to-vehicle communications," *Wireless Communications, IEEE Transactions on*, vol. 8, no. 7, pp. 3646–3657, Jul. 2009.
20. L. Greenstein, S. Ghassemzadeh, V. Erceg, and D. Michelson, "Ricean – factors in narrow-band fixed wireless channels: Theory, experiments, and statistical models," *Vehicular Technology, IEEE Transactions on*, vol. 58, no. 8, pp. 4000–4012, Oct. 2009.
21. L. Ahumada, R. Feick, R. Valenzuela, and C. Morales, "Measurement and characterization of the temporal behavior of fixed wireless links," *Vehicular Technology, IEEE Transactions on*, vol. 54, no. 6, pp. 1913–1922, Nov. 2005.

22. P. Karadimas, E. D. Vagenas, and S. A. Kotsopoulos, "On the scatterers' mobility and second order statistics of narrowband fixed outdoor wireless channels," *Wireless Communications, IEEE Transactions on*, vol. 9, no. 7, pp. 2119–2124, Jul. 2010.

23. G. Durgin, T. Rappaport, and D. de Wolf, "New analytical models and probability density functions for fading in wireless communications," *Communications, IEEE Transactions on*, vol. 50, no. 6, pp. 1005–1015, Jun. 2002.

24. A. Abdi, "On the utility of laguerre series for the envelope pdf in multipath fading channels," *Information Theory, IEEE Transactions on*, vol. 55, no. 12, pp. 5652–5660, Dec. 2009.

25. M. Slack, "The probability distributions of sinusoidal oscillations combined in random phase," *Electrical Engineers - Part I: General, Journal of the Institution of*, vol. 93, no. 66, p. 278, Jun. 1946.

26. I. Ryzhik, A. Jeffrey, and D. Zwillinger, *Table of integrals, series and products*. Academic Press, 2007.

27. A. Abdi, H. Hashemi, and S. Nader-Esfahani, "On the pdf of the sum of random vectors," *Communications, IEEE Transactions on*, vol. 48, no. 1, pp. 7–12, Jan. 2000.

28. P. Bello, "Characterization of randomly time-variant linear channels," *Communications Systems, IEEE Transactions on*, vol. 11, no. 4, pp. 360–393, Dec. 1963.

29. G. Matz, "On non-wssus wireless fading channels," *Wireless Communications, IEEE Transactions on*, vol. 4, no. 5, pp. 2465–2478, Sep. 2005.

30. T. Willink, "Wide-sense stationarity of mobile mimo radio channels," *Vehicular Technology, IEEE Transactions on*, vol. 57, no. 2, pp. 704–714, Mar. 2008.

31. D. Cox and R. Leck, "Distributions of multipath delay spread and average excess delay for 910-mhz urban mobile radio paths," *Antennas and Propagation, IEEE Transactions on*, vol. 23, no. 2, pp. 206–213, Mar. 1975.

32. A. Algans, K. Pedersen, and P. Mogensen, "Experimental analysis of the joint statistical properties of azimuth spread, delay spread, and shadow fading," *Selected Areas in Communications, IEEE Journal on*, vol. 20, no. 3, pp. 523–531, Apr. 2002.

33. G. Calcev, D. Chizhik, B. Goransson, S. Howard, H. Huang, A. Kogiantis, A. Molisch, A. Moustakas, D. Reed, and H. Xu, "A wideband spatial channel model for system-wide simulations," *Vehicular Technology, IEEE Transactions on*, vol. 56, no. 2, pp. 389–403, Mar. 2007.

34. E. Chiavaccini and G. Vitetta, "Gqr models for multipath rayleigh fading channels," *Selected Areas in Communications, IEEE Journal*, vol. 19, no. 6, pp. 1009–1018, Jun. 2001.

35. F. Garber and M. Pursley, "Performance of differentially coherent digital communications over frequency-selective fading channels," *Communications, IEEE Transactions on*, vol. 36, no. 1, pp. 21–31, Jan. 1988.

36. M. Clark, L. Greenstein, W. Kennedy, and M. Shafi, "Matched filter performance bounds for diversity combining receivers in digital mobile radio," *IEEE Trans. Veh. Technol.*, vol. 41, no. 4, pp. 356–362, Nov. 1992.

37. R. Feick, R. Valenzuela, and L. Ahumada, "Experimental results on the level crossing rate and average fade duration for urban fixed wireless channels," *Wireless Communications, IEEE Transactions on*, vol. 6, no. 1, pp. 175–179, Jan. 2007.

38. A. Domazetovic, L. Greenstein, N. Mandayam, and I. Seskar, "Estimating the doppler spectrum of a short-range fixed wireless channel," *Communications Letters, IEEE*, vol. 7, no. 5, pp. 227–229, May 2003.

39. A. Kuchar, J.-P. Rossi, and E. Bonek, "Directional macro-cell channel characterization from urban measurements," *Antennas and Propagation, IEEE Transactions on*, vol. 48, no. 2, pp. 137–146, Feb. 2000.

40. J. Fuhl, J.-P. Rossi, and E. Bonek, "High-resolution 3-d direction-of-arrival determination for urban mobile radio," *Antennas and Propagation, IEEE Transactions on*, vol. 45, no. 4, pp. 672–682, Apr. 1997.

41. K. Kalliola, H. Laitinen, P. Vainikainen, M. Toeltsch, J. Laurila, and E. Bonek, "3-d double-directional radio channel characterization for urban macrocellular applications," *Antennas and Propagation, IEEE Transactions on*, vol. 51, no. 11, pp. 3122–3133, Nov. 2003.

42. P. Karadimas and J. Zhang, "A generalized analysis of three-dimensional anisotropic scattering in mobile wireless channels-part i: Theory," in *Vehicular Technology Conference (VTC Spring), 2011 IEEE 73rd*, May 2011, pp. 1–5.

43. R. Clarke and W. L. Khoo, "3-d mobile radio channel statistics," *Vehicular Technology, IEEE Transactions on*, vol. 46, no. 3, pp. 798–799, Aug. 1997.

44. J. Parsons and A. Turkmani, "Characterisation of mobile radio signals: model description," *Communications, Speech and Vision, IEE Proceedings I*, vol. 138, no. 6, pp. 549–556, Dec. 1991.

45. S. Qu and T. Yeap, "A three-dimensional scattering model for fading channels in land mobile environment," *Vehicular Technology, IEEE Transactions on*, vol. 48, no. 3, pp. 765–781, May 1999.

46. Y. Ebine and Y. Yamada, "A vehicular-mounted vertical space diversity antenna for a land mobile radio," *Vehicular Technology, IEEE Transactions on*, vol. 40, no. 2, pp. 420–425, May 1991.

47. T. Taga, "Analysis for mean effective gain of mobile antennas in land mobile radio environments," *Vehicular Technology, IEEE Transactions on*, vol. 39, no. 2, pp. 117–131, May 1990.

48. A. Ando, T. Taga, A. Kondo, K. Kagoshima, and S. Kubota, "Mean effective gain of mobile antennas in line-of-sight street microcells with low base station antennas," *Antennas and Propagation, IEEE Transactions on*, vol. 56, no. 11, pp. 3552–3565, Nov. 2008.

49. K. Kalliola, K. Sulonen, H. Laitinen, O. Kivekas, J. Krogerus, and P. Vainikainen, "Angular power distribution and mean effective gain of mobile antenna in different propagation environments," *Vehicular Technology, IEEE Transactions on*, vol. 51, no. 5, pp. 823–838, Sep. 2002.

50. P. Karadimas, E. Vagenas, and S. Kotsopoulos, "A small scale fading model with sectored and three dimensional diffuse scattering," in *Consumer Communications and Networking Conference, 2008. CCNC 2008. 5th IEEE*, Jan. 2008, pp. 943–947.

51. P. Karadimas and S. A. Kotsopoulos, "A modified loo model with partially blocked and three dimensional multipath scattering: Analysis, simulation and validation," *Wirel. Pers. Commun.*, vol. 53, pp. 503–528, June 2010. Available: http://dx.doi.org/10.1007/s11277-009-9698-z

52. K. Mammasis and R. Stewart, "The fisher–bingham spatial correlation model for multielement antenna systems," *Vehicular Technology, IEEE Transactions on*, vol. 58, no. 5, pp. 2130–2136, Jun. 2009.

53. K. Mammasis, R. Stewart, and J. Thompson, "Spatial fading correlation model using mixtures of von mises fisher distributions," *Wireless Communications, IEEE Transactions on*, vol. 8, no. 4, pp. 2046–2055, Apr. 2009.

54. W. Lee, "Effects on correlation between two mobile radio base-station antennas," *Communications, IEEE Transactions on*, vol. 21, no. 11, pp. 1214–1224, Nov. 1973.

55. J. Salz and J. Winters, "Effect of fading correlation on adaptive arrays in digital mobile radio," *Vehicular Technology, IEEE Transactions on*, vol. 43, no. 4, pp. 1049–1057, Nov. 1994.

56. F. Adachi, M. Feeney, J. Parsons, and A. Williamson, "Crosscorrelation between the envelopes of 900 mhz signals received at a mobile radio base station site," *Communications, Radar and Signal Processing, IEE Proceedings F*, vol. 133, no. 6, pp. 506–512, Oct. 1986.

57. A. Abdi, J. Barger, and M. Kaveh, "A parametric model for the distribution of the angle of arrival and the associated correlation function and power spectrum at the mobile station," *Vehicular Technology, IEEE Transactions on*, vol. 51, no. 3, pp. 425–434, May 2002.

58. T. Sarkar, Z. Ji, K. Kim, A. Medouri, and M. Salazar-Palma, "A survey of various propagation models for mobile communication," *Antennas and Propagation Magazine, IEEE*, vol. 45, no. 3, pp. 51–82, 2003.

59. C. Balanis, "Antenna theory analysis and design, 1997," John Wiley & Sons Inc.

60. E. Damosso and L. M. Correia, "Digital mobile radio towards future generation systems," *Commission of the European Communities*, 1996. [Online]. Available: http://www.lx.it.pt/cost231/

61. N. Shabbir, M. T. Sadiq, H. Kashif, and R. Ullah, "Comparison of radio propagation models for long term evolution (lte) network," *CoRR*, vol. abs/1110.1519, 2011.

62. G. Turin, F. Clapp, T. Johnston, S. Fine, and D. Lavry, "A statistical model of urban multipath propagation," *Vehicular Technology, IEEE Transactions on*, vol. 21, no. 1, pp. 1–9, Feb. 1972.

63. F. Hansen and F. Meno, "Mobile fading–rayleigh and lognormal superimposed," *Vehicular Technology, IEEE Transactions on*, vol. 26, no. 4, pp. 332–335, Nov. 1977.

64. J. Salo, L. Vuokko, H. M. El-Sallabi, and P. Vainikainen, "An additive model as a physical basis for shadow fading," *Vehicular Technology, IEEE Transactions on*, vol. 56, no. 1, pp. 13–26, Jan. 2007.

65. A. Abdi and M. Kaveh, "A comparative study of two shadow fading models in ultrawideband and other wireless systems," *Wireless Communications, IEEE Transactions on*, vol. 10, no. 5, pp. 1428–1434, 2011.

5

Outdoor-Indoor Channel

Andrés Alayón Glazunov[1], Zhihua Lai[2] and Jie Zhang[3]
[1]*KTH Royal Institute of Technology, Sweden*
[2]*Ranplan Wireless Network Design Ltd., UK*
[3]*University of Sheffield, UK*

5.1 Introduction

The major part of the data traffic takes place indoors and this trend is expected to accelerate in the near future, [1]. Therefore, one of the main goals of new-coming wireless networks such as the LTE and LTE-A is to extend the indoor coverage of wireless broadband services. The design strategy of a wireless system is dictated by, among other factors, the limitations posed by the physical propagation environment. For example, the interference levels entering the link budgets that define the maximum data rates and coverage areas must be based on accurate propagation computations. The latter can also be used for planning location-based services requiring an accurate prediction of the user's position based on received signal strengths. Furthermore, in spite of advances in computing technology, the propagation predictions are usually rather time-consuming. Hence, producing accurate yet reasonably fast propagation models still remains at the core of wireless network planning and optimization. This is further accentuated by the introduction of Multiple Input Multiple Output (MIMO) technologies that employ multiple antennas at both the transmit and the receive sides of the wireless communication link.

It is well-known that building structures have an attenuating impact on the radio waves propagating into buildings. The radio frequencies covered by LTE networks are diverse, they comprise several frequency bands between 700 MHz and 2.6 GHz, which pose additional propagation modelling challenges. Indeed, the dominating propagation mechanism at different frequencies is different since the electrical properties of the building materials vary with the frequency. For example, at lower frequencies, for example, 700 MHz the buildings are more transparent to radio waves than at the higher frequency bands, for example, 2.6 GHz. On the other hand, at higher frequencies the transmitted signals will reach the receiver mainly due to the waveguiding and the diffraction mechanisms

---

LTE-Advanced and Next Generation Wireless Networks: Channel Modelling and Propagation, First Edition.
Edited by Guillaume de la Roche, Andrés Alayón Glazunov and Ben Allen.
© 2013 John Wiley & Sons, Ltd. Published 2013 by John Wiley & Sons, Ltd.

depending on the building materials and the propagation scenario. In this context, different propagation conditions, such as Line Of Sight (LOS) and Non Line Of Sight (NLOS), have a large impact on the signal traveling from the outdoor environment to the indoor environment. Here, the LOS and NLOS condition can not be related to the presence or the absence of a direct path between the transmit and receive antennas, respectively. Rather, the direct path between the outdoor antenna and the immediate exterior environment to the indoor antenna is considered. Indeed, here, the LOS corresponds to the situation when the outdoor transmit antenna has a clear path to the exterior walls of the building where the receive antenna is located. In the opposite case, when the direct path to the building of interest is blocked, for example, by another building we have a NLOS propagation scenario.

This chapter focuses on the transition between the outdoor channel and the indoor channel described in Chapters 3 and 4, respectively. Unavoidably, a comprehensive characterization of the outdoor-to-indoor channel, and by reciprocity[1] the indoor-to-outdoor channel too, imposes an overlap with those chapters. However, we have tried to keep it to a minimum while focusing on the main aspects of the propagation path loss modelling at the transition between the outdoor and the indoor environments. In order to gain a clear understanding of the propagation modelling principles, the necessary background will be given with special focus on established models as well as the latest developments in this field of research. The presented models have been devised from measurements obtained in a wide range of frequency bands from 400 MHz to 8 GHz.

5.2 Modelling Principles

The "instantaneous" propagation path loss that includes the effects of both large- and small-scale fading is obtained as

$$L_i = P_{tx} - P_{rx,i} + G_{tx} + G_{rx}, \tag{5.1}$$

where L_i is given in dB; P_{tx} and $P_{rx,i}$ are the transmit power and the "instantaneous" receive power (usually given in dBm), respectively; G_{tx} and G_{rx} are the transmit and receive antenna gains (in dBi), respectively. In the above equation, we have assumed that any other system losses have been compensated for, that is, the data is calibrated and we have only the effects of the radio propagation channel.

The mean received power P_{rx} is obtained by averaging the "instantaneous" absolute receive power $p_{rx,i}(d)$ (in Watts) over the interval Δ_λ

$$P_{rx} = 10 \log \left(\frac{1}{\Delta_\lambda} \int_{d-\frac{\Delta_\lambda}{2}}^{d+\frac{\Delta_\lambda}{2}} p_{rx,i}(x) dx \right), \tag{5.2}$$

where d is the separation distance between the TX and the RX and Δ_λ is usually chosen between 20λ and 40λ depending on the measurement data, λ is the wavelength corresponding to the transmit frequency. Δ_λ is chosen to remove the small-scale fading variation from the data, while keeping the large-scale fading variation as intact as possible, that is,

---

[1] Reciprocity here is assumed to be valid in the sense that interchanging positions between the TX and the RX antennas, while keeping the same transmit power, will result in the same receive power.

the fast-fading is filtered out. In practice, the integral in (5.2) is replaced by the running sample mean of the discrete power data.

The instantaneous fading signal can then be expressed as a function of the mean path loss

$$P_{rx,i} = P_{tx} + G_{tx} + G_{rx} - L_{tot} - X_\sigma + F, \tag{5.3}$$

where L_{tot} is the *mean path loss*, that is, it represents the deterministic component of the channel. X_σ stands for the large-scale (or shadow) fading and F stands for the small-scale fading, that is, both model the stochastic variation of the channel. The large-scale fading is attributed to obstacles blocking the direct path between the receive and the transmit antennas and its variation may be over a hundred wavelengths. The large-scale fading is usually treated as the residual remaining after the propagation law L_{tot} have been fitted to data. It has been found from experiments that X_σ, in dB-scale, can be accurately modeled by a zero-mean Gaussian variable with standard deviation σ, or equivalently, by a lognormally distributed variable in linear scale. On the other hand, the small-scale fading F is a result of interaction of different multipath components and therefore deals with signal variations over a few wavelengths. It can be typically modeled in linear scale by either the Rayleigh, Rician or Nakagami distributions.

The generally accepted approach is to average the received power according to (5.2). Thus, the path loss can be written as

$$L = P_{tx} - P_{rx} + G_{tx} + G_{rx}, \tag{5.4}$$

and therefore we can write

$$L = L_{tot} + X_\sigma. \tag{5.5}$$

It is worthwhile noticing that for the ideal free-space channel, $L = L_{tot} = L_{fs}$, that is, there are no fading effects and the mean path loss coincides with Friis' law

$$L_{fs} = 20\log(d) + 20\log(f) + 32.45, \tag{5.6}$$

where d and f are in km and in MHz, respectively.

Until now, the exposed modelling principle is common to the outdoor, the indoor and the outdoor-to-indoor propagation. The main goal is to obtain meaningful models for L_{tot} that minimize X_σ based on estimates of L. For the specific case of outdoor-to-indoor propagation modelling, L_{tot} can be further expanded into three terms

$$L_{tot} = L_{out} + L_{tw} + L_{in}, \tag{5.7}$$

where L_{out} is the path loss measured at some point just outside the building where the indoor antenna is located, L_{tw} is the transition path loss from outdoors to indoors and L_{in} is the indoor propagation path loss measured from some point located just inside the building. As we are going to see in the following sections, the expansion (5.7) is a fundamental modelling concept.

Sometimes, a modelling approach that relies on the so-called *excess path loss*, L_e, is preferable to the direct modelling approach (5.5). The excess path loss is defined as the loss in excess of the free-space propagation loss in dB scale

$$L_e = L - L_{fs} \tag{5.8}$$

where L and L_{fs} have been defined in (5.4) and (5.6), respectively.

Here, we can further introduce the *building penetration loss* (or building shielding loss) L_{bl} to obtain an expression similar to (5.5)

$$L = L_{bl} + L_{fs} + X_\sigma. \tag{5.9}$$

where

$$L_{bl} = \langle L_e \rangle_{X_\sigma}, \tag{5.10}$$

where $\langle \rangle_{X_\sigma}$ denotes the mean operator acting on the realization space of the random variable X_σ. The large-scale fading term X_σ in (5.9) is, in general, different from the corresponding term in (5.5).

The approach represented by (5.8)–(5.9) is especially suited for LOS conditions. However, in many practical situations the NLOS condition is more likely to happen, for example, in urban environments where the main propagation mechanism is diffraction over the roof tops of buildings and/or around buildings. In those cases the free-space path loss in (5.8) and (5.9) can be substituted by the reference path loss, L_{out}, defined above.

Hence, the excess path loss can be written as

$$L_e = L - L_{out}, \tag{5.11}$$

and the path loss can, then, be written as

$$L = L_{bl} + L_{out} + X_\sigma. \tag{5.12}$$

where the large-scale fading term X_σ is, in general, different from the corresponding terms in (5.5) and (5.10). Hence, L_e in (5.8) and (5.11) and L_{bl} in (5.9) and (5.12), respectively, denote the path loss in excess to the reference path loss outside a building. Furthermore, we see from (5.12), (5.5), (5.7) and (5.10) that

$$L_{bl} = L_{tw} + L_{in}. \tag{5.13}$$

Now, we can obtain the *mean building penetration loss* (or mean building shielding loss) by averaging L_{bl} over different positions of the antenna located indoors

$$L_{abl} = \langle L_{bl} \rangle_{r_{in}} = L_{tw} + \langle L_{in} \rangle_{r_{in}}, \tag{5.14}$$

where $\langle \rangle_{r_{in}}$ denotes the mean operator acting on the positions r_{in} of the indoor antenna. Hence, L_{abl} is obtained as average over the realizations of X_σ and the positions of the indoor antenna inside the building r_{in}.

In practice, it is not always possible to accurately model all aspects of the received signal according to (5.3); in particular, the variation due to fast fading is the hardest to predict since it depends on relatively many field fluctuations that interact to form the small-scale interference pattern. Moreover, the accuracy of the propagation models depends heavily on the level of details of the input data, for example, the geometry and electrical parameters of the buildings. Hence, the propagation models are usually constrained by trade-offs between accuracy and computation speed. Usually, more accuracy yields a higher computational burden.

The propagation path loss models we consider here can be subdivided into the following three categories:

- *Empirical models*: These are models based on extensive measurement data. The goal is to devise a function that captures the main traits of the propagation environment. This function often depends on parameters of the experiment and the measured data; the fitting is performed to minimize the error, for example, in the least squares sense. These parameters (or variables) are usually the transmit frequency of the signal, the distance between the TX and the RX, their heights above ground, the number of walls and/or floors passed by the signal on its way to the receiver, and so on. Clearly, the more the parameters are correctly taken into account, the better the accuracy of the model. The choice of these functions is dictated by the basic propagation mechanisms as well as the geometry of the problem and site-specific data in general. The resulting model is interpreted in the statistical sense, that is, a statistical distribution is ascribed to some of the input parameters and the output variables.
- *Deterministic models*: On the one hand, there are models based on the numerical re-solution of the fundamental equations governing macroscopic electromagnetic wave propagation, that is, the Maxwell's equations. On the other hand, there are models based on the Geometrical Optics (GO) principles and therefore are ray-based, among them, the ray-tracing and the ray-launching techniques are the most widespread in use. The resolution of Maxwell's equations requires an accurate modelling of the geometry of the propagation scenario as well as the correct definition of the electrical proper-ties of the building materials and any other objects present in the environment. It is worthwhile noticing, that even if the methods are deterministic, given the complex-ity of the propagation environment, the electrical parameters of the environment must be calibrated against measurement data, that is, "effective electrical parameters" are used instead of the actual parameters. Hence, the obtained predictions must also be interpreted in the statistical sense.
- *Hybrid models*: These models combine the best features of the above two modelling approaches to produce computationally efficient and accurate propagation path loss models.

5.3 Empirical Propagation Models

Empirical models offer the simplest and most computationally efficient prediction approach of the propagation path loss, but often at the expense of accuracy. The empirical propagation models are usually derived from a great amount of measurements, that is, mathematical relationships that describe the path loss dependence upon various parameters are constructed based on the data fitting. The simplest models rely on a minimal number of factors, for example, the transmit frequency and the distance between the transmitters and receivers. Empirical models are easy to use and they are often based on a few sets of formulas and parameters. However, by adding site-specific information such as the building structure and geometry, that is, number of walls, floors, ceilings, the complexity of the model is increased, but so is the accuracy. Sometimes, models that include extensive site-specific information are denominated *semi-empirical models*.

Empirical propagation models are usually valid for specific propagation environments since they are based on measurements. In addition, only a limited number of values of the parameters are applicable to the models which define the *validity region* of the model.

5.3.1   Path Loss Exponent Model

The simplest path loss model is the path loss exponent model

$$L = L_{\text{tot}}(d_0) + 10n \log \left(\frac{d}{d_0} \right) + X_\sigma, \tag{5.15}$$

where $L_{\text{tot}}(d_0)$ is the mean path loss at the reference distance d_0, n is the path loss exponent and X_σ is as defined above. The path loss exponent n is obtained empirically and accounts for the site-specific factors, that is, it dictates the average increase rate of the path loss with the distance d between the TX and RX. The dependence of the path loss on the transmit frequency is implicitly assumed in (5.15). The path loss exponent model has been widely used since the Okumura-Hata model was first presented, [2, 3], for outdoor and indoor propagation path loss modelling of macrocell systems. However, the model has no obvious applicability limitations to other cellular environments, such as, microcells, picocells or femtocells, [4].

An extensive measurement campaign in a residential area at the 5.85 GHz band shows, [5], as expected, that n increases as the receiver goes from the outdoor to the indoor environments. For that specific case, n was found to be 3.42 and 2.93, for the RX antenna indoors and outdoors, respectively, with corresponding standard deviations of the large-scale fading, σ, equal 8.01 dB and 7.85 dB, respectively. In [5] n and σ were extracted for various TX-RX configurations in the same experiment with similar results. It was also shown there that the height of the receive antenna had no statistically significant effect on the path loss exponent in both indoor and outdoor scenarios. The outdoor transmitters were placed at a height of 5.5 m at distances between 30 and 210 m from the homes.

Another example of the implementation of the path loss exponent model is shown in Figure 5.1(a). The coverage prediction for three transmitters operating at 2.14 GHz was carried out using the Ranplan Radiowave Propagation Simulator (RRPS) [6]. As we can see from the figure, the indoor building structure (e.g. the walls) does not have any impact on the predicted mean signal strength and therefore the mean path loss transition between any two positions on the figure tends to be smooth regardless of the building geometry and materials.[2] This is a fundamental drawback of the path loss exponent model in addition to the fact that it doesn't differentiate between the outdoor and the indoor propagation. Hence, the path loss exponent model is a result of averaging over all TX-RX distances.

5.3.2   Path Loss Exponent Model with Mean Building Penetration Loss

As we explained above, the building penetration loss can be used to account for the transition between the outdoor to the indoor environments and vice versa. A model that combines the path loss exponent model and the mean (aggregated) building penetration loss was investigated in [5]

$$L = L_{\text{out}} + L_{\text{abl}} + X_\sigma, \tag{5.16}$$

$$L_{\text{out}} = L_{\text{out}}(d_0) + 10n \log \left(\frac{d}{d_0} \right), \tag{5.17}$$

---

[2] In wireless network planning and optimization, a fading margin is added to account for the large-scale fading variation.

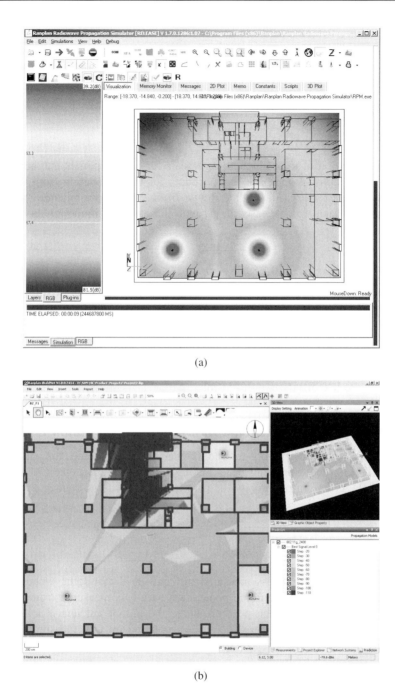

(a)

(b)

Figure 5.1 Indoor coverage prediction. (a) The indoor One-Slope-Model coverage prediction using RRPS and (b) the indoor Multi Wall coverage prediction using Ranplan iBuildNet®.

where n is obtained from the path loss measured just outside the building where the indoor antenna is located and the mean building penetration loss L_{abl} is a constant obtained according to (5.14). In [5], the reported values for L_{abl} were estimated at the 5.85 GHz band and varied between 7.2 dB and 21.1 dB with an average of 16.3 dB for the residential houses.

5.3.3 Partition-Based Outdoor-to-Indoor Model

The standard deviation of the estimated large-scale fading component can be reduced by applying a semi-empirical approach that accounts for the attenuation from the different obstructions that intersects a direct line drawn from the outdoor antenna to the indoor antenna. This approach was proposed in [5] with the path loss model given by

$$L = L_{fs} + \sum_{k=1}^{N} L_k + X_{\sigma}, \qquad (5.18)$$

where L_{fs} is the free-space path loss that would be obtained without the presence of any obstacles and L_k is the attenuation due to obstacle k that could be a tree, an exterior wall, an interior wall, and so on, N is the number of obstacles for a given TX-RX direct path. The standard deviation of the large-scale fading term could be reduced to ≈ 3 dB in average. The estimated attenuation values at 5.85 GHz were $8 - 34$ dB for home exterior, $3.6 - 5.6$ dB for home interior and $3.5 - 16.4$ dB for foliage. The model was found to work well for TX-RX distances up to 50 m.

5.3.4 Path Loss Exponent Model with Building Penetration Loss

A model for the picocellular indoor-to-outdoor channel that combines the path loss exponent model for L_{out} and a building penetration loss model for L_{bl} (see (5.12)) has been proposed in [7]. Assuming reciprocity the path loss model can be written as

$$L = L_{out} + L_{bl} + X_{\sigma}, \qquad (5.19)$$

$$L_{out} = L_{out}(d_0) + 10n \log\left(\frac{d}{d_0}\right), \qquad (5.20)$$

where L_{out} is the mean path loss measured just outside the building where the indoor antenna is located, L_{bl} is the building penetration path loss measured from the position at which L_{out} is measured to the position where the indoor antenna is located and d_0 is a reference distance. The model (5.19)-(5.20) is based on CW measurements in two residential areas at the 900 MHz, 2 GHz, 2.5 GHz and 3.5 GHz frequency bands.

The authors propose to model the path loss exponent as a function of the transmit frequency, that is, $n = n(f)$ for which a fitting function was proposed (see [7] for exact dependence). However, an important observation is that $n \approx 3.2$ for 900 MHz, 2 GHz and 2.5 GHz and $n \approx 3.8$ for 3.5 GHz. Thus, for 900 MHz to 2.5 GHz frequencies the path loss exponent is approximately constant. Furthermore, the transition loss or building penetration path loss is modelled by two terms, that is, $L_{bl} = A(f) + pL_w$, where $A(f)$ is a quadratic function of the frequency, while the second term is a linear function of the wall losses, $L_w = 5.8\,\text{dB}$. A closer examination of the results presented in [7] reveals that the frequency dependence (for 900 MHz, 2 GHz and 2.5 GHz) can be replaced by $A(f) \approx \kappa f$ without loss of accuracy, where $\kappa = 6.5\,\text{dB/GHz}$.

Hence, for typical LTE/LTE-A frequencies (between 900 MHz and 2.5 GHz), the following new model can be considered

$$L_{\text{bl}} = \kappa f + p L_w, \tag{5.21}$$

where p is the number of walls obstructing the direct path between the position where L_{out} is measured just outside the exterior wall of the building, and the position of the indoor antenna. L_w is an average for the attenuation through the exterior wall and the interior walls.

Hence, combining (5.19), (5.20) and (5.21) we obtain the following path loss model

$$L = 40.79 + 32 \log(d) + 6.5 f + p 5.8 + X_\sigma, \tag{5.22}$$

where $p = \{0, 1, 2\}$ denotes the number of walls from the exterior wall to the indoor position, $0.9 < f < 2.5\,\text{GHz}$ is the transmit frequency, $5 < d < 100\,\text{m}$ is the outdoor antenna distance to the exterior wall of the building where the indoor antenna is located, and X_σ is a zero-mean Gaussian variable with standard deviation $\sigma \approx 10\,\text{dB}$. To obtain (5.22) we have used the parameters provided in [7].

5.3.5 COST 231 Building Penetration Loss Model

5.3.5.1 COST 231 Building Penetration Loss at the LOS Condition

A comprehensive outdoor-to-indoor propagation model was developed within the COST (COperation Europeenne dans le domaine de la recherche Scientifique et Technique) project 231, COST 231 for short, and presented in [8]. The proposed model expanded the total mean path loss between isotropic antennas L_{tot} into three terms according to (5.7)

$$L_{\text{tot}} = L_{\text{fs}} + L_{\text{tw}} + L_{\text{in}}, \tag{5.23}$$

$$L_{\text{fs}} = 32.45 + 20 \log(S + d_{\text{in}}) + 20 \log(f), \tag{5.24}$$

$$L_{\text{in}} = \max\{p W_{\text{i}}, \alpha(d_{\text{in}} - 2)(1 - \sin(\theta))^2\}, \tag{5.25}$$

$$L_{\text{tw}} = W_{\text{e}} + W G_{\text{e}}(1 - \sin(\theta))^2. \tag{5.26}$$

where L_{fs} is the free-space path loss measured at the distance $S + d_{\text{in}}$ in metres, S is the distance from the outdoor antenna to the the exterior wall of the building where the indoor antenna is located, d_{in} is the distance measured perpendicularly from the exterior wall to the actual position of the indoor antenna according to Figure 5.2, f is the transmit frequency in GHz. L_{in} is the mean path loss measured from the interior side of the exterior wall and describes the indoor propagation path loss, p is the number of internal walls obstructing the direct path between the position where L_{fs} is measured and the position of the indoor antenna, W_{i} is the loss in the internal walls, α represents the indoor path loss coefficient and θ is the grazing angle to exterior wall. L_{tw} is the transition propagation path loss, it can be divided into two parts: the perpendicular loss W_{e} and the parallel penetration $W G_{\text{e}}$.

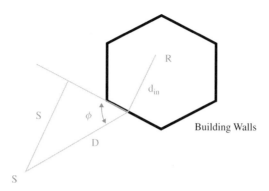

Figure 5.2   Geometrical parameter definition for the COST 231 building penetration loss at LOS.

The model is based on microcell measurements in the frequency range $900 - 1800\,\text{MHz}$ and distances S up to $500\,\text{m}$ and for base station heights below $30\,\text{m}$. Typical values for the model parameters are the following: $W_e \in [4, 20]\,\text{dB}$, for example, $10 - 20\,\text{dB}$ for concrete walls without windows, $7\,\text{dB}$ for concrete walls with windows and $4\,\text{dB}$ for wooden walls; $W_i \in [4, 10]\,\text{dB}$, for example, $7\,\text{dB}$ for concrete walls with windows and $4\,\text{dB}$ for wooden and plaster walls; $WG_e \sim 20\,\text{dB}$, $\alpha \sim 0.6\,\text{dB/m}$. Typical values for the large-scale fading standard deviation due to floor losses are $\sigma \in [5, 15]\,\text{dB}$ depending on the scenario (see [8] for further reference).

It is worthwhile noticing that, in practice, if the resolution of the building database is low additional mean building penetration losses can be added where applicable. In addition, if the the average internal wall losses are known instead of each specific value, and also the distance between internal walls are known then pW_i in (5.25) can be substituted by the equivalent law βd_{in}. This can also be applied to the case when no internal walls are present, obviously with a different β.

In [9, 10], the building penetration loss (building-shielding loss) in office-building areas at the $5.1\,\text{GHz}$ band has been studied with a focus on the azimuth and elevation dependence for satellite transmitters. The predictions provided by a model based on the COST 231 penetration loss model was found to be in good agreement with measurements. The measured building penetration loss was $10 - 45\,\text{dB}$ with a harmonic average of $19.1\,\text{dB}$ for the elevation measurements and $22.3\,\text{dB}$ for the azimuth measurements.

5.3.5.2   COST 231 Building Penetration Loss at the NLOS Condition

For the NLOS condition typically observed in macrocells, the mean path loss can be expanded into three terms according to (5.7), [11, 12]

$$L_{tot} = L_{out} + L_{tw} + L_{in}, \tag{5.27}$$

$$L_{out} = L_{out,(1 or 2)}, \tag{5.28}$$

$$L_{in} = \max\{pW_i, \alpha d_{in}\}, \tag{5.29}$$

$$L_{tw} = W_e + W_{ge} - G_{FH}. \tag{5.30}$$

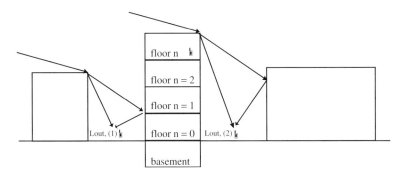

Figure 5.3 Building penetration loss v.s. floor number.

where the total mean path loss L_{tot} can be related to the outdoor path loss according to two different scenarios resulting in two different reference values, $L_{out,(1)}$ and $L_{out,(2)}$, respectively, as shown in Figure 5.3. In addition, three new terms are introduced: 1) αd_{in} models the indoor propagation in (5.28) (see (5.25) for a comparison), 2) the constant W_{ge} is a correction factor, introduced to tune the penetration loss of the external wall W_e and 3) a height gain term that takes into account the fact that the diffraction path loss decreases as the indoor antenna height relative the ground floor increases

$$G_{FH} = nG_n \quad \text{or} \quad G_{FH} = hG_h, \tag{5.31}$$

where two options are given: 1) the floor height gain G_n given in dB/floor is known and 2) the height gain G_h given in dB/m is known. In the first case n indicates the floor number, while the antenna height h is used in the second case. The parameters p, W_i, d_{in}, α are similar to the corresponding definitions in the LOS condition given above. Typical values for the correction factor W_{ge} are $3 - 5$ dB at 900 MHz, and $5 - 7$ dB at 1800 MHz. The height gain G_h is $1.5 - 7$ dB/floor at both 900 and 1800 MHz depending on the building type and height, for example, larger values should be used with taller buildings. The height gain $G_h \sim 1.1 - 1.6$ dB/m at both 900 and 1800 MHz. The other parameters take on similar values as in the COST 231 LOS model above.

5.3.6 Excess Path Loss Building Penetration Models

Combining (5.11) and (5.12) we obtain

$$L_e = L_{bl} + X_\sigma. \tag{5.32}$$

where L_{bl} is the building penetration loss (mean excess path loss) obtained by averaging the excess path loss over the large scale fading. Hence, the excess path loss depends on large-scale fading X_σ and therefore conditioned on $L_{e,tot}$, it is also Gaussian distributed variable.

An extensive building penetration loss (excess path loss) measurement campaign in an office-building environment is given in [13] for frequencies between 812 MHz, 2.2 GHz, 4.7 GHz and 8.45 GHz. The data analysis applied a modified COST 231 model to compute the mean building penetration path loss

$$L_{bl} = W + \alpha d_{in} + \alpha_f \log(f) + \alpha_{LOS}\mu - G_h h, \tag{5.33}$$

where $W = W_e + W_{ge}$ as compared to the COST231 model, α is the indoor path loss coefficient introduced above, α_f is a frequency path loss coefficient, α_{LOS} is the LOS coefficient used to include data for LOS condition, that is, $\mu = 1$ for LOS data and $\mu = 0$ for NLOS data. Multiple regression data fitting was applied to all the measurement data, which resulted in the following values of the model parameters: $W = 11.5$ dB, $\alpha = 0.6$ dB/m, $\alpha_f = -2.1$, $\alpha_{LOS} = -0.8$, $G_h = 0.5$ dB/m. Mean building penetration loss was found to be constant for all the measured frequency as in [14]. The mean building penetration loss was found to be independent from frequency and equal $L_{abl} = 10$ dB.

Another measurement campaign investigated an outdoor-to-indoor macrocellular office-building environment at frequencies: 460 MHz, 880 MHz, 1860 MHz and 5.1 GHz [14]. The outdoor antenna was placed on the roof of a 29 m tall building to cover distances within $20 - 620$ m from the antenna. The estimated L_e was confirmed to be Gaussian distributed with mean $L_{abl} = 30$ dB. It was also found that L_{abl} did not vary significantly with frequency between 460 MHz and 1860 MHz. However, it was 5 dB higher at 1.8 GHz 5.1 GHz as compared to the other frequencies. At all frequencies, the estimated standard deviation of the measured excess path loss was found to be 8 dB. It was further found that the building penetration loss decreases as the floor height increases as reported in previous studies, [11, 12].

Based on an empirical building penetration loss model based on the transition between floor levels for which the building is in NLOS and in LOS condition with respect to the outdoor antenna is proposed in [14]

$$L_{bl} = L_{bl,NLOS} + L_{bl,LOS}, \tag{5.34}$$

$$L_{bl,NLOS} = \frac{10}{a} \log \left(1 + \exp(-0.23a(n - n_{LOS}))\right) G_n, \tag{5.35}$$

where $L_{bl,LOS}$ is the building penetration loss for the LOS conditions that will prevail at higher floors, $L_{bl,NLOS}$ is the building penetration loss in the NLOS conditions, which decreases with the floor number n, n_{LOS} is the lowest floor number for which there is LOS towards the transmitter, a is a parameter used to tune the size of the model transition zone between LOS and NLOS conditions. The obtained values for G_n are $2 - 4$ dB/floor and for $L_{bl,LOS}$ are within $20 - 35$ dB, which fall in the range found in literature (e.g. see [15, 16]). Figure 5.4 shows building penetration loss as a function of the floor number for three different sets of model parameters where $a = 1$ was used. As we can see from the plots, the building penetration loss approaches the LOS building penetration loss value as the floor number increases.

5.3.7 Extended COST 231 WI Building Penetration at the LOS Condition

In [17], the COST 231 building penetration loss at the LOS condition (5.23)–(5.26) was extended to cover frequencies in the range $800 - 2000$ MHz for macrocells. The proposed

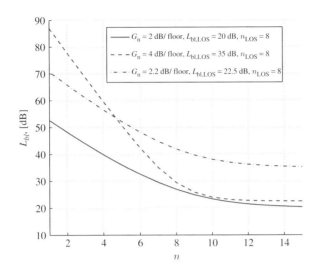

Figure 5.4 Building penetration loss v.s. floor number according to (5.34) and (5.35).

model is given by

$$L_{\text{tot}} = L_{\text{out}} + L_{\text{tw}} + L_{\text{in}}, \tag{5.36}$$

$$L_{\text{out}} = 42.6 + 26\log(S) + 20\log(f), \tag{5.37}$$

$$L_{\text{in}} = \alpha d_{\text{in}}, \tag{5.38}$$

$$L_{\text{tw}} = W_{\text{e}} + WG_{\text{e}}(1 - \sin(\theta))^2. \tag{5.39}$$

Two main modifications were introduced as compared to the COST 231 building penetration loss at the LOS condition: 1) the free-space path loss (5.24) was replaced by the LOS COST231 Walfisch-Ikegami model (5.37), [8] and 2) the indoor propagation path loss L_{in} (5.25) was replaced by the simpler expression in (5.38), [18]. In (5.37), f is the transmit frequency in MHz and S is given in km. Since (5.37) only considers the distance between the outdoor antenna and the exterior wall S, it could be more accurate than (5.24) due to reduced uncertainties. The simplification (5.38) implies that the only information required is the distance d_{in}; it is also assumed that there is only one internal wall per 10 metres distance. The outdoor antenna heights can be $4 - 50$ m and the distance range is $S \in [0.02, 5]$ km. The model parameters are $\alpha = 0.6$ dB/m, $W_{\text{e}} = 8$ dB and $WG_{\text{e}} = 23$ dB.

5.3.8 WINNER II Outdoor-to-Indoor Path Loss Models

The WINNER II (Wireless World Initiative New Radio phase II) pursued a comprehensive outdoor-to-indoor propagation modelling effort to cover frequencies between 2 GHz and 6 GHz, [19]. The developed models are a necessary complement to the COST 231 building penetration models.

5.3.8.1   WINNER II B4 LOS Model

The proposed model for the total mean path loss L_{tot} in a microcell follows the three terms approach (5.7) as presented in [19]

$$L_{tot} = L_{out} + L_{tw} + L_{in}, \tag{5.40}$$

$$L_{out} = 41 + 22.7 \log(S + d_{in}) + 20 \log\left(\frac{f}{5}\right), \tag{5.41}$$

$$L_{in} = \alpha d_{in}, \tag{5.42}$$

$$L_{tw} = W_e + WG_e(1 - \sin(\theta))^2. \tag{5.43}$$

where L_{out} is evaluated at the distance $S + d_{in}$ according to Figure 5.2. The model is based on urban/suburban microcell measurements in the frequency range $2 - 5$ GHz (suggested to be valid for up to 6 GHz) and distances $3 < S + d_{in} < 1000$ m, for outdoor antenna heights of 10 m and mobile station height $3n + 1.5$ m, where n is the floor number. The model parameters are the following: $W_e = 14$ dB, $WG_e = 15$ dB, $\alpha = 0.5$ dB/m, and $\sigma = 7$ dB. It is worthwhile noticing that L_{tw} keeps the same angle dependence as in the COST 231 LOS model.

5.3.8.2   WINNER II C4 NLOS Model

The proposed total mean path loss model L_{tot} in a macrocell follows the three terms expansion (5.7), [19]

$$L_{tot} = L_{out} + L_{tw} + L_{in}, \tag{5.44}$$

$$L_{out} = 26.46 + [44.9 - 6.55 \log(h_{BS})] \log(S + d_{in}) + \tag{5.45}$$

$$20 \log\left(\frac{f}{2}\right) + 5.83 \log(h_{BS}),$$

$$L_{in} = \alpha d_{in}, \tag{5.46}$$

$$L_{tw} = W_e' - hG_h. \tag{5.47}$$

The model is based on urban macrocell measurements in the frequency range $2 - 5$ GHz and distances $50 < S + d_{in} < 5000$ m, for outdoor antenna heights of 25 m and mobile station height $3n + 1.5$ m, where n is the floor number. The model parameters are the following: $W_e' = 17.4$ dB, $G_h = 0.8$ dB/m, $\alpha = 0.5$ dB/m, and $\sigma = 10$ dB. It is worthwhile noticing that L_{tw} keeps a similar antenna height gain as in the COST 231 NLOS model.

5.3.8.3   An Example of Semi-Empirical Model Implementation

The accuracy of the empirical models is evaluated by the standard deviation of the large-scale fading component, which is evaluated as the fitting residual of the model to the measurement data. Typical values are within the interval $5 - 15$ dB. This is usually not a problem for system level simulations since empirical models provide fast computations and the outcome of the simulations is interpreted in terms of statistical figures of merit.

Usually, the poor accuracy is primarily a result of not handling site-specific information on the propagation environment well. To improve the accuracy, some environmental information can be considered, which yields the semi-empirical models. The semi-empirical models offer a higher level of accuracy than the empirical models and their execution time is still acceptable within minutes, even seconds, on a standard PC. For example, the COST231-Multi Wall model is a typical semi-empirical propagation model designed for indoor scenarios given below, [20]

$$L_{\text{multiwall}} = L_{\text{fs}} + \sum_{i=1}^{N_{\text{walls}}} \alpha_i W_i + \sum_{j=1}^{N_{\text{floors}}} \beta_j F_j + C, \tag{5.48}$$

where the attenuation of transmissions through floors and/or walls/doors is accumulated, that is, it depends on the cumulative losses of the number N_{walls} and N_{floors} of walls and floors, respectively, that are encountered between the transmitter and receiver (note the similarity with the partition model (5.18)). L_{fs} is the free space path loss (5.6) based on the TX-RX separation distance and transmit frequency. α_i and β_j denote the coefficients for transmission through walls and floors, respectively. W_i and F_i are the losses for each wall i and floor j, respectively. The constant C is used for calibration purposes. The multiwall model is fast as it only requires the computation of the number of wall or floor penetrations. An example of a coverage prediction obtained by Ranplan iBuildNet®, [6] is shown in Figure 5.1(b). The plots represent the prediction of Multi Wall model for three transmitters at 2.4 GHz for WiFi applications.

However, the Multi Wall model may produce incorrect predictions under different circumstances. For example, in the near field, where the receiver is separated a few metres away from the transmitter or if there is only one or a few walls/floors between, the path loss computation from the Multi Wall model may seem reasonable. On the contrary, in the far field, where the direct ray between the receiver and the transmitter penetrates a considerable number of walls or floors, the prediction given by the Multi Wall model tends to be pessimistic. In this case, the dominant contribution to the signal strength comes from multiple reflections and/or diffractions, which the Multi Wall model fails to predict. To improve the accuracy of the Multi Wall model, the transmission coefficients can be tuned according to the distance separation, that is, α and β decrease as the distance increases.

5.4 Deterministic Models

The physical properties of the environment have a large influence on the signal propagation. In order to accurately capture these characteristics, deterministic approaches can be adequate. They often rely on numerical computations of the solutions to Maxwell's equations or computations based on the Geometrical Optics theory [21]. The deterministic models can offer a higher level of prediction accuracy as compared to empirical path loss models at the expense of high computational burden. However, these models need to be calibrated against measurement data to perform with good accuracy. Indeed, deterministic models require an accurate characterization of the electrical and geometrical properties of the physical propagation environment. Lately, the deterministic models have gained an increase in popularity due to the advancement of computational technology such as Graphics Processing Unit (GPU) technology that allows for efficient parallel code implementation, [22].

Deterministic models considered here are divided into two categories:

- Finite-Difference Time-Domain (FDTD);
- Ray-based methods.

5.4.1 FDTD

FDTD is a differential equation solver for Maxwell's equations. The first FDTD algorithm can be tracked back to a paper by Kane Yee in 1966 [23]. FDTD is a discrete space-time method, that is, the computational domain is discretised in both time and space. At each discrete spatial point, the electric field E and magnetic field H are computed forward in the time domain. An important feature of the FDTD method for propagation channel predictions is that it implicitly considers the reflections and diffractions that result in both multipath and large-scale fading. However, it is known that the grid size has to be less than one tenth of wave length. Hence, computations of large volumes are still not feasible with FDTD due to memory limitation since each cell requires about 30 bytes storage. Another important feature of the FDTD algorithm is that multiple sources can be computed simultaneously without extra computational efforts and suitable for broadband calculation required, for example, for UWB channel predictions. A fundamental issue of the implementation of the FDTD algorithm is the numerical boundary condition required to limit the computational domain. Appropriate Absorbing Boundary Conditions (ABC) need to be defined for the FDTD computation in order to decrease spurious reflections to the computational domain. Another important feature of the FDTD is that it is straight-forwardly parallelizable. Many previous efforts have developed parallel FDTD models to accelerate the computation, for example, via the use of GP-GPU (General Purpose Graphic Processing Unit) [24]. More details on Finite-Difference Time-Domain methods will be provided in Chapter 11.

5.4.2 Ray-Based Methods

The geometry-based methods rely on the path finding algorithms that utilize GO principles to handle the different paths that a transmitted signal takes under its way to the receiver. These paths or rays undergo reflections, diffractions, transmission attenuation through materials and scattering before they reach the receivers. These methods can be divided into three categories:

- Ray launching;
- Ray tracing;
- Dominant path.

The ray launching model is a ray-sampling method. Ray launching emits the rays in all directions with a separation angle as shown in Figure 5.5(a). It can be proved that no matter how small the angle is, rays disperse as they propagate, which leads to missed pixels, especially in the far field. To compensate for this, usually a reception sphere is employed to capture the nearby rays. Ray launching is suitable for point-to-area coverage prediction. The computation of the path loss of a pixel can usually be referenced to its neighboring pixel on the same ray. The complexity of the ray launching is almost independent of the size of the computational domain (scenario) or the number of objects in the scene. However, the complexity grows as the number of ray configurations, for example, reflections or diffractions, is incremented.

The ray tracing model computes the rays in a backward manner. It computes the deterministic path between the transmitter and receiver. For example, the reflection point can be computed by intersecting the reflected object and the line between the transmitter's mirror point and the receiver as shown in Figure 5.5(b). This also ensures the shortest path between the transmitter and the receiver. Therefore, compared to the ray launching method, the ray tracing provides a more accurate result. However, the execution time of ray tracing algorithms is usually longer than ray launching algorithms. The ray tracing is suitable for point-to-point predictions, where exact paths can be computed. However, for full coverage predictions, ray tracing becomes rather time consuming. Many previous studies, such as the one presented in [25] and [26] have proposed methods to accelerate the computation of the ray tracing algorithms. For example, the "Intelligent Ray Tracing" method employs the *one-time pre-processing* which stores the visibility relations between buildings used to accelerate the computations.

The ray-based methods are generally time-consuming since they require complex ray-object intersection tests. However, methods have been developed to accelerate these computations. For example, in [27] and [28], the authors proposed a discrete ray launching model which eliminates the need of the ray-object intersection tests. In [29], the authors presented a method to preprocess the environment. Namely, according to this method, the walls are discretised into tiles and the visibility relations are built and stored only once. This way, the speed of the ray-object intersection tests runtime can be considerably increased. Figure 5.6 shows the tree searching algorithm based on the preprocessed visibility tree. Ray-object intersection tests can be accelerated by looking up the tree.

In most cases, the signal strength is made of just one or few dominant rays. Therefore, the dominant path model was developed based on this idea and presented in [30]. First,

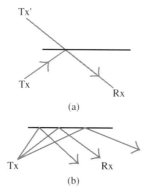

(a)

(b)

Figure 5.5   Ray-based methods. (a) Ray launching principle and (b) Ray tracing principle.

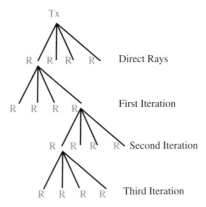

Figure 5.6   Tree searching using the preprocessing.

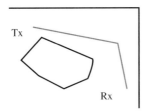

Figure 5.7   Dominant path model principle. From [30].

the scenario is analyzed and then the multiple dominant paths are computed and merged into one or few dominant paths. The path loss computation is then based on the identified dominant paths (Figure 5.7). The wave guiding effect can also be included in the path loss computation. The computation of the path loss (in dB) along an analyzed path can be formulated as

$$L = 20 \log \left(\frac{4\pi}{\lambda} \right) + 20 p \log(d) + \sum_{i=0}^{n} \alpha(\phi, i) - \frac{1}{c} \sum_{k=0}^{c} w_k, \tag{5.49}$$

where λ is the wavelength corresponding to the transmit frequency. The path loss exponent p depends on the visibility state between the current pixel and the transmitter, it can be adaptively changed to reflect the current state, that is, whether the LOS or the NLOS condition is prevalent and then change it accordingly. The $\alpha(\phi, i)$ expresses the changes of the ray interaction with the environment, that is, change of direction of the ray. The parameter w_k is the wave guiding factor. The clutter and terrain information is also considered in the dominant path model as this affects the visibility p.

The main two advantages of the dominant path model are: 1) obviously, the computation of a few dominant rays is faster and less time consuming than the ray tracing or ray launching model and 2), the dominant path model is less sensitive to inaccuracies of the building database, which in turn may influence the accuracy of prediction of the ray tracing and the ray launching models. Indeed, a slight change of the positions of the obstacles will lead to totally different prediction results. Unfortunately, accurate building databases are usually not available in most cases. Chapter 10 will fully focus on ray based methods. However in the next paragraph a few details about one particular implementation are given.

5.4.3   Intelligent Ray Launching Algorithm (IRLA)

In [27], the authors have proposed a highly computationally efficient ray launching algorithm (computations are performed within a few minutes) designed for outdoor scenarios and denominated IRLA. The standard deviation of the error was approximately 8 dB when applied to the COST-Munich scenario, [31]. The ray launching algorithm was optimized to capture most dominant rays. Later, the authors extend the model to indoor coverage computations, [28, 32]. Various acceleration techniques have been employed in combination with the ray launching algorithm, among others, code parallelization have been shown to greatly shorten computation times, [33–36]. Several combined models have been implemented to cover the outdoor, indoor, outdoor-to-indoor and indoor-to-outdoor propagation predictions, [36–40].

In the IRLA, the specified environment is discretised (rasterised) into a large number of cubes. Each cube is associated with information about the environment, that is, whether it is a wall cube, a ground cube or rooftop cube. This information is gathered prior to computations. The simulation algorithm consists of three main components that take into account different propagating phenomena:

- Collecting LOS cubes;
- Intelligent Vertical Diffraction (VD);
- Intelligent Horizontal Reflection and Diffraction (HRD).

The LOS cubes step is responsible for computing the direct paths visible to the transmitter and collecting the secondary cubes to launch reflections and/or diffractions. The HRD component recursively traces the rays until the signal strength falls below a threshold or the ray iteration has reached a limit. The VD component is based on a fast pixel checking procedure that computes the number of rooftop diffractions. Finally, the HRD component launches 3D rays undergoing reflections and diffractions. The 3D ray launching model is based on a discrete data set of size (N_x, N_y, N_z) that represents the number of cubes in X, Y, and Z dimensions, respectively. Hence, the total number of pixels is

$$N_{\text{total}} = N_x N_y N_z. \tag{5.50}$$

However, the total number of discrete rays launched from the transmitter to the cubes at the periphery is

$$N_{\text{rays}} = 2N_x N_y + 2(N_z - 2)(N_x + N_y - 2), \tag{5.51}$$

where $N_z > 1$. N_{rays} gives the total number of rays required to launch at start to speed up computations in a parallelization code implementation.

5.5   Hybrid Models

Hybrid propagation models combine two or more propagation modelling approaches. Great performance improvement can be obtained when applied to complex scenarios, such as the indoor-to-outdoor or outdoor-to-indoor propagation scenarios. For example, in [37], the authors proposed a hybrid propagation model that uses the IRLA algorithm for outdoor path loss predictions and a FDTD-like method, the Multi Resolution Frequency Domain Parflow (MR-FDPF), to predict the indoor propagation path loss. Figure 5.8 shows a schematic representation of the hybrid propagation method for outdoor-to-indoor path loss prediction. As we can see from Figure 5.8, the outdoor rays hitting the floor level of a target building are converted into indoor flows for the MR-FDPF propagation model. The advantage of using the hybrid model in this case is that it combines the best features of the models used: 1) the ray launching model is suitable for outdoor scenarios, where FDTD-like methods may not be computationally efficient due to memory limitation, 2) the ray launching model does not need the details of the indoor structure of every building since only the details of the indoor structure of the target building are required, and 3) the MR-FDPF provides accurate predictions of the indoor path loss based on the flows converted from the outdoor rays.

5.5.1   Antenna Radiation Pattern

Transmit and receive antennas have radiation patterns that need to be taken into account in path loss computations, for example, as given by (5.4). The antenna gains G_{tx} and G_{rx}

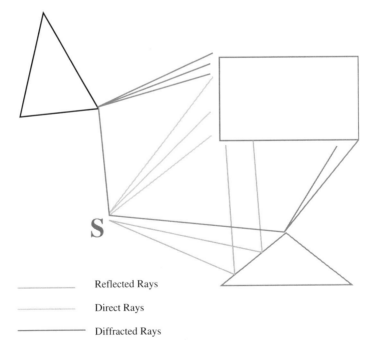

Figure 5.8   The hybrid scheme of outdoor-to-indoor propagation modelling.

of the transmit and the receive antenna, respectively, take on different values at different directions. At low frequencies, for example, 900 MHz, the radiation patterns of mobile handsets are more omnidirectional and the maximum gain can be -3 to 2 dBi, while at higher frequencies the mobile antennas may show distinct lobes at some discrete directions and also higher directivity. Wireless Fidelity (WiFi) access points usually have 3 to 5 dBi or sometimes higher. On the other hand macrocell and microcell base station antennas are usually much more directive and have considerably higher gain, for example, in the order of 18 dBi.

The antenna patterns are usually given in two 2D cuts: one horizontal plane contains gains for roughly 360° in the azimuth and similarly for the gain in the vertical plane. In this case, a full 3D pattern reconstruction is necessary. Several methods have been proposed in the past to reconstruct the full 3D pattern from the 2×2D antenna gain matrix, for example, by nonlinear interpolation methods [41]. A classical method to interpolate the full 3D pattern is to add the antenna gain in the horizontal and vertical cuts. However, a comparison with field measurements has shown the limitation of such a method, especially for directive antennas that are either electrically or mechanically tilted. The RRPS employs an enhanced nonlinear interpolation method that fits optimally the measurements, [42]. The improvements obtained were 4 dB in mean error and 3 dB in standard deviation.

In practice, the antenna pattern gains depend on direction and the path loss budget (5.4) can be written as $P_{rx} = P_{tx} - L + G_{rx}(\phi_{tx}, \theta_{tx}) + G_{tx}(\phi_{rx}, \theta_{rx})$, where $G(\phi, \theta)$ is the antenna gain in the azimuthal direction ϕ and the elevation angle θ. In particular, the full radiation pattern of the antenna should be used to dynamically adjust the ray gains in geometry-based models, that is, the gains are added to each ray launched or traced.

5.5.2 Calibration

There are many unknowns and uncertainties related to propagation path loss modelling. For example, it is not always possible to obtain the exact material properties of the buildings. Other factors can be moving objects in the measurement environment whose effect on the measured data has to be minimized. In order obtain realistic simulations a calibration must be performed based on measurement data of the simulated environment. Outcomes of the calibration process are, for example, the electrical parameters of the building materials or path loss model parameters, which can be used for further simulations in the same scenario. Hence, the measurements are compared to the simulations in order to minimize the Root Mean Square Error (RMSE)

$$\text{RMSE} = \sqrt{\frac{1}{K} \sum_{k=1}^{K} (M_k - P_k)^2}, \qquad (5.52)$$

where K is the number of measurement points; M_k denotes the measured signal and P_k denotes the predicted signal. The mean error can also be computed and is interpreted as the offset between the measurement and the model

$$\Delta\Psi = \frac{1}{K} \sum_{k=1}^{K} (M_k - P_k). \qquad (5.53)$$

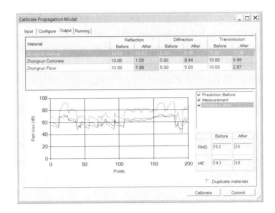

Figure 5.9   Simulated-annealing based calibration.

The calibration process seeks to minimize RMSE within the solution domain (finite or infinite). For example, the ray launching model presented in [28] employs a meta-heuristic approach based on the Simulated-Annealing. Figure 5.9 shows a calibration example that was implemented by Ranplan iBuildNet(R) using the simulated annealing approach for the ray launching model presented in [27, 28].

5.5.3   IRLA Case Study: INSA

This section presents simulations performed at the 2300 MHz LTE band on the INSA campus (Lyon, France) building database. The size of the scenario is roughly 800 × 560 m². The indoor coverage for the target building is also simulated, and covers an area of approximately 110 × 100 m² at 11 m above ground. In Figure 5.10, the dot gives the position of the transmit antenna, which is pointing towards the building in the direction of the arrow. The antenna is attached to the window, which is roughly 5.6 m above ground. The transmit power is 20 dBm and maximum antenna gain is 10 dBi, while the receive antennas are assumed to be omni-directional with a gain of 3 dBi. The performance of various outdoor-to-indoor propagation models is compared based on the signal strength level of points inside the building illuminated by the transmit antenna.

Three simulations were carried out on a standard PC (4GB RAM, AMD 64 Dual): 1) One Slope model, 2) Multi Wall model, and 3) Ray Launching model. The running time is listed in Table 5.1. The One Slope model is the fastest of the three because it is the simplest empirical propagation path loss (see 5.15). The next fastest was the Multi Wall model which needs to compute the number of transmissions between the transmitter and the receivers. Then came the Ray Launching model due to complex computation of combined reflections and diffractions, where the maximum number of reflections and diffractions was set to 5 and 5, respectively. The FDTD(MR-FDPF) took roughly 6 minutes including the one-time calibration preprocessing. Due to memory limitations, an artificial frequency of 120 MHz was used instead of a real frequency 2300 MHz. A calibration is thus necessary in this case to obtain a reasonable prediction accuracy using FDTD. It is worthwhile noticing that only 2-D predictions were possible using the MR-FDPF method in this scenario.

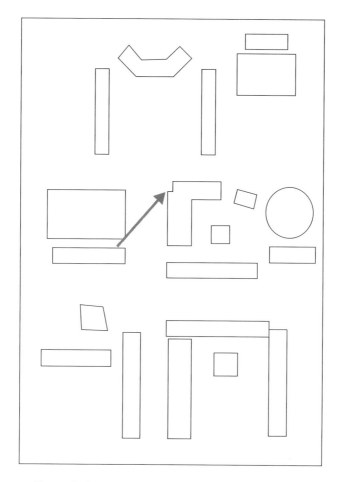

Figure 5.10 Coverage prediction using FDTD model.

Table 5.1 Running Time (s)

One Slope	Multi Wall	Ray Launching	FDTD
< 1	< 10	< 60	< 360

The coverage predictions for the One Slope model, the Multi Wall, the Ray Launching model and the MR-FDPF method are shown in Figure 5.11(a), 5.11(b), 5.11(c) and 5.12, respectively. From the figures, it can be seen that the One Slope model does not take buildings into account and therefore there is no attenuation caused by the buildings and hence no large-scale fading may be predicted. The predictions are obviously optimistic. The Multi Wall model provides a reasonable prediction in the near field when the rays undergo a small number of transmissions. However, in the far field, the prediction seems too pessimistic since other contributions from reflections and diffractions are

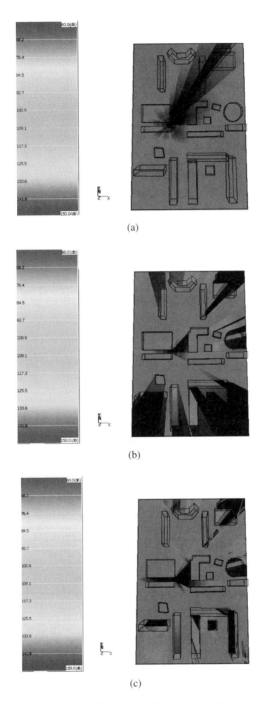

Figure 5.11   Scenario of INSA, Lyon, France. (a) Coverage prediction using One Slope model, (b) coverage prediction using Multi Wall model and (c) coverage prediction using Ray Launching model.

Figure 5.12   Coverage prediction using MR-FDPF model.

not considered. The Ray Launching model in this case overcomes this problem and manages to provide a reasonable prediction at the cost of increased computational burden. Finally, in Figure 5.12, the prediction using the MR-FDPF is given, which shows a finer resolution of the prediction, that is, the large-scale fading and small-scale fading effects can be clearly seen. The coverage area prediction is clearly affected by the reflections and diffractions of the multipath components. However, as the MR-FDPF

Figure 5.13   3D Xinghai coverage scenario.

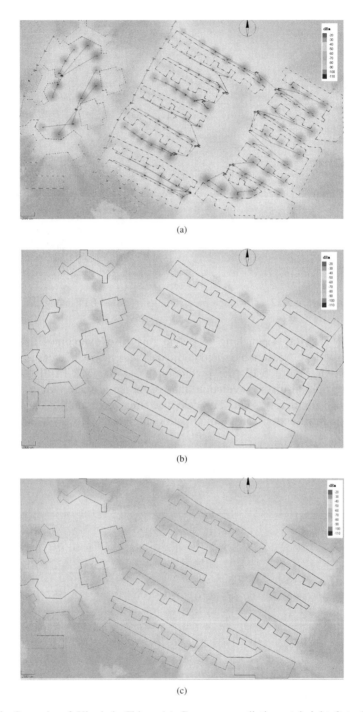

Figure 5.14 Scenario of Xinghai, China. (a) Coverage prediction at height 0 to 3 metre, (b) coverage prediction at height 6 to 9 metre and (c) Coverage prediction at height 15 to 18 metre.

currently is only implemented in 2D, the vertical rays traveling over roof-tops are not present. Moreover, a large measurement campaign should be undertaken to investigate the accuracy of these methods in more detail. More details on MR-FDPF and this combined model will be provided in Chapter 11.

5.5.4   IRLA Case Study: Xinghai

This section presents simulations of the outdoor-to-indoor scenario for LTE at 2300 MHz. The objective was to study the coverage of a DAS (Distribute Antenna Systems) placed on street lamps to buildings at different floors and on the roof top of buildings. Figure 5.13 shows the scenario of Xinghai, China, where the size is approximately $500 \times 500 \, m^2$. The transmit antennas are installed in the street lamps at 2.4 m above ground. The transmit power level is 25 dBm.

Using Ranplan iBuildNet® and RRPS, [6], a coverage prediction is carried out on a PC (4GB, AMD Athon 4600+). The computation time takes approximately 30 seconds for each antenna, which is by far more computationally efficient than other deterministic methods, for example, ray-tracing or FDTD models.

Coverage predictions at heights from 0 to 3 metres, from 6 to 9 metres and from 15 to 18 metres are plotted in Figure 5.14(a), Figure 5.14(b) and Figure 5.14(c), respectively. The attenuation for each antenna due to the first penetration into the building ranges from 9 dB to 20 dB depending on the material properties and incident angle of rays. There is a clear trend of path loss increase with the height, which indicates that there might not be sufficient indoor coverage at higher floors from the outdoor street antennas. This is due to the longer distance and due to the penetration losses caused by the building walls. There are several methods to resolve these issues:1) at first, increasing the transmit power might be an easy solution. However, this might also cause interference to other users such as femtocells. In addition it may increase issues related to health concerns, 2) instead of using omni-directional antenna, one can use directional antenna which targets a particular building of interest. A directional antenna, compared to an omni-directional antenna, can provide better coverage given the direction of interests, and 3) indoor networks such as femto cells or indoor DAS can provide better indoor coverage. All these issues require reliable and accurate path loss predictions. Therefore, research on outdoor-to-indoor propagation and vice versa is at its beginning.

Acknowledgements

The work is supported by the iPLAN project. Special thanks have to be given to Dr. Guillaume De La Roche, Professor Jean-Marie Gorce and Meiling Luo.

References

1. J. Zhang and G. de la Roche, *Femtocells:Technologies and Deployment*. John Wiley & Sons, New York, NY, USA, 2010.
2. Y. Okumura, E. Ohmori, T. Kawano, and K. Fukuda, "Field Strength and Its Variability in VHF and UHF Land-Mobile Radio Service," *Review of the Electrical Communication Laboratory*, vol. 9-10, pp. 825–873, September-October 1968.

3. M. Hata, "Empirical formula for propagation loss in land mobile radio services," vol. 29, no. 3, pp. 317–325, 1980.

4. A. F. Molisch, *Wireless Communications*, 2nd ed. IEEE Press - Wiley, 2011.

5. G. Durgin, T. S. Rappaport, and H. Xu, "Radio path loss and penetration loss measurements in and around homes and trees at 5.85 ghz," in *Proc. IEEE Antennas and Propagation Society Int. Symp*, vol. 2, 1998, pp. 618–621.

6. U. Ranplan Wireless Network Design Ltd., " http://www.ranplan.co.uk," ranplan iBuildNet and RRPS.

7. A. Valcarce and J. Zhang, "Empirical indoor-to-outdoor propagation model for residential areas at 0.9-3.5 GHz," *IEEE TRANSACTIONS ON ANTENNAS AND PROPAGATION*, vol. 9, pp. 682–685, 2010.

8. E. C. in the Field of Scientific and T. Research, "Digital mobile radio towards future generation systems, COST231 final report," 1999, http://www.lx.it.pt/cost231/.

9. A. A. Glazunov, L. Hamberg, J. Medbo, and J.-E. Berg, "Building shielding loss measurements and modelling at the 5 ghz band in office building areas," in *Proc. 52nd Vehicular Technology Conf. IEEE VTS-Fall VTC 2000*, vol. 4, 2000, pp. 1874–1878.

10. A. A. Glazunov and J.-E. Berg, "Building-shielding loss modelling," in *Proc. IEEE 51st VTC 2000-Spring Tokyo Vehicular Technology*, vol. 3, 2000, pp. 1835–1839.

11. J. Berg, *Digital Mobile Radio Toward Future Generation Systems*, 1997.

12. J. Berg, "4.6 building penetration: Digital mobile radio toward future generation systems," in *COST Telecom Secretariat, Commission of the European Communities*, Brussels, Belgium, 1999, pp. 167–174.

13. H. Okamoto, K. Kitao, and S. Ichitsubo, "Outdoor-to-indoor propagation loss prediction in 800-mhz to 8-ghz band for an urban area," vol. 58, no. 3, pp. 1059–1067, 2009.

14. J. Medbo, J. Furuskog, M. Riback, and J.-E. Berg, "Multi-frequency path loss in an outdoor to indoor macrocellular scenario," in *Proc. 3rd European Conf. Antennas and Propagation EuCAP 2009*, 2009, pp. 3601–3605.

15. E. F. T. Martijn and M. H. A. J. Herben, "Characterization of radio wave propagation into buildings at 1800mhz," vol. 2, no. 1, pp. 122–125, 2003.

16. A. F. De Toledo, A. M. D. Turkmani, and J. D. Parsons, "Estimating coverage of radio transmission into and within buildings at 900, 1800, and 2300mhz," *IEEE Personal Communications*, vol. 5, no. 2, pp. 40–47, 1998.

17. E. Suikkanen, A. Tolli, and M. Latva-aho, "Characterization of propagation in an outdoor-to-indoor scenario at 780MHz," in *2010 IEEE 21st International Symposium on Personal Indoor and Mobile Radio Communications*, Istanbul, Turkey, September 2010.

18. M. Alatossava, E. Suikkanen, J. Meinila, V. Holappa, and J. Ylitalo, "Extension of COST 231 path loss model in outdoor-to-indoor environment to 3.7ghz and 5.25ghz," in *Int. Symp. Wireless Pers. Multimedia Commun*, Sarriselka, Finland, September 2008.

19. P. Kyösti, J. Meinilä, L. Hentilä, X. Zhao, T. Jämsä, C. Schneider, M. Narandzić, M. Milojević, A. Hong, J. Ylitalo, V.-M. Holappa, M. Alatossava, R. Bultitude, Y. de Jong, and T. Rautiainen, "WINNER II channel models," EC FP6, Tech. Rep., Sep. 2007. [Online]. Available: http://www.ist-winner.org/deliverables.html

20. M. Lott and I. Forkel, "A multi-wall-and-floor model for indoor radio propagation," in *IEEE Vehicular Technology Conference*, Rhodes, Greece, May 2001.

21. A. Valcarce, G. de la Roche, L. Nagy, J.-F. Wagen, and J.-M. Gorce, "A new trend in propagation prediction," vol. 6, no. 2, pp. 73–81, 2011.

22. D. De Donno, A. Esposito, L. Tarricone, and L. Catarinucci, "Antennas and propagation magazine, ieee," *Issue: 3*, vol. 52, no. 3, pp. 116–122, 2010.

23. K. Yee, "Numerical solution of initial boundary value problems involving maxwell's equations in isotropic media," *IEEE Transactions on Antennas and Propagations*, vol. 14, no. 3, pp. 302–307, 1966.

24. A. Valcarce, G. De La Roche, and J. Zhang, "A GPU approach to FDTD for radio coverage prediction," in *IEEE 11th International Conference on Communication Systems*, Guangzhou, China, November 2008.

25. G. Wolfle, B. Gschwendtner, and F. Landstorfer, "Intelligent ray tracing - a new approach for the field strength prediction in microcells," in *IEEE Vehicular Technology Conference*, Phoenix, Arizona, USA, May 1997, pp. 790–794.

26. A. Cavalcante, M. De Sousa, J. Costa, C. Frances, and G. Dos Santos Cavalcante, "A parallel approach for 3D ray-tracing techniques in the radio propagation prediction," in *Journal Of Microwaves and Optoelectronics*, June 2007.

27. Z. Lai, N. Bessis, G. De La Roche, H. Song, J. Zhang, and G. Clapworthy, "An intelligent ray launching for urban propagation prediction," in *The Third European Conference On Antennas and Propagation EUCAP*, Berlin, Germany, March 2009, pp. 2867–2871.
28. Z. Lai, N. Bessis, G. De La Roche, P. Kuonen, J. Zhang, and G. Clapworthy, "On the use of an intelligent ray launching for indoor scenarios," in *The Fourth European Conference On Antennas and Propagation EUCAP*, Barcelona, Spain, April 2010.
29. R. Hoppe, G.Wolfle,, and F. M. Landstorfer, "Fast 3-D ray tracing for the planning of microcells by intelligent preprocessing of the data base," in *3rd European Personal and Mobile Communications Conference (EPMCC)*, Paris, France, March 1999.
30. R. Wahl, G. Wolfle, P. Wertz, P. Wildbolz, and F. Landstorfer, "Dominant path prediction model for urban scenarios," in *14th Ist Mobile and Wireless Communications Summit*, Dresden, Germany, 2005.
31. "COST231 urban micro cell measurements and building data," http://www2.ihe.uni-karlsruhe.de/forschung/cost231/cost231.en.html.
32. Z. Lai, G. De La Roche, N. Bessis, P. Kuonen, G. Clapworthy, D. Zhou, and J. Zhang, "Intelligent ray launching algorithm for indoor scenarios," *Radioengineering: Towards EuCAP 2012: Emerging Materials, Methods, and Technologies in Antenna & Propagation*, vol. 20, no. 2, June 2011.
33. Z. Lai, N. Bessis, P. Kuonen, G. De La Roche, J. Zhang, and G. Clapworthy, "A performance evaluation of a grid-enabled object-oriented parallel outdoor ray launching for wireless network coverage prediction," in *The Fifth International Conference On Wireless and Mobile Communications*, Cannes/La Bocca, French Riviera, France, August 2009, pp. 38–43.
34. Z. Lai, N. Bessis, G. De La Roche, P. Kuonen, J. Zhang, and G. Clapworthy, "A new approach to solve angular dispersion of discrete ray launching for urban scenarios," in *2009 Loughborough Antennas and Propagation Conference*, Burleigh Court Conference Centre, Loughborough University, United Kingdom, November 2009, pp. 133–136.
35. Z. Lai, N. Bessis, G. De La Roche, P. Kuonen, J. Zhang, and G. Clapworthy, "The development of a parallel ray launching algorithm for wireless network planning," *International Journal of Distributed Systems and Technologies, IGI*, vol. 2, no. 2, 2010.
36. Z. Lai, H. Song, P. Wang, H. Mu, L. Wu, and J. Zhang, "Implementation and validation of a 2.5d intelligent ray launching algorithm for large urban scenarios," in *The Sixth European Conference on Antennas and Propagation EUCAP*, Prague, Czech Republic, March 2012.
37. G. De La Roche, P. Flipo, Z. Lai, G. Villemaud, J. Zhang, and J. Gorce, "Combination of geometric and finite difference models for radio wave propagation in outdoor to indoor scenarios," in *The Fourth European Conference On Antennas and Propagation EUCAP*, Barcelona, Spain, April 2010.
38. G. De La Roche, P. Flipo, Z. Lai, G. Villemaud, J. Zhang, and J. Gorce, "Combined model for outdoor to indoor radio propagation," in *10th COST2100 Management Meeting, TD(10)10045*, Athens, Greece, February 2010.
39. G. D. Roche, P. Flipo, Z. Lai, G. Villemaud, J. Zhang, and J. Gorce, "Implementation and validation of a new combined model for outdoor to indoor radio coverage predictions," *Hindawi Publishing Corporation EURASIP Journal on Wireless Communications and Networking*, 2010.
40. D. Umansky, G. D. Roche, Z. Lai, G. Villemaud, J. Gorce, and J. Zhang, "A new deterministic hybrid model for indoor-to-outdoor radio coverage prediction," in *The Fifth European Conference On Antennas and Propagation EUCAP*, Rome, Italy, April 2011.
41. Y. Corre and Y. Lostanlen, "Methods to extrapolate 3D antenna radiation pattern ensuring reliable radio channel predictions," in *8th COST2100 Management Meeting, TD(08)426*, Wroclaw, Poland, February 2008.
42. T. G. Vasiliadis, A. G. Dimitriou, and G. D. Sergiadis, "A novel technique for the approximation of 3-d antenna radiation patterns," vol. 53, no. 7, pp. 2212–2219, 2005.

6

Vehicular Channels

Laura Bernadó[1], Nicolai Czink[1], Thomas Zemen[1], Alexander Paier,
Fredrik Tufvesson[2], Christoph Mecklenbräuker[3] and Andreas F. Molisch[4]

[1]*Forschungszentrum Telekommunikation Wien, Austria*
[2]*Lund University, Sweden*
[3]*Vienna University of Technology, Austria*
[4]*University of Southern California, USA*

6.1 Introduction

Nowadays, mobility has become a need and has brought a large number of vehicles circulating on the streets. Unfortunately, it has also brought an increase in deaths on the roads, as well as an increase of CO_2 emissions, both far from ideal side effects. Communications from Vehicle-to-Vehicle (V2V) and Vehicle-to-Infrastructure (V2I), where the infrastructure is controlled by the road operator, is envisioned to provide information about the traffic flow such that the traffic accidents rate could be reduced, and vehicles could drive in a more environmentally friendly way.

Long Term Evolution Advanced (LTE-A) supports high mobility and this makes the vehicular environment a good candidate for implementing V2V and V2I communications. However, some applications, such as hazard notification or accident mitigation, impose demanding time constraints, which are difficult to meet in cellular communication systems. Based on this consideration, ad-hoc networks are considered as the main communications technology.

The radio channel observed in V2V and V2I links is characterized by being highly time-varying and it is described through a *non-stationary* fading process. The low antenna position, the high velocities of Transmitter (TX) and Receiver (RX), and the large number of scattering objects make the propagation characteristics of the radio waves very peculiar. As a result, the deployment of a trustworthy and highly reliable system for vehicular communications is a challenging task.

A deep understanding of the underlying radio propagation channel is needed for setting up an efficient vehicular communication system. In general, computer simulations

LTE-Advanced and Next Generation Wireless Networks: Channel Modelling and Propagation, First Edition.
Edited by Guillaume de la Roche, Andrés Alayón Glazunov and Ben Allen.
© 2013 John Wiley & Sons, Ltd. Published 2013 by John Wiley & Sons, Ltd.

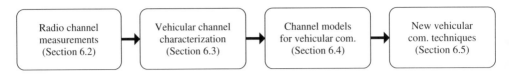

Figure 6.1   Chapter outline.

allow system design, transceiver testing, and so on, in realistic settings without the need for expensive, and difficult to reproduce, field tests. However, such simulations can only be realistic if all their components, including the channel model, reflect the realities of deployment. Hence, system engineers must rely on good channel emulators, for their simulations. Trustworthy channel models can only be derived after observing and investigating the radio channel itself, that is, measuring it. This flow of information from the behaviour of the radio channel to the system design is also the outline of this chapter, as depicted in Figure 6.1.

Vehicular radio channel measurements (Section 6.2) are needed to investigate and characterize (Section 6.3) the propagation of electromagnetic waves. The information gained is then used for deriving proper channel models (Section 6.4) used for developing new communication techniques suitable for vehicular channels (Section 6.5).

6.2   Radio Channel Measurements

Realistic and trustworthy channel models need to be based on radio channel measurements. For the class of Geometry-based Stochastic Channel Model (GSCMs) [1], see Section 6.3, as well as those of purely stochastic models, the parameters have to be obtained from measurement data [2–6]. In the case of deterministic Ray Tracing (RT) models [1], the model parameters need to be carefully chosen to represent the environment correctly. All models have in common that their quality has to be validated by comparison to measurements.

Measuring the radio channel is also known as *channel sounding*: A TX sends out a known training signal that excites (i.e., sounds) the channel. The RX stores the received signal and can thus estimate the radio channel from the known transmit signal [7]. Since communication systems have become more and more complex, also the need for more sophisticated channel models and therefore more sophisticated channel sounding is necessary. In the 1960s, it was sufficient to just measure the field strength. Later on, by the introduction of wideband radio systems, it was necessary that the channel sounder is able to record the Channel Impulse Response (CIR) (i.e., delay dispersion). In the 1990s the focus was on directional propagation properties, which affected the channel sounders in a way to be able to measure double directional CIRs using multiple transmit and receive antennas and very sophisticated antenna arrays.

The requirements for a channel sounder are determined by the considered measurement environment. Before the 1990s, measurement campaigns were usually performed in macro cell outdoor scenarios. Accommodating for indoor users, ad-hoc networks, and smaller cell sizes, measurements in micro cells and also in indoor environments became more popular. Due to the increased bandwidth of the systems, a finer delay resolution was required.

For vehicular communication systems the channel is rapidly time-varying. Therefore, current channel sounders have to be able to measure and store fast fluctuations of the time-varying, wideband, double-directional propagation channel. This section will provide insight into this challenging task. Further information about channel sounding can be found in [8] and [9].

6.2.1   Channel Sounders

There exist different types of channel sounders [8]. Their common property is that the usual output at the RX side is the time-varying CIR $h(t, \tau)$ or the time-varying frequency response $H(t, f)$, which are related via the Fourier transform [10]. The radio channel is completely described by either one of these functions within the time span and bandwidth used for sounding.[1] The parameters of the radio channel, discussed in Section 6.3, as well as the validation of channel models make use of these functions.

6.2.1.1   Narrowband Versus Wideband Channel Sounding

Overall one can distinguish between *narrowband measurement systems* and *wideband measurement systems*.

With *narrowband measurements* one measures the channel gain and Doppler shift experienced by a narrowband (sinusoidal) signal. It is not possible to investigate wideband parameters like frequency selectivity or delay spread. Possible practical realizations for narrowband systems are a sine wave generator at the TX side and a vector signal analyzer [11–13] or a spectrum analyzer [14] at the RX side.

Wideband measurement systems can be grouped in the most popular form of correlative channel sounder (time domain measurements) and multitone sounding (frequency domain measurements). With wideband sounders the CIR or channel frequency response is measured. Correlative sounders transmit a Pseudorandom Noise (PN) sequence, generated in the time domain, which is correlated with the received signal at the RX. It can be shown that the result of the RX correlation is the convolution of the autocorrelation function of the PN sequence with the CIR, in the case when the channel is time-invariant for the duration of the PN sequence, see Equation (6.2). Different types of correlative sounders are used, for example, a dedicated device as in [15], or using a arbitrary-waveform generator as TX in [16–19]. At the RX side either a signal analyzer or a sampling scope can be used. The correlation process at the RX side is often done offline in a postprocessing step. A problem of PN sequences is that their spectral power significantly decreases towards the band edge. By that, the Signal to Noise Ratio (SNR) is not constant over the whole sounding band.

Another principle of wideband sounding is multitone sounding which is similar to the principle of Orthogonal Frequency Division Multiplexing (OFDM) systems [20]. The signal is composed at the TX with the goal of having a flat spectral shape, while maintaining a low Peak-to-Average Power Ratio (PAPR). Some devices even include pre-amplification of higher frequencies to accommodate for frequency-dependent cable losses. The RX just samples the signal and can perform all calculations in the frequency domain. Multitone measurements need a more complex measurement equipment, but provide the channel characteristics in the same quality over the whole bandwidth.

---

[1] We consider that the radio channel includes the antennas, with a single antenna at TX and RX.

It has to be mentioned that channel sounders based on vector network analyzers (performing frequency sweeping), which were widely used for indoor channel measurements, can not be used for vehicular radio channels, because their channel acquisition rate is too low, as will be discussed in the section on *Channel sounding of rapidly time-varying channels* below.

6.2.1.2   Multi-Antenna Channel Sounding

Multiple Input Multiple Output (MIMO) systems with multiple antennas at the TX and RX are very important, particularly for vehicular communications systems, where reliable communication is necessary (e.g., for safety applications). In principle there are three types of MIMO channel sounders:

- the virtual array principle;
- the use of multiple parallel Radio Frequency (RF) chains at the TX and RX;
- switched-array principle.

The virtual array principle is very simple. Only one antenna is used at the TX and RX side. The antennas are physically moved to their wanted positions before each sounding of the channels. The drawback is that it works only in very slow time-varying channels, which makes it not usable for vehicular channel sounding.

In the second case the test signals are transmitted and received simultaneously at the TX and RX side. The different transmit signals need to be distinguished for example, using specific PN sequences [21]. A drawback of this principle is the high effort in multiple RF chains and their calibration.

The switched-array principle uses only a single RF chain which is sequentially switched to different antenna elements at the TX and RX side. Compared to the one maintaining multiple RF chains, this approach is much cheaper and easier to calibrate. The drawback is that the sounder needs more time to acquire one channel snapshot (switching over all transmit and receive antenna elements). Still, this approach is the most common way to sound radio channels at the current time.

Figure 6.2 shows the concept of the switched-array principle. The TX and RX are synchronized using rubidium clocks. After modulation at the TX, the test signal is sequentially

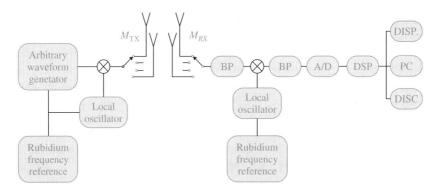

Figure 6.2   Block diagram of a wideband channel sounder based on the switched-array principle.

switched to the different antennas. At the RX the antennas are also switched sequentially. After BandPass (BP) filtering, the signal is downconverted, Analog/Digital (AD) converted and processed in the Digital Signal Processor (DSP). Afterwards the result is stored on a disc for further processing.

Even for vehicular channels, these devices provide acceptably low switching times, such that the stringent timing requirements (see, section on *Channel sounding of rapidly time-varying channels*) imposed by rapidly fluctuating channels can be met.

6.2.1.3 Channel Sounding of Rapidly Time-Varying Channels

The temporal variability of radio channels has a big impact on the measurement principle. In the case of a band-limited time-invariant radio channel the only requirement is that the RX fulfills the Nyquist sampling theorem [22] in the delay domain. In time-varying radio channels, such as occurring in vehicular communications systems, the repetition period t_s of the sounding pulse $p(t)$ is the key parameter.

For sounding these highly time-varying radio channels it has to be ensured that the channel is *underspread*. The radio channel is called underspread when the following criterion is fulfilled [23]

$$2\nu_{max}\tau_{max} \leq 1. \tag{6.1}$$

This criterion can be derived from the two following inequalities. Firstly the repetition time t_s of the sounding pulse $p(t)$ must be shorter than the coherence time of the channel, that is, the channel is not changing significantly during the repetition time, which can be described by

$$t_s \leq \frac{1}{2\nu_{max}}. \tag{6.2}$$

The repetition time is, in the case of a single antenna system, simply the test signal length of the channel sounder $p(t)$ plus a possible guard interval. For multiple antenna systems, the repetition time is the time interval over all measured single antenna links plus possible guard intervals.

Secondly, it has to be ensured that consecutive sounding signals are not overlapping, that is, that the repetition time of the sounding signal is longer than the maximum delay of the channel

$$t_s \geq \tau_{max}. \tag{6.3}$$

Combining these two inequalities in Equation (6.2) and Equation (6.3), yields Equation (6.1).

6.2.2 Vehicular Antennas

For vehicular communications one needs to distinguish between V2I and V2V links. In the case of V2V communication, the main propagation is happening in the horizontal plane (elevation $\vartheta = 90°$) because the height of the antennas on the vehicles is quite similar. For V2I communication, a fixed station, called RoadSide Unit (RSU), is installed next to the road or above the road. The preferred antenna position of the RSU is above the road [24], and therefore the main propagation link is at an elevation $\vartheta < 90°$, depending on the distance between the vehicle and the RSU ($\vartheta = 0°$, if the vehicle is below the RSU). Both situations shall be covered by the use of the same antenna on the vehicle.

The position of the antenna on the vehicle is of crucial importance and it has a big impact on the performance for both V2I and V2V links [25, 26]. Since V2V communication is expected to operate omnidirectionally in the horizontal plane, a rooftop antenna installation is preferable. The antennas can be placed in conventional rooftop antenna modules, which are usually placed near the back of the roof. In this case, V2V communication is a challenge, since the antenna pattern has to be omnidirectional in the horizontal plane. Further, the main gain of the antenna shall be in the horizontal plane ($\vartheta = 90°$). The following effects have a large impact on this desired antenna pattern:

- roof size (comparable to a finite ground plane);
- shadowing to the left and right of the vehicle from railings on the roof;
- inclination of the roof that can shadow the Line Of Sight (LOS) to the front;
- mutual coupling with other antenna elements enclosed in the rooftop antenna module;
- coupling with the dielectric housing of the antenna module.

A detailed description including measurement results for each of these challenges are provided in [27]. In general, mounting an antenna on a vehicle completely changes its three-dimensional antenna pattern. Thus, antennas have to be designed taking the roof and its mounting position into account.

For channel sounding, especially for directional analysis using multiple antenna elements, the knowledge of the exact three-dimensional antenna pattern is necessary. Therefore antenna calibration measurements have to be done after mounting, that is, including the whole vehicle, which is high effort and time consuming, see [28]. Only a few measurement facilities for such calibration measurements exists in Europe.

6.2.3   Vehicular Measurement Campaigns

Vehicular channel measurement campaigns pose additional challenges compared to measurements in other scenarios (e.g., cellular environments). Challenges concerning the channel sounder and antennas were already discussed in the last two subsections. In the following the focus is on performing vehicular channel measurement campaigns.

6.2.3.1   Choice of Measurement Scenarios

For vehicular channel measurements, it is reasonable to concentrate on application-specific scenarios, rather than on the classical environments (urban, suburban, and so on).

Application-specific scenarios include collision avoidance, emergency vehicle warning, co-operative merging assistance, or lane change assistance. Especially in these difficult situations, a highly reliable communication link has to be ensured. In the measurement campaign described in [29], the following application specific scenarios were defined, based on the European Telecommunications Standards Institute (ETSI) [30]:

- different types of road crossings;
- LOS on highways;

- merging lanes;
- traffic congestion (approaching a traffic congestion, overtaking in a traffic congestion);
- in-tunnel.

The focus of measurements, the scenarios, and the respective locations need to be already selected in the planning phase of channel measurements.

6.2.3.2 Setup of Measurement Equipment

Radio channel sounders are delicate yet heavy devices. Careful mounting of the devices in the vehicles, possibly on vibration-reducing platforms, is vital. Large vans or transporters have an abundance of loading space, which makes it simpler to load and fix the equipment. However, these vehicles may not fit the planned scenarios. Loading equipment into a station wagon is more challenging, but also provides more realistic conditions/environments. An example picture from a channel sounder, in this case the RX part, installed in a car is shown in Figure 6.3.

Another important aspect of vehicular channel measurements is the power supply of the whole equipment (channel sounder, cameras, laptops). Depending on the type of the channel sounder this can be a difficult issue. Since the vehicle battery provides only a limited amount of energy, the channel sounder usually needs separate batteries. The capacity of these batteries also limits the duration of the measurements, which has to be taken into account when planning the time schedule for the measurements on the road.

6.2.3.3 Conducting the Measurements

When conducting the measurements we have to distinguish between V2I and V2V measurements, where the V2V measurements are more challenging, because at least two

Figure 6.3   Channel sounder in the trunk of a car.

vehicles have to be coordinated and steered. Particularly in V2V communications, LOS and Non Line Of Sight (NLOS) situations can occur, because of the low height of the antennas. Thus, both situations (LOS and NLOS) need to be investigated. Measurements have to be carried out in real traffic scenarios, which is a big challenge for the drivers in both vehicles. They have to try to repeat the measurements in the same scenario several times with similar conditions (e.g., both vehicles have to enter the crossing synchronized, which sometimes poses a great challenge in real traffic). Therefore, the measurement teams need to be in constant contact for coordinating the measurements. It is recommended that there are two persons in addition to the driver in each vehicle. One person is controling the channel sounder equipment and the second person is handling all the documentation and the communication to the other vehicle(s).

Writing down important notes (measurement parameters and starting time, incidents during the measurement, speed of the vehicles, and so on) is absolutely necessary to keep track of the measurements at a later stage.

Video documentation is very important for vehicular measurements. The videos shall optimally cover the full 360° view around the vehicle (one camera looking to the front and one camera looking to the back are coming close to fulfilling this requirement). When evaluating the measurements, it is possible to analyze the environment and setup during the measurement run using the videos footage (e.g., if there was LOS or NLOS, how many vehicles were in the vicinity, positions of static scatterers, and so on). In this way, the measurement results can be mapped to the actual environment conditions (e.g., reflections from physical objects). Information from video data can be greatly enhanced by using a Global Positioning System (GPS) logging of positions and speeds of the vehicles, which is important for later extraction of channel model parameters. It has to be noted that GPS logging is not always possible (e.g., in urban environment, street canyons, tunnels); in these situations handwritten notes or other documentation needs to be used.

For targeted measurements of obstruction of LOS by another vehicle, it is helpful if the measurement team operates a third vehicle (e.g., a truck), in order to have control over the specific situation. When needed, the truck can change lane during one measurement run. To capture this situation with a truck from the usual traffic is much more complicated.

6.3   Vehicular Channel Characterization

In order to construct realistic channel models, we need to analyze the empirical data collected during measurement campaigns. In this section we will present the most important parameters to be extracted from measurements, which will be used for channel modeling purposes. We also motivate and demonstrate the importance of considering time variability of the channel properties in this analysis.

6.3.1   Time-Variability of the Channel

In vehicular communications, the environment changes rapidly, which results in a time-varying CIR. In order to evaluate the dispersion of the measured channel in the delay and Doppler domains, we calculate the Power Delay Profile (PDP) and the Doppler Power Spectra Density (DSD) based on an ensemble of CIRs.

In the well studied cellular scenario, the statistical properties of the channel usually vary slowly with time. In these cases, the Wide Sense Stationary Uncorrelated Scattering (WSSUS) assumption holds with good accuracy. A process is Wide Sense Stationary (WSS) when its statistical properties do not change with time. Similarly, the process is Uncorrelated Scattering (US) when its statistical properties do not change with frequency. This is commonly formulated as: *contributions with different Doppler frequencies are uncorrelated, and contributions at different delays are uncorrelated, respectively*.

However, due to the rapidly changing environment in vehicular communications, the second order statistics of a fading process stay constant for a finite region in time and frequency only, which we call the stationarity region. Figure 6.5 depicts the PDP and the DSD of a crossing scenario in an urban environment. It clearly demonstrates the time-variability of these two dispersion functions. Therefore, we need to use a tool for calculating the second order statistics of the fading process locally for each stationarity region. Hence, we define the scattering function per stationarity region and call it the Local Scattering Function (LSF) [23].

6.3.1.1   The LSF Estimator

Consider a (measured) channel frequency response $H(t, f)$, whose sampled version reads

$$H[m, q] := H(t_{\mathrm{s}}m, f_{\mathrm{s}}q), \tag{6.4}$$

where $m \in \{0, \ldots, M_{\mathrm{s}} - 1\}$ represents time, with M_{s} being the total number of recorded snapshots, and $q \in \{0, \ldots, Q - 1\}$ denotes frequency, with Q being the number of measured frequency bins. We define the sampling time t_{s}, and the frequency resolution $f_{\mathrm{s}} = B/Q$, where B represents the measurement bandwidth.

When estimating the power spectrum of a process using measurement data, it is very difficult to obtain statistically independent realizations of the same process. By tapering the measurement data using orthogonal windows, and by estimating the spectrum of each individual resulting windowed data, we obtain multiple independent spectral estimates from the same sample. The total estimated power spectrum is calculated by averaging over all tapered spectra. Through this averaging, the resulting estimate presents low variance. Nevertheless, by using multiple tapers on the data, the bias of the estimate increases. Therefore, the number of tapers used is a key parameter determining the performance of the estimator.

This technique is known as multitaper spectral estimation [31, 32], and we use it on the time-varying frequency response $H[m, q]$ for estimating the LSF [23].

We assume that the fading process is locally stationary within the stationarity region with dimension $M \times N$, where M denotes the number of samples in time and N the number of samples in the frequency domain, respectively. In Figure 6.4 we depict this relationship in detail. One should be careful in defining the length of M and N, taking into account the trade-off between resolution and the probability of violating the local stationarity assumption.

We calculate the LSF for consecutive stationarity regions in time and frequency indexed by (k_t, k_f). The relative time index within each stationarity region is denoted by the variable $m' \in \{-M/2, \ldots, M/2 - 1\}$, and the relative frequency index by

$q' \in \{-N/2, \ldots, N/2 - 1\}$. The relationship between the relative and absolute time index is given by

$$m = \left[(k_t - 1)\Delta_t + m'\right] + M, \tag{6.5}$$

where $k_t \in \left\{1, \ldots, \left\lfloor \frac{M_s - M}{\Delta_t} \right\rfloor\right\}$ and Δ_t denotes the time shift between consecutive stationarity regions. In the frequency domain the relationship is as follows:

$$q = \left[(k_f - 1)\Delta_f + q'\right] + N, \tag{6.6}$$

where $k_f \in \{1, \ldots, \lfloor \frac{Q-N}{\Delta_f} \rfloor\}$, and Δ_f denotes the frequency shift, see Figure 6.4.

We compute an estimate of the discrete LSF [33, 34] as

$$\hat{C}[k_t, k_f; n, p] = \frac{1}{IJ} \sum_{w=0}^{IJ-1} \left|\mathcal{H}^{(G_w)}[k_t, k_f; n, p]\right|^2 \tag{6.7}$$

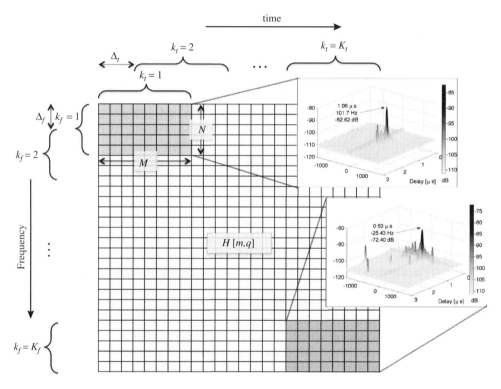

Figure 6.4 Schematic representation of the time-frequency sliding stationarity region window used for the LSF estimation. The sampling time of the data used for this figure is $t_s = 307.2\,\mu s$. The total bandwidth of $B = 240\,Hz$ consists of $Q = 769$ frequency bins. The stationarity region dimensions are chosen to be $M = 128$ samples in time and $N = 256$ samples in frequency. For simplicity, in the figure $K_t = \lfloor \frac{M_s - M}{\Delta_t} \rfloor$ and $K_f = \lfloor \frac{Q-N}{\Delta_f} \rfloor$.

where $n \in \{0, \ldots, N-1\}$ denotes the delay index, and $p \in \{-M/2, \ldots, M/2 - 1\}$ the Doppler index, respectively. The number of tapers used in the time domain is denoted by I, and J denotes the number of tapers used in the frequency domain. The LSF at k_t, k_f corresponds to the center value of the time-frequency stationarity region. The windowed frequency response reads

$$\mathcal{H}^{(G_w)}[k_t, k_f; n, p] = \sum_{m'=-M/2}^{M/2-1} \sum_{q'=-N/2}^{N/2-1} H[m' - k_t, q' - k_f] G_w[m', q'] e^{-i2\pi(pm'-nq')}.$$

(6.8)

The window functions $G_w[m', q']$ shall be well localized within the support region $[-M/2, M/2 - 1] \times [-N/2, N/2 - 1]$. We apply the discrete time equivalent of the separable frequency response used in [33], $G_w[m', q'] = u_i[m' + M/2]\tilde{u}_j[q' + N/2]$ where $w = iJ + j$, $i \in \{0, \ldots, I-1\}$, and $j \in \{0, \ldots, J-1\}$. The sequences $u_i[m']$ are chosen as the discrete prolate spheroidal sequences (DPSS) [35] with concentration in the interval $\mathcal{N}_t = \{0, \ldots, M-1\}$ and bandlimited to $\mathcal{W}_t = [-I/M, I/M]$. The DPSS are the solutions to the eigenvalue equation [32, 35]

$$\sum_{\ell=0}^{M-1} \frac{\sin\left(2\pi \frac{I}{M}(\ell - m')\right)}{\pi(\ell - m')} u_i[\ell] = \lambda_i u_i[m'].$$

(6.9)

The sequences $\tilde{u}_j[q']$ are defined similarly with concentration in the interval $\mathcal{N}_f = \{0, \ldots, N-1\}$ and bandlimited to $\mathcal{W}_f = [-J/N, J/N]$ as

$$\sum_{\ell=0}^{N-1} \frac{\sin\left(2\pi \frac{J}{N}(\ell - q')\right)}{\pi(\ell - q')} \tilde{u}_j[\ell] = \lambda_j \tilde{u}_j[q'].$$

(6.10)

The DPSSs are orthonormal on $\mathcal{N}_t = [0, M-1]$ in the time domain, and orthonormal on $\mathcal{N}_f = [0, N-1]$ in the frequency domain. The eigenvalues λ_i determine the order of the sequences as $1 > \lambda_0 > \lambda_1 > \ldots > \lambda_{M-1} > 0$ in time. The order of the sequences in frequency is defined by the eigenvalues λ_j in the same way.

The eigenvalues λ_i and u_j, and the sequences u_i and $\tilde{u}_j$ are a function of the time-bandwidth product defined by $\mathcal{N}_t\mathcal{W}_t = 2M(I/M)$ in time and $\mathcal{N}_f\mathcal{W}_f = 2N(J/N)$ in frequency. The first $2\mathcal{N}\mathcal{W}$ eigenvalues are considerably close to 1, and their respective sequences have the greatest fractional energy concentration in the interval $\mathcal{W}$. Afterwards, the eigenvalues decay rapidly to 0, this is why we only need to consider $I \leq \mathcal{N}_t\mathcal{W}_t$ and $J \leq \mathcal{N}_f\mathcal{W}_f$.

The number of tapers IJ used for the LSF estimation controls the bias-variance trade-off of the estimator. It is sufficient to select up to 3 tapers in each dimension, based on investigations performed using vehicular channel measurements [36].

In Figure 6.4 we can observe the LSF estimated at different stationarity regions. The delay and Doppler shift resolutions are given by $\tau_s = 1/(Nf_s)$ and $\nu_s = 1/(Mt_s)$.

The length of the stationarity region in time is going to determine whether the WSS assumption is violated. Similarly, the length in the frequency domain relates to the US condition. In vehicular channels, the observed fading process shows a much stronger

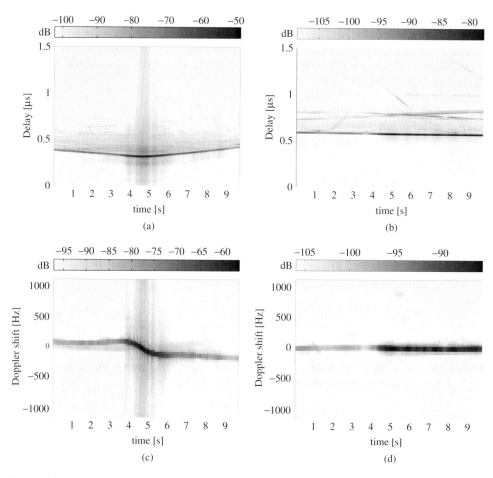

Figure 6.5 Time-varying PDP and DSD from two different measurement scenarios: (a,c) urban intersection scenario, velocities of approximately 2.78 m/s; (b, d) highway convoy measurement with temporally OLOS, constant velocities of 33.33 m/s.

violation of the WSS assumption than the US [36]. Hence, the length of the stationarity region in time is restricting its size.

If we set $Mt_s < t_{stat}$ the LSF is calculated within a region for which the WSS assumption is fulfilled. Similarly in the frequency domain by setting $Nf_s < f_{stat}$ we ensure the US assumption to hold.

In [36] the authors show that the LSF does not significantly change with frequency when M is selected such that WSS is fulfilled. This can also be seen in Figure 6.5 (a) and (b), where the PDP is plotted. For large values M, we would select a time span too large for the estimation of the LSF which would contain time-variations of the same Multipath Component (MPC). This means that the contribution of the same scatterer is spread over more than one delay component, thus violating the US assumption. Based on that, we obtain a simplified version of the LSF estimator, where we can use the whole measured frequency range for the LSF estimation as long as it does not exceed f_{stat}. Since wireless communications for vehicular applications are intended to use a bandwidth of 10 MHz (which is $< f_{stat}$ as reported in [37]), we drop the index k_f from Equation (6.7), and the simplified LSF estimator reads as

$$\hat{C}[k_t; n, p] = \frac{1}{IJ} \sum_{w=0}^{IJ-1} \left| \mathcal{H}^{(G_w)}[k_t; n, p] \right|^2. \tag{6.11}$$

6.3.1.2 The Physical Interpretation of the LSF

The LSF generalizes the spreading function [8] for a limited stationarity region. It is composed of several spectral peaks at different positions in the delay-Doppler plane. These peaks correspond to physical objects, or groups of objects, which give rise to MPCs with different power and delay-Doppler shifts. The first peak in obstruction free scenario is generally the strongest one corresponding to the LOS path from TX to RX.

Later peaks relate to other objects, such as other cars or trucks, or traffic signs and big structures. Generally, peaks with negative Doppler indicate objects moving away from the TX-RX pair, whereas a positive Doppler shift indicates approaching objects. We understand approaching/leaving with respect to the TX-RX link, that is, an object can be static but due to the TX and RX movement, it appears as approaching. The delay of the peaks indicates the distance which the transmitted wave has traveled.

Furthermore, the peaks move in the delay-Doppler plane with time, which corroborates the high time-variability of the channel and the possible violation of the WSSUS assumption. Note the position of the first peak in the two LSFs plotted in Figure 6.4. They correspond to two different stationarity regions for a highway opposite directions measurement.

The peak in the first plot presents a Doppler shift of 101.7 Hz and delay of 1.06 μs. Here, the TX and RX are approaching each other with a relative speed[2] of 5.45 m/s and they are separated by 480 m. At this point there are only a few scattering components represented by the peaks beside the LOS peak. On the lower LSF figure, the peak of the LOS has coordinates -25.43 Hz and 0.53 μs. The negative Doppler shift tells us that the TX and RX are leaving with a relative speed of 1.36 m/s, and the delay indicates that the distance between them is 159 m. In this second LSF plot, there are more spectral peaks, indicating a richer scattering environment. Note that the power of the LOS peak also varies with time.

6.3.1.3 Power Delay Profile, PDP, and Doppler Power Spectral Density, DSD

The LSF $\hat{C}[k_t; n, p]$ as defined in Equation (6.11) is a function of three variables, time index k_t, delay index n, and Doppler shift index p. Based on $\hat{C}[k_t; n, p]$, the time-varying

---

[2] We obtain an approximate relative speed v_{rel} directly from the Doppler shift component if the distance between TX and RX is larger than the maximum road width, which we consider to be at most 20 m. In that case, the angle Θ_{TX-RX} between the speed vectors of both vehicles is small enough so that $\cos(\Theta_{TX-RX}) \approx 1$, and therefore $v_{rel} = vc/f_c$, where v is the Doppler shift, c is the speed of light, and f_c is the carrier frequency.

PDP and time-varying DSD can be defined as

$$\hat{P}_\tau[k_t; n] = E_p\{\hat{C}[k_t; n, p]\} = \frac{1}{M} \sum_{p=-M/2}^{M/2-1} \hat{C}[k_t; n, p], \tag{6.12}$$

and

$$\hat{P}_v[k_t; p] = E_n\{\hat{C}[k_t; n, p]\} = \frac{1}{N} \sum_{n=0}^{N-1} \hat{C}[k_t; n, p], \tag{6.13}$$

where $E_x\{\cdot\}$ denotes expectation over variable x.

In Figure 6.5 we show two examples for the time-varying PDP and DSD. The PDP is plotted in the upper row (figures (a) and (b)), in the lower row (figures (c) and (d)) the time-varying DSD is shown. The plots on the left hand side of the figure correspond to a crossing scenario in an urban environment. The ones on the right hand side correspond to measurements taken on the highway with TX and RX driving in the same direction.

We should also point out the effect of the diffuse components, which is more pronounced in the urban scenario. When the cars are driving in opposite directions, the PDP and DSD experience stronger time-variability. Furthermore we observe some late components resulting from reflections on other objects on the road.

6.3.2 Time-Varying Vehicular Channel Parameters

As shown in the previous sections, the statistical properties of the channel are time-varying. When characterizing the channel this should be taken into account. Typically it has been assumed that a radio channel can be statistically fully characterized by its pathloss, its Root Mean Square (RMS) delay spread and RMS Doppler spread. However, as we have shown previously, the fading process in vehicular channels is a non-WSSUS process with local stationarity within a finite stationarity region. Therefore, we introduce the stationarity time as a new parameter into the characterization parameter set for these wireless channels. The channel parameters are derived from the time-varying frequency response, as well as from the time-varying PDP and DSD, hence the channel parameters for vehicular channels will be time-varying as well.

6.3.2.1 Pathloss

The pathloss describes the signal attenuation due to the propagation distance and shadowing from objects like other vehicles, buildings and so on. Vehicular measurements have shown a constant offset between the pathloss observed for approaching (forward pathloss) and leaving (reverse pathloss) situations for on-coming measurements [2].

Therefore, for modeling the pathloss, we introduce a correction term PL_c which accounts for the offset between forward and reverse pathloss. We also define a variable ζ having different values depending on the measurement conditions, and defined as

$$\zeta = \begin{cases} 1 & : \quad \text{for reverse pathloss} \\ -1 & : \quad \text{for forward pathloss} \\ 0 & : \quad \text{for convoy pathloss.} \end{cases} \tag{6.14}$$

The magnitude of the pathloss in dB is calculated from the channel frequency response $H[m, q]$ as in [2]

$$PL[k_t] = PL_0 + 10\beta \log_{10}(d[k_t]/d_0) + X_\sigma + \zeta PL_c, \quad d[k_t] > d_0, \quad (6.15)$$

where PL_0 is the pathloss at a reference distance d_0, β is the pathloss exponent, and X_σ is a zero-mean normally distributed random variable with standard deviation σ. The time-varying distance is denoted by $d[k_t]$.

6.3.2.2 Large-Scale Fading

The large-scale fading, or shadow fading, varies with time depending on whether there is LOS between the two communicating vehicles or not. Measurements have shown that the mean of the large scale fading component can be around 10 dB higher when the LOS is obstructed by another vehicle and that this also leads to an increase in the standard deviation of the fading process, which usually is described by a log-normal distribution. A 10 dB extra path loss roughly translates to a 3-fold reduction in range. It should be noted that the coherence time, or coherence distance, can be quite large for the large-scale fading process since this depends on the relative positions of vehicles. If a vehicle is shadowed by other vehicles driving in the same direction it will usually remain shadowed for quite some time. In Figure 6.6 we show an example of a scatter plot of the measured channel gain in an urban scenario where we separate samples from LOS and OLOS. As seen in the figure, the path loss exponents are nearly identical, but there is a large offset (7 dB in this particular case) between the regression lines. We can also note the large deviations from the regression lines due to the large scale fading.

6.3.2.3 RMS Delay and Doppler Spread

The second central moments are important parameters for the description of the fading process. The RMS delay spread indicates the severity of the frequency selectivity, while

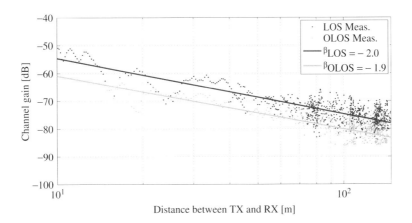

Figure 6.6 Example of a scatter plot of the channel gain for LOS and OLOS between the two vehicles in a measured urban scenario.

the time selectivity is reflected by the RMS Doppler spread. The time-varying RMS delay spread

$$\sigma_\tau[k] = \sqrt{\frac{\sum\limits_{n=0}^{N-1} (n\tau_s)^2 \hat{P}_\tau[k;n]}{\sum\limits_{n=0}^{N-1} \hat{P}_\tau[k;n]} - \left(\frac{\sum\limits_{n=0}^{N-1} (n\tau_s) \hat{P}_\tau[k;n]}{\sum\limits_{n=0}^{N-1} \hat{P}_\tau[k;n]}\right)^2}, \qquad (6.16)$$

and the time-variant RMS Doppler spread

$$\sigma_\nu[k] = \sqrt{\frac{\sum\limits_{p=-M/2}^{M/2-1} (m\nu_s)^2 \hat{P}_\nu[k;p]}{\sum\limits_{p=-M/2}^{M/2-1} \hat{P}_\nu[k;p]} - \left(\frac{\sum\limits_{p=-M/2}^{M/2-1} (m\nu_s) \hat{P}_\nu[k;p]}{\sum\limits_{p=-M/2}^{M/2-1} \hat{P}_\nu[k;p]}\right)^2} \qquad (6.17)$$

are calculated using the PDP and DSD, respectively.

Thresholding of the estimated LSFs has to be performed in order to eliminate spurious components, which would lead to erroneous results. The thresholding is done separately for each stationarity region. *Noise-thresholding* is needed for eliminating noise components that could be mistaken as MPC. *RX-sensitivity-thresholding* eliminates non-relevant components at the RX due to the receiver sensitivity.

6.3.2.4 Stationarity Time

In order to characterize for how long a fading process can be considered to be stationary, we introduce the stationarity time as an important parameter in vehicular channels. The stationarity time is calculated using the collinearity metric [38], which compares the LSF at different time instances. The collinearity is a bounded spectral metric $\gamma_t \in [0, 1]$ and it is calculated as

$$\gamma_t[k_t, k_t'] = \frac{\sum\limits_{n=0}^{N-1} \sum\limits_{p=-M/2}^{M/2-1} \hat{C}^{(t)}[k_t; n, p] \odot \hat{C}^{(t)}[k_t'; n, p]}{\|\hat{C}^{(t)}[k_t; n, p]\| \|\hat{C}^{(t)}[k_t'; n, p]\|}, \qquad (6.18)$$

where $\odot$ stands for element wise multiplication, and $\| \cdot \|$ is the norm L_2 operator. A collinearity close to 1 indicates two similar scattering functions.

The more similar the two compared spectral densities are, the closer to 1 their collinearity is. In order to define a stationarity region, we set a threshold $\alpha_{th} = 0.9$ [4, 34]. Using this threshold, the length of the stationarity region T_{stat} is obtained by

$$T_{stat}[k_t] = t_s \left(M - \Delta_t \right) + t_s \Delta_t \sum\limits_{k_t'=1}^{\lfloor \frac{M_S-M}{\Delta_t} \rfloor} \alpha[k_t, k_t'], \qquad (6.19)$$

where $\alpha[k, k']$ is an indicator function defined as

$$\alpha[k, k'] = \begin{cases} 1 & : & \gamma[k, k'] > \alpha_{\text{th}} \\ 0 & : & \text{otherwise.} \end{cases} \tag{6.20}$$

6.3.3 Empirical Results

In order to get a proper understanding of the channel behavior several measurement campaigns have been conducted providing research groups with a large amount of data from where channel parameters are extracted.

Figure 6.7 depicts the time-varying RMS delay and Doppler spreads as well as the stationarity time for two different measurement scenarios, which are the same as used in Figure 6.5. The highway same direction scenario corresponds to a convoy measurement with two vehicles driving at high speed on the highway, experiencing intermittent LOS obstruction. The measurement in the urban scenario was taken on a street intersection with open surroundings, with two vehicles approaching and leaving the crossing from perpendicular streets. For easier understanding of the plots in Figure 6.7 we recommend the reader to refer to Figure 6.5.

Larger RMS delay spreads can be observed in highway scenarios, due to some distinct scatterers with large delays and scattering from other cars. The time-variability is notable, depending on the instantaneous contributions from different scatterers.

High variations on the RMS Doppler spread are observable during the time interval where the drive-by between two cars occurs, as one can easily see between seconds 4 and 6 for the street intersection measurement. In measurements where the LOS component is strong, the RMS Doppler spread experiences time-variation only when a further MPC is strong enough, such as seen in the highway measurement at 5.5 s, Figure 6.5 (b) and (d). It is a general observation that stationarity times are larger in convoy measurements, whereas for opposite directions they remain below 0.5 s.

For channel modeling we are interested in the distribution of these parameters, whereas for system design we are more interested in determining critical values, which could affect the system performance, such as maximum RMS Doppler spread.

As we have seen in Section 6.2.3, besides the classical scenario classification, in vehicular communications we also need to define application oriented scenarios. Tables 6.1 and 6.2 summarize typical values of the channel parameters extracted from different measurement campaigns found in the literature.

Looking at the parameters listed in Table 6.1, we conclude that high RMS delay spreads are mostly observed in urban environments with rich scattering, in OLOS, and in tunnel. High RMS Doppler spreads occur in drive-by scenarios, and in situations where late Doppler components are significant, mainly caused by good reflecting objects, such as in in-tunnel environments.

Stationarity times for same direction measurements are found to be always longer than 1 s, whereas for crossing scenarios, where the two cars do not drive in the same direction, it remains below 0.5 s. Nevertheless, the stationarity time can also reach very small values, down to 40 ms for short periods.

The pathloss exponent reported in Table 6.2 shows values between 1.59 in urban environment, and 4 in rural environment. Noteworthy is that values around 1.8, below the

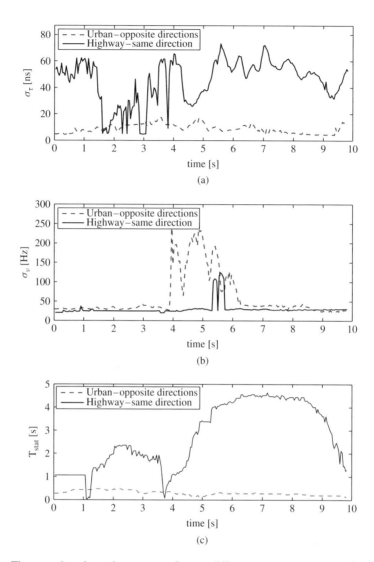

Figure 6.7 Time-varying channel parameters for two different measurement scenarios. The dashed line corresponds to urban intersection scenario, with velocities of approximately 2.78 m/s; the black line corresponds to the highway convoy scenario with temporally OLOS, with constant velocities of 33.33 m/s: (a) time-varying RMS delay spread, (b) time-varying RMS Doppler spread and (c) time-varying stationarity time.

free-space pathloss exponent, are often observed. This indicates that a larger amount of energy is available due to multipath propagation in addition to the LOS component, leading to a smaller pathloss than for free-space propagation conditions.

Regarding the RMS delay spread, the shortest value is obtained in rural environments, and the largest in urban environments, in agreement with the results in Table 6.1. Even

Table 6.1 Time-varying parameters for the new definition of scenarios [29].[1] single/multiple lane,[2] slow traffic/approaching traffic jam

Time-varying parameters	Environment	RMS delay spread σ_τ [ns]	RMS Doppler spread σ_ν [Hz]	Stationarity time T_{stat} [s]
Road crossing	Rural	34.53	87.32	0.18
	Suburban	17.42	87.61	0.55
	Urban	36.64/50.51[1]	131.89/53.01[1]	0.14/0.18[1]
Obstructed LOS	Highway	47.03	41.09	1.70
Merging lanes	Rural	16.84	69.89	1.16
Traffic congestion	Highway	19.85/11.90[2]	46.21/38.70[2]	1.44/1.74[2]
In-tunnel	Tunnel	72.00	112.66	1.11

though different measurement campaigns have reported more homogeneous results regarding the pathloss exponent, we see larger differences for the RMS delay spread results. Even fewer results have been covered in the literature with respect to the RMS Doppler spread, showing again large discrepancies among them, basically due to the measurement set-up. As mentioned before, on-coming measurements present a larger RMS Doppler spread than in convoy measurements.

6.4 Channel Models for Vehicular Communications

Channel models are important for system and algorithm design, and in order to assess the performance by computer simulations before deploying a new system. An accurate channel model will deliver reliable system simulation results. In order to derive accurate channel models, parameters extracted from measurement campaigns are used as an input so that the channel model generates representative CIRs (or transfer functions) for different environments. There are different channel modeling approaches as described in the next section together with their suitability for V2V channels.

6.4.1 Channel Modeling Techniques

The channel models are used to generate a sequence of CIRs or transfer functions, both representations are equivalent and related by the Fourier transform. In wireless communications there are basically three channel modeling approaches in use for vehicular applications:

- *Deterministic*: The CIRs are the result of solving the Maxwell equations for a specific site. It is therefore very important to correctly define the physical objects present in the environment to be simulated as well as their precise electromagnetic properties. The resulting CIRs can be very accurate at expenses of a highly demanding computational process. These channel models are known as Ray Tracing RT and have been used in vehicular communications for investigation of antenna position on the vehicle or placement of RSUs [44, 45]. Furthermore, one can also use CIRs collected in measurement campaigns as a direct input for the channel in system simulations.

The advantage of this method is its repeatability and accuracy, but only for very specific sites and antenna arrangements, namely the ones used where and when the measurements were taken [14, 46].

- *Stochastic*: The main characteristic of the stochastic approach is to generate CIRs with specific statistical properties of the fading process, without assuming an underlying geometry. The properties are defined in the model through channel parameters, such as RMS delay and Doppler spreads, Angle-of-Departure (AOD) and Angle of Arrival (AOA), or power spectra and their distributions. We can classify the stochastic channel models into two sub-groups depending on whether they are narrow- or wide-band. The narrow-band stochastic approach models the time-selectivity of the fading process by defining its Doppler spectrum [47]. The wide-band stochastic approach models also the frequency-selectivity (or delay dispersion) of the fading process and the generated CIRs are time-frequency dependent. The most used channel model of this kind is the Tapped-Delay Line (TDL) model. The CIR is built up by a finite number of delay taps, each one of them fades independently following a given Doppler spectrum. The TDL channel model is based on the WSSUS assumption [10]. Even though its complexity is fairly low, it does not always capture realistic properties for the V2V channel. Nevertheless, it has been adopted by the Institute of Electrical & Electronics Engineers (IEEE) 802.11p standard [48] as a reference channel model due to its flexibility and straightforward implementation [17, 49].

- *Geometry-based Stochastic Channel Model (GSCM)*: The GSCM is a combination of the two aforementioned approaches. The generation of the CIRs is done using simplified RT on scatterers stochastically placed in the physical environment to be simulated. Each of the scatterers has associated fading properties chosen from parameters distributions obtained from measurement campaigns. Within the GSCM approach we can distinguish two kind of models depending on where the scatterers are placed: *(i) ring-models*, and *(ii) physically-motivated models*. In the *ring-models* the scattering points are placed around the TX and RX in a shape of a circle or ellipse [50–53]. In the *physically-motivated models* the scattering points are only placed at physically realistic positions, for example, scatterers like other vehicles driving beside TX and RX are placed on a parallel position with respect to them [3, 52, 54, 55]. The GSCMs are in general accurate and can represent the properties of the V2V channels well. The channel model generates CIRs for different environments by using the right parameter set, thus it is flexible. However, the computational complexity is relatively high, especially for the *physically-motivated models*, where a large number of scatterers are needed to reproduce the contribution of diffuse components. In Section 6.4.3, we present a method for reducing the complexity of this calculation. Noteworthy is that these channel models inherently include the non-stationarities of the fading process, since they include the movements of objects.

In the following, we are going to compare the methodologies presented. We discuss the most adequate channel model to be used for V2V system performance evaluation.

We demand from a channel model that it generates *accurate* CIRs that represent *realistic V2V properties*. The channel model should be *easy to implement and compute*. A *flexible* channel model will offer the system engineer more possibilities when testing algorithms for different situations. Furthermore, we have seen in the previous Section

6.3 that the V2V fading process is non-stationary, therefore, the channel model has to include this *non-stationarity* particular of V2V channels. Hence, a good vehicular channel model should be *accurate, realistic, low-complex*, and have the ability to represent *non-stationarities*.

In Table 6.3 we summarize the main properties of the channel modeling approaches previously presented, stressing their significance for V2V channels. We list the advantages for V2V of each approach with the sign "+" and the drawbacks with "−". We conclude that the GSCM approach seems to be the most appropriate for modeling vehicular channels. The main negative aspect is the high complexity. However, in the next section, we present a complexity reduction method, with that, the GSCM becomes definitely the most suitable approach for modeling vehicular channels. We dedicate the entire following subsection to describing the GSCM.

6.4.2 Geometry-Based Stochastic Channel Modeling

The CIR is composed of contributions stemming from different scatterers that are summed according to their respective delay [3, 29, 56]. These contributions in vehicular channels can be classified into four groups depending on the class of object originating them:

- Line Of Sight contribution (LOS): the direct link between TX and RX.
- Mobile-Discrete (MD) scatterers: such as other cars or trucks driving beside the TX and RX.
- Static-Discrete (SD) scatterers: such as traffic signs, street lamps, or large metallic structures in general.
- Diffuse (DI) scattering: produced by noise barriers or vegetation on both sides of the road.

Each one of them has different statistical properties and should be modeled separately.[3] The CIR of a V2V GSCM can be written in a generic form as

$$h(t, \tau) = h^{(\text{LOS})}(t, \tau) + \sum_{p=1}^{N_{(\text{MD})}} h_p^{(\text{MD})}(t, \tau_p) + \sum_{l=1}^{N_{(\text{SD})}} h_l^{(\text{SD})}(t, \tau_l) + \sum_{r=1}^{N_{(\text{DI})}} h_r^{(\text{DI})}(t, \tau_r), \quad (6.21)$$

where t and τ denote time and delay, respectively, and $N_{(\cdot)}$ denote the numbers of MD, SD and DI scatterers, respectively. By performing a Fourier transform of the CIR we can obtain the channel frequency response

$$H(t, f) = H^{(\text{LOS})}(t, f) + \sum_{p=1}^{N_{(\text{MD})}} H_p^{(\text{MD})}(t, f) + \sum_{l=1}^{N_{(\text{SD})}} H_l^{(\text{SD})}(t, f) + \sum_{r=1}^{N_{(\text{DI})}} H_r^{(\text{DI})}(t, f),$$

$$(6.22)$$

---

[3] The channel model described assumes perfect omnidirectional antennas in the xy-plane. The model can be easily extended by including the antenna gain functions at the TX and the RX to each individual contribution x as $h_x(t, \tau) \times \delta(\varphi - \varphi_{\text{TX},x})\delta(\theta - \theta_{\text{RX},x})g_{\text{TX}}(\varphi_{\text{TX}})g_{\text{RX}}(\theta_{\text{RX}})$, being φ and θ the AOD, and AOA respectively [8].

Table 6.2 Time-varying parameters using the classical definition of scenarios.[1] breakpoint model,[2] TX-RX separation of $300 - 400\,\text{m}$,[3] low/high traffic density,[4] median value,[5] TX-RX separation of $200 - 600\,\text{m}$,[6] antenna inside/outside the car,[7] on-comming/convoy traffic,[8] $n/PL_0/\sigma/PL_c$ from Equation (6.15),[9] LOS/discrete scatterer contributions

Time-varying parameters	Pathloss exponent β	RMS delay spread σ_τ [ns]	RMS Doppler spread σ_v [Hz]	Reference
Highway	–	–	120	[14]
	1.8	247	–	[39]
	1.85	41	$92^{(4)}$	[40]
	–	$141 - 398^{(2)}$	$761 - 978^{(2)}$	[18]
	$1.9/4\,0^{(1)}$	–	–	[41]
	–	165	–	[42]
	–	$53/127^{(3)}$	–	[15]
	$3.59/2.21^{(7)}$	–	–	[6]
	$1.77/63.3/3.1/3.3^{(8)}$	–	–	[2]
	$1.8/3.08^{(9)}$	–	–	[43]
Rural	1.79	52	$108^{(4)}$	[40]
	–	22	782	[18]
	$2.3/4.0^{(1)}$	–	–	[41]
	$3.18/–^{(7)}$	–	–	[6]
Suburban	2.5	–	–	[11]
	$2.1/3\,9^{(1)}$	–	–	[11]
	–	104	–	[42]
	$3.5/–^{(7)}$	–	–	[6]
	$1.59/64.6/2.1/\text{N/A}^{(8)}$	–	–	[2]
	$1.8/3.00^{(9)}$	–	–	[43]
Urban	–	–	86	[14]
	1.61	47	$33^{(4)}$	[40]
	–	$158 - 321^{(5)}$	$263 - 341^{(5)}$	[18]
	–	373	–	[42]
	–	$126/236^{(6)}$	–	[15]
	$2.88/1.83^{(7)}$	–	–	[6]
	$1.68/62.0/1.7/1.5^{(8)}$	–	–	[2]

where f denotes frequency. We can expand Equation (6.22) as

$$H(t, f) = a^{(\text{LOS})}(t)e^{-j2\pi f \tau_{\text{LOS}}(t)} + \sum_{p=1}^{N_{(\text{MD})}} a_p^{(\text{MD})}(t)e^{-j2\pi f \tau_p(t)}$$

$$+ \sum_{l=1}^{N_{(\text{SD})}} a_l^{(\text{SD})}(t)e^{-j2\pi f \tau_l(t)} + \sum_{r=1}^{N_{(\text{DI})}} a_r^{(\text{DI})}(t)e^{-j2\pi f \tau_r(t)}, \tag{6.23}$$

where a_x denotes the complex-valued amplitude coefficient of path x. The amplitudes of the LOS, MD, and SD paths are fading, in contrast to the conventional GSCM. Using this approach one can describe the fading behavior of the combined contribution from

Table 6.3 Summary of channel models approaches and their main characteristics. Desirable properties for vehicular communications are listed with "+", on the contrary, the drawbacks of each model are listed with "−".[2] A technique for reducing computational complexity in GSCM is discussed in Section 6.4.3

Type	Technique	Characteristics
Deterministic	Stored CIR	+ Very good accuracy − Large amount of data has to be stored − Only valid for one specific site Used for validation
	RT	+ Very good accuracy + Non-stationarities well represented − Very high computational complexity Used for analyzing antenna and road side unit placements
Stochastic	TDL	+ Flexible: different environments are easy to parametrize + Relatively low computational complexity − It does not include the non-stationarities of the channel Used in simulations for system performance evaluation
GSCM		+ Good accuracy + Non-stationarities well represented + Flexible: different environments are easy to parametrize − Relatively high computational complexity[2] Used in simulations for system performance evaluation

several unresolvable paths by just modeling one. By doing so, the model can be written using two dimensions in space only [3]. Nevertheless, the complex amplitudes of the DI components are calculated as in the standard GSCM.

In order to get a clearer idea of how the GSCM works, we list its steps as follows:

1. *Define geometry and place TX, RX, and scatterers:*

The number of MD, SD, and DI scatterers is drawn from a given Probability Density Function (PDF) χ_{MD}, χ_{SD}, and χ_{DI}, respectively. The TX, the RX, and the scatterers are then placed randomly in a plane following the geometry of the road specified in the model. In Figure 6.8 we show the geometry for a highway with two lanes in each driving direction.

The MD scatterers are placed only at coordinates corresponding to realistic positions for vehicles, that is, on the road. Different mobility models can be applied, here we describe a simplistic one. The y-coordinate is drawn from a discrete uniform distribution as $y_p \sim \mathcal{U}[0, N_{lanes}] \cdot W_{lane}$, where N_{lanes} is the total number of lanes, and W_{lane} is the width of one lane in m. The x-coordinate is modeled as a continuous uniform distribution as $x_p \sim \mathcal{U}[x_{min}, x_{max}]$. Each scatterer moves with constant velocity along the x-axis obtained from a truncated Gaussian distribution. The truncation is needed to avoid negative velocities and limit the maximum driving speed. The xy-coordinates of the MD scatterers are defined only once when deploying them in the geometry, and they are updated with the movement of the scatterers.

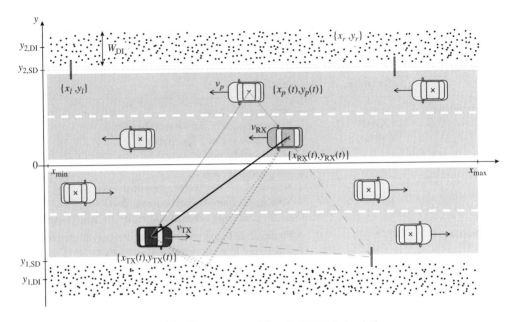

Figure 6.8 Geometry used for the GSCM simulation.

The SD scatters representing traffic signs are placed at the outer borders of the road. Their y-coordinates are determined using Gaussian distributions at each side of the road as $y_l \sim \mathcal{N}[y_{1,SD}, \sigma_{y,l,SD}]$ and $y_l \sim \mathcal{N}[y_{2,SD}, \sigma_{y,l,SD}]$. Their position along the x-axis follows a uniform distribution as $x_l \sim \mathcal{U}[x_{min}, x_{max}]$.

Finally, the DI scatterers are also uniformly distributed along the x-axis as $x_r \sim \mathcal{U}[x_{min}, x_{max}]$. Their y-coordinates are drawn from uniform distribution in a region of width W_{DI} m on both sides of the road as $y_r \sim \mathcal{U}[y_{1,DI} - W_{DI}/2, y_{1,DI} + W_{DI}/2]$ or $y_r \sim \mathcal{U}[y_{2,DI} - W_{DI}/2, y_{2,DI} + W_{DI}/2]$.

2. *Determine propagation distance:*

Calculate the propagation distance, AOA, and AOD for the LOS component, and for each single bounce path between TX, scatterer, and RX, for every time instance one intends to obtain the CIR from.

3. *Calculation of complex-valued amplitudes:*

For each individual path, calculate the time-varying fading coefficients. The complex-valued amplitude for the LOS, MD, SD paths is composed by a deterministic, distance-decaying part, and a stochastic part as

$$a_x(t) = g_{S,x} e^{j\phi_x} G_{0,x}^{1/2} \left(\frac{d_{ref}}{d_x(t)} \right)^{\beta_x/2}, \tag{6.24}$$

where the variable x denotes either the LOS, MD or SD components. The real-valued, slowly-varying, stochastic amplitude gain of component x is represented by $g_{S,x}$.

The time-varying total distance travelled by each x component is determined by $d_x(t) = d_{\text{from TX to }x}(t) + d_{\text{from }x\text{ to RX}}(t)$, which for the LOS path it is simplified as $d_{\text{LOS}}(t) = d_{\text{from TX to RX}}(t)$. The reference power $G_{0,x}$ is a fixed value for the LOS path, and it is a function of the pathloss exponent β_x for the MD, and SD paths.

The pathloss exponent β_x, and the received power $G_{0,x}$ at a reference distance d_{ref} are assigned individually to each path x. The pathloss exponent is fixed for the LOS path, and $\beta_x \sim \mathcal{U}(0, \beta_{x,\text{max}})$ for the paths from the discrete scatterers. Even though the provided model is the same for the LOS, MD and SD paths, each kind of propagation mechanism has its own set of parameters.

The phase of the complex amplitude ϕ_x is modeled as a uniformly distributed random variable as $\phi_x \sim \mathcal{U}[0, 2\pi)$, as in the classical GSCM approach.

The path gain $g_{S,x}$ can be described by a correlated log-normal stationary variable with a distance autocorrelation function

$$r_d(\Delta d) = \sigma_{S,x}^2 e^{-\frac{\ln 2}{d_{c,x}^2}(\Delta d)^2},\tag{6.25}$$

where $d_{c,x}$ is the 0.5-coherence distance defined as $\rho_d(d_{c,x}) = 0.5$, and variance $\sigma_{S,x}^2$. The variance $\sigma_{S,x}^2$ is uncorrelated with $d_{c,x}$ and is exponentially distributed as $\sim \mu_\sigma \exp\{-\mu_\sigma \sigma_{S,x}^2\}$. For obtaining the path gain $g_{S,x}$ we generate data from $\sim \mathcal{N}(0,1)$ and correlate it using the distance autocorrelation function in Equation (6.25). The coherence distance $d_{c,x}$ is given by an exponential distribution with a non-zero minimum value, determined as $d_{c,x} = d_{c,x}^{\text{min}} + d_{c,x}^{\text{rand}}$, with $d_{c,x}^{\text{rand}}$ distributed following a PDF as $\mu_{c,x} \exp\{-\mu_{c,x} d_{c,x}\}$.

The GSCM approach is used for modeling the DI path amplitudes

$$a_r^{(\text{DI})}(t) = G_{0,\text{DI}^{cr}} \left(\frac{d_{\text{ref}}}{d_r(t)}\right)^{\beta_{\text{DI}}/2},\tag{6.26}$$

where again $d_r(t) = d_{\text{from TX to }r}(t) + d_{\text{from }r\text{ to RX}}(t)$. The pathloss exponent β_{DI} and the reference power $G_{0,\text{DI}}$ are constant for all DI scatterers.

4. *Calculation of the CIR:*

We apply Equation (6.23) with the results obtained in the previous steps in order to obtain the CIR of the generated channel.

All the parameters have to be extracted from measurements. The authors in [3] present a detailed list of them, extracted from the measurements in [39]. The generated CIRs using the described GSCM is shown in Figure 6.9 (b), next to a measured channel under the same conditions in Figure 6.9 (a). Note that the properties described previously can be seen in both subfigures: *(i)* strong time-varying LOS component, *(ii)* discrete MPCs, and *(iii)* DI scattering.

6.4.3 Low-Complexity Geometry-Based Stochastic Channel Model Simulation

As mentioned previously, the accuracy of the GSCM results is very good at expenses of a long simulation time. The simulation effort grows with the number of exponential

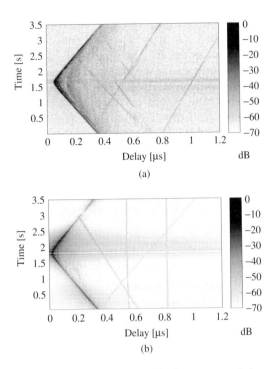

Figure 6.9 Normalized magnitude squared of the CIRs for a measured channel (a), and a simulated channel (b). The parameters chosen for the simulation are chosen from the ones extracted from the measurements. They correspond to a highway opposite direction measurement, where both cars TX and RX drive at about 33.33 m/s.

functions to be summed up in Equation (6.23), that is, with the number of scatterers. Typical values of N_{MD}, N_{SD}, and N_{DI} are 5 to 10 for the discrete scatterers, and around 500 for the diffuse ones for a $x_{max} = 500$ m. One can reduce simulation time using low-complexity techniques for modeling the DI paths [57, 58].

The authors in [58] presented a low-complexity channel simulation method based on the use of multi-dimensional DPSS as basis functions to approximate the sum of complex exponentials, as needed in Equation (6.23).

Then, we can rewrite the DI part of Equation (6.23) as

$$\sum_{r=1}^{N_{(DI)}} a_r^{(DI)}(t) e^{-j2\pi f \tau_r(t)} = \sum_{r=1}^{N_{(DI)}} \tilde{a}_r^{(DI)} e^{-j2\pi \Delta f \tau_r^{(DI)}} e^{j2\pi t \nu_r^{(DI)}}, \tag{6.27}$$

where $\tau_r^{(DI)}$ is the delay at the first simulated time instance, $t = 0$, within the WSS region. The frequency is defined by $f = f_c + \Delta f$. The complex-valued path amplitude reads $\tilde{a}_r^{(DI)} = a_r^{(DI)}(t = 0)e^{-j2\pi f_c \tau_r^{(DI)}}$, and the Doppler shift

$$\nu_r^{(DI)} = \frac{f_c}{c_0}[v_{TX}\cos(\varphi_r(t = 0)) + v_{RX}\cos(\theta_r(t = 0))], \tag{6.28}$$

where φ and θ are the AOD and AOA, respectively.

This approach demands the fulfillment of two requirements:

- the channel has to be WSS for the time-frequency region for which it is going to be simulated, that is, $(t, f) \in \{0, \ldots, T\} \times \{-B/2, \ldots, B/2\}$, with a time span of duration T and bandwidth B. The WSS assumption means that the complex-valued path amplitudes do not significantly change, and the delay paths can be modeled by a Doppler shift, thus they are changing linearly.

  It might seem that by using the low-complexity approach we are losing one of the most important properties of the V2V channels, the *non-stationarities*. However, a channel simulator using this approach can be implemented in a way such that the generated CIRs are stored in blocks or chunks, and one can assume the WSS of the complex amplitude for the DI paths to hold within each chunk.
- the channel needs to be band- and time-limited. The limits are defined in the transformed domain, that is, the frequency limitation is determined by the maximum delay, and the maximum Doppler shift determines the time limitation.

The major complexity reduction can be achieved by exploiting the time-, band-limitation of the generated CIRs which span a subspace best represented by multi-dimensional DPSS [35, 58]. We exploit the fact that the subspace dimension D is much smaller than the number of DI scatterers, $D \ll N_{(\text{DI})}$. As a result, Equation (6.28) can be approximated by

$$\sum_{d=1}^{D} g_d \omega_d(t, f), \qquad (6.29)$$

where $\omega_d(t, f)$ are the two-dimensional DPSS for the index set $(t, f) \in \{0, \ldots, T\} \times \{-B/2, \ldots, B/2\}$.

The bandlimiting region is determined by $(\tau_r^{(\text{DI})}, v_r^{(\text{DI})}) \in [0, \tau_{\max}^{(\text{DI})}] \times [-v_{\max}^{(\text{DI})}, v_{\max}^{(\text{DI})}]$, with $\tau_{\max}$ and $v_{\max}$ the maximum over all DI contributions. The basis coefficients g_d can be approximated by [58]

$$g_d = \sum_{r=1}^{(\text{DI})} \tilde{a}_r^{(\text{DI})} \lambda_d(\tau_r^{(\text{DI})}, v_r^{(\text{DI})}), \qquad (6.30)$$

being $\lambda_d(\tau_r^{(\text{DI})}, v_r^{(\text{DI})})$ the approximate projection of a complex exponential function on the basis, see Equation (29) in [58].

Figure 6.10 (a) shows the good agreement between the CIR generated with the low-complexity implementation, in thin line, and the CIR obtained with the *traditional* GSCM approach, in bold line. The generated CIR in both cases is stored in a matrix of dimensions 16320×256, representing the time and the delay dimensions. The required simulation time using the low complexity approach is 55.6 s, in contrast to the 225.1 s needed using the *tradicional* GSCM approach, 4.1 times slower.

In order to be able to compare them, we show in Figure 6.10 (b) the approximation error defined as

$$e(t, \tau) = \left| \sum_{r=1}^{N_{(\text{DI})}} a_r^{(\text{DI})}(t) e^{-j2\pi f \tau_r(t)} - \sum_{r=1}^{N_{(\text{DI})}} \tilde{a}_r^{(\text{DI})} e^{-j2\pi \Delta f \tau_r^{(\text{DI})}} e^{j2\pi t v_r^{(\text{DI})}} \right|^2, \qquad (6.31)$$

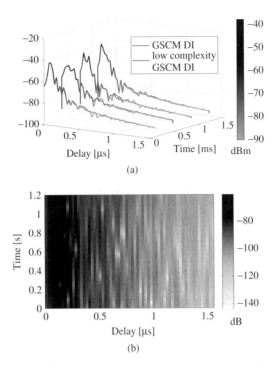

(a)

(b)

Figure 6.10 Generated CIRs only for the DI components using the traditional GSCM approach and the low-complexity implementation, subfigure (a). Approximation error in the time-delay plane, subfigure (b). Simulation performed with carrier frequency set to 5.9 GHz, bandwidth of 40 MHz and sampling time of 25 ns.

plotted in logarithmic scale. We obtain an error below -70 dB over all the time-delay plane. By using the low-complexity implementation, the simulation time becomes independent from the number of propagation paths. The simulation time reduction increases with the dimensions of the stored CIR matrix. For instance, when generating a CIR matrix two times larger than the one used in the example, the simulation time is now reduced by a factor of 5.3.

6.5   New Vehicular Communication Techniques

Key characteristics of vehicular channels are their non-stationarity, shadowing by other vehicles, and high Doppler shifts [27]. All have a major impact on link reliability and latency. For future dependable vehicular networking, improvements in coverage, reliability, scalability, and delay are required, calling for evolutionary enhancements in the IEEE 802.11p standard and inter-working with mobile telecommunication networks: 3G and LTE-A.

6.5.1   OFDM Physical (PHY) and Medium Access

Currently, the dominating standard for V2V communications is IEEE 802.11p, which is derived from the IEEE 802.11a Wireless Fidelity (WiFi) standard. Its PHY is based on

OFDM which is inherited from IEEE 802.11a with a minor modification: the symbol duration is doubled and, thus, the bandwidth is reduced to 10 MHz. Further, it integrates the Medium Access Control (MAC) layer from IEEE 802.11e, which is based on Carrier-Sense Multiple Access with Collision Avoidance (CSMA/CA) [59]. We note that both IEEE 802.11a and IEEE 802.11e were designed for Wireless Local Area Network (WLANs) in indoor environments with nomadic user mobility in mind.

There is some potential for future improvements of the PHY in IEEE 802.11p. Firstly, the currently specified pilot pattern requires the implementation of fairly sophisticated channel estimators at the RX to yield satisfactory performance in high-mobility scenarios [27]. An enhanced design of the pilot pattern would allow to reduce receiver complexity.

Secondly, the available frequency diversity within the Intelligent Transportation System (ITS) Band at 5.9 GHz is very limited. Multiple antennas at the OnBoard Unit (OBUs) and RSUs are recommended for exploiting both spatial and polarization diversity for mitigating small-scale fading and increasing link reliability. Such employment of multiple antennas also provides a reduction in the delay and Doppler spreads of the channel. On the RX side, the use of multiple antennas is compliant with the existing specifications. Transmitters with multiple antennas, on the other hand, have not yet been specified for IEEE 802.11p. Further, multi-antenna transmitters and receivers could mitigate co-channel interference by beamforming and/or increase data rate through spatial multiplexing. Various trade-offs between SINR gains from beamforming and increased data rates are achievable.

Thirdly, evolutionary improvements in the MAC are required to yield dependable systems in scenarios with a large number of vehicles crammed into a small area. Such scenarios feature many vehicles at low-mobility: for example, intersections, parking lots, and traffic jams. The current MAC layer is based on CSMA/CA which is a stochastic approach and does not guarantee bounds on the access delay.

Extensive simulations have indicated that the Carrier-Sense Multiple Access (CSMA) algorithm scales badly with the number of broadcasted Cooperative Awareness Message (CAMs) and is both unpredictable and unfair [59, 60]. On the other hand, deterministic MAC methods (e.g., self-organizing time division multiple access) are capable of guaranteeing worst case delays regardless of the number of competing nodes. The predictability of the worst case delay and scalability comes at the cost of network synchronization.

6.5.2 Relaying Techniques

The link budget for the roll-out of an IEEE 802.11p network can be relaxed by introducing relaying techniques. Although the primary benefit of relaying is pathloss mitigation, it can also exploit spatial diversity due to the broadcast nature of the shared radio channel as well as temporal diversity. Relaying techniques are capable of mitigating both small-scale *and* large-scale fading by spatial diversity. Several relaying techniques have been proposed in the literature: for example, Amplify-and-Forward (AF), Decode-and-Forward (DF), and compress-and-forward. AF is largely transparent to the employed transmission standard, but suffers from instability when the RX-side and TX-side of the relay occupy the same radio resource. DF is readily supported within the IEEE 802.11 family of specifications.

It has been recently shown for Infrastructure-to-Vehicle (I2V) communications using IEEE 802.11p that the throughput within the coverage range of the RSU is increased by 10%−20% by using a single vehicular OBU relay in-between the RSU and the OBU destination [61] in an I2V2V configuration. Depending on the position of the relay relative to the destination and the associated distances, it was possible to further increase the total throughput and extend the reliable communication range up to 40% [61].

6.5.3 Cooperative Coding and Distributed Sensing

Beyond relaying techniques as previously discussed, some individual nodes may engage in a tighter cooperation at the transmission and the reception for deploying a cooperative communication scheme [62]. Such cooperative communication networks, in which wireless nodes cooperate with each other for data aggregation, joint encoding, joint transmission, and joint decoding of information, promise significant gains in network throughput and reliability. This will be of much interest for the large-scale usage of floating car data which requires the energy-efficient transfer of small data packets. With the same diversity gain, the performance of a relaying technique is inferior to vehicular cooperative space-time coding techniques [62]. Therefore, the energy consumption for relaying is higher than for cooperative space-time coding.

6.5.4 Outlook

Following the progress in various standardization bodies, the International Telecommunication Union (ITU) has designated a 5.9 GHz band for ITS. For short and medium range vehicular connectivity, Wireless Access in Vehicular Environments (WAVE), also known as IEEE 802.11p, is released. It features increased transmission power and low-delay authentication mechanisms. For long-range vehicular connectivity which, is required for the management of traffic flow, third generation LTE-A will be beneficial. However, LTE-A will induce higher latencies which exclude its use from safety-critical applications with hard realtime constraints.

These aspects indicate that WAVE and LTE-A are complementary and motivate the implementation of a use-case dependent handover between different wireless access technologies. We therefore propose to study hybrid networking solutions for dependable vehicular connectivity which are based on WAVE, LTE-A, and other technologies. Such hybrid networking solutions may further mitigate shadowing by other vehicles by vertical handovers and relaying, as well as provide increased frequency-diversity through cooperative techniques as a measure against adverse small-scale fading.

References

1. P. Almers, E. Bonek, A. Burr, N. Czink, M. Debbah, V. Degli-Esposti, H. Hofstetter, P. Kyösti, D. Laurenson, G. Matz, A. F. Molisch, C. Oestges, and H. Özcelik, "Survey of channel and radio propagation models for wireless MIMO systems," *EURASIP Journal on Wireless Communications and Networking*, vol. 2007, 2007.
2. J. Karedal, N. Czink, A. Paier, F. Tufvesson, and A. F. Molisch, "Path loss modeling for vehicle-to-vehicle communications," *Vehicular Technology, IEEE Transactions on*, vol. 60, no. 1, pp. 323−328, January 2011.

3. J. Karedal, F. Tufvesson, N. Czink, A. Paier, C. Dumard, T. Zemen, C. Mecklenbräuker, and A. F. Molisch, "A geometry-based stochastic MIMO model for vehicle-to-vehicle communications," *Wireless Communications, IEEE Transactions on*, vol. 8, no. 7, pp. 3646–3657, July 2009.

4. L. Bernadó, A. Roma, A. Paier, T. Zemen, N. Czink, J. Karedal, A. Thiel, F. Tufvesson, A. F. Molisch, and C. F. Mecklenbräuker, "In-tunnel vehicular radio channel characterization," in *Vehicular Technology Conference (VTC Spring), 2011 IEEE 73rd*, May 2011, pp. 1–5.

5. O. Renaudin, V.-M. Kolmonen, P. Vainikainen, and C. Oestges, "Non-stationary narrowband MIMO inter-vehicle channel characterization in the 5-GHz band," *Vehicular Technology, IEEE Transactions on*, vol. 59, no. 4, pp. 2007–2015, May 2010.

6. P. Paschalidis, K. Mahler, A. Kortke, M. Peter, and W. Keusgen, "Pathloss and multipath power decay of the wideband car-to-car channel at 5.7GHz," in *Vehicular Technology Conference (VTC Spring), 2011 IEEE 73rd*, May 2011, pp. 1–5.

7. "MEDAV GmbH," http://www.medav.de.

8. A. F. Molisch, *Wireless Communications*. 2nd ed. John Wiley & Sons Ltd., 2011.

9. J. Karedal, "Measurement-based modeling of wireless propagation channels - MIMO and UWB," Ph.D. dissertation, Lund University, 2009.

10. P. A. Bello, "Characterization of randomly time-variant linear channels," *IEEE Transactions on Communications*, vol. 11, pp. 360–393, 1963.

11. L. Cheng, B. Henty, D. Stancil, F. Bai, and P. Mudalige, "Mobile vehicle-to-vehicle narrow-band channel measurement and characterization of the 5.9GHz Dedicated Short Range Communication (DSRC) frequency band," *Selected Areas in Communications, IEEE Journal on*, vol. 25, no. 8, pp. 1501–1516, October 2007.

12. L. Cheng, B. Henty, D. D. Stancil, F. Bai, and P. Mudalige, "A fully mobile, GPS enabled, vehicle-to-vehicle measurement platform for characterization of the 5.9GHz DSRC channel," in *Proc. IEEE Antennas Propagation Soc. Int. Symp.*, 2007, pp. 2005–2008.

13. L. Cheng, B. Henty, D. Stancil, and F. Bai, "Doppler component analysis of the suburban vehicle-to-vehicle DSRC propagation channel at 5.9GHz," in *Proc. IEEE Radio and Wireless Symp.*, 2008, pp. 343–346.

14. J. Maurer, T. Fugen, and W. Wiesbeck, "Narrow-band measurement and analysis of the inter-vehicle transmission channel at 5.2GHz," in *Vehicular Technology Conference, 2002. VTC Spring 2002. IEEE 55th*, vol. 3, 2002, pp. 1274–1278.

15. I. Sen and D. Matolak, "Vehicle-vehicle channel models for the 5-GHz band," *Intelligent Transportation Systems, IEEE Transactions on*, vol. 9, no. 2, pp. 235–245, June 2008.

16. L. Cheng, B. Henty, R. Cooper, D. Stancil, and F. Bai, "Multi-path propagation measurements for vehicular networks at 5.9GHz," in *Proc. IEEE Wireless Commun. Networking Conf.*, 2008, pp. 1239–1244.

17. G. Acosta and M. A. Ingram, "Model development for the wideband expressway vehicle-to-vehicle 2.4 GHz channel," in *IEEE Wireless Communications and Networking Conference (WCNC) 2006*, 3–6 April 2006.

18. I. Tan, W. Tang, K. Laberteaux, and A. Bahai, "Measurement and analysis of wireless channel impairments in DSRC vehicular communications," in *Communications, 2008. ICC '08. IEEE International Conference on*, May 2008, pp. 4882–4888.

19. P. Paschalidis, M. Wisotzki, A. Kortke, W. Keusgen, and M. Peter, "A wideband channel sounder for car-to-car radio channel measurements at 5.7 GHz and results for an urban scenario," in *Proc. IEEE Veh. Technol. Conf. 2008 Fall*, September 2008.

20. A. Paier, J. Karedal, N. Czink, H. Hofstetter, C. Dumard, T. Zemen, F. Tufvesson, C. F. Mecklenbräuker, and A. F. Molisch, "First results from car-to-car and car-to-infrastructure radio channel measurements at 5.2 GHz," in *International Symposium on Personal, Indoor and Mobile Radio Communications (PIMRC 2007)*, 3–7 September 2007, pp. 1–5.

21. G. F. Pedersen, J. B. Andersen, P. C. F. Eggers, J. O. Nielsen, H. Ebert, T. Brown, T. Yamamoto, T. Hayashi, and K. Ogawa, "Small terminal MIMO channels with user interaction," in *Proc. European Conf. Antennas Prop.*, 2007, pp. 1–6.

22. J. G. Proakis and M. Salehi, *Digital Communications*. McGraw Hill, 5th edition, 2005.

23. G. Matz, "On non-WSSUS wireless fading channels," *Wireless Communications, IEEE Transactions on*, vol. 4, no. 5, pp. 2465–2478, September 2005.

24. A. Paier, R. Tresch, A. Alonso, D. Smely, P. Meckel, Y. Zhou, and N. Czink, "Average downstream performance of measured IEEE 802.11p infrastructure-to-vehicle links," in *IEEE International Conference on Communications, Workshop on Vehicular Connectivity*, Cape Town, South Africa, 23-27 May 2010.

25. D. Kornek, M. Schack, E. Slottke, O. Klemp, I. Rolfes, and T. Krner, "Effects of antenna characteristics and placements on a vehicle-to-vehicle channel scenario," in *IEEE International Conference on Communications, Workshop on Vehicular Connectivity*, Cape Town, South Africa, 23-27 May 2010.

26. L. Reichardt, C. Sturm, and T. Zwick, "Performance evaluation of SISO, SIMO and MIMO antenna systems for car-to-car communications in urban environments," in *Intelligent Transport Systems Telecommunications, (ITST), 2009 9th International Conference on*, October 2009, pp. 51–56.

27. C. Mecklenbräuker, A. F. Molisch, J. Karedal, F. Tufvesson, A. Paier, L. Bernadó, T. Zemen, O. Klemp, and N. Czink, "Vehicular channel characterization and its implications for wireless system design and performance," *Proceedings of the IEEE*, vol. 99, no. 7, pp. 1189–1212, July 2011.

28. A. Thiel, O. Klemp, A. Paier, L. Bernadó, J. Karedal, and A. Kwoczek, "In-situ vehicular antenna integration and design aspects for vehicle-to-vehicle communications," in *4th European Conference on Antennas and Propagation*, Barcelona, Spain, April 2010.

29. A. Paier, L. Bernadó and, J. Karedal, O. Klemp, and A. Kwoczek, "Overview of vehicle-to-vehicle radio channel measurements for collision avoidance applications," in *Vehicular Technology Conference (VTC 2010-Spring), 2010 IEEE 71st*, May 2010, pp. 1–5.

30. ETSI TR 102 638, "Intelligent transport sytems (ITS); vehicular communications; basic set of applications; definitions," V1.1.1, June 2009.

31. D. Percival and A. Walden, *Spectral analysis for physical applications: multitaper and conventional univariate techniques*, ser. Cambridge University Press, 1993.

32. D. Thomson, "Spectrum estimation and harmonic analysis," *Proceedings of the IEEE*, vol. 70, no. 9, pp. 1055–1096, September 1982.

33. G. Matz, "Doubly underspread non-WSSUS channels: Analysis and estimation of channel statistics," Rome, Italy, June 2003, pp. 190–194.

34. A. Paier, T. Zemen, L. Bernadó, G. Matz, J. Karedal, N. Czink, C. Dumard, F. Tufvesson, A. F. Molisch, and C. F. Mecklenbräuker, "Non-WSSUS vehicular channel characterization in highway and urban scenarios at 5.2 GHz using the local scattering function," in *Smart Antennas, 2008. WSA 2008. International ITG Workshop on*, February 2008, pp. 9–15.

35. D. Slepian, "Prolate spheroidal wave functions, Fourier analysis, and uncertainty - V: The discrete case," *The Bell System Technical journal*, vol. 57, no. 5, pp. 1371–1430, May-June 1978.

36. L. Bernadó, T. Zemen, A. Paier, J. Karedal, and B. Fleury, "Parametrization of the local scattering function estimator for vehicular-to-vehicular channels," in *Vehicular Technology Conference Fall (VTC 2009-Fall), 2009 IEEE 70th*, September 2009, pp. 1–5.

37. L. Bernadó, T. Zemen, A. Paier, G. Matz, J. Karedal, N. Czink, F. Tufvesson, M. Hagenauer, A. F. Molisch, and C. F. Mecklenbräuker, "Non-WSSUS Vehicular Channel Characterization at 5.2 GHz - Spectral Divergence and Time-Variant Coherence Parameters," in *Assembly of the International Union of Radio Science (URSI)*, August 2008, pp. 9–15.

38. M. Herdin, N. Czink, H. Ozcelik, and E. Bonek, "Correlation matrix distance, a meaningful measure for evaluation of non-stationary MIMO channels," in *Vehicular Technology Conference, 2005. VTC 2005-Spring. 2005 IEEE 61st*, vol. 1, May-June 2005, pp. 136–140.

39. A. Paier, J. Karedal, N. Czink, H. Hofstetter, C. Dumard, T. Zemen, F. Tufvesson, A. F. Molisch, and C. Mecklenbräuker, "Car-to-car radio channel measurements at 5 GHz: Pathloss, power-delay profile, and delay-Doppler spectrum," in *Wireless Communication Systems, 2007. ISWCS 2007. 4th International Symposium on*, October 2007, pp. 224–228.

40. J. Kunisch and J. Pamp, "Wideband car-to-car radio channel measurements and model at 5.9 GHz," in *Vehicular Technology Conference, 2008. VTC 2008-Fall. IEEE 68th*, September 2008, pp. 1–5.

41. L. Cheng, B. Henty, F. Bai, and D. Stancil, "Highway and rural propagation channel modeling for vehicle-to-vehicle communications at 5.9 GHz," in *Antennas and Propagation Society International Symposium, 2008. AP-S 2008. IEEE*, July 2008, pp. 1–4.

42. O. Renaudin, V.-M. Kolmonen, P. Vainikainen, and C. Oestges, "Wideband MIMO car-to-car radio channel measurements at 5.3 GHz," in *Vehicular Technology Conference, 2008. VTC 2008-Fall. IEEE 68th*, September 2008, pp. 1–5.

43. O. Renaudin, V.-M. Kolmonen, P. Vainikainen, and C. Oestges, "Wideband measurement-based modeling of inter-vehicle channels in the 5 GHz band," in *Antennas and Propagation (EUCAP), Proceedings of the 5th European Conference on*, April 2011, pp. 2881–2885.

44. L. Reichardt, J. Maurer, T. Fugen, and T. Zwick, "Virtual drive: A complete V2X communication and radar system simulator for optimization of multiple antenna systems," *Proceedings of the IEEE*, vol. 99, no. 7, pp. 1295–1310, July 2011.

45. J. Pontes, L. Reichardt, and T. Zwick, "Investigation on antenna systems for car-to-car communication," *Selected Areas in Communications, IEEE Journal on*, vol. 29, no. 1, pp. 7–14, January 2011.

46. J. Maurer, T. Fugen, T. Schafer, and W. Wiesbeck, "A new inter-vehicle communications (IVC) channel model," in *Vehicular Technology Conference, 2004. VTC2004-Fall. 2004 IEEE 60th*, vol. 1, September 2004, pp. 9–13.

47. F. Vatalaro and A. Forcella, "Doppler spectrum in mobile-to-mobile communications in the presence of three-dimensional multipath scattering," *Vehicular Technology, IEEE Transactions on*, vol. 46, no. 1, pp. 213–219, February 1997.

48. "IEEE P802.11p: Part 11: Wireless LAN Medium Access Control (MAC) and Physical Layer (PHY) Specifications: Amendment 6: Wireless Access in Vehicular Environments," July 2010.

49. G. Acosta-Marum and M. Ingram, "Six time- and frequency – selective empirical channel models for vehicular wireless LANs," *Vehicular Technology Magazine, IEEE*, vol. 2, no. 4, pp. 4–11, December 2007.

50. M. Pätzold, B. Hogstad, and N. Youssef, "Modeling, analysis, and simulation of MIMO mobile-to-mobile fading channels," *Wireless Communications, IEEE Transactions on*, vol. 7, no. 2, pp. 510–520, February 2008.

51. A. Zajic and G. Stuber, "Three-dimensional modeling, simulation, and capacity analysis of space-time correlated mobile-to-mobile channels," *Vehicular Technology, IEEE Transactions on*, vol. 57, no. 4, pp. 2042–2054, July 2008.

52. X. Cheng, C.-X. Wang, D. Laurenson, S. Salous, and A. Vasilakos, "An adaptive geometry-based stochastic model for non-isotropic MIMO mobile-to-mobile channels," *Wireless Communications, IEEE Transactions on*, vol. 8, no. 9, pp. 4824–4835, September 2009.

53. L.-C. Wang, W.-C. Liu, and Y.-H. Cheng, "Statistical analysis of a mobile-to-mobile rician fading channel model," *Vehicular Technology, IEEE Transactions on*, vol. 58, no. 1, pp. 32–38, January 2009.

54. A. Chelli and M. Pätzold, "The impact of fixed and moving scatterers on the statistics of MIMO vehicle-to-vehicle channels," in *Vehicular Technology Conference, 2009. VTC Spring 2009. IEEE 69th*, April 2009, pp. 1–6.

55. P. Petrus, J. Reed, and T. Rappaport, "Geometrical-based statistical macrocell channel model for mobile environments," *Communications, IEEE Transactions on*, vol. 50, no. 3, pp. 495–502, March 2002.

56. T. Abbas, J. Karedal, F. Tufvesson, A. Paier, L. Bernadó, and A. Molisch, "Directional analysis of vehicle-to-vehicle propagation channels," in *Vehicular Technology Conference (VTC Spring), 2011 IEEE 73rd*, May 2011, pp. 1–5.

57. N. Czink, F. Kaltenberger, Y. Zhou, L. Bernadó, T. Zemen, and X. Yin, "Low-complexity geometry-based modeling of diffuse scattering," in *EUCAP 2010, 4th European Conference on Antennas and Propagation*, April 2010, pp. 1–5.

58. F. Kaltenberger, T. Zemen, and C. W. Ueberhuber, "Low-complexity geometry-based MIMO channel simulation," *EURASIP J. Adv. Sig. Proc.*, vol. 2007, 2007.

59. E. G. Ström, "On medium access and physical layer standards for cooperative intelligent transport systems in Europe," *Proceedings of the IEEE*, vol. 99, no. 7, pp. 1183–1188, July 2011.

60. K. Bilstrup, E. Uhlemann, E. G. Ström, and U. Bilstrup, "On the ability of the 802.11p MAC method and STDMA to support real-time vehicle-to-vehicle communication," in *EURASIP Journal on Wireless Communications and Networking*, vol. 2009, Article ID 902414.

61. V. Shivaldova, T. Paulin, A. Paier, and C. F. Mecklenbräuker, "Performance measurements of multi-hop communications in vehicular ad hoc network," in *IEEE ICC 2012 Workshop on Intelligent Vehicular Networking: V2V/V2I Communications and Applications*, Ottawa, Canada, June 2012, submitted.

62. T.-D. Nguyen, O. Berder, and O. Sentieys, "Energy-efficient cooperative techniques for infrastructure-to-vehicle communications," vol. 12, no. 3, 2011, pp. 659–668.

7

Multi-User MIMO Channels

Fredrik Tufvesson[1], Katsuyuki Haneda[2] and Veli-Matti Kolmonen[2]
[1]*Lund University, Sweden*
[2]*Aalto University, Finland*

7.1 Introduction

Multiple Input Multiple Output (MIMO) technologies are nowadays used in the major cellular and wireless standards such as LTE, and the IEEE 802.11 family. For single-link MIMO, correlation effects between antennas are of high interest since they will determine the ability to use spatial multiplexing in an efficient way. In addition, the received signal strength for each link is of course also of major interest since it ultimately determines the possibility to detect and decode the data transmitted. Usually the antennas used for a single link MIMO are located quite close to each other, though there are concepts like distributed MIMO, cooperative communication and Coordinated Multipoint (CoMP) communication as will be discussed in the next chapter. A close proximity of the antennas means that the power levels usually are quite similar, though variations might be present due to polarization effects, variations in antenna patterns and so on. Shadowing by larger objects typically affects all antennas simultaneously.

In multi-user MIMO those properties of course still hold for the individual links, but in addition to this it is important to consider the *joint* behavior of the individual MIMO links. It has been shown that there can be correlation between the links even if the units (different base stations/access points, or different user terminals) are widely separated from each other. For joint processing it is important to remember that the received power levels might differ significantly for different links, primarily due to differences in distance between units, but also if some links are shadowed and some are not.

Basically all cellular systems or Wireless Local Area Networks (WLANs) can be seen as multi-user systems, and hence if they also use MIMO as multi-user MIMO systems. However, in this chapter we mainly focus on the situation where some kind of joint processing is made to recover the different data streams.

LTE-Advanced and Next Generation Wireless Networks: Channel Modelling and Propagation, First Edition.
Edited by Guillaume de la Roche, Andrés Alayón Glazunov and Ben Allen.
© 2013 John Wiley & Sons, Ltd. Published 2013 by John Wiley & Sons, Ltd.

7.2   Multi-User MIMO Measurements

7.2.1   General Information About Measurements

Single user MIMO measurements have been extensively published in the scientific literature after the introduction of MIMO [1]. The MIMO channel characterization was further increased by the introduction of the first directional measurement equipment [2, 3] and the following double-directional channel modelling principle [4].

Due to the above mentioned steps the MIMO measurement techniques have been studied and developed intensively and highly complicated measurement configurations have been built. This measurement equipment is very accurate and produces measurement data which can be utilized in many ways, not only for wireless communication but also propagation channel investigations and modelling. On the other hand, simple configurations for investigating particular communication systems have also been developed and utilized successfully. The price and the accuracy of the simplified equipment are not on a similar level as the more complicated equipment, but they can provide sufficient information for a particular study.

Regardless of the overall system complexity, the measurement characterization comes down to measurement parameters and specifications which have been known for a long while. Although post-processing techniques have been developed extensively and they can be used to correct minor errors, the RF related hardware still should be of sufficient quality. For example, some local oscillator (LO) drifting can be corrected afterwards[1] but, for example, saturation caused by radio frequency (RF) analog circuitry is much more complicated to correct, and often cannot be corrected. Therefore, these RF related challenges exist in SISO, SU-MIMO, and MU-MIMO measurements.

We should note that MIMO often refers to communication technique where multiple antennas are utilized. In this section, the term MIMO is also used to refer to measurement system configuration where multiple antennas are used. By appropriately placing the multiple antennas in the measurement system we can use, for example, the phased array principle to estimate the directions of departure and arrival using post-processing techniques. Extensive literature on the direction finding studies exists and is thus not discussed further in this section.

In the measurements, the channel between each transmit and receiver antenna have to be distinguished. In single-link MIMO measurements three different approaches have been used; 1) virtual array principle, 2) switched based, and 3) individual RF chain based approaches. In the virtual array principle the measurement antennas are moved to different places using, for example, linear scanners and measurements for each location used. With this method the antenna configuration can be easily changed. This however increases the overall measurement time. In the switched based approach, different antenna elements in the transmitter and receiver are connected together using a predefined order and timing [2, 3]. In this approach, the antenna configuration is pre-designed, and hence it cannot be changed easily. With this method even large MIMO channel matrices can be measured in a short period of time. Finally, in the individual RF chain based approach different distinguishable signals can be transmitted from each of the transmitter elements. At the receiver, for example, switching, can be used to create a MIMO configuration.

---

[1] As long as sufficient data set is collected during the measurements.

In [5] different codes were used in the transmission, whereas, in [6] slightly different frequencies were used.

For MU-MIMO measurements the transmit and receiver antenna elements between different *links* have to be known. This complicates the overall system significantly.

7.2.2 Measurement Techniques

The single link measurement techniques can be extended to support multi-link measurements. This, however, requires more time to conduct the measurements (virtual approach) or very complicated measurement equipment (real time measurements).

Generally the MU-MIMO measurement techniques can be divided as

- Virtual measurement,
- Distributed measurement,
- Real time measurements.

Figure 7.1 illustrates the different measurement techniques. In the **virtual measurement** illustrated in Figure 7.1(a), a single-link measurement system (channel sounder, VNA, and so on) is used to measure the multi-user channel in a consecutive way. The obvious benefit of this is that already existing measurement equipment can be used. This method, however, requires that the environment remains constant over the whole measurement period. There exist only few studies which considers the applicability of the virtual measurement approach.

The **distributed measurement** illustrated in Figure 7.1(b), utilizes a single-link measurement system–often a channel sounder with fast switches – is used along with long cables to enable multinode measurements. As in the previous method, existing single-link hardware is also sufficient here, but no stability of the environment is required. However, the measurement locations should remain within the cable length.

And finally, in the **real-time measurements** illustrated in Figure 7.1(c) multiple MIMO links are measured simultaneously. This approach requires multiple synchronized nodes. Since multidimensional channel sounding equipment is very expensive, so often simplified measurement equipment is used in real-time measurements. Therefore, utilizing multiple sounders is not feasible in practice and hence no studies using a multidimensional channel sounders for multi-link studies exist to date. Only one study using a single transmitter and two receivers utilizing a real time measurement approach with two channel sounders has been reported in [7].

7.2.2.1 Validity of Virtual Measurement Approach

To enable fast multidimensional channel measurements the measurement equipment has used the switching measurement principle [2, 3]. This approach is very convenient for single-link operation but requires very accurate synchronization. This requirement, however, complicates the extension to multi-link scenario as in that case all the nodes have to be synchronized to the switching pattern as the measurements cannot be performed in a asynchronous manner due to interference problems.

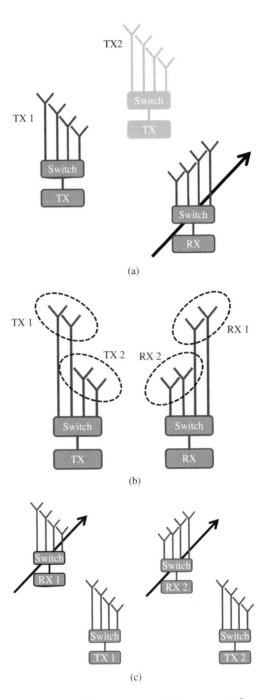

Figure 7.1 MU-MIMO measurement techniques. (a) Measurement of same routes with different TX locations in concecutive manner creates a virtual multi-link scenario. (b) In distributed measurements different antenna locations are measured within one measurement sequence. (c) In real-time measurements multiple links are measured at the sametime.

As the virtual measurement approach is feasible in practice it has been used in many campaigns. However, the validity of the virtual MU-MIMO measurement approach has not been thoroughly investigated. Two studies concerning this particular aspect have been done in [8, 9] and [10]. In [9] an extensive outdoor-to-indoor measurement campaign was conducted in Stanford University in 2008 in a regular office environment. The transmitter was located in two outdoor positions and three different directions were used in both locations while the receiver was used to measure five routes for each TX location *and* rotation inside the building. The TX was lifted with a crane above the surrounding objects while the receiver was moved using a trolley. In the measurements application specific antennas without antenna calibration were used. Measurement equipment parameters are listed in Table 7.1.

In the measurement campaign all the indoor routes were measured several times. In the campaign several precautions were taken to ensure that "the same route" would be used in the consecutive measurements:

1. The measurement route was marked into the floor and every attempt was made to follow this route.
2. A distance trigger wheel was used to trigger the measurement of the channel.
3. On each measurement run all the people were in the same places.

In [8], the measurement results from the above measurement campaign were analyzed with respect to how similar the channels remain between consecutive measurement runs. In the analysis two similarity metrics, both described in the next section, were used: 1) matrix collinearity and 2) similarity of spatial correlation matrices. In [8], the so-called correlation matrix distance (CMD) was used to evaluate the differences in the channel between consecutive measurements. In that analysis it was found that the channel characteristics at the TX side remain similar between the measurements. Furthermore, the characteristics at the RX side also seemed to remain very similar. Although the full correlation matrices indicated good similarity between the measurements, the full correlation matrices are more sensitive to the changes. Hence, in that analysis it was concluded that the general trends and the statistics of the multi-user channels can be captured using a virtual measurement approach.

In [10], a similar study to the above was conducted. There outdoor measurements were conducted in an urban environment with sophisticated antenna arrays capable of multipath parameter extraction. In this study, the same measurement route was driven using a car multiple times. By utilizing advanced parameter estimation algorithm [11] the parameters of the multipath components (MPCs) were extracted. After that, the evaluation

Table 7.1 Measurement equipment specifications for the validation of the virtual measurement technique

Parameter	Value [8]	Value [10]
Carrier frequency [GHz]	2.45	2.53
Bandwidth [MHz]	240	100
Transmit power [dBm]	27	46
Number of channels ($N_{TX} \times N_{RX}$)	4×8	16×58

of the similarity between different measurement runs were inspected in the MPC domain using delay and angular spreads. In the analysis it was found that in most cases different measurement runs provided comparable values. Only angular spread in elevation was found to have a large decorrelation level. Hence, that study also concluded that large scale characterization of multi-link propagation channels can be performed in a virtual way.

7.2.3   Phase Noise

The potential gain achieved by utilizing the MIMO communications technology can be evaluated by the MIMO capacity [12] which is directly dependent on the eigenvalue spectrum of the MIMO channel matrix. This spectrum is affected by the random properties of the channel matrix. One measurement system error is produced by the local oscillators which are used in the channel sounding equipment. In particular, the measurement equipment utilizing high accuracy LOs in the transmitter and the receiver to achieve long measurement distance is especially vulnerable to these errors. The LOs can drift with respect to each other and have small phase or frequency variation, that is, phase noise.

The LO drift can be – to some extent – be detected and corrected in post-processing in [7] and [13], but the phase noise cannot be corrected. Hence, the phase noise can randomize the channel matrix and hence increase the channel capacity.

This problem has been noted in an early phase and many studies have been conducted for the evaluation of the phase noise in MIMO measurement systems. The phase noise was theoretically re-examined in [14], where the effect of phase noise in particular in switched measurement systems to MIMO capacity was analyzed. There, it was shown that the phase noise can lead to 100% errors in capacity estimation for high signal to noise ratio (SNR) scenarios. However, in [15–18] effect of phase noise was analyzed for single link channel sounders, and it has been confirmed that the capacity estimation error is clearly smaller in practical channel sounders due to correlated phase noise. In [7] the effect of phase noise to dual-link capacity calculation was analyzed and found to be insignificant.

7.2.4   Measurement Antennas

In general the measurement antennas used in the MU-MIMO measurements can be divided into three categories

- system specific antennas,
- advanced antenna arrays,
- synthetic antennas.

A very commonly used method for the measurements is to utilize system specific – and often – commercial antennas. Examples of these are commercial base station antennas, dipoles, and USB dongles for the UE [9, 13, 19–21]. These studies can be used for radio channel evaluations [22] and the benefit of this approach is that it is very cost efficient. However, not very elaborate propagation analysis can be done with these antennas.

Only a few studies exist which utilize sophisticated antenna arrays for real-time MU-MIMO channel measurements. In the literature, however, multiple measurement campaigns have been conducted by utilizing virtual antenna arrays at the nodes. With the virtual antenna approach "any" antenna configuration can be generated on-site and hence it is not restricted to any predefined array configuration. As noted earlier, a downside of this method is that not only does it require the environment to be static for single MIMO link measurement, but the environment has to be static for multiple single MIMO link measurements.

Antenna arrays using the switched measurement principle have been used in [10, 23], and in particular in [7] for dual-link real-time measurements. The utilization of sophisticated antennas enable advanced post-processing of the measurement data so that the propagation channel characterization can be performed. In the study on the virtual measurement principle in [10], the RIMAX algorithm [11] was used whereas in [7] spherical and cylindrical antenna configurations were used along with detailed radiation pattern characterization and efficient post-processing algorithm [24].

The utilization of the parameter estimation algorithms provide a very detailed knowledge of the propagation channel. Furthermore, the MPC parameters defined by the double-directional channel model can be extracted and used for multi-link propagation channel characterization and modelling. This approach provides a very detailed understanding on the relevant phenomena in the multi-link scenario and it has also been adopted in the COST2100 model, explained in Section 8.4. We should note, that this approach enables the usage of other antenna configurations than the ones used in the measurements [25].

An optional approach is to modify the radiation pattern of the measurement array by appropriately weighting the received signals. In [23] MU-MIMO measurements were conducted with sophisticated antenna arrays. Although the measurements were conducted with a sophisticated antenna ray, the polarimetric antenna radiation pattern information was used to form simple omnidirectional antenna configuration at the RX. This way the measurement data could be utilized without demanding post-processing calculations.

7.2.5   Measurement Campaigns

7.2.5.1   Real Time Measurements

In [7] Two-channel sounder equipment was combined to a real-time dual-link propagation channel measurement system. The measurement system is based on channel sounder constructed in Aalto University[2] [16] and commercial RUSK wideband MIMO channel sounder from Lund University manufactured by MEDAV GmbH [2, 26]. As the synchronization between the transmitters could not be achieved, the orthogonality between transmitted signals cannot be preserved. Therefore, the TKK *transmitter* (TX) could not be used, and hence, the channel sounder consists of LU TX and LU and TKK receivers (RX). Both sounders use the switched array principle.

The fast RF switches at the TX and RX are synchronized to each other and to the transmit signal so that the antenna switching occurs at the beginning of the signal period. The antenna switching period is synchronized with the sampling unit, hence the correspondence between channel switching and channel samples is known. All signal processing of the received signal is performed in the post-processing.

---

[2] Formerly known as Helsinki University of Technology (TKK).

The cylindrical antenna structure used for the LU RX consists of 4 rings of 16 dual polarized antenna elements, so the antenna has 64 dual polarized (horizontal and vertical) antenna elements in total. Out of these, 16 dual-polarized elements were used so that 8 elements from each of the two middle rows were selected in an alternating fashion.

These antennas, along with the detailed antenna radiation pattern measurement, enable the dynamic directional characteristics of the two links to be evaluated. For the directional evaluation extended Kalman Filter (EKF) based algorithm was used [24].

In [27, 28] the Eurecom MIMO Openair Sounder (EMOS) [13] has been used to evaluate MU-MIMO measurements conducted in outdoor environments. The measurement system is a real-time measurement system consisting of an ordinary laptop computer.

The transmitter antenna was a Powerwave 3G antenna with four antenna elements whereas the terminal antennas were Panorama 3G laptop mountable antennas. The transmitter continuously transmits orthogonal frequency division multiplexing (OFDM) sounding sequence. The time synchronization is done based on the received signal using the OFDM synchronization symbols. The synchronization between different links is done based on the frame number dictated by the transmitter. Furthermore, the phase noise – or drift – is compensated using a phase rotation during the post-processing. As in the receivers the received signal along with the frame number is recorded and the frame synchronization is performed in the post-processing it is easy to deploy one or more receivers and hence enable MU-MIMO measurements.

In [20] multi-site measurements were conducted in Kista, Stockholm, Sweden. In the measurements three base station sites were used with a single antenna at each site, whereas four antennas were used in the receiver. The overall system is based on the Ericsson LTE testbed. In the measurements, standard base station antennas were used at the base stations and the inter-site distance was from 350 to 600 m with a transmit power of 36 dBm. The receiver antennas were two electrical and two magnetic antennas mounted on top of the measurement vehicle. The measurement frequency is 2.66 GHz and the bandwidth is 20 MHz.

In [6, 29] MU-MIMO indoor and outdoor-to-indoor measurements were conducted around 1.8 GHz with a bandwidth of 9.6 kHz. In the measurements various antenna configurations were used. At the transmitter dual-polarized planar antennas were used along with Powerwave dual-polarized antenna at some measurement locations and on the receiver side, two 4-element arrays were used; a 4-element monopole and 4-element PIFA array. The developed measurement equipment consists of separate transmitter RF chains for each antenna. These transmitters used slightly different frequencies to distinguish the different transmitter antennas. Similarly, at the receiver individual RF chains were used for each receiver antenna. The same LO signal was used for each receiver RF chain.

The measurement system specifications for the different real-time measurement campaigns are summarized in Table 7.2.

7.2.5.2 Distributed Measurements

In addition to measurements explained earlier from [9, 30] in those studies distributed measurements were conducted in outdoor-to-indoor and indoor-to-indoor scenarios. In order

Table 7.2 Real-time measurement system specifications

Parameter	[7]	[28]	[20]	[6, 29]
Carrier frequency [GHz]	5.3	1.9	2.66	1.8
Bandwidth [MHz]	120	5	20	9.6 kHz
Transmit power [dBm]	27	34	36	−5/25
Number of channels ($N_{TX} \times N_{RX}$)	$30 \times 32, 30 \times 30$	$N \times 4 \times 2$	$3 \times 1 \times 4$	$2 \times 2 \times 4$

to achieve comprehensive measurements, distributed measurements combined with virtual measurement approaches were used. For example, first a measurement from outdoor to indoor receiver locations were measured. After this, the equipment was transported indoors and the transmitting elements placed in "exactly" the same positions as some of the indoor receivers in the previous step, and then indoor-to-indoor measurements were conducted in a distributed manner.

In [19] similar measurements to the abovementioned were carried out using Electrobit PropSound [31] in indoor and outdoor-to-indoor scenarios. The measurement frequency range was 3.8 GHz with 50 MHz bandwidth. The user terminal antennas were dipole antennas whereas at the TX dipole antennas (indoor) as well as dual-polarized patch antennas (outdoor-to-indoor) were used.

In [21] a real-time *relaying* measurement campaign was conducted. The measurements were conducted in an indoor environment where the channels from two access points to two user equipment were measured. In addition, the link between the two user equipments was measured. In the measurements four monopole arrays were used in the APs and four patch elements in the UEs. In this measurement system, the channel between the APs was transmitting, the UE1 was receiving while the UE2 had TX and RX functionality.

The measurement setup described in [23] has been used for outdoor MU-MIMO measurements by combining a distributed and a virtual measurement technique. In those measurements, several basestation sectors were used in each basestation location. These basestation sector antennas were connected to same switching procedure and hence enabling (distributed) multi-sector transmission to be used. After measuring the surrounding measurement routes a new basestation site was used in a similar manner, hence creating a virtual measurement approach. In the transmitter planar array and commercial basestation antennas were used, whereas in the receiver different patch antenna arrays were used. The operation frequency was around 2 GHz.

The measurement system specifications for the different real-time measurement campaigns are summarized in Table 7.3.

Table 7.3 Distributed measurement system specifications

Parameter	value [19]	value [21]	value [23]
Center frequency [GHz]	3.8	5.2	2.53
Bandwidth [MHz]	50	200	20
BS Tx power [dBm]	23		45
Number of channels ($N_{TX} \times N_{RX}$)	$4/8 \times 8$	4×4	$16/6 \times 10/16^3$

7.3   Multi-User Channel Characterization

As in any other wireless system the ultimate performance that can be achieved in a particular situation is determined by the radio channel and its properties. In single link MIMO the possible performance is mainly determined by the SNR and the correlation of the signals at the antennas. The SNR at a specific spot is in turn determined by the distance dependent path loss and the large-scale fading. Since different users typically are located with various distances to the central unit (base station, access point and so on) and they will have different large scale fading realizations it is important to consider that their transmitted signals will arrive with different signal strengths at the receiver. This is important for joint processing, where the signal levels of some users might be so bad that they are hidden in the signals of other users.

When it comes to multiuser MIMO and the difference compared to single-link MIMO we pay special attention to correlation effects between the users. There are two kinds of correlation that are of interest: correlation of the large-scale fading between users and correlated small scale fading properties. The latter can be described as "correlation of correlation" and is attributed to correlated angular spreads for the users; if one MIMO link has high antenna correlation is there then also a large probability that a second MIMO link has high antenna correlation?

In order to characterize the similarity between the channels for different MIMO links various measures can be used. Those measures are typically characterizing:

1. The similarity between the channel matrices of two MIMO link.
2. The similarity between the correlation matrices for the two MIMO links, or
3. The alignment of the sub-spaces used by the two links.

The matrix collinearity is used to characterize the similarity of two matrices. It is defined in [32] as

$$c\left(\mathbf{H}_1, \mathbf{H}_2\right) = \frac{\left|\mathrm{tr}\left(\mathbf{H}_1\mathbf{H}_2^{\mathrm{H}}\right)\right|}{\left\|\mathbf{H}_1\right\|_{\mathrm{F}}\left\|\mathbf{H}_2\right\|_{\mathrm{F}}} \tag{7.1}$$

where $\mathbf{H}_1$ and $\mathbf{H}_2$ are the matrices to be compared; $(\cdot)^{\mathrm{H}}$ denotes conjugate transpose; and $\|\cdot\|_{\mathrm{F}}$ is the Frobenius norm. This metric can be used to evaluate the similarity of the matrix subspaces and hence it can be used to evaluate the similarity of a channel matrix. For the collinearity metric $c \in [0, 1]$.

In [33] the matrix collinearity is further developed to a correlation matrix distance (CMD). This metric is a special case of the matrix collinearity. It is based on the collinearity of the correlation matrices and is defined as

$$d_{\mathrm{corr}}\left(\mathbf{R}_1, \mathbf{R}_2\right) = 1 - \frac{\mathrm{tr}\left(\mathbf{R}_1\mathbf{R}_2^{\mathrm{H}}\right)}{\left\|\mathbf{R}_1\right\|_{\mathrm{F}}\left\|\mathbf{R}_2\right\|_{\mathrm{F}}} \tag{7.2}$$

where the $\mathbf{R}$ can represent the RX correlation matrix, the TX correlation matrix or the full correlation matrix. Those different correlation matrices are defined as $\mathbf{R}_{\mathrm{RX},1} = E\left\{\mathbf{H}_1\mathbf{H}_1^{\mathrm{H}}\right\}$ for the RX correlation, as $\mathbf{R}_{\mathrm{TX},1} = E\left\{\mathbf{H}_1^{\mathrm{H}}\mathbf{H}_1\right\}$ for the TX correlation, and as $\mathbf{R}_{\mathrm{full},1} = E\left\{\mathrm{vec}\left(\mathbf{H}_1\right)\mathrm{vec}\left(\mathbf{H}_1\right)^{\mathrm{H}}\right\}$ for the full MIMO matrix. As illustrated in Section 7.4, this metric has also been used in MU-MIMO channel modelling.

Another metric to calculate the similarity of the correlation matrices is the so-called geodesic distance [34], defined as

$$d_{\text{geod}}\left(\mathbf{R}_1, \mathbf{R}_2\right) = \left(\sum_k \left|\log\left\{\lambda_{\mathbf{R}_1^{-1}\mathbf{R}_2}\right\}_k\right|\right)^{1/2}$$ (7.3)

where $\left\{\lambda_{\mathbf{R}_1^{-1}\mathbf{R}_2}\right\}_k$ is the kth eigenvalue of $\mathbf{R}_1^{-1}\mathbf{R}_2$.

The capacity of a single MIMO link under interference in a multiuser system is determined by the mutual information between the transmitted and received signals. The mutual information is given by [35]

$$I\left(\mathbf{x}; \mathbf{y}, \mathbf{H}\right) = E\left\{\log_2\left[\det\left(\mathbf{I} + \mathbf{H}_1\mathbf{H}_1^H\mathbf{R}_2^{-1}\right)\right]\right\}$$ (7.4)

where $\mathbf{R}_2$ is the correlation matrix of the interfering channels.

In cases where the transmitter has perfect side information about the interference at the receiver the optimal sum rate can be achieved by so called Dirty Paper Coding. The technique is, however, very complex and in practice less complex, often linear, pre-coders have to be used.

Next we give an example [36] on how the correlation and power imbalance affect the sum rates for different pre-coding methods: DPC, zero forcing (ZF) and minimum mean squared error (MMSE) pre-coders in a two-user MIMO system. We consider the downlink of a multiuser-MIMO system where the base station has M antennas, and serves K single-antenna users. The received $K \times 1$ vector $\vec{y}$ at the user sides can be described as

$$\vec{y} = \sqrt{\rho}\vec{H}\vec{z} + \vec{n},$$ (7.5)

where $\vec{H}$ is the $K \times M$ channel matrix, $\vec{z}$ is the transmitted vector across the M antennas, and $\vec{n}$ is a noise vector. We have further $\vec{z} = \vec{U}\vec{x}$, where $\vec{U}$ is the $M \times K$ pre-coding matrix.

The Gram matrix associated with $\vec{H}$ can now be written as

$$\vec{G} \triangleq \vec{H}\vec{H}^H = \begin{bmatrix} 1 + g & \delta \\ \delta^* & 1 - g \end{bmatrix},$$ (7.6)

where g $(0 \leq g < 1)$ denotes the power imbalance between the two users, and δ measures the correlation between the two channels; the correlation between the channels to the two users can be expressed as $|\delta|/\sqrt{1 - g^2}$.

The DPC capacity can be calculated as

$$C_{\text{DPC}} = \begin{cases} \log_2\left[1 + \rho + \dfrac{\rho^2\left(1 - g^2 - |\delta|^2\right)^2 + 4g^2}{4\left(1 - g^2 - |\delta|^2\right)}\right], & |\delta|^2 \leq \delta_{\text{th}} \\ \log_2\left[1 + \rho\left(1 + g\right)\right], & |\delta|^2 > \delta_{\text{th}}. \end{cases}$$ (7.7)

If $|\delta|^2$ is higher than a certain threshold δ_{th}, all power will be allocated to the strongest channel, and the DPC capacity saturates at the single-user capacity.

The resulting sum rate for ZF pre-coding becomes

$$
C_{ZF} = \begin{cases} \log_2 \left[\dfrac{\left(2 + \rho \left(1 - g^2 - |\delta|^2\right)\right)^2}{4\left(1 - g^2\right)} \right], & |\delta|^2 \le \delta_{th} \\[3mm] \log_2 \left[1 + \dfrac{\rho \left(1 - g^2 - |\delta|^2\right)}{1 - g} \right], & |\delta|^2 > \delta_{th}. \end{cases} \tag{7.8}
$$

The ZF interference cancellation has significant signal power penalty if the two user channels are highly correlated. From (7.8) we can see that the capacity goes to zero when the channel correlation is high (low orthogonality), that is, when $|\delta|$ approaches $\sqrt{1 - g^2}$.

The sum rate of the MMSE pre-coding scheme, subject to $P_1 + P_2 = 1$, is given by

$$
C_{MMSE} = \max_{P_1, P_2} \sum_{i=1}^{2} \log_2(1 + SINR_i), \tag{7.9}
$$

where

$$
SINR_1 = \frac{\rho P_1 \left[(1 + g)(1 - g + \alpha) - |\delta|^2\right]^2}{\rho P_2 \alpha^2 |\delta|^2 + \gamma \left[(1 + g + \alpha)(1 - g + \alpha) - |\delta|^2\right]^2} \tag{7.10}
$$

and

$$
SINR_2 = \frac{\rho P_2 \left[(1 - g)(1 + g + \alpha) - |\delta|^2\right]^2}{\rho P_1 \alpha^2 |\delta|^2 + \gamma \left[(1 + g + \alpha)(1 - g + \alpha) - |\delta|^2\right]^2}. \tag{7.11}
$$

Closed form expressions of optimal power allocation and maximized sum rate can be reached but are far too long and complicated to be given here, but in the case of $g = 0$, a simple expression of the sum rate is obtained as

$$
C_{MMSE}|_{g=0} = 2 \log_2 \left[1 + \frac{\rho}{2} \left(1 - \frac{\rho}{\rho + 2} |\delta|^2 \right) \right]. \tag{7.12}
$$

In Figure 7.2 we show the sum rates for the different pre-coding methods when varying the user correlation factor $|\delta|^2$ for $\rho = 10\,dB$ and $g = 0.3$. When the two user channels are orthogonal the sum rates of the ZF and MMSE pre-coders are the same as the DPC capacity. When the channel correlation grows high, ZF capacity decreases rapidly to zero and the DPC capacity decreases to single-user capacity.

In Figure 7.3 we show the sum rates for the different pre-coding methods when varying the power imbalance. As g becomes large, the ZF sum rate decreases rapidly while the MMSE sum rate and DPC capacity decrease first and then achieve the single-user capacity. Furthermore, the DPC capacity and linear pre-coding sum rates are low when the power imbalance g is large.

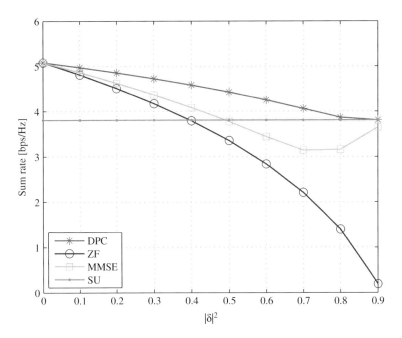

Figure 7.2 Sum rates for the different pre-coding methods when varying the user correlation factor, from [36].

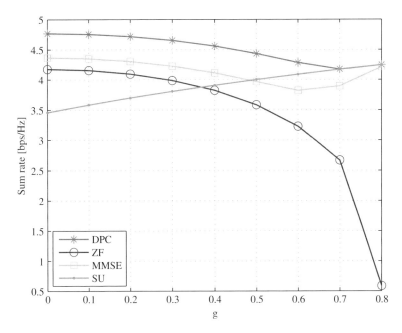

Figure 7.3 Sum rates for the different pre-coding methods when varying the power imbalance, from [36].

7.4   Multi-User Channel Models

Channel models are important tools for design, testing, and standardization of physical layer radio technologies. Though there are several multi-user channel models available, *not* all the models are capable of *controlling* correlation between multiple links. Some models introduce a numerical and physical mechanism to control the inter-link correlation, that is, the subspace alignment in analytical model [37] and shared (common) clusters in physically motivated radio propagation models [38]. The other models simulate multiple links in a random and independent way, and hence, the inter-link correlation is generated only in an implicit manner. It is also important to note that the explicit control mechanism is brought to channel model at the expense of additional complexity of the model structure and parameters compared to the implicit model. We make clear distinction between these two types of models.

Discussions also concern applicability, complexity, and accuracy of the models. Since the multi-user channel models are developed as an extension of a single-user channel model in many cases, the main features of the channel model resulting in their own advantages and limitations are determined by their inherent single-user structure. Hence this section introduces the single-user channel model structure only in a brief manner, and focuses on the capability and limitations of multi-user features. Readers interested in the details of the single-user channel models are directed to previous chapters. In this subsection, implications of the channel models on multi-user system performance are discussed. For some multi-user channel models, comparison of channel model outputs and measurements is also given to validate the channel model.

We start introducing multi-user channel models by analytical models that manipulate correlation structure between multi-user links. We then cover general cluster-based channel models as a physically motivated means of multi-user channel modelling. Finally, particular implementation of cluster-based channel models is introduced, namely, the IEEE802.11TGac [39], WINNER II [40], and the COST2100 [41] models.

7.4.1   Analytical Model

Analytical channel models are a widely used and commonly accepted method to simulate isolated (single-user) MIMO links. The Kronecker channel model [42] assumes correlation between the transmit and receive antenna correlation properties, and the Weichselberger model [43] exploits eigenstructure of channel correlation matrices to generate MIMO channel responses. Relative simplicity of the model and its well defined structure parameters makes a practical use straightforward, whereas a shortcoming of the model is the implicit inclusion of antenna effects, and hence it is difficult to reflect changes of antenna properties into channel model outputs.

In order to create correlation between multiple MIMO channels, Czink et al. [37] introduced an explicit mechanism of controlling signal subspaces of multiple MIMO links. By aligning the signal subspaces of multiple links, it is possible to control the inter-link correlation in a well defined manner. A metric closely related to the mutual information of MU-MIMO channels is introduced to quantify the subspace alignment,

$$J(\boldsymbol{R}_0, \boldsymbol{R}_1) = \log_2 \det\left(\boldsymbol{I} + \boldsymbol{R}_0(\boldsymbol{R}_1 + \sigma^2 \boldsymbol{I})^{-1}\right), \qquad (7.13)$$

where $\boldsymbol{R}_0$ and $\boldsymbol{R}_I$ are receive correlation matrices of signal of interests and aggregation of interference, σ^2 is variance of complex Gaussian noise at the receiver, and $\boldsymbol{I}$ is an identity matrix. Denoting eigenvalue decomposition of the correlation matrices as $\boldsymbol{R}_0 = \boldsymbol{U}\boldsymbol{\Lambda}\boldsymbol{U}^H$ and $\boldsymbol{R}_I = \boldsymbol{V}\boldsymbol{\Gamma}\boldsymbol{V}^H$ where $\boldsymbol{U}$ and $\boldsymbol{V}$ are eigenvectors and $\boldsymbol{\Lambda}$ and $\boldsymbol{\Gamma}$ are diagonal matrices composed of sorted eigenvalues in descending order, the metric gives possible largest and smallest values when $\boldsymbol{V} = \overleftarrow{\boldsymbol{U}}$ and $\boldsymbol{V} = \boldsymbol{U}$, respectively; $\overleftarrow{\boldsymbol{U}}$ means the reversed order of columns of matrix $\boldsymbol{U}$. The smallest value of the metric corresponds to the worst case of the mutual information when the eigenspaces of signal and interference are linearly dependent, thereby the strongest eigenmode of the interference affects that of the signal of interests. The largest value of the metric is equivalent to the best mutual information metric as a result of the strongest eigenmode of the interference aligned with the weakest one of the signal of interests. Using these notations, it is possible to further elaborate the mutual information metric to have a range [0, 1] by

$$\tilde{J}(\boldsymbol{R}_0, \boldsymbol{R}_I) = \frac{J(\boldsymbol{R}_0, \boldsymbol{R}_I) - J(\boldsymbol{R}_0, \boldsymbol{U}\boldsymbol{\Gamma}\boldsymbol{U}^H)}{J(\boldsymbol{R}_0, \overleftarrow{\boldsymbol{U}}\,\boldsymbol{\Gamma}\,\overleftarrow{\boldsymbol{U}}{}^H) - J(\boldsymbol{R}_0, \boldsymbol{U}\boldsymbol{\Gamma}\boldsymbol{U}^H)}. \tag{7.14}$$

In (7.14), any signal subspace of the interference, $\boldsymbol{V}$, fulfilling

$$J(\boldsymbol{R}_0, \boldsymbol{U}\boldsymbol{\Gamma}\boldsymbol{U}^H) \le J(\boldsymbol{R}_0, \boldsymbol{V}\boldsymbol{\Gamma}\boldsymbol{V}^H) \le J(\boldsymbol{R}_0, \overleftarrow{\boldsymbol{U}}\,\boldsymbol{\Gamma}\,\overleftarrow{\boldsymbol{U}}{}^H) \tag{7.15}$$

is guaranteed. Therefore, for a given target subspace alignment $\tilde{J}_{\text{target}} \in [0, 1]$, an eigenvector of the interference correlation matrix, $\boldsymbol{V}$, can be derived as follows.

1. Assume that the correlation matrix of the signal of interests, $\boldsymbol{R}_0 = \boldsymbol{U}\boldsymbol{\Lambda}\boldsymbol{U}^H$, and the eigenvalue profile of the interference, $\boldsymbol{\Gamma}$, and the target mutual information metric for the desired subspace alignment, $\tilde{J}_{\text{target}}$, are given.
2. Draw a unitary matrix $\boldsymbol{Z}$ at random, for example, by a singular value decomposition of an identical and independently distributed Gaussian matrix. There is an infinite number of solutions of $\boldsymbol{V}$ that fulfill $\tilde{J}_{\text{target}}$, and hence, generating multiple $\boldsymbol{V}$ makes it possible to obtain different realizations of the interference subspace that achieve the same alignment with the subspace of signals of interests.
3. Evaluate the normalized subspace alignment metric, $\tilde{J}_{\boldsymbol{Z}}(\boldsymbol{R}_0, \boldsymbol{Z}\boldsymbol{\Gamma}\boldsymbol{Z}^H)$.
4. Consider a curve

$$\boldsymbol{V}(s) = \begin{cases} P_{\boldsymbol{U} \to \boldsymbol{Z}}(s), & \tilde{J}_{\text{target}} \le \tilde{J}_{\boldsymbol{Z}} \\ P_{\boldsymbol{Z} \to \overleftarrow{\boldsymbol{U}}}(s), & \tilde{J}_{\boldsymbol{Z}} < \tilde{J}_{\text{target}} \end{cases}, \tag{7.16}$$

where

$$P_{\boldsymbol{Z}_1 \to \boldsymbol{Z}_2}(s) = \boldsymbol{Z}_1 \cdot \boldsymbol{W}e^{js\boldsymbol{\Phi}}\boldsymbol{W}^H, \tag{7.17}$$

$\boldsymbol{W}$ and $\boldsymbol{\Phi}$ result from eigenvalue decomposition of $\boldsymbol{Z}_1^{-1}\boldsymbol{Z}_2 = \boldsymbol{W}e^{j\boldsymbol{\Phi}}\boldsymbol{W}^H$. Corresponding mutual information function is given by $\tilde{J}(s) = \tilde{J}(\boldsymbol{R}_0, \boldsymbol{V}(s)\boldsymbol{\Gamma}\boldsymbol{V}(s)^H)$. The function is not necessarily increasing monotonically as s increases, but is continuous.
5. Find $s^* \in [0, 1]$ such that $\tilde{J}(s^*) = \tilde{J}_{\text{target}}$ using the bisection method [44] and return $\boldsymbol{V}(s^*)$. Now it is possible to derive the correlation matrix of the interference, $\boldsymbol{R}_I = \boldsymbol{V}(s^*)\boldsymbol{\Gamma}\boldsymbol{V}(s^*)^H$.

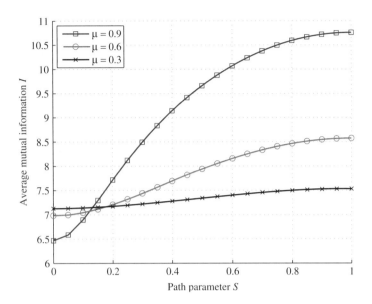

Figure 7.4 Average mutual information of multi-user interference channels with different alignments of signal and interference sub-space. For $s = 0$, signal and interference subspace are completely overlapping, while for $s = 1$, the strongest interference eigenvector couples into the weakest signal eigenvector [37].

According to [37], the interference alignment causes 65% of differences in mutual information of a 4×4 MIMO channel when the *intra-link* correlation of the signal of interests and interference is 0.9 as shown in Figure 7.4. Furthermore, the paper showed validity of the model by indoor peer-to-peer and outdoor-to-indoor MIMO radio channel measurements. The subspace alignment metric, $\tilde{J}$, took values in $[0, 1]$ in actual measurements and was found to fit well with the Beta distribution,

$$p_\beta(\tilde{J}) = \frac{\Gamma(a + b)}{\Gamma(a)\Gamma(b)} \tilde{J}^{a-1}(1 - \tilde{J})^{b-1}, \tag{7.18}$$

where $\Gamma(\cdot)$ is the Gamma function. For example, indoor peer-to-peer measurements revealed $a = 1.51$ and $b = 0.98$ with the Kolmogorov-Smirnov test at a significance of 5%. It was pointed out that a possible range of the mutual information metric depends on the similarities of radio wave propagation of the signal of interests and of the interference. When the same propagation mechanism, for example, a diffraction from a door of the building, is shared by the two links, it leads to smaller mutual information metric. Eigenvalue profiles of the links, Λ and Γ, were also parametrized by the measurements.

7.4.2 General Cluster Model

Contrary to the analytical channel model that focuses on the signal correlation, cluster models are based on physical reality of multipath propagation, that is, a cluster, an entity that is composed of multipath components having angular and delay characteristics close to each other at the base stations (BSs) and mobile stations (MSs). While cluster models are in general more complex than analytical models, they can integrate antenna effects, for example, antenna patterns, efficiencies, polarizations, orientations, and antenna array configurations because the models are based on the double-directional approach [4] as introduced in Chapter 2. Here a general cluster model that determines the multi-user correlation property explicitly, so called a common (shared) cluster, is first introduced. Two examples of implementing such clusters into channel simulations are described to show implications of the common cluster on the multi-user MIMO system performance.

7.4.2.1 Common Clusters

Correlation between two user links occurs when they share the same clusters that convey energy. The most obvious case is illustrated in Figure 7.5(a), where two MSs are almost co-locating and accessing the same BS. As we know, the spatial correlation rapidly decreases against increasing distance between two MSs. Large separation of MSs can also lead to spatially de-correlated multiple links on the other side of the link when the Kronecker structure of MIMO channels are invalid. However, in an extreme case of waveguiding propagation as shown in Figure 7.5(b), which we can find in street canyons and indoor corridors, multipaths comes from limited directions of clusters. They tend to stay at similar angles even though a MS moves over a long distance as shown in Figure 7.6(a). This easily ends up with a high spatial and shadowing correlation of multiple links where the benefit of the spatial division multiple access and the multi-user MIMO would be weakened. Similarly, when a MS sees signals from two BSs only in the same direction as illustrated in Figure 7.6(b), it could also result in correlated shadow fading between the links that deteriorate the effectiveness of macro-diversity and BS cooperation. The worst case is that multiple BSs see the same cluster both on the BS and MS sides that leads to a keyhole channel and hence the benefit of utilizing spatial degree of freedom vanishes. Such clusters are so called common clusters. Multi-user MIMO channel measurements in indoor scenarios [38, 45] revealed that energy conveyed by the common cluster was 30% at most

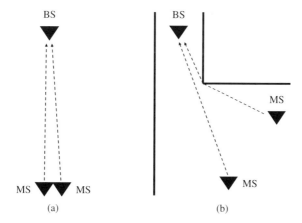

Figure 7.5 Two possible scenarios where two links would have spatial correlation on the BS side [47].

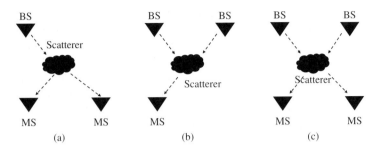

Figure 7.6 Clusters visible from (a) multiple MSs, (b) mutiple BSs, and (c) from multiple BSs and MSs [47].

in a hall environment, while it could reach up to 100% in a corridor environment where waveguiding propagation was dominant. Furthermore, the common cluster was a main reason for shadow fading correlation between multiple links, according to a measurement reported in [46].

7.4.2.2   Implementation Examples 1 – Evaluation of Multi-User Channel Capacity

Implications of the common cluster on multi-user MIMO system performance was evaluated by a simple geometry-based model illustrated in Figure 7.7(a) [38]. The model consists of a MS, two BSs, and six clusters *per link* randomly placed within the cell range. One of the two links between the BS and MS conveys a signal of interests, while another serves as an interfering link. One out of six clusters is seen from the two BSs simultaneously behaving as a common cluster, and the rest is visible only to a single BS. The BS and MS are equipped with an four-element half-the-wavelength dipole arrays along the x-axis of Figure 7.7(a). The antenna arrays have inter-element spacings of

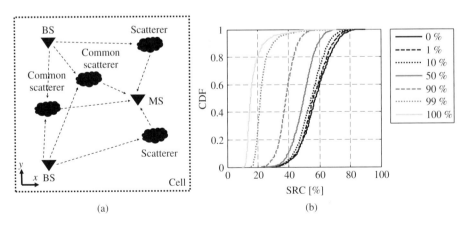

Figure 7.7   (a) Layout of a simple multi-user MIMO model and (b) Effects of common cluster the sum-rate capacity when five uncommon and one common clusters are considered. The CDFs include data from 1000 random combinations of cluster locations with each value of the S_c [38].

half-the-wavelength. The cluster geometries were generated 1000 times to see large-scale behavior of the effect of the common cluster. A sum-rate channel capacity was derived for each cluster geometry with the system signal-to-noise ratio of 10 dB for the two links, yielding 0 dB signal-to-interference ratio. The interference was cancelled using a zero-forcing algorithm [48] for each link, thereby reduction of the sum-rate channel capacity relative to the case of two isolated links was evaluated as a function of power conveyed by the common cluster S_c.

The cumulative distribution functions (CDFs) of the capacity are derived for $S_c = 0, 1, 10, 50, 90, 99$, and 100% as shown in Figure 7.7(b). The degradation of the capacity was apparent when $S_c > 50\%$, indicating the significance of the common clusters on the system performance. In the median level, the capacity decreased by 10% relative to the case without the common cluster. It was furthermore found that the impact of the common clusters is more significant when the environment contains a smaller number of clusters and when the antennas have larger electrical size because they lead to more distinguishable clusters in the angular domain. When there are two clusters in an environment and one of them is common among two links carrying $S = 50\%$ of the total energy, the deterioration of the capacity was 50% in the median level relative to a case without the common cluster in the median level.

7.4.2.3 Implementation Examples 2 – Evaluation of Multiplexing Gain in Multi-User Mimo Systems

A generic channel model was also exploited to evaluate effects of the common clusters on multiplexing gain in multi-user MIMO systems. An uplink multiuser MIMO system where n_U mobile users equipped with n_T antennas communicating with a BS having n_R antennas is considered. Theoretically, the upper bound on the multiplexing gain in multi-user MIMO systems is given by $\min[n_R, n_T n_U, n_S]$ where n_S is the number of distinct clusters involved in the multi-user MIMO channel [49]. The channel model is described in Figure 7.8(a). Each of the n_U users is surrounded by n_S clusters visible only to the particular user, as well as by n_{Sh} clusters visible from multiple MSs simultaneously. Departure angles of multipath clusters from all mobiles are uniformly distributed over 2π, while the arrival angle of multipath clusters at the BS are uniformly distributed over a total angular spread ϕ_{spr} at the BS, which is assumed to be the same for all MSs. The departure and arrival angles of multipath clusters are assumed to be uncorrelated with each other. The amplitudes of the multipaths are independent and identically Rayleigh distributed. It is assumed that the fading of the common clusters from the two (or more) users is uncorrelated. Uniform linear arrays were considered in the BS and MS.

Figure 7.8(a) shows multiplexing gain for a pair of users, with the same direction from the BS; the number of antennas at the BS and MSs is $n_R = 16$ and $n_T = 2$. The total number of visible clusters from the two mobiles is $n_U(n_S + n_{Sh})$, but the number of distinct clusters having different arrival angles at the BS is $n_U n_S + n_{Sh}$. Figure 7.8(b) shows that the number of distinct clusters determines the asymptotic multiplexing gain as evident from the curves with $n_S = 0$ and $n_{Sh} = 2$ that approaches the multiplexing gain of 2 as the signal-to-noise ratio goes into infinity. Furthermore, the curves with $n_S = 2, n_{Sh} = 0$

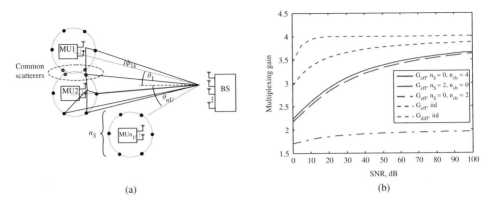

(a)   (b)

Figure 7.8   (a) Layout of a finite cluster channel model including common clusters and (b) multiplexing gain for two users with different numbers of independent and common clusters [49].

and $n_S = 0, n_{Sh} = 4$ are almost identical, supporting the theoretical derivation. According to [49], extending the scenario to an 8 user system shows that effective multiplexing gain is largely determined by the total number of distinct clusters.

7.4.3   Particular Implementation of Cluster Models

Because of the scalability of the cluster model for multi-user scenarios, its particular implementation is available in several forms. Representative implementations are the IEEE802.11TGac [39], WINNER II [40], and the COST2100 [41] models. The TGac model provides the angles of multipath clusters deterministically, while the WINNER II [40] and the COST2100 [41] models draw multipath cluster angles in a stochastic way. The latter models are particularly called geometry-based stochastic channel models, where a virtual map is used to define propagation channels. In the WINNER II and COST2100 model, the antenna effects are fully decoupled from the radio wave propagation model, hence it is possible to integrate antenna effects into the generated propagation channel responses using the concept of double-directional channels [4]. In the TGac model, antenna configurations, for example, linear antenna arrays with a specific inter-element spacing, can be taken into account to create radio channels, but it is difficult to consider characteristics of antenna element itself, for example, cross-polarization discrimination. The cluster models are general and applicable for various environments ranging from cellular, indoor, and *ad-hoc* scenarios. It is possible to realize channels corresponding to different environments by adapting a different set of model parameters of the clusters. The parameters of the TGac model support indoor pico-cell scenarios only, while those of the COST2100 models are available for indoor pico-cell as well as outdoor macro- and micro-cell scenarios. The WINNER II model supports a vast range of scenarios, for example, outdoor and indoor hotspots, outdoor-to-indoor scenarios, outdoor static feeder links, suburban and rural scenarios in addition to metropolitan outdoor and indoor scenarios. The cluster models are capable of integrating the common cluster to allow explicit control of multi-user correlation as the COST2100 model does. The TGac and

WINNER II channel model relies on implicit control mechanism of multi-user correlation since they do not use the common cluster approach.

7.4.3.1 IEEE802.11TGac Model

The IEEE802.11TGac channel model [39] is capable of a multi-user MIMO channel simulation below 6 GHz by extending the IEEE802.11TGn [50] channel model that supports only an isolated MIMO link. The TGac model also supports more antennas in a base and MS and wider system bandwidths on top of the TGn channel model. The 11ac model is designed so that the basic structure of the TGn channel model is unaltered, and the new features are given as model add-ons and corresponding model parameters that can be integrated with the 11TGn channel model.

The TGn channel model is a tapped delay line model with multipath clusters assigned to each delay tap. The interval between consecutive delay taps is determined by the system bandwidth. Properties of the clusters in the angular and delay domains, for example, the number of clusters, cluster power, mean angles, angular spreads at the base or mobile are deterministically given for various environments revealing different multipath richness [50]. Power angular spectrum of the clusters is defined by the Laplacian distribution. Antenna correlation can be derived based on the cluster angular characteristics with specific antenna configurations at the base and mobile. Analytical formula are available for truncated Laplacian distributions of the power angular spectrum with simple and ideal antenna arrays, for example, uniform linear arrays with isotropic antenna elements [51], but numerical simulations are required in more realistic and complex antenna arrays. Channel matrices are then realized by multiplying i.i.d. complex Gaussian random seeds with the antenna correlation at each delay tap. The first delay tap also includes the effect of LOS, defined by the Ricean K-factor and the array steering vector at the base and mobile.

The TGac channel model realizes multi-user MIMO scenarios by generating each base-to-mobile link independently, where departure and arrival angles of multipath clusters are subject to variation around the baseline cluster angles specified in the TGn channel model. The variation in cluster departure and arrival angles in a downlink scenario is given as follows [39].

1. The LOS cluster has the departure and arrival cluster angle variation of $\mathcal{U}[-\pi, \pi]$ at the base and mobile, where $\mathcal{U}[u_{\min}, u_{\max}]$ denotes a uniform distribution between $u_{\min}$ and $u_{\max}$.
2. The NLOS clusters have the departure and arrival cluster angle variation of $\mathcal{U}[-\pi/6, \pi/6]$ and $\mathcal{U}[-\pi, \pi]$ at the base and mobile, respectively.

Because of the independent generation of multiple base-to-mobile links, it is impossible to control inter-link correlation if there is any. In fact, the report [39] discusses a possibility that signals from multiple mobiles arrive at a base from identical angles because of common clusters shared among multiple links. However, the reports do not refer to the necessity of including such clusters into multi-user MIMO channel model because of lack of empirical evidence on existence of the clusters.

7.4.3.2 WINNER II Model

The WINNER II channel model was originally developed as an extension of the 3GPP/3GPP2 spatial channel model [52], and has been used in a wide variety of system simulations and standardization. For example, a subset of the WINNER model is adapted as in the ITU IMT-Advanced channel model [53]. This is because of remarkable features of the WINNER II model, that is, a wide range of supported environments and faithfulness to physical realities with relatively low implementation complexity.

The WINNER II model defines locations of BSs and MSs on a geometrical map, and then generates multipath clusters for each BS-MS link. Properties of the multipath clusters are determined by large-scale parameters, for example, angular and delay spreads, shadow fading, and the Ricean K-factor of clusters. The large-scale parameters are correlated random variables depending on a radio environment of interests, and are used to generate departure and arrival angles, delays, and power of clusters. The model supports a multi-user and multi-site channel simulation by "dropping" multiple BSs and mobiles on a geometrical map. Inter-link correlation between multiple links is defined solely by the correlation of large-scale parameters. When two mobiles are communicating with the same BS, the large-scale parameters of the two links have the correlation $\rho = \exp\left(-\frac{d}{\Delta}\right)$, where d is a separation distance of the two mobiles, and Δ is a correlation distance defined for each large-scale parameter and supported environment. For example, $\Delta = 15$ and 50 m for angular spreads of LOS and NLOS urban macro-cell scenarios. Beyond the correlation distance, the large-scale parameters of the two links are roughly uncorrelated. Correlation of large-scale parameters for inter-site scenario, that is, for two BSs accessing the same mobile, is set to 0 because no decisive conclusion was drawn from their measurements. When generating multipath clusters from the large-scale parameters, it is not possible to ensure that the clusters are departing or arriving from the same angles for multiple links, since the multipath angles are drawn independently for different links. Therefore the model does not allow explicit control of multi-user and multi-site correlation.

7.4.3.3 COST2100 Model

The COST2100 channel model adopts the concept of the common scatterers to be able to control the multi-user correlation in an explicit manner. We describe how the concept of the common scatterers is integrated into the model, followed by validation of the channel model by comparing outputs of channel model implementation and multi-user MIMO channel measurements.

Model Structure: The COST2100 model is expressed as an imaginary map consisting of base and MSs, and scatterers. Lines connecting the three elements represent radio wave propagation paths, leading to a virtual multipath environment. By a stochastic control of scatterer locations and attenuation of each multipath, it is possible to approximate angular, delay, and power characteristics of actual propagation environment. The scatterer locations are classified into single and multiple bounces as shown in Figure 7.9, where the latter is also called twin clusters. The scatterer locations are derived essentially by power angular profile at the base and mobile. The power and delay of multipaths are determined by total length of lines connecting BSs, scatterers, and mobiles. An extra delay sometimes needs to be added between multiple scatterers in order to create long delayed multipaths observed in measurements. One of the distinct characteristics of the COST2100 model is that it

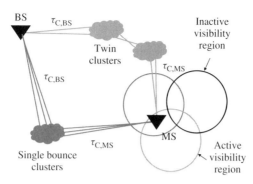

Figure 7.9 Basic structure of the COST2100 channel model: BS, MS, activated and inactivated VR and clusters [47].

allows a dynamic channel simulation by means of a concept of visibility region (VR) that originated from the COST259 macrocell channel model [54] and was further elaborated in the COST273 channel model [55]. As illustrated in Figure 7.9, each scatterer is always associated with a VR. The scatterers produce multipaths only when the mobile comes into the VR. When the mobile goes outside the VR, the corresponding scatterer is inactivated and stops contributing multipaths. In this way, VRs emulate scatterer death/birth effects that lead to non-wide-sense stationary simulations.

The concept of the common scatterers was integrated into the COST2100 channel model. The scatterers (clusters is almost the same terminology) are associated with VRs and therefore the common scatterers are attributed to VRs that activate the same scatterer, that is, mobile-side common scatterer. Furthermore, a scatterer can also be visible from more than one BS, that is, base-side common scatterer. The COST2100 channel model implements this by generating VRs on an imaginary map, and then a set of VRs that activates the same scatterer are selected (MS-side common scatterers). At the same time, a subset of scatterers is connected to more than a single BS (BS-side common scatterers). It was found from a corridor measurement where waveguiding effects are dominant [38] that 56 percent of the scatterers are visible for two BSs, while a single scatterer is connected with 1.4 VRs on average. Allocation of VRs for mobile- and/or base-side common scatterers is managed by establishing a VR assignment table in the channel model implementation [41]. The implementation is available for the public at [56].

Validation of the Model: It is important to show validity of the model in order to make sure that the model output emulates characteristics of measured propagation channels. Validity of the model is discussed in terms of the inter-link correlation by comparing COST2100 channel model outputs and corresponding measurements from which parameters of the model were derived. The parameters are available [41] for indoor office and hall scenarios at 5.3 GHz radio frequency, but still further parameterization works are required for other scenarios and frequencies. As a measure for the inter-link correlation, a correlation matrix co-linearity (CMC) [32] was calculated for two cases. The first case is inter-link correlation at a BS communicating with two mobiles represented by the configuration in Figure 7.6(a). The second case considers inter-link correlation at a mobile communicating with two BSs like the scenario in Figure 7.6(b). Correlation matrices were derived for two

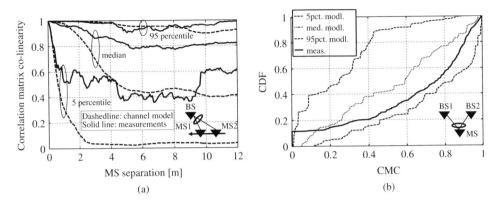

Figure 7.10   Inter-link correlation (a) in a multi-MS scenario on the BS side and (b) in a multi-BS scenario on the MS side. Median as well as 5 and 95 percentile curves are shown [47].

base-to-mobile links to calculate the CMC. CDFs of the CMC were considered for the first case, while variation of the CMC against distances between two MSs was evaluated in the second case. The scenario with two mobiles was emulated by choosing channel samples at two different MS locations on a measurement route, and hence is *not* a true multi-mobile environment. However, variation of surrounding radio environment could be regarded negligibly small because of fast channel acquisition capability of a multi-user MIMO sounder used for the measurements. Square four-element vertically polarized half-the-wavelength dipole antenna arrays with the inter-element spacings of the half-the-wavelength were considered at bases and mobiles. Figure 7.10(a) shows the correlation at the base communicating with two mobiles. The curves represent the median values as well as 5 and 95 percentile values of the CDF. Since the measured curves fall within the COST2100 model output, it was proven that the COST2100 channel model encloses measured characteristics in it. Because of a specific characteristic of a corridor scenario where the waveguiding propagation is prevailing, the correlation remains high even if the distance between two mobiles is large. Figure 7.10(b) shows the inter-link correlation at a mobile communicating with two bases. The measured CMC falls into the interval that the model provides, demonstrating credibility of the model for the multi-base cases too.

References

1. G. J. Foschini and M. J. Gans, "On limits of wireless communications in fading environments when using multiple antennas," *Wireless Personal Communications*, vol. 6, no. 3, pp. 311–335, March 1998.
2. R. Thomä, D. Hampicke, A. Richter, G. Sommerkorn, A. Schneider, U. Trautwein, and W. Wirnitzer, "Identification of time-variant directional mobile radio channels," *IEEE Trans. Instrum. Meas.*, vol. 49, no. 2, pp. 357–364, April 2000.
3. K. Kalliola, H. Laitinen, L. I. Vaskelainen, and P. Vainikainen, "Real-time 3-D spatial-temporal dual-polarized measurement of wideband radio channel at mobile station," *IEEE Trans. Instrum. Meas.*, vol. 49, no. 2, pp. 439–448, Apr. 2000.
4. M. Steinbauer, A. Molisch, and E. Bonek, "The double-directional radio channel," *Antennas and Propagation Magazine, IEEE*, vol. 43, no. 4, pp. 51–63, Aug 2001.

5. J. Nielsen, J. Andersen, P. Eggers, G. Pedersen, K. Olesen, and H. Suda, "Measurements of indoor 16 times; 32 wideband MIMO channels at 5.8 GHz," in *Spread Spectrum Techniques and Applications, 2004 IEEE Eighth International Symposium on*, Sydney, Australia, Aug. 30–Sep. 2 2004.

6. L. García, N. Jaldén, B. Lindmark, P. Zetterberg, and L. de Haro, "Measurements of MIMO indoor channels at 1800 MHz with multiple indoor and outdoor base stations," *EURASIP J. Wirel. Commun. Netw.*, vol. 2007, pp. 58–58, January 2007.

7. V.-M. Kolmonen, P. Almers, J. Salmi, J. Koivunen, K. Haneda, A. Richter, F. Tufvesson, A. F. Molisch, and P. Vainikainen, "A dynamic dual-link wideband MIMO channel sounder for 5.3 GHz," *IEEE Trans. Instrum. Meas.*, vol. 59, no. 4, pp. 873–883, 2010.

8. N. Czink, B. Bandemer, G. Vazquez-Vilar, L. Jalloul, and A. Paulraj, "Stanford July 2008 radio channel measurement campaign," in *Proc. COST2100 6th Management Committee Meeting*, 2008, October 6-8, 2008, Lille, France, TD(08)620.

9. N. Czink, B. Bandemer, G. Vazquez-Vilar, L. Jalloul, and A. Paulraj, "Can multi-user MIMO measurements be done using a single channel sounder?" in *Proc. COST2100 6th Management Committee Meeting*, 2008, October 6-8, 2008, Lille, France, TD(08)621.

10. M. Narandzic, M. Kaske, G. Sommerkorn, S. Jackel, C. Schneider, and R. S. Thoma, "Variation of estimated large-scale MIMO channel properties between repeated measurements," in *Proc. IEEE 73rd Vehicular Technology Conf. (VTC Spring)*, 2011, pp. 1–5.

11. A. Richter, "Estimation of Radio Channel Parameters: Models and Algorithms," Ph.D. dissertation, Technischen Universität Ilmenau, Ilmenau, Germany, May 2005.

12. I. E. Telatar, "Capacity of multi-antenna Gaussian channels," *Technical Memorandum, Bell Laboratories, Lucent Technologies*, October 1998, published in *European Transactions on Telecommunications*, vol. 10, no. 6, pp. 585–595, November/December 1999.

13. R. de Lacerda, L. S. Cardoso, R. Knopp, D. Gesbert, and M. Debbah, "Emos platform: Real-time capacity estimation of MIMO channels in the umts-tdd band," in *Proc. 4th Int. Symp. Wireless Communication Systems ISWCS 2007*, 2007, pp. 782–786.

14. D. Baum and H. Bolcskei, "Impact of phase noise on MIMO channel measurement accuracy," in *Proc. IEEE Vehicular Technology Conference, VTC Fall*, vol. 3, Los Angeles, CA, USA, 26–29 Sept. 2004, pp. 1614–1618.

15. J. Kivinen and P. Vainikainen, "Phase noise in a direct sequence based channel sounder," in *The 8th IEEE International Symposium on Personal, Indoor and Mobile Radio Communications, 1997. 'Waves of the Year 2000'. PIMRC '97.*, vol. 3, 1-4 Sept. 1997, pp. 1115–1119.

16. V.-M. Kolmonen, J. Kivinen, L. Vuokko, and P. Vainikainen, "5.3-GHz MIMO radio channel sounder," *IEEE Trans. Instrum. Meas.*, vol. 55, no. 4, pp. 1263–1269, Aug. 2006.

17. P. Almers, S. Wyne, F. Tufvesson, and A. F. Molisch, "Effect of random walk phase noise on MIMO measurements," in *Proc. IEEE VTC-Spring*, vol. 1, 2005, pp. 141–145.

18. A. Taparugssanagorn, J. Ylitalo, and B. Fleury, "Phase-noise in TDM-switched MIMO channel sounding and its impact on channel capacity estimation," in *Proc. IEEE Global Telecommunications Conference GLOBECOM '07*, 2007, pp. 4559–4564.

19. P. Chambers, P. Castiglione, L. Liu, F. Mani, F. Quitin, O. Renauding, F. Sanchez-Gonzales, N. Czink, and C. Oestges, "PUCCO Radio Measurement Campaign," in *COST 2100 11th Management Committee Meeting*, 2010, Jun. 2–4, 2010, Aalborg, Denmark, TD(10)11015.

20. J. Medbo, I. Siomina, A. Kangas, and J. Furuskog, "Propagation channel impact on lte positioning accuracy: A study based on real measurements of observed time difference of arrival," in *Proc. IEEE 20th Int Personal, Indoor and Mobile Radio Communications Symp*, 2009, pp. 2213–2217.

21. P. Kyritsi, P. Eggers, R. Gall, and J. Lourenco, "Measurement based investigation of cooperative relaying," in *Proc. 64th IEEE Vehicular Technology Conference, VTC Fall*, Montreal, Quebec, Canada, Sep. 2006, pp. 1–5.

22. M. Steinbauer, A. F. Molisch, and E. Bonek, "The double-directional radio channel," *IEEE Antennas Propag. Mag.*, vol. 43, no. 4, pp. 51–63, Aug. 2001.

23. S. Jaeckel, L. Thiele, A. Brylka, L. Jiang, V. Jungnickel, C. Jandura, and J. Heft, "Intercell interference measured in urban areas," in *Proc. IEEE Int. Conf. Communications ICC '09*, 2009, pp. 1–6.

24. J. Salmi, A. Richter, and V. Koivunen, "Detection and tracking of MIMO propagation path parameters using state-space approach," *IEEE Trans. Signal Process.*, vol. 57, no. 4, pp. 1538–1550, Apr. 2009.

25. K. Sulonen, P. Suvikunnas, L. Vuokko, J. Kivinen, and P. Vainikainen, "Comparison of MIMO antenna configurations in picocell and microcell environments," *IEEE J. Sel. Areas Commun.*, vol. 21, no. 5, pp. 703–712, 2003.

26. http://www.channelsounder.de/ruskchannelsounder.html.

27. F. Kaltenberger, M. Kountouris, L. Cardoso, R. Knopp, and D. Gesbert, "Capacity of linear multi-user MIMO precoding schemes with measured channel data," in *Proc. IEEE 9th Workshop Signal Processing Advances in Wireless Communications SPAWC 2008*, 2008, pp. 580–584.

28. F. Kaltenberger, D. Gesbert, R. Knopp, and M. Kountouris, "Performance of multi-user MIMO precoding with limited feedback over measured channels," in *Proc. IEEE Global Telecommunications Conf. IEEE GLOBECOM 2008*, 2008, pp. 1–5.

29. L. Garcia-Garcia, B. Lindmark, N. Jalden, and L. de Haro, "Multi-user MIMO capacity from measurements in indoor environment with in- and outdoor base stations," in *Proc. Second European Conf. Antennas and Propagation EuCAP 2007*, 2007, pp. 1–7.

30. C. Oestges, N. Czink, B. Bandemer, P. Castiglione, F. Kaltenberger, and A. J. Paulraj, "Experimental characterization and modeling of outdoor-to-indoor and indoor-to-indoor distributed channels," *IEEE Trans. Veh. Technol.*, vol. 59, no. 5, pp. 2253–2265, 2010.

31. "Electrobit homepage," http://www.elektrobit.com/.

32. G. Golub and C. V. Loan, *Matrix Computations*, 3rd ed. Johns Hopkins, 1996, p. 694.

33. M. Herdin, N. Czink, H. Ozcelik, and E. Bonek, "Correlation matrix distance, a meaningful measure for evaluation of non-stationary MIMO channels," in *Proc. IEEE Vehicular Technology Conference, VTC Spring*, Jun. 2005, pp. 136–140.

34. A. Perez-Neira, M. Lagunas, M. Rojas, and P. Stoica, "Correlation matching approach for spectrum sensing in open spectrum communications," *Signal Processing, IEEE Transactions on*, vol. 57, no. 12, pp. 4823–4836, Dec. 2009.

35. R. Blum, "MIMO capacity with interference," *Selected Areas in Communications, IEEE Journal on*, vol. 21, no. 5, pp. 793–801, June 2003.

36. X. Gao, O. Edfors, F. Rusek, and F. Tufvesson, "Linear pre-coding performance in measured very-large MIMO channels," in *Vehicular Technology Conference (VTC Fall), 2011 IEEE*, Sept. 2011, pp. 1–5.

37. N. Czink, B. Bandemer, C. Oestges, T. Zemen, and A. Paulraj, "Analytical multi-user MIMO channel modeling: Subspace alignment matters," *Wireless Communications, IEEE Transactions on*, vol. 11, no. 1, pp. 367–377, January 2012.

38. J. Poutanen, F. Tufvesson, K. Haneda, V. Kolmonen, and P. Vainikainen, "Multi-link MIMO channel modeling using geometry-based approach," *Antennas and Propagation, IEEE Transactions on*, vol. 60, no. 2, pp. 587–596, Feb. 2012.

39. G. Breit et al., " TGac channel model addendum supporting document," May 2009.

40. P. Kyösti et al., "IST-WINNER d1.1.2 WINNER II channel models, https://www.ist-winner.org/WINNER2-Deliverables/D1.1.2v1.1.pdf."

41. R. Verdone ed., *Pervasive Mobile & Ambient Wireless Communications, The COST Action 2100*, 1st ed. Springer, 2011.

42. D.-S. Shiu, G. Foschini, M. Gans, and J. Kahn, "Fading correlation and its effect on the capacity of multielement antenna systems," *Communications, IEEE Transactions on*, vol. 48, no. 3, pp. 502–513, Mar 2000.

43. W. Weichselberger, M. Herdin, H. Ozcelik, and E. Bonek, "A stochastic MIMO channel model with joint correlation of both link ends," *Wireless Communications, IEEE Transactions on*, vol. 5, no. 1, pp. 90–100, Jan. 2006.

44. R. Hamming and R. Hamming, *Numerical Methods for Scientists and Engineers*, ser. Dover Books on Mathematics. Dover, 1973.

45. J. Poutanen, "Geometry-based radio channel modeling: Propagation analysis and concept development," Ph.D. dissertation, Aalto University, School of Electrical Engineering, Mar. 2011.

46. J. Poutanen, K. Haneda, V.-M. Kolmonen, J. Salmi, and P. Vainikainen, "Analysis of correlated shadow fading in dual-link indoor radio wave propagation," *Antennas and Wireless Propagation Letters, IEEE*, vol. 8, pp. 1190–1193, 2009.

47. K. Haneda, J. Poutanen, F. Tuvfesson, L. Liu, V. Kolmonen, P. Vainikainen, and C. Oestges, "Development of multi-link geometry-based stochastic channel models," in *Antennas and Propagation Conference (LAPC), 2011 Loughborough*, Nov. 2011, pp. 1–7.

48. R. Blum, "MIMO capacity with interference," *Selected Areas in Communications, IEEE Journal on*, vol. 21, no. 5, pp. 793–801, June 2003.

49. A. Burr, "Multiplexing gain of multiuser MIMO on finite scattering channels," in *Wireless Communication Systems (ISWCS), 2010 7th International Symposium on*, Sept. 2010, pp. 466–470.

50. V. Erceg et al., "TGn channel models," May 2004.

51. L. Schumacher, K. Pedersen, and P. Mogensen, "From antenna spacings to theoretical capacities-guidelines for simulating MIMO systems," in *Personal, Indoor and Mobile Radio Communications, 2002. The 13th IEEE International Symposium on*, vol. 2, Sept. 2002, pp. 587–592 vol.2.

52. 3GPP, "Spatial channel model for multiple input multiple output (MIMO) simulations (release 10), http://www.3gpp.org/ftp/Specs/html-info/25996.htm."

53. ITU-R, "Report ITU-R m.2135-1, guidelines for evaluation of radio interface technologies for IMT-advanced, http://www.itu.int/pub/R-REP-M.2135-1-2009."

54. H. Asplund, A. Glazunov, A. Molisch, K. Pedersen, and M. Steinbauer, "The COST 259 directional channel model-part ii: Macrocells," *Wireless Communications, IEEE Transactions on*, vol. 5, no. 12, pp. 3434–3450, December 2006.

55. L. M. Correia, Ed., *Mobile Broadband Multimedia Networks*. Elsevier, UK, 2006.

56. L. Liu, "A public COST 2100 channel model matlab source code, http://ftp.COST2100.org/WG2.3 Model/," 2010.

8

Wideband Channels

Vit Sipal[1], David Edward[1] and Ben Allen[2]

[1]*University of Oxford, UK*
[2]*University of Bedfordshire, UK*

This chapter describes wave propagation relating to wideband channels. It links the propagation properties in an indoor environment to receiver performance and receiver architecture. The conclusions here are relevant to Long Term Evolution (LTE) systems but the chapter approaches the channel behaviour from the propagation properties of Ultra Wide Band (UWB) channels. The consideration of UWB channels is justified for two reasons. Firstly, UWB systems are considered as a complement to LTE systems for short range communication at extremely high data-rates. Secondly, analysis of the wireless channel over the extreme bandwidth associated with UWB provides an unprecedented insight and understanding into the fundamentals of wave propagation. This understanding can then be translated into narrowband channels which are more specific than the general UWB case.

UWB wireless communication addresses the impending spectrum gridlock, which is one of the main issues current wireless technology is addressing. Over the last century of wireless communication, the usable frequency bands have been fully allocated and there is little room left for new services [1]. Even though the spectrum is almost fully allocated, it is not fully used [2]. Henceforth, re-use of the frequency spectrum has been recognized as an opportunity and UWB is one of the approaches which seek to exploit the unused bands in the spectrum. It re-uses the spectrum elegantly by transmitting an extremely wideband signal (bandwidths over 500 MHz) with a low power spectral density. For the primary licensed narrowband users, a UWB transmission appears as white noise. Conversely, the narrowband user represents only a narrowband interference to the UWB system and this interference can be resolved using the inherent frequency diversity of the UWB system as long as the front end of the UWB receiver is not overloaded.

The extremely wide bandwidth and the low power spectral density provide the UWB wireless channel with unique properties that make it very different from the classical narrowband channel. For instance, the US Federal Communications Commission (FCC) limits

---

LTE-Advanced and Next Generation Wireless Networks: Channel Modelling and Propagation, First Edition.
Edited by Guillaume de la Roche, Andrés Alayón Glazunov and Ben Allen.
© 2013 John Wiley & Sons, Ltd. Published 2013 by John Wiley & Sons, Ltd.

the UWB radiation for wireless communication in the frequency band 3.1–10.6 GHz, with Equivalent Isotropically Radiated Power (EIRP) spectral density of −41.3 dBm/MHz. Due to the large bandwidth, many individual multipath components can be resolved which reduces the impact of fading. The low power spectral density limits the dynamic range of the system and practically confines the UWB communication to one room only [3].

In this chapter, the fundamental properties of the UWB channel and their impact on the performance of UWB communication systems are introduced. These properties are confined to the FCC 3.1–10.6 GHz bandwidth. Some of the conclusions are applicable in general but caution is required. For instance, as the frequency decreases the penetration properties of electromagnetic wave improve and the confinement of the communication and associated interference is no longer valid.

The path-loss in UWB channels is discussed in Section 8.1. This section is based mainly on the IEEE802.15.4a UWB channel model [4]. The impulse response of UWB channel and the impact of antennas on it is presented in Section 8.2. Section 8.3 investigates the frequency selective fading and its impact on the performance and complexity of UWB systems. The last sections discuss practical challenges and opportunities associated with Multiple Input Multiple Output (MIMO) UWB, and the impact of the wideband propagation on LTE-Advanced systems.

8.1 Large Scale Channel Properties

The most significant large scale channel property is undoubtedly the path-gain/loss. This defines the expected received energy and is therefore traditionally one of the main parameters defining the range of a wireless communication system. The path gain, G_p, is typically defined as follows [5, 6]:

$$G_p(r, f) = E\left\{ \int_{f-B/2}^{f_c+B/2} |H(r, f)|^2 df \right\} \tag{8.1}$$

where r represents range; f stands for frequency; f_c is the centre frequency of the channel; B is the channel bandwidth; and $H(r, f)$ represents the transfer function of the wireless channel.

Path-gain and path-loss are fully equivalent quantities. Their relationship is inversion. As a result, path-gain decreases with range whilst path-loss increases. In this chapter, we choose path-gain because of the lack of inversion in (8.1), but the conclusions are fully equivalent.

Equation (8.1) sums the energy carried by all multipath components. As a result, in indoor environments the path gain will be typically larger than in free-space where path gain decreases with the square of range. The mean $E\{\bullet\}$ in (8.1) reflects the fact that the superposition of multipath components can be constructive or destructive which varies the received energy. Such an effect is called "fading" and is studied in detail in Section 8.3. For path-loss measurements, averaging over a local area is performed to eliminate the impact of fading.

Another conclusion from (8.1) is that the path gain is a function of range and frequency. Whilst literature typically agrees on the range dependency of a channel's path-gain, the frequency dependency of path gain in a UWB channel is a more controversial topic because it strongly reflects the impact of antennas. The IEEE802.15.4a channel model

[4] chooses the compromise approach and defines path gain as a product of range and frequency [4]:

$$G_p(r, f) = G_p(r) \cdot G_p(f) \tag{8.2}$$

Both dependencies will be discussed in separate subsections.

8.1.1   Path Gain – Range Dependency

The distance part of the Equation 8.2 can be modeled with the conventional formula known for narrowband wireless propagation [4]:

$$G_p(r)|_{\text{dB}} = G_0|_{\text{dB}} - 10n \cdot log_{10}\left(\frac{r}{r_0}\right) \tag{8.3}$$

where $G_0|_{\text{dB}}$ represents the path gain at reference distance r_0 and n is path gain coefficient. For UWB, the choice of r_0 is 1 m in the IEEE802.15.4a channel model. This choice is based on the standard communication ranges expected in indoor channels.

The value of the path gain coefficient n depends on the environment and on the condition Line Of Sight (LOS) or Non Line Of Sight (NLOS). For LOS free space, $n = 2$ which is also a path gain coefficient applicable when individual rays are observed. As shown in (8.1) the path gain is defined as a sum over all multipath. Thus, for a dense multipath indoor environment the actual path gain is higher than in free space, in other words the path gain coefficient is typically lower than 2. Its actual value depends on the type of environment. For instance, an industrial environment with a large number of metallic scatterers differs from a residential area. Furthermore, there is an impact of the relative positions of the transmitter and receiver to walls and scatterers. As a result, reported values of path gain coefficient from different reports may vary significantly as illustrated below.

In the **office areas**, the actual value of path gain exponent varies from measurement to measurement. For the LOS scenario, the following results are reported as 1.9 in [7] or as 1.3 in [8]. For the NLOS scenarios, the reported values are 3.6 in [7] or 2.3–2.4 in [8].

The variations between the individual measurements can be explained with results from [9–11], where **residential areas** were investigated and it was found that the path gain exponent, n, is in fact a Gaussian random variable with standard deviation σ_n. References [9–11] propose to use n = 1.7 with standard deviation 0.3 in the case of LOS and $n = 3.5$ with standard deviation 0.97 in the case of NLOS.

The random nature of path gain coefficient, n, as well as G_0 is generally accepted and mentioned in the IEEE802.15.4a UWB channel model [4]. If the random nature is considered in simulations, the computational complexity increases, which removes one of the main advantages of empirical models over complex deterministic models. Henceforth for the sake of low computational complexity, the IEEE802.15.4a UWB channel model [4] proposes values for path gain parameters as summarized in Table 8.1.

8.1.2   Path Gain – Frequency Dependency

The frequency dependency of path gain in UWB wireless channels is a topic well studied. It often depends to what extent the antennas are or are not included in the channel. In [12],

Table 8.1 Path gain parameters according to IEEE802.15.4a channel model [4]

Environment	n	G_0[dB]
Office LOS	1.63	35.4
Office $NLOS$	3.07	57.9
Residential LOS	1.79	43.9
Residential $NLOS$	4.58	48.7
Industrial LOS	1.2	56.7
Industrial $NLOS$	2.15	56.7
Outdoor LOS	1.76	45.6
Outdoor $NLOS$	2.5	73.0

the waveform of a pulse transmitted in free space between two antennas is studied. It is shown that the waveform does not change as the distance increases. As a result, [12] argues that the propagation itself is not frequency selective and that the frequency selectivity is caused by the antennas. Furthermore [12] explores the Friis formula for power spectral density at the receiver $S_R(r, f)$ for free space [12]:

$$S_R(r, f) = S_T(f) \cdot G_T(f) \cdot G_R(f) \cdot \left(\frac{c}{4\pi r f}\right)^2 \qquad (8.4)$$

where $S_T(r, f)$ is the transmitted power spectral density; $G_T(f)$ and $G_R(f)$ represents the gains of transmit and receive antennas.

In (8.4), the antenna gains are presented as a function of frequency. Reference [12] argues that the standard assumption of path gain $G_p(f) \propto f^{-2}$ is valid for the assumption of constant gain of antennas $G_T(f) = G_T$ and $G_R(f) = G_R$. This is not always true for UWB antennas. In fact, if it is assumed that the effective aperture of antennas is constant, that is, $A_{eT}(f) = A_{eT}$ and $A_{eR}(f) = A_{eR}$, then (8.4) can be rewritten as [12]:

$$S_R(r, f) = S_T(f) \cdot A_{eT}(f) \cdot A_{eR}(f) \cdot \left(\frac{f}{rc}\right)^2 \qquad (8.5)$$

which yields in a completely different frequency dependency of path gain $G_p(f) \propto f^2$.

These observations make the frequency dependency of path gain quite complex. As pointed out in [3], the gain of the practical antennas for UWB communication is neither constant nor is their effective apperture constant. As a result, the frequency dependency for free space caused by the antennas will be somewhere between $G_p(f) \propto f^2$ and $G_p(f) \propto f^{-2}$. In a practical multipath channel, this frequency dependency is experienced individually by each multipath because of the angular variation of antenna gain.

Furthermore, the assumption that propagation itself is frequency independent is valid only for direct rays. In [13], the physical propagation of individual rays is studied and it is shown that if the antennas are excluded the path gain of the i-th ray, $G_p, i(f)$, depends on the physical propagation experienced by the ray [13]:

$$G_p, i(f) \propto f^{2\alpha(i)} \qquad (8.6)$$

Table 8.2 Frequency selectivity for different diffraction mechanisms according to [13]

Physical mechanism	Frequency dependence factor α
Line of sight	0
Reflection	0
Diffraction from smooth or flat surface	0
Diffraction by edge	-0.5
Diffraction by corner or tip	-1
Diffraction by axial cylinder face	0.5
Diffraction by broadside of a cylinder	1

The frequency dependence factor $\alpha(i)$ depends on the physical mechanisms that are applicable to the ith ray. Table 8.2 from [13] summarizes the frequency dependencies for typical physical mechanisms.

In a nutshell, there is a frequency dependency of the path gain for UWB channels. This is given by the combination of antenna effects (frequency and angular dependency of antenna gain) and physical propagation. These mechanisms are well understood and can be implemented in a deterministic channel model. However, for an empirical model a "typical" scenario is taken into account by the IEEE802.15.4a UWB channel model [4] to provide a low-complexity dependency:

$$G_p(f) \propto f^\kappa \tag{8.7}$$

where κ depends strongly on the propagation scenario. The measured values for various environments are listed in Table 8.3 based on data taken from [6].

8.2 Impulse Response of UWB Channel

The knowledge of the channel impulse response is important for the robust design of a wireless communication system. For instance, energy carried by multipath components can contribute to inter-symbol interference and/or it determines the frequency selectivity of a wireless channel. The standard assumption is to model the channel impulse response

Table 8.3 Frequency dependency coefficient κ according to [4]

Environment	κ for LOS	κ for NLOS
Residential	-1.12 ± 0.12	-1.53 ± 0.32
Office	-0.03	-0.71
Industrial	$+1.103$	$+1.427$
Outdoor	-0.12	-0.13

as a sequence of Dirac impulses [3], but it is important to know the statistics of wireless channel for the deployment scenario, that is, the number of multipath components, the mean delay between the components, and the attenuation coefficient and so on. In this section, we firstly introduced the channel impulse response adopted by the IEEE802.15.4a channel model [4]. This is later critically discussed and corrected.

8.2.1 Impulse Response According to IEEE802.15.4a

The IEEE802.15.4a channel model recommends the Saleh-Valenzuela model that defines the rays arriving in clusters. As a result, the channel impulse response is given as [14]:

$$h(t) = \sum_{l=0}^{\infty} \sum_{k=0}^{\infty} \alpha_{kl} e^{j\theta_{kl}} \delta(t - T_l - \tau_{kl}) \tag{8.8}$$

where index l stands for the cluster number, index k stands for ray number in the l-th cluster l. Accordingly, index T_l is the cluster arrival time and index τ_{kl} relates to the ray delay in the l-th cluster. α_{kl} represents the ray amplitude and θ_{kl} its phase.

The cluster arrival times are given by the Poisson probability density function [4]:

$$p(T_l | T_{l-1}) = \Lambda \exp[-\Lambda(T_l - T_{l-1})], \quad l > 0 \tag{8.9}$$

$T_0 = 0$ by definition, for modelling it corresponds to the delay of the direct path for LOS or to the delay of the shortest path for NLOS. For the arrival of rays within a cluster the IEEE802.15.4a channel deviates from the original Saleh-Valenzuela model and suggests a mixture of two Poisson processes [4]:

$$p(\tau_{kl} | \tau_{(k-1)l}) = \beta \lambda_1 \exp[-\lambda_1(\tau_{kl} - \tau_{(k-1)l})] + (1 - \beta)\lambda_2$$
$$\exp[-\lambda_2(\tau_{kl} - \tau_{(k-1)l})], \quad k > 0 \tag{8.10}$$

$\tau_{0l} = 0$ by definition.

This high number of multipath components in the model of channel impulse response caused by the fact that for a UWB channel, the immense bandwidth allows us to resolve more multipath components.

The mean energy of a cluster, Ω_l, has an exponential decay [4]:

$$10 \log(\Omega_l) = 10 \log(\exp(-T_l / \Gamma)) \tag{8.11}$$

where Γ is a model parameter specific for the deployment scenario. The total energy of all clusters (sum of Ω_l) is given by the total path gain as specified in (8.1). The mean power delay profile of individual rays also follows an exponential decay with a model parameter [4]:

$$E\left\{|\alpha_{kl}|^2\right\} = \Omega_l \frac{1}{\gamma_l[(1 - \beta)\lambda_1 + \beta\lambda_2 + 1]} \exp(-\tau_{kl} / \gamma_l) \tag{8.12}$$

Reference [4] notes that the γ_l might depend on the cluster number l and that the normalization in 8.12 is only approximate.

8.2.2   *Impact of Antenna Impulse Response in Free Space*

The IEEE802.15.4a channel models specifies the set of model parameters Λ, λ_1, λ_2, Γ, and γ_l for various environments. However, recent work [15, 16] points out some fundamental errors in the assumption that the Saleh-Valenzuela model from (8.8) is a suitable descriptor of the UWB channel. Firstly, it is pointed out that the mean delays between rays given by the coefficient λ_1 correspond to a path difference of 10–15 cm. These path loss differences might be justifiable for the reflected rays but there seems to be no physical reason for them for the direct path. Secondly, the model does not consider the impulse response of the antennas (e.g. [17]) that are manifested in channel measurements. Based on these discrepancies an alternative explanation is offered.

Reference [15] begins with the hypothesis that the clusters in channel impulse response are not individual random rays but they are a manifestation of antenna impulse responses. Quite simply, antennas are spatial filters. A communication between two antennas in free space, with one direct path only, can be modelled as a band pass filter. Thus an impulse response of such a channel will correspond to an impulse response of a bandpass filter which explains the peaks in the cluster. To confirm this hypothesis, [15] uses a commercial fullwave simulation software to determine the antenna impulse response of a discone antenna using algorithm introduced in [18]. The antenna impulse response for the transmit and the receive case then models the ideal free space channel with two discone antennas as described in [17]. This result is compared to a measurement of the same antennas in an anechoic chamber. The resulting impulse responses are shown in Figure 8.1.

From Figure 8.1, firstly, a good agreement of theory and measurement can be observed. The small discrepancy can be attributed to simulation errors and non-ideal measurement conditions. Secondly, even though there is only one ray present in the measurement and simulation, the channel impulse response does not resemble a single Dirac impulse, but it resembles well a manifestation of an impulse response of a bandpass filter. Therefore,

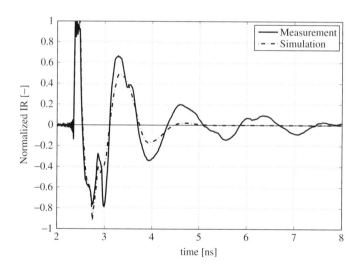

Figure 8.1   Simulated and measured normalized impulse response of a single path channel.

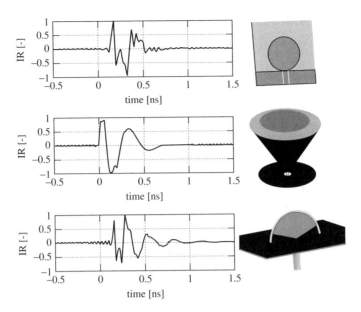

Figure 8.2 Simulated impulse response of a free space UWB channel for three different types of antennas, taken from [16]. Reproduced with permission of IEEE.

Figure 8.1 supports well the hypothesis that ray clusters are actually a manifestation of the antenna impulse responses.

It is believed that the reason why the individual non-zero values within a cluster are often attributed to individual multipath components, arises from the insufficient bandwidth of channel measurements (often less than 3 GHz [6]) and because the concept of antenna impulse response is not yet used as a standard antenna descriptor.

Additionally, the assumption that rays within a cluster originate from the antenna impulse response can also explain the differences among measurements carried out by different research groups simply because they used different antennas. To illustrate this, the simulated impulse response of the free space channels channel for three different types of UWB antennas to explore the effect of antenna type are presented in Figure 8.2.

8.2.3 Manifestation of Antenna Impulse Response in Realistic Indoor Channels

Indoor channels are described in Chapter 3. In [15], a simulation and measurement for an ideal free space channel with one multipath only was presented. Reference [16] presents measurements in a real laboratory environment with several multipath components. These show that antenna impulse responses manifest themselves not only for the line-of-sight cases but for later reflected rays as well. The distance between the transmitter and receiver was chosen to be 120 cm and the measurement bandwidth was 3–20 GHz. Small discone antennas operating above 2.5 GHz were employed. Their polarization was vertical. The output power was set to keep below the limit of −41.3 dBm/MHz so that the measurement

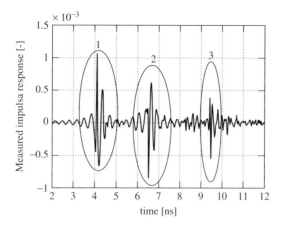

Figure 8.3 Measured channel impulse response for an example realistic wireless channel, taken from [16]. Reproduced with permission of IEEE.

is as close to a practical application as possible. An example of such a measured channel impulse response is presented in Figure 8.3.

In Figure 8.3, three detectable multipath components are highlighted. It can be seen that rays 1 and 2 are similar, apart from the sign caused by the reflection in the case of ray 2. Ray 3, however, is different – shorter. The reason for this was obtained by an analysis of the specific channel. Delays of the rays were compared to real distances in the channel setup to determine the ray paths. Ray 1 is the direct path, ray 2 represents a reflection in the horizontal plane. Due to the rotational symmetry of discone antennas its shape remains the same and the sign is determined by the reflection coefficient. Ray 3 represents a ceiling reflection with elevation angle of 60°. This is also the reason for its different waveform. As already mentioned, the antennas act as spatial filters. It is well known that for higher frequencies the elevation angle of the radiation maximum increases, for example [16, 33], which impacts the waveform of the antenna impulse responses.

To confirm this claim, [16] simulates the antenna impulse responses (transmit and receive case) for the discone antennas for direct paths corresponding to elevations of 0° and 60°. The simulated antenna impulse response for 0° is then de-convolved from ray 1 and ray 2 in Figure 8.3 whereas the simulated antenna impulse response for 60° is de-convolved from ray 3. The results are collated together to represent the channel impulse response without the impact of the antennas. This is presented in Figure 8.4 where the channel is manifested as three Dirac impulses. There is further evidence supporting the initial hypothesis that the clusters in the Saleh-Valenzuela model are in fact a manifestation of antenna impulse responses, and suggest the need for a revision of the IEEE802.15.4a channel model.

8.2.4 New Channel Model For UWB

Further comparison of Figure 8.4 with Equation (8.10) defining the probability of ray arrival in clusters reveals the reason for the use of the mixture of two Poisson processes in Equation (8.10), whilst the standard Saleh-Valenzuela model for narrowband systems uses a single Poisson process. For the indoor office environment and LOS measurements the IEEE802.15.4a model suggests $\lambda_1 = 0.19[1/ns]$ and $\lambda_2 = 2.97[1/ns]$. These values correspond to a mean ray arrival delay of 5.26 ns and 0.33 ns, respectively. Using the data presented above, the Poisson process with mean ray arrival delay of 0.33 ns can be linked to the antenna impulse response (see Figure 8.3) whereas the second Poisson process corresponds to the actual multipath propagation (see Figure 8.4). The same can be shown for the parameters for the outdoor environment, indoor residential and indoor industrial environments as listed in [4]. In other words, there are ray clusters in the model of antenna impulse response, but the number of rays in each cluster is significantly lower and it corresponds to the original Saleh-Valenzuela model. The difference is that for UWB bandwidths, the antenna impulse response is resolvable.

This leads to a proposal of the following change to the model of channel impulse response of UWB channels:

$$h(t) = \sum_{l=0}^{\infty} \sum_{k=0}^{\infty} h_{kl}(t) * \alpha_{kl} e^{j\theta_{kl}} \delta(t - T_l - \tau_{kl}) \tag{8.13}$$

where $h_{kl}(t)$ represents the antenna impulse response associated with the antenna type, launch angle and arrival angle of the multipath (given by the parameters k, l) and normalized to have unit energy; $*$ stands for convolution.

The definition of inter-cluster parameters remains the same as in Equations (8.9), (8.11). For the inter-cluster parameters the Equation (8.10) is simplified as follows:

$$p(\tau_{kl} | \tau_{(k-1)l}) = \lambda \exp[-\lambda(\tau_{kl} - \tau_{(k-1)l})], \quad k > 0 \tag{8.14}$$

where for λ, the smaller parameters out of the pair λ_1 and λ_2 offered by IEEE802.15.4a model is selected.

So far, the new model presented in Equation (8.13) seems more complex and less suitable for practical use in channel simulations, however it can be further simplified as described in the following subsection.

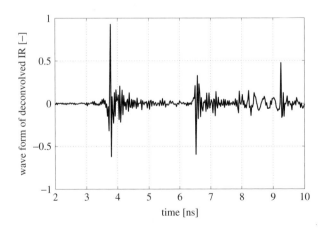

Figure 8.4 Channel impulse response after de-convolution of antenna impulse responses, taken from [16]. Reproduced with permission of IEEE.

8.2.5   UWB Channel Impulse Response – Simplified Model
for Practical Use

The first simplification of the model can be achieved if the practical dynamic range of the UWB systems is considered. This dynamic range is typically about 70 dB (EIRP limit is −41.3 dBm/MHz and the noise floor is −114 dBm/MHz). As a result of this dynamic range, rays with delays more than 100 ns do not need to be considered. Such rays must have a path length of at least 30 m. A simple path loss calculation for free space at a frequency of 3 GHz yields in path loss of 71.5 dB ($PL = (4\pi r/\lambda_0)^2$). This path loss increases for higher frequencies and whilst the antenna gain might reduce it, it does not include the reflection losses. For typical indoor environments, path lengths of 30 m correspond to multiple reflections. Such a consideration reduces the number of clusters that need to be considered. For instance, for indoor office environment the IEEE802.15.4a suggests mean inter-cluster delay $E\{T_l − T_{l-1}\}$ of 62.5 ns, and an average of 5.4 clusters. It is apparent that for practical UWB systems there is no need to consider more than two clusters.

Furthermore, for LOS scenarios in residential areas the mean inter cluster delay is $E\{T_l − T_{l-1}\}$ reported to be 21.3 ns. The mean intra-cluster delay between rays $E\{\tau_{k,l} − \tau_{k-1,l}\}$ given by λ is 6.7 ns which is about one third of the inter-cluster delay. This suggests that the clusters actually be reduced by a sequence of rays with the delay given by Equation 8.14.

If these two points are combined, the model can be simplified into following form:

$$h(t) = \sum_{k=0}^{K_{DR}} h_k(t) * \alpha_k e^{j\theta_k} \delta(t − \tau_k) \tag{8.15}$$

where K_{DR} is the number of rays that need to be considered given the dynamic range of the system. The value of K_{DR} is selected based on the system architecture. For instance, it is the lower for non-coherent impulse radio receiver with a low dynamic range, whereas it will be larger for direct sequence Code Division Multiple Access (CDMA) with good noise suppression capabilities. The ray arrival time is then given as:

$$p(\tau_k|\tau_{k-1}) = \lambda \exp[−\lambda(\tau_k − \tau_{k-1})], \quad k > 0 \tag{8.16}$$

For the energy of rays $E\{\alpha_k^2\}$, Equation (8.11) can be transformed and used as:

$$E\{\alpha_k^2\} = \exp(−\tau_k/\Gamma) \tag{8.17}$$

Alternatively, a path-loss estimate for a given frequency band can be employed.

8.2.6   UWB Channel Impulse Response – Conclusion

In a nutshell, this section presents a novel model of channel impulse response for UWB channels. It is shown that the immense bandwidth on hand allows us to resolve and identify the impulse response of an antenna which is a deterministic waveform. As a result, rays that were assumed random in the IEEE802.15.4a model are shown to be deterministic. This is useful for practical system considerations.

Firstly, the number of independent multipaths is significantly reduced. This can be utilized, for example, by a reduced number of fingers in coherent rake receivers in impulse radio receivers [16]. There are issues associated with the random nature of antenna type and ray launch-/arrival-angle, but [16] shows that the waveform of manifested antenna impulse responses $h_k(t)$ does not change for small displacements (<25 cm) of receiver or transmitter. As a result, a use of a pilot sequence enables the form of $h_k(t)$ to be determined.

Similarly for Orthogonal Frequency Division Multiplexing (OFDM) UWB systems, the manifestation of antenna impulse responses $h_k(t)$ seems to the long narrowband sub-carriers extremely short (compare WiMedia OFDM symbol duration is 312.5 ns [19] to $h_k(t)$ of about 1–2 ns) act only as Dirac impulse and causes minimum distortion. In other words, the narrowband signal passes through the bandpass filter represented by the antenna almost without any distortion.

Note that: unlike in Figure 8.1 there is only one peak in baseband representation of $h_k(t)$). This effectively removes the impact of $h_k(t)$ from Equation (8.15) resulting in reduced complexity of simulations.

8.3   Frequency Selective Fading in UWB Channels

Frequency selective fading in a wireless channel is caused by the superposition of multi-path components. As a result, the path loss of a channel may vary significantly even for a small displacement of transmitter or receiver. Consequently, the link budgets of wire-less communication systems plan a fade margin to overcome this issue [3]. The actual path loss is a random variable. The typical probability distribution functions suggested by the literature are: Rayleigh (no dominant line-of-sight), Rice (dominant line-of-sight) or Nakagami (combination of both suitable for dense multipath environments).

One of the major advantages of UWB wireless communications is the robustness against the impact of fading because the available bandwidth allows us to resolve individual multipath components in the time domain [3]. This is illustrated by Figure 8.5 which shows the variation of empirical normalized channel energy, CE, (8.18) for the same position of a receiver but different bandwidths.

$$CE(r, f) = \frac{\int_{f-B/2}^{f_c+B/2} |H(r, f)|^2 df}{E\left\{\int_{f-B/2}^{f_c+B/2} |H(r, f)|^2 df\right\}} \tag{8.18}$$

where the variable definition is the same as in 8.1.

As can be seen in Figure 8.5, there is a minimum variation of channel energy, CE, for UWB channels for displacements relatively small compared to the communication distance. In this section, we focus on the proper mathematical characterization of fading. The transition in fading severity between narrowband and wideband channel, called fade depth scaling, is studied. Then the link between the type of environment and the severity of fading is discussed and its impact on the performance of real systems is introduced. Lastly, the probability distribution function of fading is discussed and linked to the performance of UWB systems using various modulation schemes.

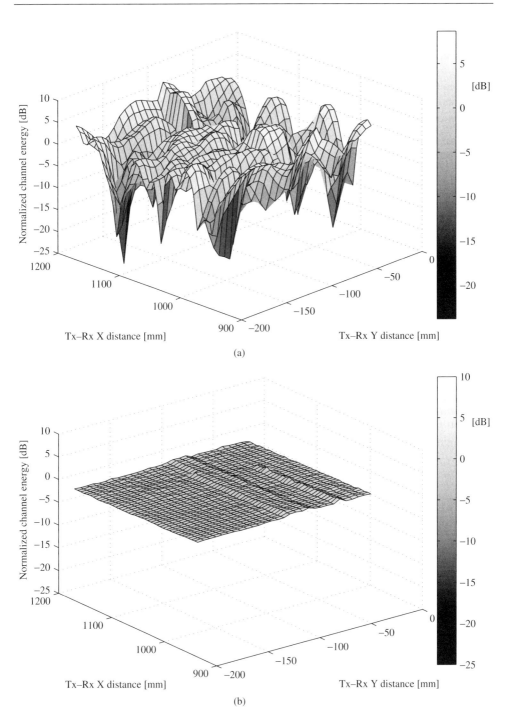

Figure 8.5 Empirical spatial variation of path loss (Equation 8.1) normalized to a local mean for a channel with centre frequency of 6 GHz (a) bandwidth of 6 MHz, (b) bandwidth of 6 GHz, taken from [20].

8.3.1   Fade Depth Scaling

8.3.1.1   Fade Depth Scaling – Cause

We begin the investigation of fade depth with the simplified channel model form Equation (8.15) and perform a Fourier transformation in order to use path loss from Equation (8.1) which is defined in frequency domain:

$$H(f) = \sum_{k=0}^{K_{DR}} H_k(f) \cdot \alpha_k e^{j\theta_k} e^{-j2\pi f \tau_k} \tag{8.19}$$

where all the variables are the same as in (8.15), $H_k(f)$ is the Fourier transform of the manifestation of antenna impulse responses of transmit and receive antennas, $h_k(t)$.

Before using (8.19) in (8.1), the impact of $H_k(f)$ is discussed. As it was shown in [16], the $H_k(f)$ are displacement-invariant for a small displacement relatively small to the distance between the transmitter or receiver. The frequency dependence is determined by the type of antenna and is similar to the discussion about the frequency dependency of path-loss. Here, we make an assumption $H_k(f) = $ const. For OFDM UWB systems, this assumption is perfectly valid as the investigation is limited only to the narrow bandwidth of individual subcarriers. It is less so for wideband impulse radio UWB systems or wideband single carrier UWB systems. However, the state-of-the art impulse radio systems only rarely consider bandwidths beyond 1 GHz and state-of-the-art antennas provide good constant performance over bandwidths of several GHz [3] and are designed to minimize the distortion of transmitted pulses. Here, the main objective to explore the impact of small changes of τ_k on the channel energy. As a result, marginal error caused by the assumption $H_k(f) = $ const can be neglected. The phase and amplitude of $H_k(f)$ become part of the θ_k and α_k with no loss of generality. Using this simplification and substituting (8.19) into (8.18) and gives:

$$CE(f_c) = \frac{\int_{f_c-B/2}^{f_c+B/2} |H(f)|^2 df}{E\left\{ \int_{f_c-B/2}^{f_c+B/2} |H(f)|^2 df \right\}} \tag{8.20}$$

$$CE(f_c) = \frac{\int_{f-B/2}^{f_c+B/2} \left(\sum_{k=0}^{K_{DR}} \alpha_k^2 + \sum_{k=0,k\neq l}^{K_{DR}} \sum_{l=0}^{K_{DR}} \alpha_k \alpha_l e^{-j(2\pi f \tau_k - \theta_k)} e^{j(2\phi f \tau_l - \theta_l)} \right) df}{E\left\{ \int_{f_c-B/2}^{f_c+B/2} |H(f)|^2 df \right\}} \tag{8.21}$$

$$CE(f_c) = \frac{B \sum_{k=0}^{K_{DR}} \alpha_k^2 + \sum_{k=0,k\neq l}^{K_{DR}} \sum_{l=0}^{K_{DR}} \int_{f_c-B/2}^{f_c+B/2} \alpha_k \alpha_l \cos\left(2\pi f (\tau_k - \tau_l) - (\theta_k - \theta_l)\right) df}{E\left\{ \int_{f_c-B/2}^{f_c+B/2} |H(f)|^2 df \right\}} \tag{8.22}$$

If the integration and additional operations with the trigonometric functions are performed, following result is obtained:

$$CE(f_c) = \frac{B \left[\begin{array}{c} \sum_{k=0}^{K_{DR}} \alpha_k^2 + \sum_{k=0,k\neq l}^{K_{DR}} \sum_{l=0}^{K_{DR}} \alpha_k \alpha_l \\ \mathrm{sinc}\left(\pi B(\tau_k - \tau_l)\right) \cos\left(2\pi f_c(\tau_k - \tau_l) - (\theta_k - \theta_l)\right) \end{array} \right]}{E\left\{ \int_{f_c-B/2}^{f_c+B/2} |H(f)|^2 df \right\}} \tag{8.23}$$

The mean operation in denominator is performed over the same integral as in the nominator. Here, only the double sum varies for small displacements. Due to the random nature of delays between multipaths $(\tau_k - \tau_l)$ and of phase components θ_k, θ_l, Equation (8.23) can be simplified as:

$$CE(f_c) = \frac{\left[\begin{array}{c} \sum_{k=0}^{K_{DR}} \alpha_k^2 + \sum_{k=0,k\neq l}^{K_{DR}} \sum_{l=0}^{K_{DR}} \alpha_k \alpha_l \mathrm{sinc} \\ \left(\pi B(\tau_k - \tau_l)\right) \cos\left(2\pi f_c(\tau_k - \tau_l) - (\theta_k - \theta_l)\right) \end{array} \right]}{\sum_{k=0}^{K_{DR}} \alpha_k^2} \tag{8.24}$$

Equation (8.24) illustrates the issue of fading. There are two terms, the single-sum represents the energy carried by the multipath components. The double-sum represents the energy fluctuations caused by the superposition of band-limited signal in time domain. For a UWB signal with large $B(\tau_k - \tau_l)$ the second term goes to zero. Hence, all multipaths are resolved and no fluctuations appear as illustrated to the wideband plot in Figure 8.5.

For a narrowband channel in dense multipath environment $B(\tau_k - \tau_l) \to 0$, the sinc-term equals one. For a small displacement change, the $(\tau_k - \tau_l)$-term in the cos-term will cause a change. Due to the large centre frequency f_c, even a small change of $(\tau_k - \tau_l)$ can induce a phase shift of π in the cos-term. As a result, a rather large variation of the energy can occur.

An empirical way to express the variation of channel energy defines a quantity called fade depth [23]. For a small displacement of antennas (compared to communication range) the fade depth is defined as three times the standard deviation of the channel energy (expressed in dB). According to Equation (8.24), the fade depth is expected to be a large number for narrowband channel and it is expected to approach 0 dB for infinite bandwidth. The changes of fade depth with bandwidth is called fade depth scaling. References, [23] studied mainly the transitions of fade depth between the wideband and UWB channel. References, [26] studied the transition from narrowband channels to UWB channels. An example of the change of fade depth with bandwidth is presented in Figure 8.6.

As well as the trend that results from empirical data, Figure 8.6 also introduces a fitted curve that can be used to model fade depth scaling. This model was introduced in [25] and is expressed as:

$$FD = A + \frac{B}{1 + C \cdot BW} \tag{8.25}$$

where A, B, and C are model parameters and BW stands for bandwidth.

The parameter A represents the minimum fade depth for a channel with infinite bandwidth. It is typically just above 0 dB due to the fact that the small changes of path loss are inevitably associated with any spatial displacement. The parameter B represents

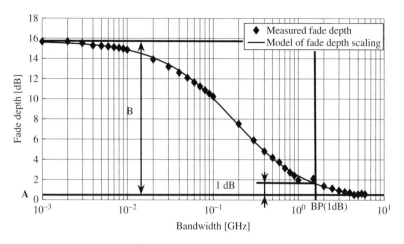

Figure 8.6 Empirical scaling of fade depth with bandwidth, taken from [24]. Reproduced with permission of IET.

the increase of fade depth for a narrowband channel. In terms of Equation (8.25), it describes the contribution of the double sum for a zero bandwidth. C describes the transition between narrowband and UWB channel. Reference [25] introduces a new descriptor, $BP(1\,\text{dB})$, which is believed to provide better insight than C. $BP(1\,\text{dB})$ is a bandwidth for which the fade depth increases 1 dB above the minimum value for UWB channel. The $BP(1\,\text{dB})$ bandwidth represents the minimum bandwidth required to resolve individual multipaths and to suppress the impact of fading on the variation of channel energy. It can be calculated as follows:

$$BP(1\,\text{dB}) = \frac{B - 1}{C} \qquad (8.26)$$

This curve fitting for the fade depth scaling is introduced because it allows us to use the parameters A, B, and $BP(1\,\text{dB})$ to quantify the changes in fade depth scaling behaviour for various conditions. For instance, reference [24] uses these parameters to investigate the changes of fade depth scaling behaviour as a function of the volume of the room.

8.3.1.2   Fade Depth Scaling in Confined Environments

The motivation for investigating how fading and fade depth changes for various volumes of confined environment is the desire to ensure robust deployment of wireless technology in environments such as airplanes, cars or even inside computer cases [27]. The considerations in [24] start with an expression similar to (8.24). [24] points out that the fade depth scaling is caused by the decrease of the sinc-term which removes the impact of the double sum (i.e. removes fading) for large products of $B(\tau_k - \tau_l)$. It is apparent that the product is the smallest for the consecutive rays, that is, $k = l \pm 1$. Furthermore, it is expected that the delays, $\tau_k - \tau_l$, between the rays decrease as the volume of the environment decreases. Thus, the bandwidth required to resolve individual multipaths increases for a smaller volume. Theoretical considerations suggest a dependency of $BP(1\,\text{dB})$ on fine volume V [24]:

$$BP(1\,\text{dB}) = V^{-\frac{1}{3}} \tag{8.27}$$

This theoretical value is confirmed in [24] where 21,200 random environment were analyzed by means of ray-tracing. The 21,200 different $BP(1\,\text{dB})$ were fitted by minimum-mean square error of the curve fitting and the results gives a dependency of

$$BP(1\,\text{dB}) = V^{-0.295} \tag{8.28}$$

which agrees well with the theoretical expectations. The ray-tracing results were validated by empirical measurements. The results – the simulation, the measurements and the minimum mean square error fitted curve according to 8.28 are presented in Figure 8.7.

The dependency of parameters A and B is also studied in [24]. It concludes that these parameters are approximately independent on the volume of the environment. In terms of fade depth scaling with bandwidth as depicted in Figure 8.6, this means that the shape of the curve remains approximately unchanged but the curve is shifted along the horizontal axis. For smaller environments, the curve moves to the right.

8.3.1.3 Impact of Fade Depth Scaling on System Performance

So far, the reasons for the existence of frequency selective fading have been introduced. The severity of fading has been quantified by means of fade depth and the scaling of fade depth with channel bandwidth has been introduced. Here, the practical impact of the

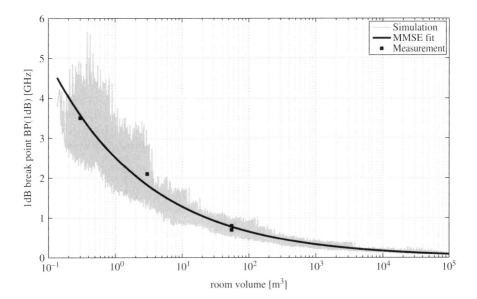

Figure 8.7 The dependence of $BP(1\,\text{dB})$ on the room volume, taken from [24]. Reproduced with permission of IET.

fading on performance of systems is explored. As mentioned before, fading is accounted for in the link budget by means of a fade margin that increases the transmit power, so that even for deep fading, the Signal to Noise Ratio (SNR) at the receiver is sufficient for the required Quality of Service (QoS) [21]. This fade margin can be determined based on the fade depth scaling curve (see Figure 8.6). Reference [24] points out that this increase in energy can also be achieved by the means of spreading the signal over frequency as performed within the WiMedia standard where, for the mode with the strongest forward error correction, each information bit is transmitted on four subcarriers.

However, notwithstanding this consideration, the change of fade depth scaling with the scaling of the volume of the environment is an important effect. One of the results might be the fact that devices designed for indoor operation might fail when operating in a smaller confined environment. For narrowband systems, the issue will be less severe because the maximum fade depth is reported to be approximately independent on the volume of the environment [24], but it impacts wideband and UWB systems.

This issue can be illustrated, for example, in reference [28] where a UWB development system is shown to fail operating at the highest-data rate (weakest forward error correction) in a small van. Similarly, reference [29] reports that a UWB development system cannot operate with packet-error-rates below 80 percent when placed in a small metallic cabinet the size of a computer server. In a nutshell, state-of-the-art wideband and UWB wireless systems designed for operation in indoor environment often cannot be directly used in smaller environments because of the increased severity of fading. This has to be addressed either by the use of a higher fade margin, stronger forward error correction or by changing the nature of the environment by introduction of absorbers blocking some multipath components and reducing the impact of fading.

8.3.2   Probability Distribution Function of Fading

8.3.2.1   Distribution Type

The preceding section has introduced the concept of fading but the fading has only been quantified by the means of fade depth which is effectively the standard deviation of channel energies. For a practical evaluation of its impact on performance of real systems, the probability distribution function needs to be known. Thus, using Equation (8.1) for mean path loss, we might define the path loss for a specific point in space using the mean path loss and a non-negative random variable representing fading, $F(B)$, as follows:

$$G_p(r, f_c) = F(B)E\left\{\int_{f_c-B/2}^{f_c+B/2} |H(r, f)|^2 df\right\} \tag{8.29}$$

The fading $F(B)$ depends on the bandwidth as presented in the preceding subsection and it can possess any non-negative value. Different authors suggest various probability distributions of $F(B)$. The most popular ones are Rice, Rayleigh, Lognormal and Nakagami. Nakagami fading is suggested by the IEEE802.15.4a model [4] and many other works, for example [5, 30]. Figure 8.8 compares empirical Cumulative Distribution Function (CDF) with CDFs of Ricean, Lognormal and Nakagami distribution. For a wideband

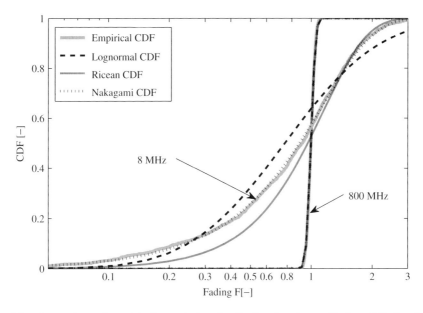

Figure 8.8 Comparison of empirical cumulative distribution functions of fading F for bandwidths of 8 MHz and 800 MHz.

channel, all three distributions correspond well to the fading. For a narrowband channel, it is apparent that Nakagami distribution is the best fit to the measured data.

Therefore, the Nakagami distribution is assumed in this work. Its probability distribution function of fading F is given by [4]:

$$p(F) = \frac{1}{\Gamma(m)} \left(\frac{m}{\Omega}\right)^m F^{2m-1} e^{-\frac{m}{\Omega}F^2} \tag{8.30}$$

where m and Ω are distribution parameters and Γ stands for Gamma-Function.

8.3.2.2 Impact of Bandwidth

Figures 8.9 and 8.10 explore the impact of bandwidth on the shape of the Probability Density Function (PDF) and CDF and the good agreement between the empirical data and the fits prove the suitability of the selection of Nakagami distribution. Figure 8.9 presents the empirical PDF obtained from measurements presented in [16] and compared it to the Nakagami distribution obtained from this data. The PDF is presented for two bandwidths: 8 MHz and 800 MHz. The changing nature of fading is documented by the change in the shape of the PDF. For narrow bandwidths, the PDF is wide and deep fades are possible, whereas for wider bandwidths of 800 MHz a deep fade (below 0.5) is extremely im-probable.

Figure 8.10 then presents the empirical and Nakagami CDF of fading for four different bandwidths. The advantage of the representation by the means of CDF is given by the fact that the CDF is more suitable for the evaluation of the goodness of the fit by a selected distribution. The Kolmogorov-Smirnov test is such a test.

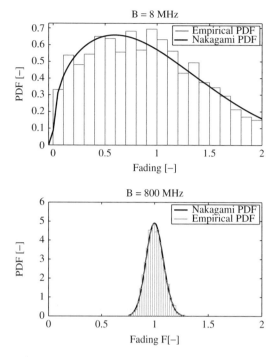

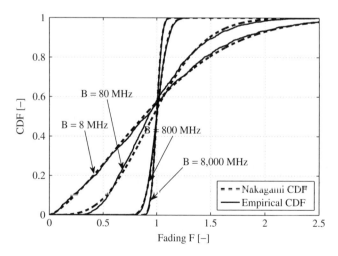

Figure 8.9 The probability distribution function of fading F for bandwidth of 8 MHz and 80 MHz – empirical values.

Figure 8.10 The cumulative distribution function of fading F for bandwidth of 8, 80, 800 and 8,000 MHz.

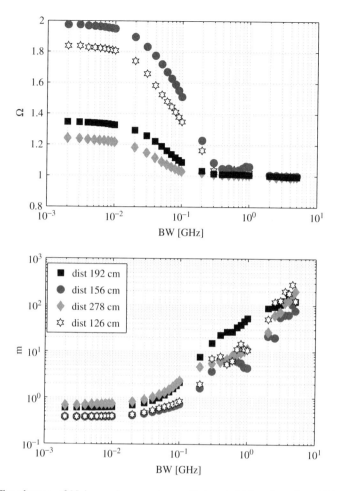

Figure 8.11 The change of Nakagami parameters with bandwidth, taken from [30]. Reproduced with permission of IET.

The impact of bandwidth is further explored in Figure 8.11 where the dependency of Nakagami parameters is plotted for four different channel measurements in a laboratory environment. Each parameter was estimated using 1,600 channel functions. Details on the measurement can be found in references [30]. Despite the differences in Ω for narrowband channels, there is an apparent change of the parameters with bandwidth. Reference [30] defines the limit for a wideband channel where multipath components can be resolved by the condition $m > 5$ & $\Omega < 1.1$ and shows that this is typically fulfilled for indoor channels with bandwidths beyond 500 MHz.

8.3.2.3 Impact of Fading on Bit Error Rate

The knowledge of the channel–the path loss and the probability distribution function of fading can be directly linked to the system performance. The Bit Error Rate (BER) can be expressed in a closed form as a function of SNR. For instance, for BPSK, BER is expressed as:

$$BER = \frac{1}{2}\text{erfc}\left(\sqrt{\frac{E_b}{N_0}}\right) \qquad (8.31)$$

Typically, the actual $\frac{E_b}{N_0}$ is not known and the mean expected $E\left\{\frac{E_b}{N_0}\right\}$ is estimated using path gain. However, the expected $E\{BER\}$ performance for a wireless channel cannot be expressed solely by the means of $E\left\{\frac{E_b}{N_0}\right\}$ using Equation (8.1). The reason is as follows. The path loss (Equation 8.1) can be used to determine the mean $E\left\{\frac{E_b}{N_0}\right\}$ but this value is the same for narrowband and wideband channels, whereas for narrowband channels the actual variation of $\frac{E_b}{N_0}$ is much larger. Thus, the fading needs to be accounted for as follows:

$$BER = \int_0^\infty \frac{1}{2}\text{erfc}\left(\sqrt{F \cdot E\left\{\frac{E_b}{N_0}\right\}}\right) p(F) dF \qquad (8.32)$$

Equation (8.32) can be used to estimate the mean expected performance of wireless communication systems as long as the fading PDF is known. Mean parameters from Figure 8.11 can be used and they describe a wide range of bandwidths. Originally, Equation (8.32) is set to be applicable for one carrier only and the integration corresponds to the uncertain level of fading in a specific point. However, it can also be used for total BER estimation of OFDM UWB systems such as those defined by the WiMedia standards [19]. For an OFDM system with N subcarriers, the total BER can be calculated as the mean of BER_n for individual subcarriers:

$$BER = \frac{1}{N}\sum_{n=1}^{N} BER_n \qquad (8.33)$$

Reference [25] shows that the nature of the fading does not significantly change with centre frequency over the UWB frequency band. This can also be demonstrated by Equation (8.24) where it is shown that the main source of fading is the delays between individual multipath components. It is noted that there is a correlation between fading experienced by adjacent subcarriers [19] but for a bandwidth over 500 MHz, which is a requirement for UWB systems, this can be averaged out. Additionally, with reference to Equation (8.24) the changes in frequency have the same impact on the cos-term as a change in $\tau_k - \tau_l$ induced by a change in position. So the changes in frequency are equivalent to the small changes in position. Thus, for an OFDM system with a large number of subcarriers, all the BER_i of individual subcarriers in Equation (8.33) can be calculated using Equation (8.32). Effectively, it removes the need for the sum in Equation 8.33, and thus Equation 8.32 can be used.

To illustrate the impact of bandwidth on the mean BER, Figure 8.12 shows the BER curve for four different bandwidths. The parameters of the fading PDFs for these bandwidths were determined from experimental channel measurements as discussed above and as presented in reference [25]. A significant performance improvement can be observed

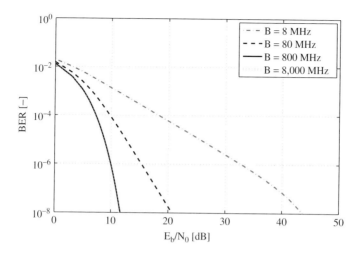

Figure 8.12 BER curves calculated using for Equation 8.32 for various bandwidths of 8, 80, 800, and 8,000 MHz.

for a change from narrowband channel to the UWB channel, but it is also noted that there is not much performance gain between the bandwidth of 800 MHz and 8,000 MHz. This is due to the fact that for 800 MHz, all multipath components can already be resolved.

To investigate the transition between the narrowband channels versus UWB bandwidth the BER versus bandwidth for $\frac{E_b}{N_0} = 10$ is plotted in Figure 8.13. As can be seen, the plot is very similar to the plot of fade depth scaling (Figure 8.6). For bandwidths below 10 MHz, the bandwidth does not play a role. An improvement follows as the systems

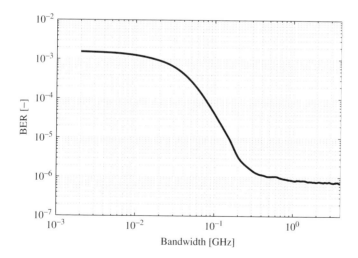

Figure 8.13 The change BER calculated using for Equation (8.32) for $\frac{E_b}{N_0} = 10$, taken from [31]. Reproduced with permission of IET.

obtain the capability to resolve individual multipath components. The BER improvement saturates for bandwidths above 300 MHz.

8.3.2.4 System Architecture and Fading

So far, we have analyzed the fading thoroughly and its impact on fading on the BER performance of systems with various bandwidths. We will now consider to what extent is it possible and practical to apply these findings into a UWB system design and the choice of suitable architecture.

Looking at Figure 8.13 it would seem that using wideband communication would be a preferable option. This would suggest the choice of Impulse Radio UWB systems which were the initial choice for UWB at the beginning of the recent technical developments [32]. However, it is not necessarily the best option as fading is not the only propagation effect to consider. Another issue is inter-symbol-interference. To avoid it, impulse radio systems can only be used as low-duty-cycle systems, which effectively limits the achievable data-rate. For instance, a guard interval of 15 ns, which might not be sufficient according to the delay spreads reported in [4], means a data-rate of 67 Mbps. Such data-rates are far from the initial promise of Gigabit Wireless that was associated with UWB. Furthermore, there are synchronization issues for coherent impulse radio systems whilst non-coherent impulse radio UWB inherently suffer from inferior performance in terms of BER performance [33]. As a result, impulse radio does not seem to be the choice for high data-rates application, even though there is a future for short-range low complexity wireless systems such as wireless sensor networks.

The alternative use of the bandwidth is the OFDM approach adopted in the WiMedia standard [19]. The apparent advantage is the fact that for longer symbols, the inter-symbol-interference is avoided whilst no guard intervals limit the data-rate. This is at the expense of increased fading. As a result, relatively strong channel coding needs to be employed. The Wimedia standards operate at a basic data-rate of 640 Mbps but due to channel coding this is reduced to data-rates of 53.3, 80, 106.6, 160, 200, 320, 400, and 480 Mbps depending on the chosen code [19]. As mentioned above, the system is generally vulnerable to fail if it operates in confined environments smaller than a typical room. For instance, for the 480 Mbps mode it is reported to fail to operate in a van [28], and reference [29] reports that the system fails completely in a confined environment with the size of a computer-server case.

A promising alternative is Single Carrier Frequency Domain Equalization (SC-FDE), which has lately received significant attention by the UWB community. This idea is based on a paper from 1970s that introduces frequency domain equalization [34]. Time domain equalization is practically a rake receiver that adds multipath components with the correct delays and phases so that the output SNR is maximized. Time domain equalization is, unfortunately, associated with high complexity [35] whereas the same operation can be performed in the frequency domain with lower complexity. As a result, blocks of symbols undergo a Fast Fourier Transform (FFT), based on channel estimation they are multiplied with an equalization matrix, and transformed back to time domain. The issue of inter-block interference is avoided by the use of periodic prefix. The advantage of such an approach is the fact that the wideband modulated single-carrier does not suffer from fading like its UWB OFDM counterpart, whilst the complexity associated with the double FFT at the receiver is the same as for the OFDM where a transceiver also needs to implement two FFTs (one for transmitter, one for receiver). SC-FDE systems do not require FFT at the transmitter. To underline the potential of SC-FDE UWB it is noted that such a system manufactured by PulseLink^{TM} has already entered the marketplace.

8.4   Multiple Antenna Techniques

The advantages of multiple antennas are well understood for narrowband wireless systems [36]. Multiple antenna techniques can be used to increase the range, data throughput or the security of the communication. In terms of UWB, the same advantages are available but the extreme bandwidth represents an additional challenge for the design of antenna arrays as well as for beamforming algorithms. Here, we focus on the description of antenna arrays and the challenges and opportunities for UWB MIMO wireless.

There are two different ways to describe wideband arrays: the narrowband description with radiation patterns suitable for OFDM systems, and the beam pattern descriptions suitable for impulse radios. These descriptors are also tightly related to the behaviour of the array and the suitable strategy for each system architecture.

A significant difference of UWB compared to systems such as LTE or IEEE802.11 is the fact that UWB is EIRP constrained rather than power constrained. As a result, different algorithms need to be used at the transmitter to ensure that the array gain does not increase the EIRP over the regulatory limits, whereas the receivers do not have any gain constraints.

8.4.1   Wideband Array Descriptors

In an arbitrary antenna array, the signals from individual antenna elements are combined. For narrowband signals, the signal from each element can pass through a delay line which introduces a phase shift. The phase shift defines the radiation pattern, mainly the direction of the main lobe. Such a description is not possible for wideband pulses as the delay line does not introduce a constant phase shift. Therefore, the description is performed based on time delays. Let us assume an N-element array with element spacing d combining a wideband signal signal $s(t)$ as depicted in Figure 8.14.

For a signal arriving from angle α, the optimum combining strategy is to introduce constant delay $\Delta\tau$ between neighbouring elements. The optimum beam steering/signal combining occurs for [37]:

$$\Delta\tau = -\frac{d}{c}\sin\alpha \tag{8.34}$$

where c is the speed of light.

This produces two main beams symmetrical around the array. So far, this does not differ from the narrowband beam steering where the main lobe is steered towards the direction of the dominant path. The difference occurs when the remaining multipath components are considered: signals $s_i(t)$ which are copies of the signal from the dominant path delayed by T_i ($s_i(t) = s(t - T_i)$) arriving from arbitrary angles α_i. For a narrowband system, these signals would be combined as they overlap in the time domain. The amplification or attenuation of these signals is determined by the radiation pattern describing the antenna

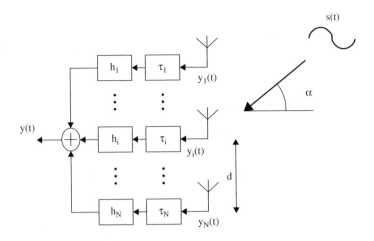

Figure 8.14 Antenna array for reception of wideband pulse $s(t)$.

array. For extremely short wideband pulses, later multipaths may not overlap in the time domain. As a result, there might be up to N non-overlapping copies of signal $s_i(t)$ at the output. To describe this situation, reference [37] introduced the concept of beam patterns for the horizontal plane of the array depicted in Figure 8.14 [37]:

$$BP(\Delta\tau, \varphi) = \frac{\max_t \left\{ \int\limits_{t-T/2}^{t+T/2} |y(t, \Delta\tau, \varphi)|^2 \right\}}{\max_t \left\{ \int\limits_{t-T/2}^{t+T/2} |s(t)|^2 \right\}} \tag{8.35}$$

where $y(t, \Delta\tau, \varphi) = \sum\limits_{i=1}^{N} h_i s \left(t - \frac{d}{c}\sin\varphi - (N-1)\Delta\tau \right)$ is the output signal, T is an integration period.

In fact, reference [37] lists three different beampatterns that differ by the integration period T and this difference corresponds to various receiver architectures assumed at the receiver.

- $T = \infty$ corresponds to a receiver that processes all the available multipath components.
- $T \to 0$ corresponds to a receiver that processes only the dominant path. This is feasible for multiple antenna impulse radio UWB systems as the energy of the dominant path is amplified by $20 \log N$ whereas the amplicition of later (non-overlapping) multipath components is significantly lower.
- arbitrary T corresponds to a system that processes only a limited number of multipaths.

The main practical conclusion of the above discussion on signal combining is the fact that for UWB antenna arrays processing short pulses grating lobes are not exhibited. In other words, the energy of the receiver or transmitter is ideally focused and only the desired dominant multipath is received and amplified (with the exception of the symmetric second main lobe). Thus as a consequence of the spatial filtering effect, inter-symbol interference is reduced which can reduce the guard interval and lead to systems with a higher data-rate. The same is applicable to inter-user-interference.

8.4.2   Antenna Arrays – UWB OFDM Systems

The preceding section highlighted the behaviour of antenna arrays for short pulses. For narrowband signals that constitute UWB OFDM systems with long pulses $s(t)$, the beam pattern in Equation 8.34 converges towards the standard description based on radiation pattern of elements and array factor. As a result, the standard beamforming algorithms can be used. It must be considered though that the antenna array factor depends on the spacing of the antennas relative to wavelength. As a result, the radiation pattern is slightly different for each OFDM subcarrier. Thus, based on system design parameters (OFDM symbol bandwidth and so on), it must be considered to what extent can one array factor, which effectively defines the channel matrix, be used for multiple subcarriers to reduce the computational complexity.

For practical WiMedia systems though, the change of radiation pattern is not significant for frequencies within a single OFDM symbol with a bandwidth of 528 MHz, but may differ significantly with the frequency hopping as illustrated in Figure 8.15. Therefore, beamforming must typically be performed individually for each frequency hop. Also, as can be seen in Figure 8.15, the issue of grating lobes might need to be considered. For a dense spacing which does not cause grating lobes at the higher end of the frequency band, the array factor at the lower end of the band is almost omni-directional which reduces the contribution of beamforming. Thus, the spacing needs to be a compromise between the existence of grating lobes for high frequencies and no beam-forming for lower frequencies.

To conclude it should be mentioned that, unlike impulse radio UWB, OFDM UWB systems can exploit spatial diversity because as presented in preceding chapters, individual subcarriers are affected by frequency selective fading. The following subsection will discuss some strategies for MIMO for OFDM UWB.

8.4.2.1   MIMO Strategies for OFDM UWB

Antenna selection for OFDM-UWB systems is introduced for example in reference [39]. The idea of antenna selection uses the fact that the OFDM symbol is combined by subcarriers from multiple antennas. In both the transmitting and receiving case, each antenna transmits/receives only a subset of subcarriers. As a result, the total system does not violate the regulatory requirements on EIRP spectral density. This approach effectively reduces fading because, if a channel from one antenna suffers from a deep fade for a particular subcarrier, based on the random nature of fading, the channel for another antenna is likely not to suffer from fading. One issue associated with antenna selection is the fact that, at the receiver, the signal will have a discontinuous phase, which can be an issue for the receiver, especially if algorithms assuming continuous phase are employed. As a result, reference [39] suggest minimum phase techniques that ensure continuity of phase and help to achieve optimum performance.

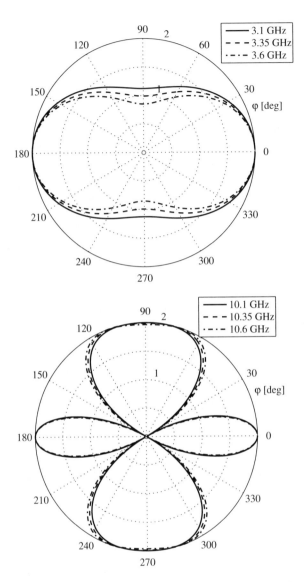

Figure 8.15 Array Factor for 2-antenna array spaced by 4 cm for frequencies from the lower end of the FCC UWB spectrum (top) and the upper end of the FCC UWB spectrum (bottom), taken from [38]. Reproduced with permission of IET.

Whilst antenna selection offers a significant improvement of BER performance, further improvement can be achieved by the use of **beamforming**. Even with antenna selection, the energy is radiated in/received from all directions. Hence, the constructive and destructive multipaths remain present at the receiver. However, a beamformer can be used to direct energy in the constructive multipath components only. This will increase the SNR at the receiver even though the total radiated energy is actually reduced because it will not be radiated/received omni-directionally. Also, the spatial selectivity of the antenna might be utilized to suppress interferers.

As mentioned in the introduction of well-known beamforming algorithms such as Eigen-beamforming can be used at the **receiver**, but will not be used at the **transmitter** because they have been developed for power constrained systems where EIRP in a certain direction can be increased. According to [39], the EIRP constrained beamforming can be solved as a convex optimization problem, which is computationally complex. Henceforth, computationally less demanding suboptimal use of Eigen-beamforming where the final result is scaled so that the EIRP limits are not violated were suggested. Unfortunately according to [39], scaled Eigen-beamforming's BER performance is inferior to antenna selection. On the otherhand, [39] shows that the lobes of antenna factor generated by scaled Eigen-beamforming have the same spatial distribution as the lobes of the optimal scheme, even though their level is lower. Therefore, reference [39] suggests a new method that uses the result of scaled Eigen-beamforming and reduces the spatial peak-to-average ratio of the array factor. Thus, a sub-optimal method with a BER performance closer to the optimal scheme than antenna selection or scaled-Eigen-beamforming is obtained. Performance of this method is shown to be significantly superior to antenna selection.

Another advantage of beamforming over antenna selection is the potentially lower complexity. WiMedia OFDM symbols consist of 128 subcarriers [19]. With frequency hopping over three frequency bands, it is 384 subcarriers [19]. Antenna selection algorithms as presented in [39], require individual processing for each subcarrier. As pointed out above, the array factor remains approximately unchanged over the bandwidth of a Wimedia OFDM symbol. Thus, beamforming for the entire bandwidth of OFDM symbol is possible even though such an approach is suboptimal.

8.5   Implications for LTE-A

This chapter has presented an analysis of wave propagation in UWB channels and links it to the performance of UWB systems, but the conclusions are applicable to any wireless system operating indoors. Systems such as LTE macrocells, however, have a higher dynamic range, therefore more multipath components need to be considered. The transmission range of the system is also larger meaning that multipath components incur more delay. Femtocells, on the other hand, usually operate indoors and at much lower powers than macrocells and so will have characteristics more akin to UWB channels, but also exhibit some fundamental differences, mainly due to their considerably narrower bandwidth signals compared to those of UWB. Here we list a few examples illustrating that the wave propagation is very similar and many conclusions from UWB channel are transferable to LTE.

One of the considerations within the LTE community is the bandwidth scaling of wireless systems from 20 MHz to 100 MHz. This is a significant increase in bandwidth which has an important impact on system performance as more multipath components will be resolvable. Figure 1.13 illustrates the change of mean BER as the bandwidth of the transmitted system increases for $E\left\{\frac{E_b}{N_0}\right\} = 10$ dB. With a change from 20 MHz to 100 MHz, the BER is reduced by a factor of more than 10. This improvement can determine the choice of modulation scheme. Due to the reduced impact of frequency selective fading, wideband modulation schemes such as CDMA or SC-FDE would then require weaker channel coding. This reduction in overheads can be used to increase the data-rate, range, number of users and to reduce complexity and hence power and cost.

LTE networks make provision for multi-link systems, where a mobile user may communicate with multiple base stations at the same time. For correct estimates of available channel capacities, the number of independent multipath components must be known. For this estimation, the conclusions relating to the manifestation of channel impulse response are crucial as they link the occurance of multiple peaks together.

With the increase of bandwidth, the duration of communication symbols in many cases corresponds to the distance between the base stations. In this case, the beam-pattern considerations mentioned in Section 8.4 can be employed to improve localization of mobile station or to enable low-complexity beam.

As well as drawing technical similarities between UWB and LTE wave propagation, we should also consider UWB as complimentary to LTE, where UWB is designed to provide high data rate, short range links. This forms part of a suite of wireless technologies that enables the provision of mobile multi-media services in a wide range of scenarios. The data may originate or be destined for around the home, about a person or within a vehicle, but require a gateway to a wider area network in order to reach a distant destination. UWB, as well as some other signalling schemes, provides a solution for these short range links requiring close to the data source or destination.

References

1. A. Goldsmith, S. Jafar, I. Maric, and S. Srinivasa, "Breaking spectrum gridlock with cognitive radios: An information theoretic perspective," *Proceedings of the IEEE*, vol. 97, no. 5, pp. 894–914, May 2009.
2. D. Cabric, S. M. Mishra, D. Willkomm, R. Brodersen, and A. Wolisz, "A cognitive radio approach for usage of virtual unlicensed spectrum," in *14th IST Mobile Wireless Communications Summit, 2005*, 2005.
3. B. Allen, M. Dohler, E. Okon, W. Malik, A. Brown, and D. Edwards, *Ultra Wideband Antennas and Propagation for Communications, Radar and Imaging*. John Wiley & Sons, 2006.
4. A. F. M. et al., "IEEE 802.15.4a channel model - final report," IEEE, Tech. Rep., 2004.
5. A. F. Molisch, "Ultrawideband propagation channels-theory, measurement, and modeling," *Vehicular Technology, IEEE Transactions on*, vol. 54; no. 5, pp. 1528–1545, 2005.
6. A. F. Molisch, K. Balakrishnan, D. Cassioli, C.-C. Chong, S. Emami, A. Fort, J. Karedal, J. Kunisch, H. Schantz, and K. Siwiak, "A comprehensive model for ultrawideband propagation channels," pp. 6–3653, 2005.
7. A. Durantini, W. Ciccognani, and D. Cassioli, "Uwb propagation measurements by pn-sequence channel sounding," pp. 3414–3418, Vol.6, 2004.
8. B. M. Donlan, S. Venkatesh, V. Bharadwaj, R. M. Buehrer, and J.-A. Tsai, "The ultra-wideband indoor channel," pp. 208–212 Vol.1, 2004.
9. S. S. Ghassemzadeh, R. Jana, C. W. Rice, W. Turin, and V. Tarokh, "Measurement and modeling of an ultra-wide bandwidth indoor channel," *Communications, IEEE Transactions on*, vol. 52; no. 10, pp. 1786–1796, 2004.
10. S. S. Ghassemzadeh, R. Jana, C. W. Rice, W. Turin, and V. Tarokh, "A statistical path loss model for in-home uwb channels," pp. 59–64, 2002.
11. S. S. Ghassemzadeh and V. Tarokh, "Uwb path loss characterization in residential environments," pp. 501–504, 2003.
12. D. R. McKinstry and R. M. Buehrer, "Uwb small scale channel modeling and system performance," vol. 58. Affiliation: Mobile/Portable Radio Res. Group, Virginia Tech, Blacksburg, VA, United States, 2003, pp. 6–10.
13. R. C. Qiu and I.-T. Lu, "Wideband wireless multipath channel modeling with path frequency dependence," pp. 277–281 vol.1, 1996.
14. A. Saleh and R. Valenzuela, "A statistical model for indoor multipath propagation," *Selected Areas in Communications, IEEE Journal on*, vol. 5; no. 2, pp. 128–137, 1987.

15. V. Sipal, B. Allen, and D. Edwards, "Analysis and mitigation of antenna effects on wideband wireless channel," *Electronics Letters*, vol. 46, no. 16, pp. 1159–1160, 2010.

16. V. Sipal, B. Allen, and D. Edwards, "Effects of antenna impulse response on wideband wireless channel," in *Antennas and Propagation Conference (LAPC), 2010 Loughborough*, Nov. 2010, pp. 129–132.

17. A. Shlivinski, E. Heyman, and R. Kastner, "Antenna characterization in the time domain," *Antennas and Propagation, IEEE Transactions on*, vol. 45; no. 7, pp. 1140–1149, 1997.

18. V. Sipal, B. Allen, and D. Edwards, "Modelling of antenna pattern descriptors for antenna performance evaluation," in *International Conference on Modelling and Simulation, 2010. MS 10 Prague*, June 2010.

19. G. Heidari, *WiMedia UWB - technology choice for wireless USB and Bluetooth*. John Wiley & Sons, 2008.

20. V. Sipal, B. Allen, and D.J. Edwards, "Enhanced Fade Depth Model for Extremely Wideband Channels," *Int. Conference Modelling and Simulation 2010*, AMSE. Prague, 22–25, Jun. 2010.

21. W. Q. Malik, B. Allen, and D. J. Edwards, "Fade depth scaling with channel bandwidth," *Electronics Letters*, vol. 43; no. 24, pp. 1371–1372, 2007.

22. W. Malik, B. Allen, and D. Edwards, "Bandwidth-dependent modelling of smallscale fade depth in wireless channels," *Microwaves, Antennas Propagation, IET*, vol. 2, no. 6, pp. 519–528, Sept. 2008.

23. W. Q. Malik, D. J. Edwards, and C. J. Stevens, "Frequency dependence of fading statistics for ultrawideband systems," *Wireless Communications, IEEE Transactions on*, vol. 6, no. 3, pp. 800–804, 2007.

24. V. Sipal, J. Gelabert, B. Allen, C. Stevens, and D. Edwards, "Frequency-selective fading of ultrawideband wireless channels in confined environments," *Microwaves, Antennas Propagation, IET*, vol. 5, no. 11, pp. 1328–1335, 2011.

25. V. Sipal, B. Allen, and D. Edwards, "Exploration and analysis of fade depth scaling," in *Antennas and Propagation Conference (LAPC), 2010 Loughborough*, Nov. 2010, pp. 125–128.

26. G. Llano, J. Reig, and L. Rubio, "Analytical approach to model the fade depth and the fade margin in uwb channels," *Vehicular Technology, IEEE Transactions on*, vol. 59, no. 9, pp. 4214–4221, Nov. 2010.

27. J. Gelabert, D. Edwards, and C. Stevens, "Experimental evaluation of uwb wireless communication within pc case," *Electronics Letters*, vol. 47, no. 13, pp. 773–775, 2011.

28. I. Garcia Zuazola, J. Elmirghani, and J. Batchelor, "High-speed ultra-wide band in-car wireless channel measurements," *Communications, IET*, vol. 3, no. 7, pp. 1115–1123, July 2009.

29. V. Sipal, J. Gelaber, C. Stevens, B. Allen, and D. Edwards, "Impact of confined environments on wimedia uwb systems," in *Antennas and Propagation Conference (LAPC), 2011 Loughborough*, Nov. 2011, pp. 125–128.

30. V. Sipal, B. Allen, and D. Edwards, "Exploration of nakagami fading in ultra-wideband wireless channels," *Electronics Letters*, vol. 47, no. 8, pp. 520–521, 2011.

31. V. Sipal, B. Allen, and D. Edwards, "Study of multi-tone frequency shift keying for ultrawideband wireless communications," *IET Communications*, 2012.

32. M. Win and R. Scholtz, "Impulse radio: how it works," *Communications Letters, IEEE*, vol. 2, no. 2, pp. 36–38, Feb 1998.

33. N. He and C. Tepedelenlioglu, "Performance analysis of non-coherent uwb receivers at different synchronization levels," *Wireless Communications, IEEE Transactions on*, vol. 5, no. 6, pp. 1266–1273, June 2006.

34. T. Walzman and M. Schwartz, "Automatic equalization using the discrete frequency domain," *Information Theory, IEEE Transactions on*, vol. 19, no. 1, pp. 59–68, Jan 1973.

35. D. Falconer, S. Ariyavisitakul, A. Benyamin-Seeyar, and B. Eidson, "Frequency domain equalization for single-carrier broadband wireless systems," *Communications Magazine, IEEE*, vol. 40, no. 4, pp. 58–66, Apr 2002.

36. C. Oestges and B. Clerckx, *MIMO Wireless Communications*. Orlando, FL: Academic Press, 2007.

37. T. Kaiser, F. Zheng, and E. Dimitrov, "An overview of ultra-wide-band systems with mimo," *Proceedings of the IEEE*, vol. 97, no. 2, pp. 285–312, Feb. 2009.

38. V. Sipal, B. Allen, D. Edwards, and B. Honary, "20 years of ultrawideband: Opportunities and challenges," *IET Communications*, 2012.

39. C. Vithanage, M. Sandell, J. Coon, and Y. Wang, "Precoding in ofdm-based multi-antenna ultra-wideband systems," *Communications Magazine, IEEE*, vol. 47, no. 1, pp. 41–47, January 2009.

9

Wireless Body Area Network Channels

Rob Edwards[1], Muhammad Irfan Khattak[2] and Lei Ma[1]
[1]*Loughborough University, UK*
[2]*NWFP University of Engineering and Technology, Pakistan*

9.1 Introduction

The 1980s saw the beginnings of what has now become a revolution in personal communication systems. Although there are several contenders, the first viable voice-only cellular system is thought to have been NMT (Nordic Mobile Telephone) which was an analogue cellular system deployed in Nordic countries, Eastern Europe and Russia. Other early starters included Total Access Communications System (TACS) in the United Kingdom and Advanced Mobile Phone System (AMPS) in the United States. Early systems tended to have less than convenient handsets with cumbersome power requirements. Over almost three decades now there has been a generally increasing demand for handsets that are smaller with increased facility and longevity of use. This demand has been satisfied by improvements in technology particularly in the field of miniaturization whereby components of a communication system are generally much reduced in size.

Long Term Evolution (LTE) wireless network technology comes to us at a time when the Apple iPhone and similar smart phones from other manufacturers have constrained this strand of devices to have a display area of roughly 90 mm across the diagonal. Nevertheless, there continues to be significant demand for ever smaller mobile communications devices. Although realization of a 4G phone can be achieved in many ways, we can divide the methods into those widely achieved and those desired. The four widely achieved methods are: the classic mobile phone; the smartphone; the integrated mobile computer phone; and the stand alone modem phone. The desired methods of realization include wearable phones, either in clothing or on the body, and more exotically, mobile phones implanted into the body with various degrees of permanence.

LTE-Advanced and Next Generation Wireless Networks: Channel Modelling and Propagation, First Edition.
Edited by Guillaume de la Roche, Andrés Alayón Glazunov and Ben Allen.
© 2013 John Wiley & Sons, Ltd. Published 2013 by John Wiley & Sons, Ltd.

Since allegorically mobile phones are now more numerous than people it is probable that all types of human phone interaction are currently happening, and therefore in this chapter we will explore the physics of how the radio frequency radiation generated by mobile phones interacts with our skin, tissue and bones. We will look at how energy from mobile phones is absorbed and how propagation occurs close to the surface of the body. The safety aspects of LTE devices which relate to specific absorption rates are not discussed in this chapter. Note that 4G hand sets are likely to deliver less energy into users than the 3G hand sets they are replacing, and that recent research indicates that the current safety advice on the power of handsets is considered safe for normal use [1].

For the purposes of clarity, Wireless Body Area Network (WBAN) and Wireless Body Sensor Network (WBSNs) will be referred to collectively as Radio Body Networks (RBNs). Typically such networks would comprise a master controller worn somewhere on the body termed a Body Central Unit (BCU) and one or more sensors wirelessly connected on the body. Note that although both ultrasonic channels and infrared channels are covered by the generic term 'wireless', neither are considered here since we look only at radio.

Germane to LTE we have two modes of operation for RBNs: that of inter-mode in which a network on the body uses a cellular network to route traffic to other RBNs; and intra-mode in which telemetry from sensors on the body is routed via a BCU using cellular technologies such as LTE as a bearer technology. Therefore there are two channels to consider. One is the cellular channel (forward/down link and reverse/up link), which may use LTE as a bearer. Then there is the on-body channel between the BCU and sensors on the body which use an arbitrary bearer.

Figure 9.1 shows two of the main example scenarios involved in RBNs, the first being a point-to-point system with two monopole antennas on the chest, and the second being a cellular mode with an antenna designed to be used on the back of the hand. Note that there are an infinite number of antenna positions and types for both modes but propagation efficiency is best achieved using antennas high off the ground and insulated from the skin.

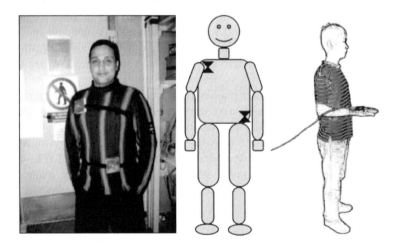

Figure 9.1 Scenarios for on-body antennas: left and centre - point-to-point monopole, right - glove mounted wearable antenna.

Propagation on the body is best achieved using transverse electric type antennas such as monopoles that see the body as a lower impedance path than their transverse magnetic counterparts [2].

For a proper understanding of radio propagation close to users, four topics need to be considered. These are:

1. Wearable antennas (that can be considered briefly).
2. How radio frequency energy interacts with human biological tissue.
3. How humans change the properties of wearable antennas.
4. How from measurements researchers have predicted radio waves behave on the surface of the body.

Inside humans, electromagnetic waves travel at differing speeds dictated by the type of tissue in which they exist. As LTE advanced has a relatively wide range of possible frequency bands, for example TR 36.913 identifies bands from as low as 450 MHz to as high as 4.99 GHz, the on-body channels can exist in dramatically different forms for this technology. This broad range of frequencies, combined with multiple-input multiple-output antenna systems and carrier aggregation makes inter-RBNs a complex problem particularly for antenna engineers.

The majority of current personal communication systems are based around wireless technologies. Typically, an antenna is used in such systems to transmute from or into guided energy in the radio into or from an electromagnetic wave for transmission and reception of the information. Antennas have the property that they work best at resonance and their efficiencies are strongly linked to physical size such that they are often made to be at least half a wavelength in at least one of their dimensions. This has meant that they are perhaps somewhat more resistant to miniaturization than other components of personal communication systems. A recent trend, therefore, has been to move the antennas out of the system and onto the human body where more space is available. We call them on-body antennas or Wearable Antenna (WAs).

By selective tuning, a good antenna design can decrease the power that a communication system needs to maintain its signal budget. This gives engineers a choice between range, battery life and improved signal-to-noise ratio margins. Modern cellular is defined by its ability to dip into the noise floor to an ever increasing degree and even improvements of less than 1 dBm are significant in low power technology.

9.2 Wearable Antennas

For wearable applications, antennas should be able to resist the effects from the human body and in addition be demonstrably safe in terms of the power levels they create in a wearer. Electrically large ground planes are usually preferred for wearable antennas since they can increase a wearable antenna's stability and partially isolate the human body from radio frequency radiation. The greater the proportion of electromagnetic flux existing in the user, the greater the loss the antenna will suffer. Radio frequency radiation is linked to Specific Absorption Rate (SAR) which should be kept low, and SAR can be reduced by shielding the body with a ground plane. However, big ground planes increase the size and rigidity of wearable antennas making them less compatible with the user.

Most traditional antennas are rigid, which makes them too uncomfortable to be embedded into skin or clothing. This has resulted in a demand for new types of antenna better suited to operating in close proximity to humans. Several types of antennas now have wearable implementations. In particular, these are microstrip antennas, printed dipoles or loop antennas, printed monopoles or Planar Inverted-F Antenna (PIFAs). Such antennas are normally flat and can loosely be described as comprising of four elements, namely: a feed; a ground plane; a dielectric substrate; and one or more radiating elements all of which have typically been rigid.

Wearable antennas have received interest due to the introduction of personal communications technology which provides momentum to make mobile phones smaller and simpler to carry and use. Clothing may now have a variety of consumer electronics integrated into it. In personal electronics, antennas play a paramount role in the optimal design of the wearable or hand-held units used in communications. Such antennas can increase range, reduce infrastructure, increase battery life and improve diversity. The antenna requirements are given by the particular specification, but common to many applications of wearables the requirements are: to be light weight, inexpensive, low maintenance and robust to wear and tear. They should also be comfortable to wear and this has resulted in demand for flexible antennas, which can be easily attached to clothing.

A very early paper written in 1968 entitled *The Effect of the Human Body on Radiation Properties of Small-Sized Communication Systems* [3] illustrates the problem of an antenna close to the body very well. Here, the author describes how it was found that radio sets achieve a better range when located away from the body even though the whip antennas the radio sets used had been designed (tuned), to be on the body. Krupka concluded

> ... the human body acts mainly as an absorbing element, reducing in principle the radiated power of an antenna situated in close proximity to it.

He was not able to analytically describe the problem and therefore produced a paper of measurements. In fact it is only in the last decade that mathematical solutions for body antennas have begun to emerge and the problem of the body's interactions with radio frequency radiation from mobile communications devices is in essence the wearable antenna problem. In a more recent paper (2000), an RF helmet antenna and RF vest antenna were put forward by the authors of [4] and [5] respectively, which could be operated over the frequency range of 500-2000 MHz. A conductive cloth and polyester interwoven with nickel and copper were used to construct the helmet antenna whilst the vest antenna was made of canvas and flectron from Laird Technologies. The main advantages of these antennas were light weight, inexpensive to manufacture and low maintenance.

Medical imaging is another notable application of wearable antennas. For example, the detection of cancer cells as early as possible is an area of concern and very promising results were put forward by the authors of [6] for the early detection of breast cancer. The idea behind this research was that the breast tumours have different electrical properties compared to healthy breast tissues at certain frequencies. The breast was illuminated with electromagnetic waves and the transmitted or reflected waves were then measured and compared with known results. Monopole antennas were used by the authors of [6] but research is also being done to develop textile antennas for medical imaging as these antennas are uncomfortable against the skin.

As a small aside it is worth noting that human tissue is dispersive and therefore antennas for 4G cellular, which tend to wider bandwidths than early cellular radio, interact with the body in a different way to previous types of antennas used for early cellular systems such as Global System for Mobile communication (GSM). As will be shown later in this chapter the electrical properties of biological matter, such as conductivity and complex permittivity (which includes losses), vary with frequency and over some bands their rates of change can be quite dramatic. Spectrum selection in 4G therefore has implications for power losses close to humans.

The concept of *Smart Clothing* may date back to as early as 1993 [7, 8]. Smart clothes are augmented clothes improved by intelligent wearable systems. In these systems new textiles, now known as technical textiles, are beginning to appear. Technical textiles may be conductive but also flexible. A term associated with technical textiles used in wearable antennas is drapeability. Drapeability is a subjective measure of how well an antenna behaves like clothing. In other words, the usual purpose of clothing can be extended to new tasks such as a shirt calling for emergency assistance or a vest transmitting continuous health telemetry. Smart clothes were recently introduced by the authors of [7, 8] for arctic environments.

Since wearable antennas can be incorporated with clothes, so the textile material can used as a substrate for these types of antennas. One of the main advantages of this type of substrate is that clothing tends to have low permittivity which means that surface wave losses are decreased and impedance bandwidth is increased. On the other hand, fabric stretch and compression changes the dimensions of the antenna and influences the antenna's characteristics. Typical changes are detuning in the frequency band and loss of Q, both of which result in reduced wireless performance.

It turns out that although smart clothing offers a simple conceptual avenue for wearable antennas it is the opinion of the authors that antennas as clothing decoration, such as badges or logos or antennas in accessories, such as hats, bags and shoes are likely to be much more popular amongst the general population. This aspect is simply a function of the diversity of humans.

Having introduced the topic we are now able to discuss the physics of how the electromagnetic fields from antennas interact with human tissue and the characteristics of the on-body channel for LTE advanced wireless networks.

9.3 Analysis of Antennas Close to Human Skin

It is widely accepted that small antennas interact strongly with human tissues [9, 10]. In fact at all of the popular frequencies used for mobile communications a typical unoptimized free space antenna would see all of its important parameters, for example directivity, radiation pattern, input impedance and efficiency, markedly changed in the proximity of biological tissue. This is mainly because the body consists of elements that have different electrical properties and is therefore inhomogeneous and thus dispersive. S. Gabriely, R. W. Lau and C. Gabriel from Kings College London have measured the dielectric properties of human body tissues. They did this by adapting a 50 ohm conical probe to interface with the target tissues. Then, using an automatic swept-frequency network and impedance analyzers, the authors obtained results which can be used to calculate the properties of human body tissue from 10 Hz to 20 GHz. Their work proves that the electrical properties of human body tissues are significantly different from those of free space. The result of this is that EM waves behave differently in free space than in the human body. For example the muscle in an arm has a dielectric constant of 53.55; the skin in an arm has the dielectric constant of 38.87, and the air that surrounds them has a dielectric constant of 1. At 1.8 GHz the effective wavelength goes from 166.67 mm in air to 26.73 mm in the skin to 22.78 mm in muscle. In other words the resonant length of an antenna inside a body may be more than seven times smaller than one outside it.

The body is also conductive which means that an antenna on the skin may be shorted out and therefore prevent the current getting to the radiating parts with sufficient magnitude to allow the antenna to perform well. It turns out to be really tricky to keep current on just one side of an antenna if the size of the antenna and its ground plane is small, as it is in most user equipment. Current from the antenna therefore will interact with the user. The study of these interactions contributes to improved design and safety of WAs. Note that since there is very little tissue with magnetic properties in the body, most researchers assume non-magnetic materials with a relative permeability set to unity, $\mu_r = 1$.

9.3.1 Complex Permittivity and Equivalent Conductivity of Medium

The permeability and permittivity of free space are $\mu_0 = 1.257 \times 10^{-6}$ H/m and $\varepsilon_0 = 8.85 \times 10^{-12}$ F/m, respectively. Assuming non-magnetic materials, the media through which electromagnetic waves pass have the same permeability as free space but different permittivity. For most materials permittivity ε is usually a complex value which consists of a real part ε_R and an imaginary part ε_I [11].

$$\varepsilon = \varepsilon_R - j\varepsilon_I \tag{9.1}$$

ε_R represents the ability of bound charges in a material to polarize in response to a time varying electric field. With a given field strength, higher values of ε_R mean stronger polarization. Since these bound charges are limited in motion by other properties of the tissue, there is always a delay before their polarization follows the time varying electric field. To overcome the motion limitation some of the energy contained in the radio wave will converted into heat. Simplistically one can see this as heat generated by friction. Therefore we see a loss mechanism which illustrates why humans present a lossy load to the mobile phone antenna. Thus the bound charges align themselves with the polarity of the field from the handset. This is more obvious at high frequencies where the polarization rate of some bound charges cannot follow the frequencies such that the value of it may in fact decrease with the increased frequency. The time delay between the time varying electric field and the polarization rate can be represented in terms of ε_I, which is a function of frequency and can be used to compute the power loss.

According to the time-harmonic form of Maxwell's equations, in free space with no source,

$$\nabla \times \mathbf{H} = j\omega \mathbf{D} = j\omega\varepsilon_0 \mathbf{E} \tag{9.2}$$

Where $\mathbf{H}$ is the magnetic field intensity (volts per meter), ω is the angular frequency (radian per second), $\mathbf{D}$ is the electric displacement field (coulombs per square meter) and $\mathbf{E}$ is the electric field intensity (volts per meter).

While in other mediums [12]

$$\nabla \times \mathbf{H} = \mathbf{J} + j\omega \mathbf{D} \tag{9.3}$$

$$= \sigma_c \mathbf{E} + j\omega\varepsilon \mathbf{E} \tag{9.4}$$

$$= \sigma_c \mathbf{E} + j\omega \left(\varepsilon_R + j\varepsilon_I\right) \mathbf{E} \tag{9.5}$$

$$= (\sigma_c + \omega\varepsilon_I)\mathbf{E} + j\omega\varepsilon_R \mathbf{E} \tag{9.6}$$

$$= \left(\sigma_c + \sigma_p\right)\mathbf{E} + j\omega\varepsilon_R \mathbf{E} \tag{9.7}$$

$$= \sigma_e \mathbf{E} + j\omega\varepsilon_R \mathbf{E} \tag{9.8}$$

where $\mathbf{J}$ is the current density (amperes per square meter), σ_c and σ_p are defined as conduction conductivity and polarization conductivity respectively and indicate the conduction current and polarization delay.

Since σ_c and $\omega\varepsilon_I$ have effects that lead to the power loss they can be lumped together to form σ_e and termed the 'equivalent conductivity'. Alternatively we can write

$$\frac{\sigma_c}{\omega} = \varepsilon_c \tag{9.9}$$

so the effective imaginary part of permittivity becomes

$$\varepsilon_{el} = \varepsilon_I + \varepsilon_c \tag{9.10}$$

A second form of Equation 9.8 can be derived as follows

$$\nabla \times \mathbf{H} = \sigma_e \mathbf{E} + j\omega\varepsilon_R \mathbf{E} \tag{9.11}$$

$$= [\sigma_e \mathbf{E} + (j\omega\varepsilon_R \mathbf{E} - j\omega\varepsilon_0 \mathbf{E})] + j\omega\varepsilon_0 \mathbf{E} \tag{9.12}$$

$$= \sigma_{\text{total}} \mathbf{E} + j\omega\varepsilon_0 \mathbf{E} \tag{9.13}$$

$$= \mathbf{J}_{\text{total}} + j\omega\varepsilon_0 \mathbf{E} \tag{9.14}$$

By comparing Equation 9.8 with Equation 9.14 it can be seen that $\mathbf{J}_{\text{total}}$ comprises all of the effects on the radio wave when it is inside tissue and that the complex value σ_{total} comprises both the conduction and polarization currents. Therefore these equations yield the amount of energy turning into heat inside human tissue as a function of complex permittivity and frequency.

9.3.2 Properties of Human Body Tissue

As a result of the analysis in the previous section and given the data relating to the permittivity of human tissue it is now possible to plot the properties of human tissue over the LTE spectrum as they change with frequency. Using the data from [13] the next three figures (Figures 9.2–9.4) show how the dielectric constant $\varepsilon_r = \frac{\varepsilon_R}{\varepsilon_0}$, loss tangent $\tan\delta = \frac{\varepsilon_I}{\varepsilon_R}$ and the conductivity vary with the frequency for blood, cortical bone (supporting

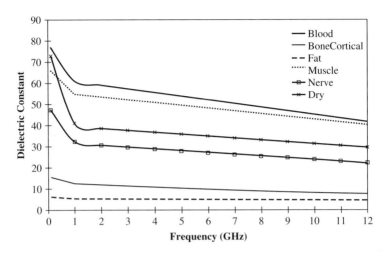

Figure 9.2 Dielectric constant versus frequency for common body tissues.

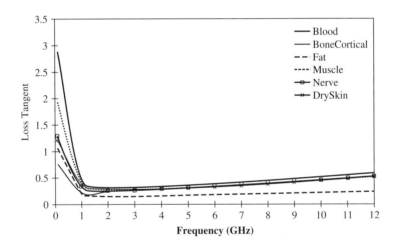

Figure 9.3 Loss tangent versus frequency for body tissues.

bone), fat, muscle, nerve and dry skin (Figure 9.3). The frequency range – dc to 12 GHz includes all proposed LTE bands.

In Figure 9.2 we see the knee of the dielectric constant versus frequency curve occurs at about 1.2 GHz, below which the rate of change of permittivity with frequency is relatively steep. This is an important consideration for wearable antennas designers at these frequencies.

In general it can be seen that the body's tissues have dielectric constants greater than that of free space. Below 7 GHz the dielectric constants of blood and muscle are more than 50 times that of air. In the range of frequencies considered here, fat is found to have the permittivity closest to that of free space. All tissue permittivity values in the range

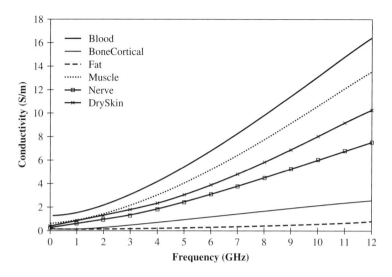

Figure 9.4   Conductivity versus frequency for body tissues.

decrease with increasing frequency. A consequence of this is that at high frequencies antennas can operate closer to the skin for constructive interference of the reflected wave which leads to improved gain.

The values of loss tangent can be divided into two sections. In the first section below 1 GHz the loss tangent rate of change steeply decreases to below 0.5 while above 1 GHz, there is a small rate of increase. This means that close-to-body antennas using the lower LTE bands will experience greater losses (maximally six times greater), than their higher frequency counterparts.

In Figure 9.4 it can be seen that the conductivity of dry skin (skin exposed to air), lies between approximately 1.75 S/m at 3 GHz and 7 S/m at 10 GHz. Whilst for blood the conductivity across the same range of frequencies varies from approximately 3 to 13 S/m. The outer surface of humans is therefore quite conductive in the range of frequencies considered here.

We know that the permittivity, ε, associated with the permeability, μ, and the frequency, f, will decide the wavelength λ of a radio wave in the body.

$$\lambda = \frac{1}{f\sqrt{\varepsilon\mu}} \tag{9.15}$$

So with a fixed frequency, the wavelength of the radio wave will be shorter in the human body than in free space. A consequence of this is that an antenna designed for free space applications will be electrically larger in biological tissue than in air.

Since the human body is conductive fields from antennas produce currents which flow in our tissues. Such currents may oppose currents on the radiating elements of the antenna. Therefore in addition to heat losses in biological matter we have currents in the body that may exist with unhelpful polarity. These effects are inversely proportional to the distance a handset antenna is from the body.

Antennas close to humans should therefore be considered as a system comprising a human and an antenna rather than just an antenna. Closeness in this context can be defined by the frequency of operation but at LSE frequencies effects become very significant for distances less than a few centimeters.

9.3.3 Energy Loss in Biological Tissue

In Equation 9.8 it was shown how σ_c and σ_p can be combined into the effective conductivity σ_e. Using this parameter along with knowledge of the electric field, the energy loss in biological tissue W_{loss} can be defined as

$$W_{loss} = \int_{v_h} \sigma_e |\mathbf{E}|^2 \, dv_h \tag{9.16}$$

where v_h is the volume of tissue. In essence, energy from the radio wave becomes heat in the tissue and therefore constitutes a loss. The field density of an antenna is typically greatest close to the antenna and thus moving a wearable antenna closer to the body increases the proportion of loss in the tissue. Note that all equipment in LTE must use power levels below the safety limit and heating from handsets is not currently thought to pose a health risk. A paper covering this aspect is published by the author in [14] which includes a great deal of discussion on power levels from mobile handsets in humans.

9.3.4 Body Effects on the Q Factor and Bandwidth of Wearable Antennas

The Quality Factor, Q is an important parameter of antennas that relates the radiated energy, P_R, and ohmic losses, P_L, of an antenna to the reactive energy P_S stored around the antennas structure.

$$Q(\omega) = \frac{\omega P_s(\omega)}{P_R(\omega) + P_L(\omega)} \tag{9.17}$$

All these parameters are functions of angular frequency, ω. Here we assume $e^{j\omega t} \to 1$ and thereafter neglect it. In [15], the Q factor was related to the voltage standing-wave ratio bandwidth FBW_V using the following equalities.

$$Q(\omega) \approx \frac{\omega}{2R(\omega)} |Z'(\omega)| \tag{9.18}$$

where

$$Z(\omega) = R(\omega) + jX(\omega) \tag{9.19}$$

is the input impedance of the antenna and

$$Z'(\omega) = R'(\omega) + jX'(\omega) = \frac{dR(\omega)}{d(\omega)} + j\frac{dX(\omega)}{d(\omega)} \tag{9.20}$$

is the frequency derivative of the complex input impedance.

If we now define β as in [15] as the defined bandwidth of an antenna close to a human we can write

$$\text{FBV}_\text{V}(\omega) \approx \frac{4\sqrt{\beta}R(\omega)}{\omega|Z'(\omega)|} \tag{9.21}$$

and now restate the Q of an on-body antenna as

$$Q(\omega) = \frac{2\sqrt{\beta}}{\text{FBW}_\text{V}(\omega)} \tag{9.22}$$

We can see from Equation 9.22 that the Q is roughly proportional to the inverse of the defined bandwidth.

The human body, is as we have seen, a lossy medium that has its greatest effect when closest to an antenna. We can calculate its effects on the Q of an antenna using the methods described in [16] and [15]. If we assume a human/antenna system consisting of lossy materials with permittivity $\varepsilon_a = \varepsilon_{aR} - j\varepsilon_{aI}$ to be non-magnetic (permeability is μ_0) we can state that

$$P_{Sf} = \frac{1}{4}\lim_{r\to\infty}\left[\int_{v(r)}\left[(\omega\varepsilon)'|\mathbf{E}|^2 + (\omega\mu)'|\mathbf{H}|^2\right]dv - 2\varepsilon_0 r\int_{4\pi}|F(\theta,\phi)|^2\,d\Omega\right] \tag{9.23}$$

$$= \frac{1}{4}\lim_{r\to\infty}\left[\int_{v(r)-v_a}\left[\varepsilon_0|\mathbf{E}|^2 + \mu_0|\mathbf{H}|^2\right]dv - 2\varepsilon_0 r\int_{4\pi}|F(\theta,\phi)|^2\,d\Omega\right]$$

$$+\frac{1}{4}\left[\int_{v_a}\left[\left|(\omega\varepsilon_{aR})' - j(\omega\varepsilon_{aI})'\right||\mathbf{E}|^2 + \mu_0|\mathbf{H}|^2\right]dv\right] \tag{9.24}$$

$$P_{Rf} = \frac{1}{2}Re\int_s[\mathbf{E}\times\mathbf{H}^*]\cdot\hat{n}\,dS = \frac{1}{2\eta}\int_{4\pi}|F(\theta,\phi)|^2\,d\Omega \tag{9.25}$$

$$P_{Lf} = \frac{\omega}{2}\int_{v_a}\varepsilon_{aI}|\mathbf{E}|^2\,dv \tag{9.26}$$

where, P_{Sf}, P_{Rf} and P_{Lf} are the reactive energy, radiated energy and ohmic losses of the antenna respectively and r is the far field position vector and is assumed to be large compared to the wavelength.

For this analysis we see that $d\Omega = dS/r^2 = \sin\theta d\theta d\phi$ and $F(\theta,\phi)$ is the electric field pattern.

$$F(\theta,\phi) = \lim_{r\to\infty} re^{jkr}\mathbf{E}(r) \tag{9.27}$$

Also in Equation 9.25, η, is the impedance of free space, $v(r)$ is the volume including the antenna and surrounding space and v_a is the volume of the antenna. So the Q factor can be obtained using

$$Q_f = \frac{\omega P_{Sf}}{P_{Rf} + P_{Lf}} \tag{9.28}$$

In the following way, the same analysis can now be extended to cover the case of an antenna close to a human which has a complex permittivity $\varepsilon(\omega) = \varepsilon_R(\omega) - j\varepsilon_{el}(\omega)$, is nonmagnetic with a volume v_h as follows

$$P_{Sh} = \frac{1}{4} \lim_{r \to \infty} \left[\int_{v(r)} \left[(\omega\varepsilon)' |\mathbf{E}|^2 + (\omega\mu)' |\mathbf{H}|^2 \right] dv - 2\varepsilon_0 r \int_{4\pi} |F(\theta, \phi))|^2 d\Omega \right] \quad (9.29)$$

$$= \frac{1}{4} \lim_{r \to \infty} \left[\int_{v(r)-v_a-v_h} \left[\varepsilon_0 |\mathbf{E}|^2 + \mu_0 |\mathbf{H}|^2 \right] dv - 2\varepsilon_0 r \int_{4\pi} |F(\theta, \phi)|^2 d\Omega \right]$$

$$+ \frac{1}{4} \left[\int_{v_a} \left[\left| (\omega\varepsilon_{aR})' - j\left(\omega\varepsilon'_{al}\right) \right| |\mathbf{E}|^2 + \mu_0 |\mathbf{H}|^2 \right] dv \right]$$

$$+ \frac{1}{4} \left[\int_{v_h} \left[\left| (\omega\varepsilon_R)' - j\left(\omega\varepsilon'_{el}\right) \right| |\mathbf{E}|^2 + \mu_0 |\mathbf{H}|^2 \right] dv \right] \quad (9.30)$$

$$P_{Rh} = \frac{1}{2} Re \int_S [\mathbf{E} \times \mathbf{H}^*] \cdot \hat{n} dS = \frac{1}{2\eta} \int_{4\pi} |F(\theta, \phi)|^2 d\Omega \quad (9.31)$$

$$P_{Lh} = \frac{\omega}{2} \int_{v_a} \varepsilon_{al} |\mathbf{E}|^2 + \int_{v_h} \sigma_e |\mathbf{E}|^2 dv \quad (9.32)$$

where P_{Sh}, P_{Rh} and P_{Lh} are the reactive energy, radiated energy and ohmic losses of an antenna close to the human body respectively. Figure 9.5 shows the extent of the volumetric integrals used in this analysis [15].

Therefore, treating a handset system as a single resonant antenna, the Q factor of the handset is

$$Q_h = \frac{\omega P_{Sh}}{P_{Rh} + P_{Lh}} \quad (9.33)$$

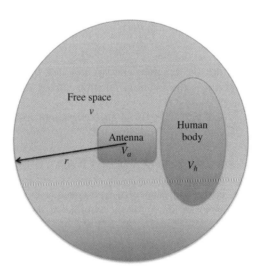

Figure 9.5 An antenna close to the body indicating the volumetric integral surfaces and position vector. Note the position vector originates at the center of the antenna.

From this discussion it can be seen that the body acts as a lossy load that increases the losses of the handset regardless of tuning. Across the range of current LTE frequencies the body acts as a series inductance and resistance to the system thereby causing a free space antenna to be detuned with associated loss of Q that increases with proximity to biological matter.

It should also be borne in mind that since the permittivity of skin is quite different to that of air, a substantial proportion (typically in excess of 50 percent), of the energy from a handset is reflected away from the body which can be a factor in antenna design. To expand on this point a little an antenna over a perfectly conducting ground plane exhibits constructive interference at $\frac{1}{4}$ of the wavelength of the operating frequency, whilst for an antenna close to humans this distance tends to occur at about $\frac{1}{6}$ of the operating frequency. Constraining the operating position of the handset either by clothing or grip design can therefore have a significant effect on performance.

Having looked at the fields inside human tissue the next section will consider models for on-body propagation.

9.4   A Survey of Popular On-Body Propagation Models

This section provides a short review of theoretical models used to predict the properties of a radio wave traveling on the surface of the body. Typically the on-body wave is thought to be most interesting in point-to-point short range body radio networks in which two or more wearable antennas are placed on the body at differing points. This channel may be considered useful since the scattered wave may be blocked by a limb or the curvature of the body and since any propagation through tissue will be severely attenuated the on-body channel may be the only one available. Thus in some circumstances the surface wave may be the most reliable form of propagation.

The term "surface wave" is used a great deal in the study of on-body communications and there are two popular forms. These are the Norton surface wave [17] and the Zenneck surface wave [18]. Both are in fact special cases of plane waves with the Poynting vector (the direction of energy flux density–in essence the direction of the power flow), being their only actual difference. In the Norton on-body surface wave the Poynting vector dips just below the epidermis and therefore suffers heavy attenuation in the tissue of the body. With the Zenneck wave the Poynting vector is aligned parallel to the surface of the body. However, current views suggest that a Zenneck wave can only be launched by antennas with large apertures which is difficult on the body. Therefore it is reasonable to suggest that the on-body surface wave is of the Norton type which is an extension of Arnold Sommerfeld's work and originally considered propagation mechanisms on the surface of the earth in radio [17].

It turns out that body-surface waves are severely attenuated by the loss in the tissue such that any scattered component in the on-body propagation between two antennas will dominate to an extent beyond the dynamic range of typical radio transceivers (greater than 40 dBm).

Work regarding the human body's influence on antenna parameters began in earnest in the late 1990s and major contributions regarding on-body propagation began to appear in 2002–2003, [19, 20]. In Europe, key research on the on-body channel has been provided by ETH Zurich [21], IMEC (NL), ULB (Brussels) and UCL (Louvain) [22–25]. More specifically, in the UK contributions have come from the University of Birmingham [26], Queen Mary University of London [27, 28], and Queens University of Belfast [29, 30]. Note, although a great deal of this research relates to ultra-wide band radio that has a 3.1 GHz to 10.6 GHz authorized unlicensed bandwidth according to the Federal Communications Authority (FCC), many authors have considered lower frequency ranges as well.

The electrical properties of the tissues of the human body make it difficult to derive a simple path loss model for a Body-Area-Network (BAN). Since the antennas used for Body Area Networks (BANs) lie close to the body, a BAN channel model has to take into account the influence of the human body.

Previously, radio nodes in on-body communication systems have been divided into three groups, namely: Implant Node (nodes placed inside the human body under the skin); On-body Node (nodes placed on the surface of the human body); and External Node (nodes which are not in direct contact with the human body surface). User equipment in LTE is therefore likely to be realized with external nodes for inter-body area networks.

To make LTE particular to the on-body channel we assume that at a future date modulation schemes used in cellular will also be used for point to point on-body communications.

For any non-stationary human on-body, channels are typically highly dynamic with

a) intermittent line of sight components,
b) complex close in scatterers/interferers,
c) highly variable dielectric constants related to the tissue types that support Norton on-body surface waves.

Many researchers have sought to isolate c) using either the absorbing boundary conditions of their simulations or the absorbing properties of anechoic chambers to attenuate the space wave which contributes to scattering. This is not trivial in measurements since the amount of energy in the on-body wave are relatively small when compared to line-of-sight. For example, the floor in most anechoic chambers will reflect waves stronger than typical surface waves and will usually need to be time gated out.

When the nodes are not in line-of-sight with one another, the distance between them will be the distance around the body and creeping wave diffraction will need to be taken into account. There are many routes around the surface of the body and therefore a form of multi-path does exist for on-body channels with the associated delay spread problems.

Body movement can be highly dynamic. Typical postures are: sitting and standing, and bending and twisting. However, all of these postures may change slowly as in a social environment, or quickly as in a sporting environment [28]. Most changes in posture will change the distance between nodes as well as the propagation path. Thus the characterization of radio wave propagation needs to not only account for these movements, but also the location of the nodes on the body surface. According to Hall et al. [28] changes in the local geometry of the environment will also affect the antenna parameters such as input match and radiation pattern, so the design of wearable transceivers should consider changes in the antenna and propagation loss to enable maximum channel capacity and minimum power consumption. This comment is important for it implies that the same type of antenna will behave differently according to its position on the body. Further, it can be inferred that different antennas may suit different parts of the body.

The on-body propagation channel around the human body is a combination of free space propagation, diffraction (creeping wave) and reflections from the environment (see Chapter 2). As radio wave propagation in free space is characterized by the Friis formula for transmission, which is

$$P_r = \frac{P_t G_r G_t \lambda^2}{16\pi^2 d^2} \tag{9.34}$$

where G_r and G_t are the gain of the receiving and transmitting antennas respectively, d is the separation between them and λ is wavelength in meters. In some cases Equation 9.34 also includes a system loss factor which does not depend on the propagation. It is clear from Equation 9.34 that this path loss model does not hold for $d = 0$ which implies a range of application is also needed.

According to the authors of [31–34] the path loss model in dB between a transmitting and receiving antenna is given by

$$PL(d) = PL_0 + 10n \log_{10}\left(\frac{d}{d_0}\right) \tag{9.35}$$

where d_0 is the reference distance which is taken as 1m for on-body measurements by many researchers, PL_0 is the path loss at some reference distance and n is the path loss exponent. Depending upon the environment, antenna positions on the body, antenna type and frequency, different authors have given different values for path loss exponent. Fort et al. [31], (3 to 10 GHz) found path loss exponents of 7.5 and 3.1 for different scenarios (around and infront of the torso that is NLOS and LOS respectively) while authors of [33], (2–12 GHz), gave a value of 4.4 for horizontal, vertical and diagonal propagation on the body. Similarly A. Alomainy et al. [34], (3 to 9 GHz), have found $n = 3.9$ and 2.6 for Horn Shaped self-Complementary Antenna (HSCA) and Planar Inverted Cone Antenna (PICA).

Path loss models for the mobile communication frequencies of 900 MHz and ISM band 2.4 GHz were presented by the authors of [35–37]. According to the authors of [35], whose measurements were done inside an anechoic chamber and a hospital room, the channel can be fitted to an on-body propagation model given by

$$PL(d)_d B = a \cdot \log_{10}(d) + b + c + N \tag{9.36}$$

where a and b are the coefficients of linear fitting (which were different for both environments and were frequency dependent), d was the distance between transmitter and receiver, N was a normally distributed random variable with a standard deviation of σ_N and c was the difference in the signal level for the different types of antennas that were used. In this case a dipole and a chip antenna were measured.

As a result of the on-body measurements done and reported in [36], it was concluded that the path loss model around the human body follows an exponential decay that can be modelled by the following equation.

$$PL(d)_{dB} = -10 \log_{10}\left(P_0 e^{-m_0 d} + P_1\right) + \sigma_p n_p \tag{9.37}$$

where m_0 is the average decay rate in dB/cm for the surface wave travelling around the perimeter of the body. This rate was found to be approximately 2.1 dB/cm. P_1 is the average attenuation of the scattered components in an indoor environment radiated away from the body and reflected back towards the receiving antenna. σ_p is the log-normal variance in dB around the mean, representing the variations measured at different body and room locations. This parameter will depend on variations in the body curvature, tissue properties and antenna radiation properties at different body locations.

Many on-body measurements were performed by the authors of [37]. Their method involved locating the transmitting antenna on different locations on the human body and fixing the receiver antenna at a position on the chest and the right hip. Three different postures were examined in this study, namely: standing; walking; and running. The main factors which caused variation in the path loss were found to be human body movements. The path loss was found to be

$$PL_{dB} = P_{tx} - P_{rx} + G_{amplifiers} - L_{cable} \qquad (9.38)$$

where P_{tx} was the transmitted power, P_{rx} was the RMS received power, $G_{amplifiers}$ was the combined amplifier gain and L_{cable} were any cable losses. Note that unlike other papers that have been discussed so far, this model includes some of the systematic losses as well. In general, for scenarios in which the subject is moving it is not yet possible to calibrate out systematic errors due to cabling.

Values for the different parameters used in Equations 9.36, 9.37 and 9.38 are of course different for differing frequency. Typical values can be found in [35], [36], [37].

Hall et al. [28] performed many on-body measurements for the modeling of path loss for different situations such as Line Of Sight (LOS) and Non Line Of Sight (NLOS) at 2.45 GHz. A path loss model for the LOS case using transmitter and receiver at the front of the body was found to be

$$PL(d)_{dB} = -5.33 - 20\log_{10} d_{cm} \qquad (9.39)$$

Where d_{cm} is the separation distance in centimeters.

A path loss model for NLOS case using a transmitter and receiver at front and back of the human body respectively for the approximation of exponential decay using a linear regression formula was

$$PL(d)_{dB} = -0.36 d_{cm} - 35 \qquad (9.40)$$

This gave a decay of 0.36 dB/cm.

Finally, according to Fort et al. [22] at 2.45 GHz when the distance between the transceivers is less than 25 cm, in line-of-sight conditions, the distribution is Rician [38] but for non-line-of-sight situation the distribution is mainly Rayleigh [38] (see Chapter 2) for details on the fading distribution). In this case the authors made their measurements at the frequency of 2.45 GHz.

In Figure 9.6 are plotted four of the most popular models for close to and on the body propagation along with some spot values (to be published research from the authors), for measured results for a volunteer wearing wet clothing. The curves show how the channel is attenuated as two antennas are moved a part on the body at a frequency of 2.45 GHz. It can be noticed that there is a wide disagreement in models which leads to a wide range of measured results.

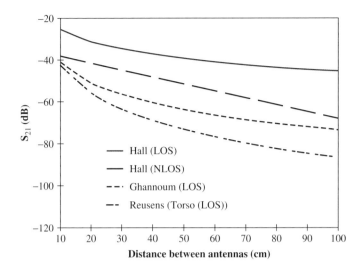

Figure 9.6 Comparison of on-body propagation models around and close to the human torso at 2.45 GHz.

It turns out that the mode of an antenna is particularly important when close to the body and that this factor is not always apparent in the literature.

Theoretically it can be proven that Transverse Electric (TE) propagating waves are supported by an inductive surface and and Transverse Magnetic (TM) waves supported by a capacitive surface. Norton proves this in [39], but a clearer treatment is given by Vaughan in [2]. In this application the plane of the wave has to be referenced to the surface of the body. A monopole type antenna on the body will produce a TE wave and a dipole parallel to the skin surface will produce a TM wave. Looking again at Figure 9.6 we see that Hall's model was obtained using a monopole type antenna (as were the the authors' spot values) and that curves of Reusens [32] and Ghannoum [33] used TM mode antennas.

9.5 Antenna Implants-Possible Future Trends

Over the past two decades, the UK has been able to benefit from its significant wealth in knowledge relating to telecommunications engineering. During this time, there has been an exponential growth in the field of mobile communications, proving beyond doubt that people love to talk. Coincident with this success there has been a massive increase in healthcare provision in the UK combined with an associated revolution in how treatment is offered to the patient. The simplicity and utility of technologies like Global System for Mobile Communications (GSM) with voice, data, 3G with streaming video and 4G with its superior resource allocation all offer much to healthcare, particularly for non-secure medical telemetry. We discuss here the future concept of 4G systems implanted into the body with bidirectional link to the cellular network. This is different from current systems that communicate with implanted devices over short range links (<410 m).

Given the right safeguards for implanted mobile phone technology, it would for example be possible to measure the properties of a heart attack in real time and perhaps monitor the effects of treatment subsequent to the event, whilst allowing the patient freedom of movement. What would be needed would be a system that could be implanted into a patient for short periods of time (perhaps several weeks) that could be used to transmit data out of the body and to a medical expert. In this context such a system would use data rather than voice, be non-real time with low isochronous application usage.

What is envisaged might be low SAR flexible antennas just beneath the skin surface for use with cellular systems and their vast networks of base stations. Such a system would comprise a small telecommunications module with integrated micro controller and power supply attached by cable to an antenna. The module, its battery and its associated sensors would lie inside the body. To minimize SAR the antenna would lie as close to the outside of the body as practical but not outside the skin. The system would be encapsulated and screened to reduce energy interactions with tissue. From the point of view of avoiding infection the proposal to have the whole system inside the body's protective skin is of clear benefit. By not breaching the skin complications arising from infection; hygiene and painful snagging would be avoided. Furthermore such in-body systems would be invisible to other people and may allow patients an extended freedom of movement and much more privacy. All of the components except the antenna are state-of-the-art. In considering the size of such antennas we are helped a great deal by the permittivity of the surrounding tissue which is generally high. Therefore, such antennas would tend to be much smaller than their free-space counterparts (for example about 25 mm long for a half wave dipole at 900 MHz).

It can be shown, for example with LTE wireless communications, that it would not currently be a problem sending 4G signals to a modem implanted inside a body cavity. However, because of the very strict legacy limits related to medical implants, the tricky part of such a system would be how to get 4G signals out of the body to a base station without exceeding SAR limits within the body. For a mobile handset power levels from a handset are limited to 2 Watts but are typically around 0.6 Watts split across several channels. However for medical implants the limit is 25 µ Watts.

The standards germane to this discussion are the Medical Device Radiocommunications Service (MedRadio) [40] and the Wireless Medical Telemetry Service (WMTS) [41]. MedRadio has a spectrum between 401 and 457 MHz. The more common devices realized have been implanted cardiac pacemakers and defibrillators, and neuromuscular stimulators for physical mobility. WMTS has spectrum at around 0.6 GHz and 1.4 GHz and has been used for sending data about such things as pulse and respiration rates to close in receiving stations. A typical application would be a cardiac monitor wirelessly linked to a nurse's station for post operative care.

The ability to communicate with an implant over a high bandwidth link would facilitate many new applications and enhance existing applications such as pacemakers, implantable cardioverter defibrillators, neuro-stimulators, hearing aids, robotic prostheses, artificial eyes, brain pacemakers to control Parkinsons disease, monitoring of blood glucose levels for diabetic patients, stimulation and recording of brain and muscle activity, swallow-able pills for traversing the gastrointestinal tract and implantable drug delivery systems. Implanted medical devices save lives, increase the quality of the user's life, reduce the number of trips a patient has to make to a hospital (increasing their freedom and reducing the risk of diseases such as MRSA) and save billions of pounds in hospital beds, resources and doctors' time. Lifesaving implants such as cardiac pacemakers, neuro-stimulators and pumps, have now become routine and do not attract the negative media attention that normally follows attempts to create so called *bionic* people. The cardiac pacemakers and defibrillators have grown into a multi-billion pound industry since the first implanted pacemaker in 1958. As implanted antennas are aimed at the same market, improving the quality of life of severely ill patients or others who are at high risk of illness, it is expected that it will be well received by both the medical community and patients alike. After all, antennas are implanted inside the human body as a method of treating tumors using hyperthermia. Previously, battery power has been a limiting factor. A typical pacemaker uses less than 10 mW and the battery lasts 10 years. However, a long range medical biotelemetry system consisting of a sensor(s), a battery, a 4G communications module and a low power subcutaneous antenna may only need to be in place for a few weeks.

9.6 Summary

It has become apparent that most of the previous generations of cellular wireless communications have not reached the potential in terms of bandwidth and reliability suggested by the specification. Modern handsets are still plagued by high rates of drop out and poor battery life. Since much of the energy from handsets is absorbed by the user's head, hand and shoulders (and thus wasted), great improvements in performance can be achieved by careful antenna design. In this chapter we have shown that humans have a significant effect on handsets across all of the proposed LTE bands and have presented methods for calculating loss in tissue the Q of antennas close to humans. In addition the current state of the art models for on-body channel have been compared and discussed. We have also provided technical advice on antenna deployment near to the skin of a user.

The chapter concluded with a brief discussion on medical implants and possible future avenues of research.

References

1. IEEE, "Ieee standard for safety levels with respect to human exposure to radio frequency electromagnetic fields, 3khz to 300 ghz amendment 2: Specific absorption rate (sar) limits for the pinna," *IEEE Std C95.1b-2004 (Amendment to IEEE Std C95.1-1991 as amended by IEEE Std C95.1a-1998)*, pp. 1–6, 2004.
2. R. Vaughan, A. Lea, P. Hui, and J. Ollikainen, "Theory of propagation for direct on-body wireless sensor communications," in *Antennas and Propagation for Body-Centric Wireless Communications, 2009 2nd IET Seminar on*, April 2009, pp. 1–5.
3. Z. Krupka, "The effect of the human body on radiation properties of small-sized communication systems," *Antennas and Propagation, IEEE Transactions on*, vol. 16, no. 2, pp. 154–163, Mar 1968.
4. J. Lebaric and A.-T. Tan, "Ultra-wideband rf helmet antenna," in *MILCOM 2000. 21st Century Military Communications Conference Proceedings*, vol. 1, 2000, pp. 591–594 vol.2.
5. R. Abramo, R. Adams, F. Canez, H. Price, and P. Haglind, "Fabrication and testing of the comvin vest antenna," *IEEE MILCOM*, 2000, vol. 1, pp. 595–598, 2000.
6. E. Fear, S. Hagness, P. Meaney, M. Okoniewski, and M. Stuchly, "Enhancing breast tumor detection with near-field imaging," *Microwave Magazine, IEEE*, vol. 3, no. 1, pp. 48–56, Mar 2002.
7. A. Smailagic, D. P. Siewiorek, R. Martin, and J. Stivoric, "Very rapid prototyping of wearable computers: A case study of vuman 3 custom versus off-the-shelf design methodologies," in *Journal on Design Automation for Embedded Systems*, 1997, pp. 217–230.

8. J. Rantanen, N. Alfthan, J. Impio, T. Karinsalo, M. Malmivaara, R. Matala, M. Makinen, A. Reho, P. Talvenmaa, M. Tasanen, and J. Vanhala, "Smart clothing for the arctic environment," in *Wearable Computers, 2000. The Fourth International Symposium on*, 2000, pp. 15–23.

9. Z. N. Chen, A. Cai, T. See, X. Qing, and M. Chia, "Small planar uwb antennas in proximity of the human head," *Microwave Theory and Techniques, IEEE Transactions on*, vol. 54, no. 4, pp. 1846–1857, June 2006.

10. A. Byndas, A. Kucharski, and P. Kabacik, "Experimental study of the interactions between terminal antennas and operators," in *Antennas and Propagation Society International Symposium, 2001. IEEE*, vol. 3, 2001, pp. 74–77 vol.3.

11. S. Gabriel, R. W. Lau, and C. Gabriel, "The dielectric properties of biological tissues: Ii. measurements in the frequency range 10hz to 20 ghz," *Physics in Medicine and Biology*, vol. 41, no. 11, p. 2251, 1996. [Online]. Available: http://stacks.iop.org/0031-9155/41/i=11/a=002,

12. V. V. Sarwate, *Electromagnetic fields and waves*. New York; Chichester: John Wiley, 1993.

13. IFAC, "Ifac italian national research council, institute for applied physics," Web Application, 2011.

14. W. Whittow, C. Panagamuwa, R. Edwards, and J. Vardaxoglou, "The energy absorbed in the human head due to ring-type jewelry and face-illuminating mobile phones using a dipole and a realistic source," *Antennas and Propagation, IEEE Transactions on*, vol. 56, no. 12, pp. 3812–3817, Dec. 2008.

15. A. Yaghjian, "Improved formulas for the q of antennas with highly lossy dispersive materials," *Antennas and Wireless Propagation Letters, IEEE*, vol. 5, no. 1, pp. 365–369, Dec. 2006.

16. A. Yaghjian and S. Best, "Impedance, bandwidth, and q of antennas," *Antennas and Propagation, IEEE Transactions on*, vol. 53, no. 4, pp. 1298–1324, April 2005.

17. K. Norton, "The propagation of radio waves over the surface of the earth and in the upper atmosphere," *Proceedings of the Institute of Radio Engineers*, vol. 24, no. 10, pp. 1367–1387, Oct. 1936.

18. J. Wait, "The ancient and modern history of em ground-wave propagation," *Antennas and Propagation Magazine, IEEE*, vol. 40, no. 5, pp. 7–24, Oct 1998.

19. P. Hall, M. Ricci, and T. Hee, "Measurements of on-body propagation characteristics," in *Antennas and Propagation Society International Symposium, 2002. IEEE*, vol. 2, 2002, pp. 310–313 vol.2.

20. T. Zasowski, F. Althaus, M. Stager, A. Wittneben, and G. Troster, "Uwb for noninvasive wireless body area networks: channel measurements and results," in *Ultra Wideband Systems and Technologies, 2003 IEEE Conference on*, Nov. 2003, pp. 285–289.

21. T. Zasowski, G. Meyer, F. Althaus, and A. Wittneben, "Propagation effects in uwb body area networks," in *Ultra-Wideband, 2005. ICU 2005. 2005 IEEE International Conference on*, Sept. 2005, pp. 16–21.

22. A. Fort, J. Ryckaert, C. Desset, P. De Doncker, P. Wambacq, and L. Van Biesen, "Ultra-wideband channel model for communication around the human body," *Selected Areas in Communications, IEEE Journal on*, vol. 24, no. 4, pp. 927–933, April 2006.

23. A. Fort, C. Desset, P. De Doncker, P. Wambacq, and L. Van Biesen, "An ultra-wideband body area propagation channel model-from statistics to implementation," *Microwave Theory and Techniques, IEEE Transactions on*, vol. 54, no. 4, pp. 1820–1826, june 2006.

24. J. Ryckaert, P. De Doncker, R. Meys, A. de Le Hoye, and S. Donnay, "Channel model for wireless communication around human body," *Electronics Letters*, vol. 40, no. 9, pp. 543–544, april 2004.

25. S. Van Roy, C. Oestges, F. Horlin, and P. De Doncker, "On-body propagation velocity estimation using ultra-wideband frequency-domain spatial correlation analyses," *Electronics Letters*, vol. 43, no. 25, pp. 1405–1406, 6 2007.

26. P. Hall, M. Ricci, and T. Hee, "Measurements of on-body propagation characteristics," in *Antennas and Propagation Society International Symposium, 2002. IEEE*, vol. 2, 2002, pp. 310–313 vol.2.

27. P. S. Hall and Y. Hao, "Antennas and propagation for body centric communications," in *Antennas and Propagation, 2006. EuCAP 2006. First European Conference on*, Nov. 2006, pp. 1–7.

28. P. Hall, Y. Hao, Y. Nechayev, A. Alomalny, C. Constantinou, C. Parini, M. Kamarudin, T. Salim, D. Hee, R. Dubrovka, A. Owadally, W. Song, A. Serra, P. Nepa, M. Gallo, and M. Bozzetti, "Antennas and propagation for on-body communication systems," *Antennas and Propagation Magazine, IEEE*, vol. 49, no. 3, pp. 41–58, June 2007.

29. S. Cotton and W. Scanlon, "A statistical analysis of indoor multipath fading for a narrowband wireless body area network," in *Personal, Indoor and Mobile Radio Communications, 2006 IEEE 17th International Symposium on*, Sept. 2006, pp. 1–5.

30. S. Cotton and W. Scanlon, "Higher order statistics for lognormal small-scale fading in mobile radio channels," *Antennas and Wireless Propagation Letters, IEEE*, vol. 6, pp. 540–543, 2007.

31. A. Fort, C. Desset, J. Ryckaert, P. De Doncker, L. Van Biesen, and P. Wambacq, "Characterization of the ultra wideband body area propagation channel," in *Ultra-Wideband, 2005. ICU 2005. 2005 IEEE International Conference on*, Sept. 2005, p. 6 pp.

32. G. V. E. Reusens, W. Joseph and L. Martens, "On-body measurements and characterization of wireless communication channel for arm and torso of human," *4TH INTERNATIONAL WORKSHOP ON WEARABLE AND IMPLANTABLE BODY SENSOR NETWORKS (BSN 2007)*, vol. 13, pp. 264–269, 2007.

33. H. Ghannoum, R. D'Errico, C. Roblin, and X. Begaud, "Characterization of the uwb on-body propagation channel," in *Antennas and Propagation, 2006. EuCAP 2006. First European Conference on*, Nov. 2006, pp. 1–6.

34. A. Alomainy, Y. Hao, X. Hu, C. Parini, and P. Hall, "Uwb on-body radio propagation and system modelling for wireless body-centric networks," *Communications, IEE Proceedings-*, vol. 153, no. 1, pp. 107–114, Feb. 2006.

35. T. Aoyagi, J. Takada, K. Takizawa, H. Sawada, N. Katayama, K. Yazdandoost, T. Kobayashi, H. Li, and R. Kohno, "Channel models for wbans," *NICT, IEEE 802.15 Working Group Document, IEEE 802.15-08-0416-03-0006*, Sept 2008.

36. G. Dolmans and A. Fort, "Channel models wban-holst centre/imec-nl," IEEE 802.15-08-0418-01-0006, 2008.

37. D. Miniutti, L. Hanlen, D. Smith, A. Zhang, D. Lewis, D. Rodda, and B. Gilbert, "Narrowband channel characterization for body area networks," IEEE P802.15-08-0421-00-0006, 2008.

38. W. C. Y. Lee, *Mobile Communications Design Fundamentals*. John Wiley and Sons, Inc. New York, NY, USA, 1992.

39. K. Norton, "The physical reality of space and surface waves in the radiation field of radio antennas," *Proceedings of the Institute of Radio Engineers*, vol. 25, no. 9, pp. 1192–1202, Sept. 1937.

40. FCC. (2012) Medical device radiocommunications service (medradio). [Online]. Available: http://www.fcc.gov/encyclopedia/medical-device-radiocommunications-service-medradio

41. FCC. (2012) Wireless medical telemetry service (wmts). [Online]. Available: http://www.fcc.gov/encyclopedia/wireless-medical-telemetry-service-wmts

Part Three

Simulation and Performance

10

Ray-Tracing Modeling

Yves Lostanlen[1] and Thomas Kürner[2]
[1]*University of Toronto, Canada*
[2]*Technische Universität Braunschweig, Germany*

10.1 Introduction

Born in computer graphics, ray tracing algorithms have been elaborated to generate an image by computing and tracing the light trajectories through pixels in a 2D image and simulating the effects of its interactions with digitized objects. The technique produces very realistic scenes but at a greater computational cost. This makes ray tracing best suited for applications where the image can be rendered slowly ahead of time, such as in still images and film and television special effects, and more poorly suited for real-time applications like video games where speed is critical. Ray tracing methods can simulate a wide variety of interactions, such as reflection and refraction (transmission), and diffraction phenomena. Those ray-based techniques have been adapted at lower frequencies of the electromagnetic spectrum and are valid in the far-field region of the transmitters. Indeed it is crucial to predict accurately the radio-wave propagation when designing a wireless system. Indeed the propagation is the first limiting factor for most wireless equipment and networks. Modeling reliably the radio-wave behavior in complex environments (urban and indoor) provides a key asset to understand better the wireless characteristics and impairment and to get optimal performance. Deterministic modeling tools are useful to study the wave propagation in specific environments and to understand physical effects observable in measurements. Thanks to advanced ray-based propagation models, specific phenomena can be isolated in simulations giving a better insight into the physics underlying the system operations. Initially confined to academic studies, the ray-based propagation models embedded in simulators are now used for operational purposes thanks to fast computing machines. For 2G systems, the determination of the path loss was generally sufficient to design and deploy wireless networks. The recent systems have a larger bandwidth which requires the prediction of enhanced channel characteristics such as the multi-path delays,

LTE-Advanced and Next Generation Wireless Networks: Channel Modelling and Propagation, First Edition.
Edited by Guillaume de la Roche, Andrés Alayón Glazunov and Ben Allen.
© 2013 John Wiley & Sons, Ltd. Published 2013 by John Wiley & Sons, Ltd.

the angles of departure and arrival. Thus, beyond the path loss estimation, a full space-time channel response can be determined by ray-based methods for each individual location of a geographic area. The assessment of multi-antenna systems performances (implemented in Long Term Evolution (LTE) and Long Term Evolution Advanced (LTE-A) wireless systems) can take full advantage of those space-time predictions.

Full-wave methods such a Finite-Difference Time-Domain (FDTD) and Method Of Moments (MOM) solve directly the Maxwell equations. These techniques model accurately the details of electromagnetic phenomena for very restricted volumes such as complex antennas in their environment, and possibly in near fields. More details on full-wave methods, also referred to as finite different methods, will be presented in Chapter 11. Although some attempts have been made to consider them in indoor and urban areas, without tremendous enhancement work, these methods are mainly seen as complements of asymptotical methods to model accurately the close surroundings of the radiating elements. The asymptotical methods are the most commonly used techniques nowadays to predict the signal propagation for operational wireless networks, especially in dense urban areas, although nothing prevents a larger use as will be shown later. These techniques rely on geometrical propagation path and approximations of the Maxwell equations to model the physical phenomena such as the reflection, the diffraction and the transmission.

This document is dedicated to the description of ray-tracing techniques that underly the most advanced site-specific propagation models. First, the physical phenomena involved in the radio-wave propagation for wireless systems are described. Then a focus is made on the geometrical path determination by ray-based techniques such as the ray-tracing and the ray-launching as they provide the best-in-class models to assess efficiently the system and network performance. Finally extensions of those methods to ultra-wideband signals are presented.

10.2 Main Physical Phenomena Involved in Propagation

Radio transmission is affected by many influencing factors and phenomena and the relevance of propagation phenomena depends on the respective environment. Important features that have to be taken into account are the distance between a base station and the mobile stations, which can vary from a few meters to some kilometers as well as the ratio between the wave length and the object size that has to be considered. Ray-optical approaches, which are subject to the methods described in this chapter are applicable, if the dimensions of all obstacles in the propagation path are much larger than the wavelength. Especially in a smaller environment, like indoor cells, where structures of objects are typically small, this has to be considered, when ray-tracing is applied. On the other hand, in large macro cells objects of relevance are buildings or mountains, whose typical dimensions allow the application of ray tracing even for wave lengths in the order of meters. Many wave propagation models are based on the ray-optical description of diffraction and scattering processes, where separate observations of single propagation mechanisms, each with its own physical and mathematical description are possible. This enables a description of the fields in the form of rays in analogy to geometrical optics. Hence multipath modeling in ray-tracing consists of a search for all relevant transmission paths and the analysis of their interaction. The latter is the subject of this section. First, in

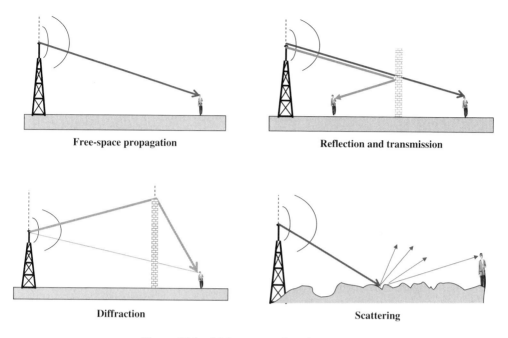

Figure 10.1 Main propagation phenomena.

Section 10.2.1 basic modeling principles are introduced followed by a short review of methods to model to the main propagation phenomena displayed in Figure 10.1.

10.2.1 Basic Terms and Principles

Geometrical Optics (GO) is a method for calculation of scattered fields in illuminated areas. This can be explained by Figure 10.2, with a light source and an object [1]. This object partitions the room illuminated by the light source in zones with and without light. The lines of separation are called Incident Shadow Boundarys (ISBs). In the scenario depicted in Figure 10.2 two ISBs exist. When considering additionally reflected rays, Reflection Shadow Boundarys (RSBs) also occur, separating the regions where reflected rays exist and regions where no reflected rays exist. Reflection points are defined by generalizing the Fermats principle. In the current scenario two RSBs exist. When applying GO to the following zones and rays exist [1]:

- zone I (visual zone): direct and reflected ray exist;
- zones II, III (visual zone, reflection shadow): only a direct ray exists;
- zone IV (shadow zone): no field.

A more detailed description of the scattered field in zone I is given by the approach from the Physical Optics (PO), which applies methods from the antenna theory. To explain this effect let us consider the case of an ideally conductive body. In this case the surface has

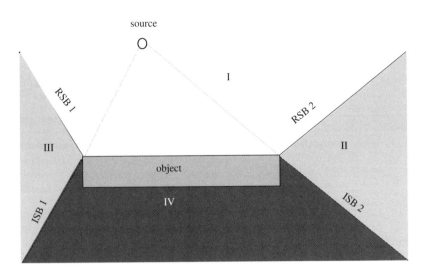

Figure 10.2 Definition of visual and shadow zones, derived from [1].

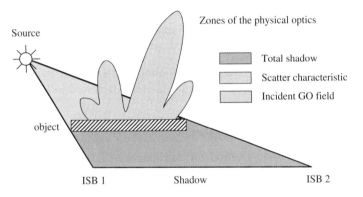

Figure 10.3 Concept of physical optics, derived from [1].

a current density induced by the incident GO field, see Figure 10.3. The re-radiated field and the scattering characteristics (comparable to antenna characteristics) can be defined. PO does not consider shadowing effects since current density in shadowed zones is equal to zero because there is no GO field. Consequently, in terms of GO, a ray exists between a starting point and an end point only if no obstacle blocks the path. This geometrical-optical approach corresponds to the extreme case for infinitely high frequencies, since the field strengths in the shadow ranges disappear here identically. This brings us to the introduction of the concept of diffraction. Diffraction can be seen as the summary of all those effects observed at finite frequencies, that is, in case of not disappearing wave lengths, where propagation of electromagnetic waves differs from the forecast of the PO.

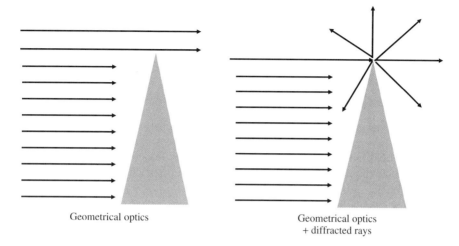

Geometrical optics

Geometrical optics
+ diffracted rays

Figure 10.4   Existing rays for the cases GO only and GO combined with diffraction.

This applies especially to shadow areas. Figure 10.4 depicts a scenario with the rays for the case of applying GO and GO combined with diffraction.

10.2.2   Free Space Propagation

A ray traveling directly from the transmitter to the receiver without interaction with any object, see Figure 10.1, is attenuated by the so-called free-space loss. For calculating the free space loss the influence of antennas on both sides of the link are not taken into account and isotropic propagation is assumed. In this case the transported power is equally spread over the surface area of a sphere with a radius of r, where r is the distance between the base station and the mobile station. This explains why the received power at the mobile stations is proportional to $1/r^2$.

10.2.3   Reflection and Transmission

Let us consider an incident plane wave impinging onto a plane boundary of infinite size between two lossless media with different dielectric properties. Two waves will result from the interaction with the boundary: one wave that will be reflected on the boundary with the reflection angle being the same as the incident angle; and a second wave that goes through the boundary, that is, the transmitted wave. The situation with the incident angle being equal to the reflecting angle is also called a specular reflection. The amount of energy that is reflected and transmitted, respectively, can be described by Fresnel's reflection and transmission coefficients. In ray-tracing applications typically one media is air, whereas the other one is for example a wall, see Figure 10.1. In the case of a lossy media an additional loss occurs proportional to the thickness of the media. If the surface of the boundary is rough, modified Fresnel coefficients can be applied for specular reflection. For the case of layered media reflection and transmission factors are computed using the wave matrix method. Further details on all these methods can be found, for example, in [2–4].

10.2.4 Diffraction

Several methods exist for calculating diffraction effects [3]:

- Field theoretical methods, for example, the Parabolic Equation Method.
- High frequency approximations, for example, Geometrical Theory of Diffraction (GTD) or Uniform Theory of Diffraction (UTD).
- Methods for calculating diffraction on absorbing knife edges (Knife Edge Diffraction).
- Empirical solution methods, where obstacles are not covered explicitly, but globally by a few correction terms.

The most frequently applied techniques in ray-tracing models are UTD/GTD and Knife Edge Models. Since Knife Edge Models are described in Chapter 12 only UTD/GTD is described here. GTD was developed in the 1960s by Keller [5]. The GTD is a completion of the GO solution, see Figure 10.4, by considering diffraction effects at surface disconti-nuities, geometrical discontinuities (peaks, edges, bendings) and dielectric discontinuities (impedance jump). The diffracted ray path is described by diffraction coefficients (accord-ing to the reflection coefficients for reflected rays). One of the drawbacks of Kellers GTD is that it contains poles for the angles Incident Shadow Boundary (ISB) and Reflection Shadow Boundary (RSB), see Figure 10.2. Hence, field calculations in angular ranges near ISB and RSB are not possible. This problem was solved by an extension valid to the full area by Kouyoumijan and Pathak in 1974 [6]. This solution, which also allows field calculations near ISB and RSB is called UTD.

The following rays in zones I to IV as defined in Figure 10.2 are now existing [1]:

- zone I (visual zone): direct, reflected and diffracted ray exist;
- zones II, III (visual zone, reflection shadow): direct and diffracted ray exists;
- zone IV (shadow zone): only a diffracted rays exists.

For the calculation of the resulting scattered fields, all ray paths have to be considered:

- rays not touching a surface at any point (direct rays);
- rays touching a surface at a point (reflected rays);
- rays touching a discontinuity at a point (edge-diffracted rays, diffraction at impedance jump);
- rays touching a surface over a distance (surface rays);
- all rays going through multiple processes (multiple reflection, multiple diffraction or combined multiple reflection and diffraction).

By adding the electric fields at the observation point occurring at the different ray paths, the full scattered field is obtained. However, in practical implementations single ray paths may be neglected according to the requirements in terms of accuracy. From literature, solutions for a couple of so-called canonical objects are known. For more details see, for example, [2].

10.2.5 Scattering

In the case of rough surfaces and at interfaces that are not infinitely large, respectively, the incident radiation energy is scattered in other directions than the specular one. For calculation of the scattering in wave propagation models, methods for estimating contributions from scattering at single area elements are required based on a statistical description of the surface. One possibility is to use the theory of bistatic radar cross section of the surface element. This concept was first introduced to ray-tracing models by Lebherz et al. in 1990 [7], who applied the Kirchhoff and small perturbation methods as given in [8–10] to a 3D propagation model in rural areas. This method allows for a full-polarimetric description of the scattering process. One drawback of the applied method is that they are only valid for certain surface properties (relating height standard variation and correlation length to the wave length). More details on these methods can be found for example, in [3].

Another popular and much more simple method is to describe scattering in analogy to the modeling of diffuse scattering of light in the field of computer graphics. Therefore, the scattering surface is assumed to be a Lambertian emitter. One of the first ray-tracing models applying this technique was published by Liebenow et al. in 1994 [11] and is also applied in a couple of other ray-tracing models, for example [12–14]. In this model the attenuation depends only on the angles of incidence and reflection as well as on an empirically found attenuation factor. The impact of polarization is not considered in this model. An alternative approach has been introduced by Lostanlen in [15] using PO.

A more advanced method based on the Lambertian model for the modeling of scattering from buildings in urban areas has been introduced by Degli-Esposti in 2001 [16]. This model allows tuning of the main parameters, which have a precise physical meaning, by measurements. A further development of this method is presented in [17]. The extension to ultra-wideband signals was done by Lostanlen in [18] for indoor environments.

10.3 Incorporating the Influence of Vegetation

When incorporating the influence of vegetation on propagation into a ray-tracing model the following effects are typically considered [19]:

- *Diffraction* of rays over the canopy of the trees.
- *Attenuation* of rays going through the tree canopy. This effect is often called coherent scattering or tree shadowing in the literature.
- *Scattering* from trees. This effect is also called incoherent or diffuse scattering.

These three effects are also visualized in Figure 10.5. In the following, a short review of methods to model these effects is given also mentioning some ray-tracing models, where these methods are applied.

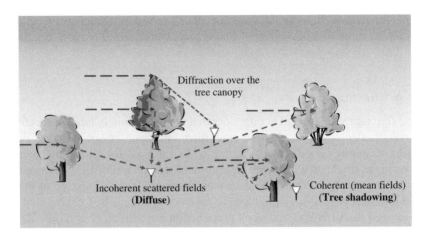

Figure 10.5   Diffraction and scattering in vegetation, derived from [19].

10.3.1   Modeling Diffraction Over the Tree Canopy

For the calculation of the additional loss observed, when a ray is diffracted over the tree canopy each tree can be modeled by a single knife edge. Examples where this effect has been applied to ray-tracing are mentioned in [12] and [19].

10.3.2   Modeling Tree Shadowing

Simple empirical approaches to model the attenuation when a ray is penetrating through the tree take into account the length of the path through vegetation which is multiplied with a specific attenuation loss [12, 20]. The specific attenuation increases with frequency and depends on the polarization for frequencies below 2 GHz. Typical values for the frequencies between 1 and 3 GHz are in the order of 0.1 to 0.4 dB/m. In [12] this approach is combined with diffraction over vegetation, see Figure 10.6 for each ray in the ray-tracing model passing through vegetation, see Figure 10.5.

More advanced methods to model coherent tree scattering have been developed for vegetated built-up areas see Figure 10.7 based on the discrete scattering theory of Foldy-Lax combined with the Kirchhoff-Huygens diffraction theory [20–22]. Based on these results [24] provides a simplified analytical approach to combine the specific attenuation of trees with multiple-screen diffraction and extends the Walfsich-Bertoni [25] model for the case of vegetated built-up areas. With additional extensions consider also roof-top-to-street diffraction in streets with vegetation, as in [26], which proposes an extension of the COST231-Walfsich-Ikegami-Model [27] to vegetated areas.

10.3.3   Modeling Diffuse Scattering from Trees

In [14] a simple model to consider diffuse scattering from vegetation is described. In this approach scattering is modeled by a Lambertian emitter and applied to a complete ray-tracing model. A more advanced tree-scattering model is proposed by de Jong in 2004 [28].

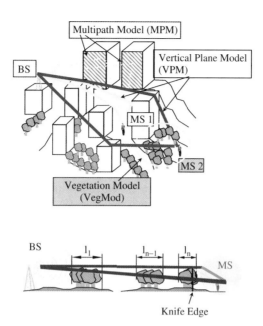

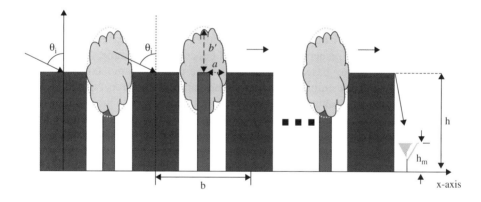

Figure 10.6 Propagation paths and vegetation submodel (top) and superposition of rays propagating through vegetation and rays diffracted over the tree canopy (bottom) (from [12]).

Figure 10.7 Vegetation in built-up areas (from [24]).

The tree canopy is modeled by randomly distributed and oriented cylinders and thin disks, which represent the branches and leaves, respectively. The expressions derived by de Jong show good agreement with measurements from single trees at 1.9 GHz. The approach is incorporated into a complete ray-tracing model and reveals a considerable improvement of the prediction accuracy in micro cellular environments. De Jong's approach is further developed by Chee [19, 29]. Chee introduces a tree group model and extends the frequency range up to 3.5 GHz. When applied to 3D ray tracing in a rural residential area at 3.5 GHz the model shows significant improvement as compared to the case when the influence of vegetation is not considered.

The main physical phenomena intervening in the electromagnetic wave propagation have been reviewed in this section. In the remaining of the chapter, a detailed description of the ray-tracing techniques is given. As those methods required are computationally intensive, some acceleration techniques of the computation times have been elaborated and will be mentioned.

10.4 Ray-Tracing Methods

The focus here is made on the techniques called *ray-tracing*. These computational methods have gained an increased interest in the last two decades as they model intuitively deterministic behaviors of the electromagnetic waves propagation in outdoor and indoor areas. The main idea consists in assuming that the energy wave-fronts radiated by a source propagates following infinitesimal conical pipes, called rays. Consequently, these rays are the trajectories of equipotent wave-fronts traveling in straight directional lines.

Rays interact with the environment along their path following three main schemes: transmission through obstacles, reflection on planar surfaces of obstacles, and diffraction on the edge of obstacles. Transmissions and reflections are modeled using Geometrical Optics while diffractions are modeled using the GTD and the UTD proposed by [5] and [6] respectively.

Ray-based methods directly follow the Fermat principles and the Snell-Descartes laws. In optics, Fermat's principle states that the ray trajectory between two points (e.g., a source and a receiver) is the path that can be traveled in the least time. The derivative Snell-Descartes law is a formula used to describe the relationship between the angles of incidence and transmission (in other words, refraction), when referring to waves passing the interface between two isotropic media of different material properties (permittivity, permeability, conductivity), such as open air and concrete for example.

The first objective of ray-based methods is to trace the trajectories taken by modeled electromagnetic waves transmitted by a source and captured by a receiver. In a digitized representation of the environment, the modeled "obstacles" (outdoor building, indoor pieces of furniture) modify the ray course creating various distinctive paths before the wave reaches the receiver. Thus, inherently the ray-tracing techniques provide a direct modeling of the multi-path occurring in the electromagnetic scene between the two communication ends, in Line Of Sight (LOS) and Non Line Of Sight (NLOS) areas.

10.4.1 Modeling of the Environment

The ray-based methods use descriptions of the environment for indoor and outdoor areas, as illustrated in Figure 10.8. The most important elements of the digitized environments are the unitary elements: segments and/or facets (in 3D) that are spatial samplings of closed polygons representing the obstacles. In indoor environments these objects are the main building walls, partition walls, windows, floor and ceiling, and pieces of furniture when required. For outdoor environments the building contours and their details (podiums, rooftop obstacles), bridges, and vegetation are represented. The environment is sampled spatially into small elements (segment or closed polygons) called facets (Figure 10.9). The nature of those obstacles is also important as it may impact the type and strength of the interactions applied on the objects (transmission loss, reflection and diffraction "strength").

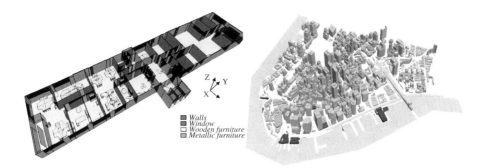

Figure 10.8   Indoor and outdoor environments.

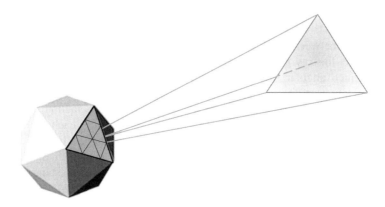

Figure 10.9   Spherical front wave out of the source and unitary facet.

The facet orientation is determinant in the accuracy of the computation because it gives the trajectory of the path after the interactions. A trade-off needs to be found when digitizing environments: on the one hand, too few details lead to missed interactions, and on the other hand too many details lead to long computation times and to some extent to erroneous computations because of the quantity of possibly inaccurate details captured, that do not necessarily have such an impact on reality.

10.4.2   Geometric Computation of the Ray Trajectories

The geometrical computation of the ray trajectories will depend on the targeted use of the rays and their application. The main assumption is to consider dimensions of the obstacles far larger than the wavelength (which holds for the electromagnetic spectrum frequency bands used in personal wireless communications). To determine a couple of individual Tx-Rx links, the image backward method may be selected, whereas when multiple links must be computed such as a coverage area then the forward method may be chosen. We now detail these techniques.

10.4.3   Direct Method or Ray-Launching

The first approach called Forward-Ray Tracing or Shooting and Bouncing Rays (SBR) consists in launching rays from the transmitter in all directions and then analyzing, following interactions on the obstacles, which rays have reached the receiver area. This method considers that reflections are specular, and not diffuse, that is, a ray reflected on a facet will create one and only one ray by reflection. This technique is the basis of the computationally efficient technique called Ray-Launching. Assuming that a source radiates (shooting) an isotropic wave, we can imagine the wave-front as a propagating sphere centered on the transmitter. The surface of the sphere is decomposed into equal area triangle patches, using the projection of a particular polygon, called icosahedron, on the sphere. The sub-division into unitary elements of the triangular faces of this polygon is realized by a decomposition technique called tessellation. The lines between the center of the sphere (Tx) and the center of the faces are the rays. Then the beam is defined by the three lines Center-Vertex of the triangulated sphere, modeling the propagating vector of the unitary wave-fronts.

The direct line between the source and the receiver is analyzed. If there is no obstacle blocking the line, this ray will be the direct path. Then rays are launched from the source with a pre-defined angular step (α). An intersection of each ray with the obstacles is tested. If a reflection or diffraction is observed (bouncing), then a new ray is created and follows the same intersection checking procedure. When an interaction occurs, then the ray interaction is stored. A tree gathering all interactions is built: each node corresponds to an interaction and each branch to a new ray. At the reception, a bundle of rays might be collected. Two main families of ray collection are introduced in the literature: the reception sphere and the illumination method. At a reception (Rx) distance d far enough from the source (Tx), it can be observed in Figure 10.10 that two adjacent rays are separated by a distance roughly equal to $d\alpha$ in 2D and $d\alpha\sqrt{3}$ in 3D. Thus, for each reception location, rays intersecting a disk in 2D or sphere in 3D with a radius of $d\alpha/2$ and $d\alpha\sqrt{3}/2$ respectively, would be considered as contributing to the total received signal. This reception sphere is built such that one and only one ray per possible interaction is collected. The second technique consists in the creation of illumination zones for each interaction of a ray beam. A ray beam is defined by the area between two adjacent rays (in 2D) or the volume inside three adjacent rays (in 3D). An illumination zone is a cone

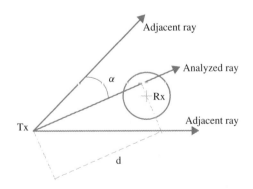

Figure 10.10   Reception sphere for ray-launching.

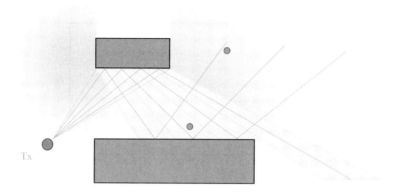

Figure 10.11   Ray-launching principle.

representing the reception area possibly reached by the ray-beam (see Figure 10.11). At the reception, if the bundle of rays is located inside the illumination area, then they are considered as contributing to the total reception signal. The accuracy of the ray-launching algorithm depends on the angular separation between adjacent wave-front propagation vectors. Decreasing the angular separation between rays will cause an increase in the total number of rays to evaluate, and thus will increase the overall simulation runtime. The major issue of this technique is the lack of accuracy when the angular separation is too large. Since the cross-section of the wave-front is growing with distance, the wave-front propagation vector could miss some interactions with small obstacles. In this case, a smaller sampling angle must be chosen, at the price of a longer simulation time. To avoid this divergence with the distance, several techniques have been proposed such as the beam-launching [30] and the coupling of angular and linear steps [31].

10.4.4   Image Method Ray-Tracing

Originally called Backward Ray Tracing and usually referred as Indirect Method or Image Method, the ray tracing starts from the receivers' location to find all the possible paths between the transmitter and the receiver. Geometrical Optics are used to determine all the interactions (transmissions, reflections and diffractions) in the digitized environment. Rays reflected by a facet are calculated using the theory of images. An image source (virtual source) is calculated by axial symmetry toward the facet of reflection, as shown in Figure 10.12 for a 1st order reflection. By applying the same process for each wall, multiple images are obtained corresponding to multiple orders of reflection. This method considers that reflections are specular, and not diffuse, that is, a ray reflected on a facet will create one and only one ray by reflection. Each image and its associated facet are stored in a tree for all possible facets and for all selected orders of reflections. An image gives a viable ray if the geometrical segment between the source and the receiver intersects the facet associated to the image. If there is no intersection, it means that the receiver is not in the reflection zone of the facet, that is, the image does not give a valid ray. If one or more

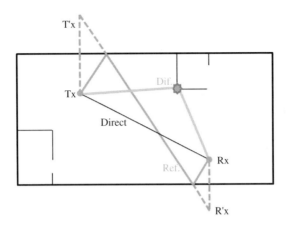

Figure 10.12   Ray-tracing principle.

intersections occur in addition to the source's facet intersection, then the ray is considered valid, but one or more transmissions will occur before the ray reaches the receiver. All the images that do not provide a valid path are therefore deleted from the tree, and only the valid branches remain. As mentioned, transmissions are determined through the tests of the valid rays constructed by reflections. These transmission interactions are saved for each branch of the tree, between a source and the receiver, or between a lower order source and its associated higher order's one. Transmissions in materials are subject to multiple internal reflections and also refractions, leading to multiple reflected and transmitted rays. Nevertheless, an approximation is generally made in ray-tracing algorithms to consider a single reflected ray and a single transmitted ray from these interactions. Reflections and transmissions coefficients are therefore calculated to gather all the energy carried by those multiple rays into a single one. Diffraction edges and wedges are considered as secondary sources from which ray tracing is performed as if it was the original source, that is, the propagation is studied in all the directions. In Figure 10.8, the propagation between two points, Tx and Rx, and the diffraction on point "Diff" is drawn. "Diff" acts as a new source, radiating energy all around, with specific diffractions coefficients depending on arrival and departure angles.

   In theory, the combination of all possible interactions should be studied. However, in complex environments with a lot of obstacles, it is not efficient to test all possible interactions. Usually, ray tracing algorithms are limited–by heuristic values–the number of interactions to reduce computation time. In addition, some speed-up techniques have been suggested to reduce the number of tested rays, which is greedy in computation time. The main speed-up techniques are presented in the following paragraphs.

10.4.5   Acceleration Techniques

Acceleration techniques usually rely on speeding up intersections tests and/or reducing the number of rays to test. A simpler ray intersection check with the object can be realized on a simplified version of the building itself: a bounding volume. Bounding volumes represent small volumes of the 3D digitized environment that contain a single object of the environment. Only if a ray intersects the bounding volume does the object itself need to be checked for intersection. Though this actually increases the computation for rays that pierce its bounding volume (two intersection tests instead of one), in a typical environment most rays closely approach only a small fraction of the objects. The simplified intersection tests on the bounding volumes result in a significant net gain in efficiency. A multi-scale technique consists in building bigger bounding volumes that enclose a few objects at once. Fewer bounding volumes are therefore tested and a computation gain is obtained. A further refinement creates first a space subdivided uniform (or not) volume matrix. The elements in the matrix are called voxels. The uniformity of the spatial subdivision simplifies the ray's tests while it may raise a little bit the number of voxels to test. Glassner [32] provides a comprehensive detailed presentation of those acceleration techniques.

We discuss thereafter two advanced methods leading to faster computations.

The main objective of the *illumination technique* is to reduce the number of reflection images to be created by using the concept of illumination zones [33–35]. For N facets, a brute force algorithm will create N first order images, N^2 second order images and so on. Then to test if a ray is valid, a backward operation (from Rx to Tx) is needed for each image obtained. This leads to a lot of wasted intersection tests. To circumvent this issue, for each image created by a reflection or a diffraction, its illumination zone is stored. The illumination zone is a cone that represents the area that is possibly reached by an image source on the corresponding facet, as described for tow base stations in Figure 10.13. Hence the image tree is built with only image sources that have a valid path from its lower order source. At the end of tree building, the backward validity test has to be performed only between the transmitter and the highest order images because we know that the paths created from lower order images are valid. Moreover, since illumination zones are stored, the validity of a ray is known by checking if the receiver is lying in the illumination zone of the image source, that is, there is no need to trace the ray joining the source to the receiver and perform intersection tests. An even more refined method has been proposed for a 2D ray-launching in [31] with a non-uniform angle separation between rays. The

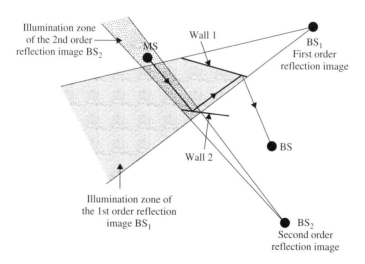

Figure 10.13 Illumination zone for a *1st* and *2nd* order reflection [33].

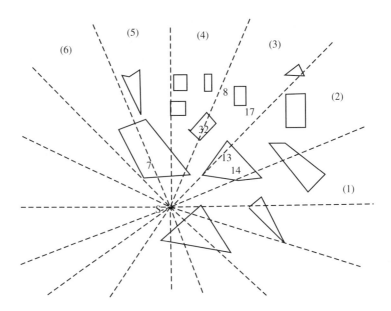

Figure 10.14   Angular Z-Buffer subdivision of an area [36].

method is suited to any environment because the ray launching makes sure that any illuminated facet will at least receive one ray. Thus the divergence usually observed for ray-launching for distant reception areas, is extremely reduced and becomes negligible. The results are as accurate as the image-based ray-tracing, with a better computational time though.

The *angular Z-Buffer* [36, 37] consists in dividing the area in a similar approach to the voxel presented above, but instead of using squared areas, this method uses angular pixels, or anxels, created around the source location (point *S* in Figure 10.14). The facets entirely or partially located into an anxel are stored and sorted from the shortest to the furthest distance to the source. By simply discarding the facets that are not in the ray-beam, this sorting order allows reducing the computation time during transmission tests. In addition, the illumination zone principle presented in the previous paragraph may also be used to process faster the reflections. For each image source created by a reflection, the illumination zone is calculated, and anxels are constructed from the image source only in the bounding box of the illumination zone. Further acceleration techniques are hardware based using for example Graphical Processing Units (GPU) [38, 39].

10.4.6 Hybrid Techniques

Even when acceleration techniques are implemented, the use of ray-based method can be computationally intensive for real electromagnetic scene, in particular in 3D. Hence hybrid techniques have been proposed [40, 41] to reduce the complexity of the volume approach by considering two planes: horizontal and vertical, that limits, among others, the intersection tests. In the horizontal plane, a ray-launching is generally performed. This part requires the greater details and accuracy as the direction of the bounced rays depends on the segment orientation. In the vertical plane, a different technique may be used: the most common are ray-tracing and multiple knife-edges.

10.4.7 Determination of the Electromagnetic Field Strength and Space-Time Outputs

Based on the identified paths between the transmitter and possible receiver locations, ray based methods predict propagation effects in mobile and personal communication environments. They are based on Geometrical Optics (GO), which is an easily applied approximate method for predicting a high-frequency electromagnetic field. The use of high-frequency electromagnetic theory allows calculating the amplitude, phase, delay and polarization of each ray.

Considering that each ray potentially interacts with the surrounding environment, the complex received field amplitude is expressed by

$$E_i = E_0 . G_{ti} . G_{ri} . \prod^{NbR} R_{NbRi} . \prod^{NbT} T_{NbTi} . \prod^{NbD} D_{NbDi} . \frac{e^{-\frac{j2\pi d}{\lambda}}}{d} \tag{10.1}$$

Where E_0 is a reference field (V/m), G_{ti}, G_{ri} are respectively the transmitter and receiver gains for the ith ray, R_{NbRi}, T_{NbTi}, D_{NbDi} are respectively the reflection, transmission and diffraction coefficients, λ is the wavelength and d is the length of the path.

10.4.8 Extension to Ultra-Wideband (UWB) Channel Modeling

The ray-based method does play an important role in characterizing the electromagnetic wave propagation in various environments including urban areas and inside buildings. The wide system bandwidth has increased this importance of using techniques relating the time delayed arrival of the paths. The multi-antenna systems have made the spatial channel response characterization a premium requirement. Hence, the ray-based methods, provided that they are well implemented to handle large operational computations areas, are widely used in the Wireless Industry to deploy and optimize mobile networks. Beside their use for the radio network planning, the ray-based techniques do also find a direct application for localization purposes in complex environments. With some adaptations that are described below, an UWB impulse response can be obtained. This UWB channel response relates to a site specific situation which is crucial for positioning applications.

A brief overview of deterministic UWB channel models is given here. A literary review of existing UWB channel models and measured UWB channel properties is detailed in [42].

UWB impulse radio communications transmit data using ultra short impulses and spreading the energy over a large bandwidth (up to several GHz) [43]. Given the very wide bandwidth and the reduced signal energy, UWB pulses are severely distorted when propagating in the environment [42]. Furthermore the large frequency band introduces an important frequency dependence of the wireless channel. This frequency-dependence has an impact on material properties, physical interaction mechanisms, and also on antenna patterns that experience significant variations over a very wide frequency range [44]. Besides, when modeling UWB channel, the phase information must capture the effects

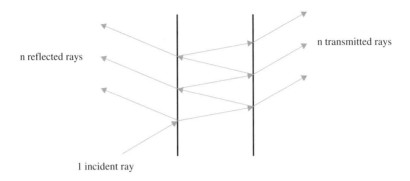

Figure 10.15   Internal reflection inside wall for UWB signals.

from propagation, antennas and materials. The modeling of this phase behavior is key to understanding transmitted signals attenuations and dispersions.

Several contributions have demonstrated the usefulness of the ray-based method to model the UWB wireless channel [45–48]. Generally, most mechanisms are processed in the frequency domain, such as the frequency-dependence of the material. An Inverse Fourier Transforms translates the signals in time-domain. The phase information behavior is carefully separated into local phenomena, such as the interactions, and longer distance phenomena, such as the free space propagation [45]. The material properties used in the reflection, transmission and diffraction coefficients are expressed as a function of the frequency. For ultra-short pulses, the transmissions inside walls need to be revisited as multiple reflections and transmissions result from the incident waves due to the internal bouncing. This phenomenon illustrated in Figure 10.15 mentioned in [46] was further detailed in [49]. Antennas are handled either in frequency domain, or in the time domain [50]. But conceptually it is to be understood that the antenna diagram and antenna distortion (including the trailer) will differ for each radiated path (as illustrated in Figure 10.16 in time domain). The diffuse scattering experienced by the waves over the large frequency bandwidth may be of paramount importance, and require specific processing. In [18] the authors have proposed an initial solution to introduce the diffuse scattering into ray-based UWB models. The wideband channel model is enriched with more phenomena inserted

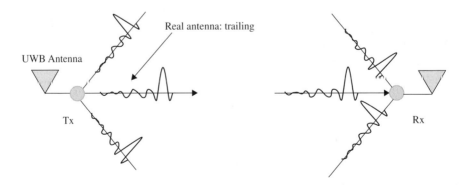

Figure 10.16   UWB antenna time domain trailing.

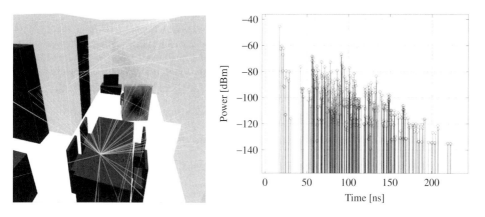

Figure 10.17   Ray-traced geometry in a room with furniture and the corresponding UWB impulse.

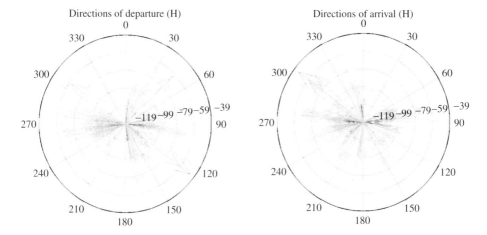

Figure 10.18   Directions of departure and arrival in the horizontal plane of the UWB impulse response.

as pseudo-stochastic parameters. This technique has demonstrated large enhancements of the prediction accuracy of wideband channel characteristics from diffuse scattering modeling. The impulse response obtained by such a UWB deterministic simulation tool in a room is represented in Figure 10.17. The corresponding direction of departure and direction of arrival in the azimuthal plane are illustrated in Figure 10.18. The UWB impulse responses can be used to give precise time-of-arrival and angle-of-arrival data that are exploited by localization algorithms to determine precisely the position of mobile devices as investigated in the European funded projects Where [51] and Where2 [52].

References

1. M. Lebherz, *Wellenausbreitungsmodelle zur Versorgungsplanung unter Bercksichtigung der Mehrwegeausbreitung (in German)*. Nomos Verlag Baden-Baden, Ph. D. Thesis Universität Karslruhe (Germany), 1990.

2. C. A. Balanis, *Advanced Engineering Electromagnetics*. Wiley, 1989.

3. N. Geng and W. Wiesbeck, *Planungsmethoden für die Mobilkommunikation* (in German). Springer, 1998.

4. S. Saunders, *Antennas and Propagation for Wireless Communication Systems*. Wiley, 1999.

5. J. B. Keller, "Geometrical theory of diffraction," *J. Opt. Soc. Amer.*, vol. 52, no. 2, pp. 116–131, 1962.

6. R. Kouyoumjian and P. Pathak, "A uniform geometrical theory of diffraction for an edge in a perfectly conducting surface," *Proceedings of the IEEE*, vol. 62, no. 11, pp. 1448–1461, Nov. 1974.

7. M. Lebherz, W. Wiesbeck, and W. Krank, "A versatile wave propagation model for the vhf/uhf range considering three-dimensional terrain," *Antennas and Propagation, IEEE Transactions on*, vol. 40, no. 10, pp. 1121–1131, Oct. 1992.

8. F. T. Ulaby, R. K. Moore, and A. K. Fung, *Microwave Remote Sensing: Active and Passice*. Addison-Wesley, 1982.

9. D. E. Barrick, G. T. Ruck, W. D. Stuart, and C. K. Krichbaum, *Radar Cross Section Handbook*, vol. 11. Plenum, 1970.

10. P. Beckmann and A. Spizzichino, *The Scattering of Electromagnetic Waves from Rough Surfaces*. Artech House, 1987.

11. U. Liebenow and P. Kuhlmann, "Theoretical investigations and wideband measurements on wave propagation in hilly terrain," in *Vehicular Technology Conference, 1994 IEEE 44th*, Jun 1994, pp. 1803–1806 vol. 3.

12. T. Kürner and A. Meier, "Prediction of outdoor and outdoor-to-indoor coverage in urban areas at 1.8ghz," *Selected Areas in Communications, IEEE Journal on*, vol. 20, no. 3, pp. 496–506, Apr. 2002.

13. T. Kürner and M. Schack, "3d ray-tracing embedded into an integrated simulator for car-to-x communications," in*Electromagnetic Theory (EMTS), 2010 URSI International Symposium on*, Aug. 2010, pp. 880–882.

14. T. Fügen, J. Maurer, T. Kayser, and W. Wiesbeck, "Capability of 3-d ray tracing for defining parameter sets for the specification of future mobile communications systems," *Antennas and Propagation, IEEE Transactions on*, vol. 54, no. 11, pp. 3125–3137, Nov. 2006.

15. Y. Lostanlen, B. Uguen, G. Chassay, and H. Griffiths, "Modelling of the air-ground interface for ultrawideband radar applications," in *Ultra-Wideband, Short-Pulse Electromagnetics 5*, P. D. Smith and S. R. Cloude, Eds. Springer US, 2002.

16. V. Degli-Esposti, "A diffuse scattering model for urban propagation prediction," *Antennas and Propagation, IEEE Transactions on*, vol. 49, no. 7, pp. 1111–1113, Jul. 2001.

17. V. Degli-Esposti, F. Fuschini, E. M. Vitucci, and G. Falciasecca, "Measurement and modelling of scattering from buildings," *Antennas and Propagation, IEEE Transactions on*, vol. 55, no. 1, pp. 143–153, Jan. 2007.

18. Y. Lostanlen and G. Gougeon, "Introduction of diffuse scattering to enhance ray-tracing methods for the analysis of deterministic indoor uwb radio channels (invited paper)," in *Electromagnetics in Advanced Applications, 2007. ICEAA 2007. International Conference on*, Sept. 2007, pp. 903–906.

19. K. L. Chee, "Fixed bordband wireless access in vegetated rural residential areas," Shaker-Verlag, Ph.D. Thesis, Technische Universiät Braunschweig (Germany), 2011.

20. *ITU-R Recommendation P833: Attenuation in Vegetation*, 1992.

21. S. Torrico and R. Lang, "A simplified analytical model to predict the specific attenuation of a tree canopy," *Vehicular Technology, IEEE Transactions on*, vol. 56, no. 2, pp. 696–703, March 2007.

22. K. Chee, F. Catalan, S. A. Torrico, and T. Kürner, "Modelling tree scattering in rural residential areas at 3.5ghz," in *12th URSI Commision-F Triennial Open Symposium on Radio wave Propagation and Remote Sensing, Garmisch-Partenkirchen, 8-11th March*, 2011.

23. S. Torrico, H. Bertoni, and R. Lang, "Modeling tree effects on path loss in a residential environment," *Antennas and Propagation, IEEE Transactions on*, vol. 46, no. 6, pp. 872–880, Jun. 1998.

24. S. A. Torrico, K. Chee, and T. Kürner, "A propagation prediction model in vegetated residential environments a simplified analytical approach," in *5th European Conference on Antennas and Propagation, Rome, Italy*, 2011.

25. J. Walfisch and H. Bertoni, "A theoretical model of uhf propagation in urban environments," *Antennas and Propagation, IEEE Transactions on*, vol. 36, no. 12, pp. 1788–1796, Dec. 1988.

26. K. Chee, S. A. Torrico, J. Baumgarten, A. Hecker, P. Zahn, M. Rohner, and T. Kürner, "Propagation prediction and measurement in vegetated moderately built-up areas," in *6th European Conference on Antennas and Propagation, Prague, Czech Repubic*, 2012.

27. E. Damosso and L. M. Correia (Eds.), *COST Action 231: Digital mobile radio towards future generation systems-Final Report*. Office for the Official Publications of the European Communities, 1999, http://www.lx.it.pt/cost231.

28. Y. de Jong and M. Herben, "A tree-scattering model for improved propagation prediction in urban microcells," *Vehicular Technology, IEEE Transactions on*, vol. 53, no. 2, pp. 503–513, March 2004.

29. K. Chee, S. A. Torrico, and T. Kürner, "Foliage attenuation over mixed terrains in rural areas for broadband wireless access at 3.5ghz," *IEEE Trans. on Antennas and Propagation*, vol. 59, no. 7, pp. 2698–2706, 2011.

30. S.-H. Chen and S.-K. Jeng, "An sbr/image approach for radio wave propagation in indoor environments with metallic furniture," *Antennas and Propagation, IEEE Transactions on*, vol. 45, no. 1, pp. 98–106, Jan. 1997.

31. Y. Corre and Y. Lostanlen, "Three-dimensional urban em wave propagation model for radio network planning and optimization over large areas," *Vehicular Technology, IEEE Transactions on*, vol. 58, no. 7, pp. 3112–3123, Sept. 2009.

32. Glassner, *An introduction to Ray-Tracing*, academic press; 1st edition ed. University of Chicago Press, 1989.

33. G. Athanasiadou, A. Nix, and J. McGeehan, "A microcellular ray-tracing propagation model and evaluation of its narrow-band and wide-band predictions," *Selected Areas in Communications, IEEE Journal on*, vol. 18, no. 3, pp. 322–335, Mar. 2000.

34. K. Rizk, J.-F. Wagen, and F. Gardiol, "Two-dimensional ray-tracing modeling for propagation prediction in microcellular environments," *Vehicular Technology, IEEE Transactions on*, vol. 46, no. 2, pp. 508–518, May 1997.

35. F. Aguado Agelet, A. Formella, J. Hernando Rabanos, F. Isasi de Vicente, and F. Perez Fontan, "Efficient ray-tracing acceleration techniques for radio propagation modeling," *Vehicular Technology, IEEE Transactions on*, vol. 49, no. 6, pp. 2089–2104, Nov. 2000.

36. M. Catedra, J. Perez, F. Saez de Adana, and O. Gutierrez, "Efficient ray-tracing techniques for three-dimensional analyses of propagation in mobile communications: application to picocell and microcell scenarios," *Antennas and Propagation Magazine, IEEE*, vol. 40, no. 2, pp. 15–28, Apr. 1998.

37. C. Saeidi and F. Hodjatkashani, "Modified angular z-buffer as an acceleration technique for ray tracing," *Antennas and Propagation, IEEE Transactions on*, vol. 58, no. 5, pp. 1822–1825, May 2010.

38. T. Rick, and R. Mathar, *Antennas, 2007. INICA'07. 2nd International ITG Conference on*, Fast Edge-Diffraction-Based Radio Wave Propagation Model for Graphics Hardware, 2007, March 15–19, 10.1109/INICA.2007.4353923.

39. M. Reyer, T. Rick, and R. Mathar, *Antennas and Propagation, 2007. EuCAP 2007. The Second European Conference on*, Graphics Hardware Accelerated Field Strength Prediction for Rural and Urban Environments, 2007, Nov. 1–5.

40. G. Liang, and H. L. Bertoni, *Vehicular Technology Conference, 1997, IEEE 47th*, A new approach to 3D ray tracing for site specific propagation modeling, 1997, May 2, 1113–1117 vol. 2, 10.1109/VETEC.1997.600503.

41. T. Kürner, D. J. Cichon, and W. Wiesbeck, *Selected Areas in Communications, IEEE Journal on*, Concepts and results for 3D digital terrain-based wave propagation models: an overview, 1993, Sep. 11, 7, 1002–1012, 10.1109/49.233213.

42. A. Molisch, "Ultrawideband propagation channels-theory, measurement, and modeling," *Vehicular Technology, IEEE Transactions on*, vol. 54, no. 5, pp. 1528–1545, Sept. 2005.

43. M. Win and R. Scholtz, "Impulse radio: how it works," *Communications Letters, IEEE*, vol. 2, no. 2, pp. 36–38, Feb. 1998.

44. R. Qiu, "A study of the ultra-wideband wireless propagation channel and optimum uwb receiver design," *Selected Areas in Communications, IEEE Journal on*, vol. 20, no. 9, pp. 1628–1637, Dec. 2002.

45. B. Uguen, E. Plouhinec, Y. Lostanlen, and G. Chassay, "A deterministic ultra wideband channel modeling," in *Ultra Wideband Systems and Technologies, 2002. Digest of Papers. 2002 IEEE Conference on*, 2002, pp. 1–5.

46. Y. Lostanlen, G. Gougeon, and Y. Corre, "A deterministic indoor uwb space-variant multipath radio channel model," in *Ultra-Wideband, Short-Pulse Electromagnetics 7*, F. Sabath, E. L. Mokole, U. Schenk, and D. Nitsch, Eds. Springer New York, 2007, pp. 796–815.

47. W. Wiesbeck, C. Sturm, W. Soergel, M. Porebska, and G. Adamiuk, "Influence of antenna performance and propagation channel on pulsed uwb signals," in *Electromagnetics in Advanced Applications, 2007. ICEAA 2007. International Conference on*, Sept. 2007, pp. 915–922.

48. J. Jemai, P. C. F. Eggers, G. F. Pedersen, and Kurner, T., *Antennas and Propagation, IEEE Transactions on*, Calibration of a UWB Sub-Band Channel Model Using Simulated Annealing, 2009, Oct., 57, 10, 3439–3443, 10.1109/TAP.2009.2028676, 0018–926X.

49. E. Plouhinec and B. Uguen, "Ray-tracing and multiple reflections inside materials applied to uwb localization," in *Electromagnetics in Advanced Applications, 2009. ICEAA '09. International Conference on*, Sept. 2009, pp. 674–677.

50. A. M. Attiya and A. Safaai-Jazi, "Simulation of ultra-wideband indoor propagation," *Microwave and Optical Technology Letters*, vol. 42, no. 2, pp. 103–108, 2004.

51. EU Seventh Framework Programme, "Wireless Hybrid Enhanced Mobile Radio Estimators" – http://www.ict-where.eu/.

52. EU Seventh Framework Programme, "Wireless Hybrid Enhanced Mobile Radio Estimators Phase 2" , – http://www.kn-s.dlr.de/where2/.

11

Finite-Difference Modeling

Guillaume la Roche

Mindspeed Technologies, France

Finite-Difference Time-Domain (FDTD) methods and similar approaches have been used for decades for radio wave simulation and field computation in small areas due to their high computational requirements. Furthermore, due to the increasing performance of personal computers, such approaches are more and more popular for the purpose of deterministic simulation of radio wave propagation. Therefore this chapter presents Finite Difference (FD) methods as an alternative to the popular ray tracing model presented in the previous chapter. We will demonstrate in this chapter that these techniques have interesting properties, mainly due to their simple implementations and high accuracy, that is why it is believed they can open the way towards very accurate simulations for LTE-advanced and beyond wireless networks. However there are also some technical challenges that will need to be addressed and will be discussed at the end of this chapter.

11.1 Introduction

FD methods, also referred to as Electromagnetic Model (EM) or full-wave equations, are aimed at solving Maxwell's equations, or other sets of equations derived from them, on a computer using some sort of domain-discretization.

With Long Term Evolution (LTE)-advanced and as detailed in Chapter 1 future wireless networks are expected to be more complex with higher numbers of cells. Therefore, software tools that aid in network optimization are necessary. When planning a wireless network in specific scenarios, it is necessary to take into account all the obstacles and their materials, which will impact the attenuation of the signal and its propagation due to the numerous reflections/diffractions. For such purpose deterministic models are usually implemented and the most common approach is the ray tracing approach described in Chapter 10. However another alternative for deterministic radio propagation is the use of full wave method techniques a.k.a. FD methods. These methods directly solve Maxwell's equations on a discrete spacetime grid, thus achieving very high accuracy, because all

LTE-Advanced and Next Generation Wireless Networks: Channel Modelling and Propagation, First Edition.
Edited by Guillaume de la Roche, Andrés Alayón Glazunov and Ben Allen.
© 2013 John Wiley & Sons, Ltd. Published 2013 by John Wiley & Sons, Ltd.

the reflections and diffractions are implicitly taken into account (unlike ray tracing which limits the number of rays to compute due to complexity reasons).

In this chapter we will give a short survey on theory and possible FD models. Then, in the following sections, we will illustrate a few results obtained with one of these models regarding path loss prediction and fading prediction. Since the use of such models is still at an early stage, we will also give perspectives regarding the use of FD models for simulating next generation wireless networks. In particular we will discuss the challenges related to 3D implementation, as well as the combination of FD methods with ray tracing, and the perspectives for simulating wide-band signals like beyond LTE-Advanced systems.

11.2 Models for Solving Maxwell's Equations

Maxwell's equations describe how electric charges and electric currents act as sources for the electric and magnetic fields. Moreover they describe how a time varying electric field generates a time varying magnetic field and vice versa. Two equations describe how the fields emanate from charges, whereas two others detail how magnetic fields circulate around time varying magnetic fields and vice versa.

The solution of Maxwell's equations can not be solved in a real world scenario for wireless communications, because real-world electromagnetic problems like scattering, radiation and wave guiding are not analytically calculable. Therefore the common approach is to use approximations of Maxwell's equation. In such approximations, the scenario is discretized in space, and the equations are solved at each pixel in the spatial grid. This discrete resolution of Maxwell's equations, also referred to as FD methods or differential equation solvers, requires paying attention to the following properties in order to be sure that the results approximate well the ones from continuous Maxwell's equations:

- The size of the pixels in the spatial grid is likely to be very small compared to the wavelength. For instance if a LTE system is simulated at the frequency of 2 GHz, the spacial step Δ must be chosen so that $\Delta \ll 15$ cm. When simulating large scenarios (like urban scenarios) this leads to very large matrices and therefore the memory requirements and simulation time duration can be high.
- Because we restrict the propagation space to the scenario under consideration, the continuous phenomena of radio wave propagation is restricted to a limited space. Hence, special attention must be paid to the space boundaries. In practice this is done by adding special absorbing materials (also referred to as Perfect Matched Layer (PML)) on the borders of the scenario which will avoid reflections on the boundaries of the scenarios.

Many FD methods have been implemented in the literature. In general these are used for antenna design applications or electronic circuits, but not for simulating large scenarios like the propagation of radio signals inside streets or buildings. The list of common methods includes (but is not restricted to):

- Finite-Difference Time-Domain. In this approach, at each pixel the electric field is solved at a given instant in time, then the magnetic field is solved at the next instant in time, and the process is repeated over and over again. This is the most popular method because it is very simple to implement, that is why this method is included in many commercial EM software. There is also a model called Multi Resolution Time Domain (MRTD) which is a modification of FDTD based on wavelet analysis.

- Finite element method (FEM) is a more complex model used to find approximate solution of partial differential equations and is well adapted to complex domains.
- Finite Integration Technique (FIT) is another approach. In time domain the numerical effort of FIT increases more slowly with the problem size as compared to other commonly employed methods, so larger structures may be analyzed.
- Pseudo Spectral methods like Pseudo Spectral Time Domain (PSTD) and Pseudo Spectral Spatial Domain (PSSD) use discrete Fourier or Chebyshev transforms to calculate the derivatives of the electric and magnetic field. Their advantage is a reduced numerical phase velocity compared to FDTD.
- Transmission Line Matrix (TLM) or ParFlow (Partial Flows) is a simplified model where the fields are approximated to scalar fields instead of vectors. ParFlow is a formulation of TLM which was applied to wireless communications, however both approaches are very similar.

It is to be noticed that many other implementations of FD methods were proposed in other applications requiring the propagation of EM field to be analyzed in particular mediums.

Due to the increasing performance of computers, these models have started to be used for larger scenarios such as radio wave propagation for wireless network planning. The most commonly used is FDTD due to its very easy implementation. For instance in [1], FDTD is used for simulating the radio coverage of Wireless Interoperability for Microwave Access (WiMAX) femtocells. ParFlow was also investigated due to its reduced complexity, making it more suitable for scenarios restricted to $2D$ [2] as will be discussed later in this chapter. However many other models were also applied like FIT in [3] which is applied to inbuilding radio propagation.

The performance of FD models in terms of speed depends for a large part on their implementation, however they also have similar general properties. That is why in the remainder of this chapter we will describe and present results about two of these models which are, to the knowledge of the author, the most commonly used for wireless network planning: FDTD and ParFlow.

11.2.1 FDTD

FDTD method was proposed by Yee in 1966 in [4] and it is today one of the most efficient computational approximations to the Maxwell equations. In this approach, if we consider the case of a vertically polarized antenna, the FDTD equations can be written as follows:

$$
H_x\big|_{i,j+\frac{1}{2}}^{n+1} = H_x\big|_{i,j+\frac{1}{2}}^{n}
$$

$$
- D_b\big|_{i,j+\frac{1}{2}} \cdot \left[\frac{E_z\big|_{i,j+1}^{n+\frac{1}{2}} - E_z\big|_{i,j}^{n+\frac{1}{2}}}{\Delta \kappa_{y_{j+\frac{1}{2}}}} + \Psi_{H_{x,y}}\big|_{i,j+\frac{1}{2}}^{n+\frac{1}{2}} \right] \tag{11.1}
$$

$$H_y\big|_{i+\frac{1}{2},j}^{n+1} = H_y\big|_{i+\frac{1}{2},j}^{n}$$

$$+D_b\big|_{i+\frac{1}{2},j} \cdot \left[\frac{E_z\big|_{i+1,j}^{n+\frac{1}{2}} - E_z\big|_{i,j}^{n+\frac{1}{2}}}{\Delta\kappa_{x_{i+\frac{1}{2}}}} + \Psi_{H_{y,x}}\big|_{i+\frac{1}{2},j}^{n+\frac{1}{2}} \right] \quad (11.2)$$

$$E_z\big|_{i,j}^{n+\frac{1}{2}} = C_a\big|_{i,j} \cdot E_z\big|_{i,j}^{n-\frac{1}{2}} + C_b\big|_{i,j} \cdot \left[\Psi_{E_{z,x}}\big|_{i,j}^{n} - \Psi_{E_{z,y}}\big|_{i,j}^{n} \right.$$

$$\left. + \frac{H_y\big|_{i+\frac{1}{2},j}^{n} - H_y\big|_{i-\frac{1}{2},j}^{n}}{\Delta\kappa_{x_i}} - \frac{H_x\big|_{i,j+\frac{1}{2}}^{n} - H_x\big|_{i,j-\frac{1}{2}}^{n}}{\Delta\kappa_{y_j}} \right], \quad (11.3)$$

where H is the magnetic field and E the electrical field in a discrete grid sampled with a spatial step of Δ. D_b, C_a, and C_b are the update coefficients that depend on the properties of the different materials inside the environment. $\Psi_{H_{x,y}}$, $\Psi_{H_{y,x}}$, $\Psi_{E_{z,x}}$ and $\Psi_{E_{z,y}}$ are discrete variables with nonzero values only in some CPML regions and are necessary to implement the absorbing boundary.

These equations are typically solved using a cellular automata approach, that is, a recursive implementation in which the E and H fields are updated at each iteration based on the fields from their neighbors. For instance, when simulating the propagation of a Base Station (BS) in a scenario, the energy in the pixel representing the emitter is recursively translated and updated to its neighbors. Then a criteria must be used to decide when the steady state is reached (can be based on a number of time iterations depending on the size and complexity of the scenario). These FDTD equations can be easily translated into 3D at the cost of high computational load. The challenges related to 3D FD models will be investigated at the end of this chapter.

11.2.2 ParFlow

The first implementation of ParFlow, a formulation of TLM was applied to wireless networks in 1998. In this approach [5] a parallel solver was used in order to reduce the simulation time. This model was also implemented into $2D$. The idea with ParFlow is that the E and H fields, are approximated to scalar fields, also referred to as flows. This has the advantage of reducing the number of variables to store which is a key issue for large scenarios, where memory is usually the main limitation to the size in pixels of the scenario to simulate. However this loss of full E and H fields information makes this method inappropriate for polarization studies. In ParFlow, the flows are driven by a local transition matrix derived from the discrete Maxwell's wave equations. The flows are shown in Figure 11.1 and are referred to as $\overleftarrow{f_d}$ and $\overrightarrow{f_d}$ for inward and outward flows respectively. The index d, ($d \in \{E, W, S, N\}$) indicates the flow direction (East, West, South, North). The frequency domain ParFlow (FDPF) algorithm bounds steady-state neighbor flows by the FD scattering equation:

$$\overrightarrow{F}(m) = \Sigma_f(m) \cdot \overleftarrow{F}(m) + \overrightarrow{S}(m) \quad (11.4)$$

where $\Sigma_f(m)$ is the local scattering matrix at frequency f, m referring to the pixel (i, j) with $m = j + i \cdot N_c$, N_c being the horizontal size of the environment. $\overleftarrow{F}(m)$ and $\overrightarrow{F}(m)$

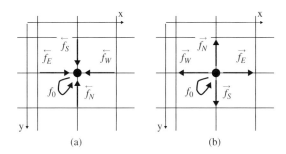

Figure 11.1   5 inward flows (a) and 5 outward flows (b) are associated with each pixel.

are respectively the inward and outward flow vectors according to:

$$\overleftarrow{F}(m) = \begin{pmatrix} \overleftarrow{f}_E(i, j) \\ \overleftarrow{f}_W(i, j) \\ \overleftarrow{f}_S(i, j) \\ \overleftarrow{f}_N(i, j) \end{pmatrix} ; \quad \overrightarrow{F}(m) = \begin{pmatrix} \overrightarrow{f}_E(i, j) \\ \overrightarrow{f}_W(i, j) \\ \overrightarrow{f}_S(i, j) \\ \overrightarrow{f}_N(i, j) \end{pmatrix} \qquad (11.5)$$

similarly, $\overrightarrow{S}(m)$ contains source flows at pixel m.

Note that the local scattering matrix contains specific coefficients adapted to the material (air, concrete, glass, and so on) located in m. This equation can be solved in a similar way as FDTD using an iterative implementation where flows are updated at each pixel until the steady state is reached. However, a special implementation called Multi Resolution Frequency Domain Parflow (MR-FDPF) was proposed by Gorce in 2003. The idea is to reformulate the ParFlow equations in the frequency domain as detailed in [6]. Using such an approach it has been shown that the problem can be solved as a multi-resolution technique. In this approach the ParFlow nodes are gathered into multi resolution (MR) nodes, where the inward and outward *exchange* flow vectors are given by:

$$\overleftarrow{F}_e(b_k) = \begin{pmatrix} \overleftarrow{f}_E(b_k) \\ \overleftarrow{f}_W(b_k) \\ \overleftarrow{f}_S(b_k) \\ \overleftarrow{f}_N(b_k) \end{pmatrix} ; \quad \overrightarrow{F}_e(b_k) = \begin{pmatrix} \overrightarrow{f}_E(b_k) \\ \overrightarrow{f}_W(b_k) \\ \overrightarrow{f}_S(b_k) \\ \overrightarrow{f}_N(b_k) \end{pmatrix} \qquad (11.6)$$

These vectors are still bounded by a local scattering equation similar to the standard scattering equation available for unitary ParFlow nodes

$$\overrightarrow{F}_e(b_k) = \Sigma_e(b_k) \cdot \overleftarrow{F}_e(b_k) + \overrightarrow{S}_e(b_k) \qquad (11.7)$$

where $\Sigma(b_k)$ is now the scattering matrix involving the exchange flows only. This matrix can be divided into 16 blocks, each one involving flows from one side to another one. For instance, $\sigma^e_{WS}(k)$ relates west outward and south inward flows. Compared to the usual ParFlow scattering matrix, the only difference stems from the fact that scalar flows are now replaced by vectorial ones. Using this formulation it is shown in [6] that a whole environment to simulate can be recursively divided into MR-nodes. Based on this binary tree of MR-Nodes, the scattering matrices can be built during a preprocessing step which

Table 11.1 ParFlow: comparison between time domain and frequency domain approaches

Time domain parflow	MR-FDPF
straightforward basic implementation	involves matrices computations
easy massively parallel implementation	parallel implementation not straighforward
no preprocess	preprocess needed only once
easy for time simulations (impulse response)	steady state is reached implicitly
slightly faster for a single set of sources	faster investigation of independent sources

does not depend on the location of the source to simulate. Then, once the preprocessing is computed the propagation of one source only requires an upward and downward phase, which has a very reduced complexity compared to simulating the propagation of one source using a standard time domain ParFlow approach, to be performed. The focus of this chapter being more on the application of FD models than their implementation, readers are referred to [6] for more details on the ParFlow equations. Also, as a summary concerning the time domain and the frequency domain, [7] summarizes well the advantages of both approaches, as represented in Table 11.1.

11.3 Practical Use of FD Methods

In the previous paragraphs we introduced the models able to solve Maxwell's equation and gave more details about two implementations that have started to be used for wireless network planning. However, one major question to be answered is: what are the advantages of such FD models for simulating LTE advanced systems and what are the practical use cases?

11.3.1 Comparison with Ray Tracing

When dealing with deterministic radio propagation simulators, that is, propagation tools which provide accurate radio coverage in a given scenario, where the obstacles (typically the buildings and walls) and their positions are taken into account, most of the current works are based on ray tracing as described in Chapter 11. Now, due to high performance of recent computers, FD models can be used as well for radio coverage prediction of wireless networks. Therefore choosing an approach would require a performance comparison of both techniques. Giving a general comparison regarding the accuracy or simulation time duration of these models is not an easy task, because usually the performance of propagation models tools depend for a large part on the implementation itself. That is why only a few works such as [8] try to compare FDTD and ray tracing in the same scenario. Let us also take note that the accuracy of the models can also have a dependency with how the model was calibrated (which will be discussed in the following sections), making the comparison even more difficult.

However, based on a general understanding of these models it is possible to give general characteristics of both methods. A major difference between them is how the complexity varies. On the one hand, for ray tracing models complexity depends directly

Table 11.2 Main properties of ray tracing and finite difference methods

X	Ray Tracing	Finite difference
Speed	varies with the number of obstacles	varies with the size of the scenario
Accuracy	depends om the number of reflections	depends on how Δ is small
Diffraction	requires to use UTD or GTD	implicit
Programming	time consuming	easy

on the number of obstacles, because more obstacles will lead to more reflections. On the other hand, complexity for FD models depend mainly on the size of the spatial grid, that is, the size of the scenario. The main difference between ray tracing and FD methods are summarized in Table 11.2.

One advantage of FDTD is its easy implementation, that is, the programming of the method itself is an easy task due to the simplicity of the equations (in practice, with matlab only a few lines of code are needed). Furthermore, as summarized in Table 11.1, FD models appear as an interesting alternative to ray tracing in very complex scenarios where many diffractions and reflections occur, which may be difficult to simulate with ray tracing. For instance in indoor scenarios with many obstacles (walls, furniture, and so on) or dense urban scenarios with complex buildings, FD models are expected to provide very high accuracy. In LTE-Advanced it is expected that in order to reach high capacity, cellular operators will have to deploy more small cells. It is believed that such small cells like femtocells and relays (see Chapter 2) will be very efficiently simulated based on FD models. That is why there are already a few published papers covering this kind of application [1].

Currently, due to high memory requirements, these FD models are not preferred for large scenarios where ray tracing is preferred. That is why finding optimal implementations, where the complexity is reduced, has been studied for a few years by researchers, as will be investigated in the following section.

11.3.2   Complexity Reduction

When implementing FD methods for radio coverage prediction, two main techniques can be used in order to reduce the computational load. The first approach is to use parallel implementation, and the second approach is to slightly reduce the size of the environment by increasing the spatial step Δ.

• *Parallel implementation*: One of the advantages of FDTD and other finite difference methods is the facility with which they are implemented on parallel architectures. Indeed, the main task of such algorithms is the update of the E and H fields at each grid cell. These depend on the field values of the neighboring cells, which have to be computed until the steady-state condition is reached. Therefore, since different cells can be independently computed, a parallel implementation in which each thread computes one cell usually results in a higher performance than implementations on traditional linear processors. The first of such parallel-computing FDTD implementations for EM purposes was carried out on high-performance supercomputers, which have the draw-back of being very expensive. Other alternatives include the use of computer clusters, which lead to strong bandwidth requirements. More recently, there is a growing inter-est in the implementation of FDTD algorithms on graphical processing units (GPUs). Indeed, it has been shown that today GPUs are powerful enough for this purpose, and their prices have significantly decreased since their conception. For instance, the Computed Unified Device Architecture from the NVIDIA Corporation provides a C language-based interface that allows an easy access to the parallel capabilities of the underlying hardware. Furthermore, advanced cards including more than one GPU that are dedicated to the tasks of intense scientific computing are now available. However, the gain in performance is very implementation dependent, and it is therefore important, when implementing the FDTD algorithm on a GPU, to carefully distribute the tasks among the multiprocessors.

- *Low frequency simulation*: Another obvious approach to reduce the computation time is to discretize the environment with a larger grid than strictly required. This is equivalent to performing the simulations at a lower frequency than the frequency of the physi-cal system. Of course, this low frequency simulation approach leads to approximations. However, although this may seem to lead to unphysical results such as spurious waveg-uiding effects in the simulation of narrow streets, accuracy losses can be compensated for by calibrating the electrical properties of the materials used in the simulation. In fact, it is shown that a when using the lower frequency approximation and a proper calibration, only diffraction phenomena are still subject to some attenuation errors. Fur-thermore, it was also shown in [9] that reducing the simulation frequency by a factor of ten with respect to the physical frequency has a negligible effect on the error of the lower frequency simulation as long as the material properties are carefully calibrated. However, very low simulation frequencies will lead to substantial accuracy degradations owing to the low spatial resolution that decreases the accuracy of the scenario.

In practical indoor environments, the lower frequency approximation is often a simple yet effective approach for estimating the coverage when the hardware limitations do not allow the simulation of a large scenario or when inaccuracies in floor plan, wall types, and other important building characteristics compromise the accuracy of more precise computations. However it is to be remembered that simulating propagation at a lower frequency is an approximation, which requires, in order to be effective, doing a proper calibration of the model.

11.3.3 Calibration

Because FD methods solve Maxwell's equations, they are very accurate only if the physical properties of the materials are known, which translates into assigning proper coefficients to the pixels in the matrix representing the scenario. Hence FD simulations usually require calibrating many parameters. In fact each material is characterized by numerous coefficients. For instance ParFlow has two parameters (refraction and attenua-tion which is related to permittivity) per material. It is usually hard to find accurate values for these parameters due to the following reasons:

- In a practical scenario the constituting materials of a wall may not be perfectly known.
- The width of the walls is usually not known.
- The database of the scenario may not be up to date.
- When simulating a scenario some elements like furniture or humans are usually not included.

A solution to overcome this issue is to calibrate the parameters of the materials based on on-site measurements. In practice a cost function which minimizes the difference between the simulation and the measurements is used. Then, during the optimization process the algorithm tries many possible values for each parameter. The most common cost function for this purpose is the Root Mean Square Error (RMSE) defined as:

$$RMSE = \sqrt{\frac{1}{N} \sum_{p=0}^{N-1} |P_{PRED}(p) - P_{MES}(p)|^2} \qquad (11.8)$$

where N is the number of measured points, $P_{PRED}(p)$ is the predicted received power in dBm at point p and $P_{SIM}(p)$ the measured value.

Higher accuracy is expected when N is higher. The minimization problem is a hard optimization problem. First, it is not convex and presents several local minima. Second, the computational load to evaluate one set of parameters is high. Since generally an exhaustive search of all possible settings is not affordable, a heuristic approach adapted to this problem must be used. Because the search space is continuous and the derivatives of the cost function cannot be analytically assessed, gradient based methods are excluded. In practice techniques like simulated annealing and Tabu search are shown to perform well. These methods are efficient to avoid local minima. The algorithm can be stopped when an acceptable solution is reached or when a maximal number of iterations has been reached. It is to be noted that the optimal solution is often valid only for the environment under test and for the configuration used for calibration. For instance in a given building, after calibration of all the materials, it is usually possible to reach an accuracy of 3 to 5 dB between simulations and predictions, which can be helpful when studying the performance of wireless networks in this specific scenario.

11.3.4 Antenna Pattern Effects

Although omnidirectional antennas like dipoles or whips are widely used for indoor access points such as Wireless Fidelity (WiFi), directional antennas are sometimes preferred to increase the range. For instance relays which cover point to point communications or to extend range (see Chapter 1) may be extensively used in Long Term Evolution Advanced (LTE-A) and beyond. Simulating such directive sources requires being able to accurately take into account the antenna pattern of the BS. Including a radiation pattern in ray tracing is easier than in FDPF approaches where point sources are essentially omnidirectional. This issue can be overcome by radiation pattern synthesis exploiting arrays of point sources, in both receiving and transmitting modes. This approach was proposed by Villemaud et al. and is detailed in [10]. In this chapter a directive antenna is simulated using 3×3 pixels sources spaced by $\Delta = \lambda/6$. Then an optimization method was

Figure 11.2 Antenna pattern simultion, onmidirectional source (a), and directional sector antenna (b) obtained by combination of 3 × 3 sources.

implemented so that the powers (amplitude and phase) to be applied to each source were properly allocated. The method was evaluated with indoor access points, showing a 1.4 dB improvement in the accuracy of the radio coverage. As an illustration, an example of typical results for a 10 × 10 m scenario at the 1.9 GHz frequency is plotted in Figure 11.2. It is seen that, even if not as simple as compared to ray tracing, sector antennas can be efficiently simulated using FD models.

11.3.5 3D Approximation

In many scenarios, it is necessary to perform 3D simulations. For instance, in urban scenarios where antennas are located on the roofs of the buildings and where end users are moving in the streets, a flat $2D$ FD simulation can not provide acceptable results. The 3D implementation of ray tracing models is quite straightforward [11] and even if it leads to high computational load, restricting the number of reflections to typically two or three reflections can provide good results in a short time. There is not such flexibility in FD models to reduce complexity. Therefore, even if techniques from Section 11.3.2 can be used, they have their limitations when simulating large scenarios in full 3D. However, there has been some attempts to use 3D FD models for radio coverage predictions, and it is expected that they will be more and more common due to the hardware constraints being less and less an issue.

In [12] a full 3D ParFlow method was implemented. However, due to complexity reasons its application is limited to simulating only one room of a building, making it not really suitable for larger scenarios like LTE advanced. In [13], Austin proposes using a 3D FDTD in a whole building (18 × 18 × 9 m) using the frequency of 1 GHz and a spatial step $\Delta = 1$ cm. This approach seems to be very interesting for simulating wireless networks because having a full 3D radio coverage of a building at a 1 cm resolution can help to identify very fine propagation effects. However, their implementation based on a 64-node computer cluster (using Intel Xeon 2.66 GHz processors) still requires 48 hours to reach the steady state, which may not be acceptable for practical application.

Hence, due to the very high complexity, FD models which approximate 3D propagation may be an interesting alternative. This kind of model, also referred to as 2.5D models, can only be applied in flat scenarios where main propagation phenomenas occur in the horizontal plane. This is typically the case for multi-floor buildings, where the height of each floor (2 to 3 meters) is very small compared to the other dimensions. 2.5D models usually compute the propagation at each floor using a 2D method. Then the 2D radio coverages are projected among floors using different techniques. In [13], a 2.5D FDTD approach is proposed. This simplified model is based on the dominant propagation mechanisms and has shown good agreement against experimental measurements. In [14] a 2.5 ParFlow model is presented. In order to develop the model three different projection techniques were proposed and are summarized here:

- *Direct field projection*: This approach is the easiest and fastest one. The true field propagation is computed for the source floor only with the 2D standard approach. It thus exploits the knowledge of the walls of the source floor only. Starting from the source floor coverage prediction, the coverage at the n + 1 floor is computed at each pixel by applying a constant ceil attenuation factor.
- *Direct source projection*: Another simple approach tackles the multi-floored problem by firstly computing an equivalent source at each floor. More precisely, a constant attenuation factor is applied to the source amplitude for each virtual source as a function of the number of cells between the virtual and the real sources. The penetration through ceils is thus taken into account only at the source coordinates. Secondly each virtual source is propagated in each floor, no matter what happens on other floors.
- *Combined flows projection and propagation*: Because both previous approaches cannot deal with the obstacles at all floors, a combined approach is alternatively proposed. The steady-state and associated FDPF flows are computed at the source floor firstly. Instead of computing the mean power, exact flows are stored and are used as source flows at the receiver floor, after applying a constant attenuation factor. All of these source flows are then propagated at the receiver floor, leading to a new coverage map.

Based on experimental measurements in a three-floor building, it was shown that the third approach (combined flows) leads to a RMSE of 4.4 dB between measurements and simulation, which is in the same order of accuracy as the full 3D approach. This demonstrates that in multi-floor buildings 2.5D methods can be sufficient. The two other techniques, that is, direct field projection and direct source projection, obtained an accuracy of 8.2 dB and 6.1 dB respectively. The results using the combined flows approach are plotted in Figure 11.3. If 2.5 methods appear as a good compromise for multi floor buildings, there is no doubt that full 3D methods are needed for large open space buildings like airports, stations, and so on, where the 2,5D projection technique cannot be applied.

11.4   Results

11.4.1   Path Loss Prediction

In this section, MR-FDPF method has been firstly assessed with calibrated radio measurements in harmonic mode. The measurement platform consisted of an arbitrary waveform

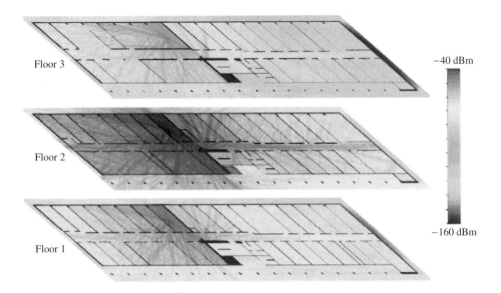

Figure 11.3   2.5D Radio coverage of one antenna located at floor 2.

generator (ESG4438C of Agilent Technology c) and a Vector Signal Analyser (VSA 89641 of Agilent Technology c), both equipped with 2.4 GHz, 4 dBi, omnidirectional antennas. A pure sinusoidal waveform was generated and the span of the receiver was fixed at 2 kHz allowing measurements from .40 dBm down to .110 dBm. For each referred position (cf. red crosses on Figure 11.4), about 200 measurement samples were collected during 60s with slow displacements and rotations to obtain a mean value compensated for flat fading. The source was located at floor 2, in room E7. Measurements were taken at 80 locations at the source floor and respectively at 11 and 15 locations for the upper and lower floors. Each floor is made of three different types of walls: *concrete walls* for the main load-bearing walls, *plaster walls* corresponding to standard walls between rooms and *glazed walls* corresponding to the external walls made of windows. So three couples (n, α) were optimised during the calibration process. The test building was made of three floors. The source was located in room E7 at intermediate level. Measurement points are indicated by red crosses with an attenuation coefficient of walls having a low impact on

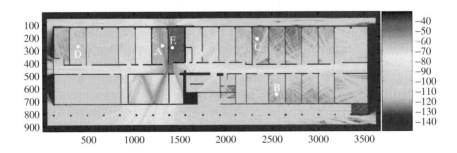

Figure 11.4   Computed signal power (dBm), and fading measurement points.

Table 11.3 k parameter estimated values

Scenario	Size	Frequency	Simulation time	Accuracy
Indoor	$20 \times 80\,\text{m}$	2 GHz	6.5 s	3.88 dB
Outdoor	$3.4 \times 2.4\,\text{km}$	900 MHz	3 min	8 dB

the simulations. This result can be assigned to the thin thickness of walls in comparison to the simulation wavelength. On the contrary, the air attenuation factor impacts strongly on the predictions because it modifies the attenuation law. However, the best predictions were obtained with $\alpha_{air} = 1.0$. This corresponds to a logarithmic attenuation corresponding to a 2D free-space propagation model. This can be charged to the fact that the transmitter was situated in a regular multifloor building in which waves are bounded by the floor and the ceiling. After calibration, we found the following values for parameter n: 5.4 for concrete walls, 2.4 for plaster walls and 1.3 for glazed walls. Only slight variations were observed as a function of the resolution step. These values were eventually close to experimental values provided in other papers.

In [15] the same method was applied to an urban scenario based on the Global System for Mobile communication (GSM) measurement campaign in Munich. The accuracy was of course not so high compared to indoor environment, because this time the low frequency approximation was used and a spatial step of 1 m had to be used for memory reduction reasons.

The performance of the model for both *indoor* and *outdoor* scenarios are summarized in Table 11.3.

11.4.2 Fading Prediction

In this section, we are interested with the local fading strength estimation. Indeed, if FD methods simulate Path Loss well, it is also interesting to investigate to simulate fading. Indeed, due to the high accuracy and resolution (few centimeters), fast fading due to multi path could also be investigated. For this purpose, the power distribution from homogeneous regions was analyzed and fitted with a Rice's law, the k parameter indicating the fading strength. $k < 1$ indicates a Rayleigh like channel while $k > 10$ an AWGN channel. Figure 11.4 represents the radio coverage of an access point (Point E on the figure, emitted power = 17 dBm), from blue (-140 dBm) to red (-40 dBm) computed with our simulator. We see that the simulator we developed provides us with high resolution. In this chapter, we are interested with the local fading strength estimation. For this purpose the power distribution from homogeneous regions is analyzed and fitted with a Rice's law, the k parameter indicating the fading strength. $k < 1$ indicates a Rayleigh like channel while $k > 10$ an AWGN channel.

11.4.2.1 Indoor Channel Measurements

The measurement platform consists of an arbitrary waveform generator (ESG4438C of Agilent Technology ©) and a vector signal analyzer (VSA 89641 of Agilent Technology ©), both equipped with 2,4 GHz, 4 dBi, omnidirectional antennas.

For each referred position in the same test environment, about 200 measurement samples were collected during 60 s with slow displacements and rotations. From these measurements, the experimental pow r distribution associated with each referred position is achieved. These distributions are then fitted with the Rice's laws and the k parameter is obtained.

11.4.2.2   Estimation of k Parameter

Rician model estimation of k parameter has been described in [16]. The k parameter Rice law supposes that a dominant direct path exists between the emitter and the receiver. In this case k parameter corresponds to the following ratio:

$$k = \frac{s^2}{2.\sigma^2} \tag{11.9}$$

with s the magnitude of the direct signal and σ^2 the variance of the received signal amplitude. When there is no dominant path, s is null and the power signal distribution follows a Rayleigh law.

k parameter can be computed from measurement or simulation by:

$$k = \frac{\sqrt{1 - \frac{\sigma^2}{\omega^2}}}{1 - \sqrt{1 - \frac{\sigma^2}{\omega^2}}} \tag{11.10}$$

with $\omega = E(R^2)$ and R the magnitude of the received signals. The k parameters estimated from simulation and measurements can then be used for fading strength comparison.

11.4.2.3   Results

To compute the simulated distributions, square zones from 30×30 pixels, that is, 60×60 centimeters are chosen around the considered point. Then the distribution of the points in this area is considered.

To present the results, two points (A and B on Figure 11.4) have been chosen and analyzed. The coverage in dBm of these areas are represented in Figure 11.5. Measured distributions are built using the measurement samples collected during the measure.

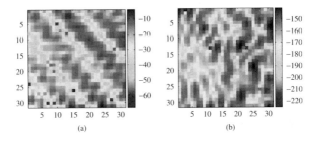

Figure 11.5   Simulated power in dBm centered on point A (a) and point B (b).

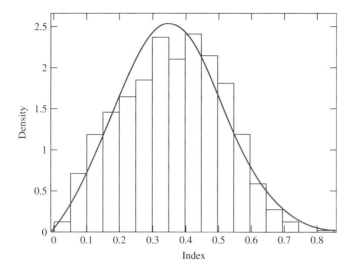

Figure 11.6   Power signal distribution on simulated point A.

Point A is located in the room next to the emitter. The simulated coverage (Figure 11.6) and the measurement (Figure 11.7) have been found to best fit to a Rice law. The k parameter has been computed using Equation 11.9. Measured values for point A where 1.689 for the simulation and 1.697 for the measurements.

We compared the simulated and measured power signal repartitions and the k values were respectively 0.624 and 0.482. Is this case the fading is more Rayleigh like.

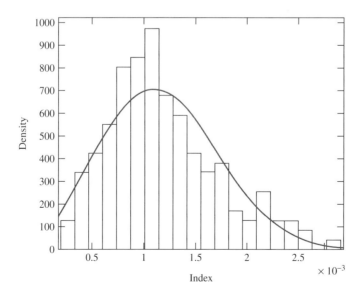

Figure 11.7   Power signal distribution on measured point A.

Table 11.4 k parameter estimated values

point	A	B	C	D
measured	1.697	0.624	0.396	1.55
simulated	1.689	0.482	0.533	0.356

Other measurements (point B, C and D) have been conducted to verify that simulation fits the measurements. They are presented in Table 11.4.

It is to be noted that there is a high level of variation in the fading phenomenas, so it is difficult to give exact values for parameter k. To see if the differences between simulated and measured k parameters were significant or not we represented the theoretical Rice functions corresponding to the parameters obtained in Table 11.3.

We see that except for point D where there is a difference, the distributions are approximately the same. This is a good match, indicating the simulator predicts well the fast fading statistics at different positions in the environment.

11.5 Perspectives for Finite Difference Models

11.5.1 Extension to 3D Models

There has been a few attempts to implement full 3D versions of FD models. If the implementation itself is quite straightforward and easy to perform (the main concern is to add the two directions corresponding to the vertical axis), it usually leads to very high memory requirements due to the size of the problem [12]. Most papers using FDTD or ParFlow are able to simulate propagation inside buildings with good accuracy. Figure 11.8 illustrates the results obtained with the MR-FDPF method in a ($18 \times 18 \times 9$ m) building (same scenario as Section 11.3.5). During the preprocessing the 3D multi resolution nodes are computed (preprocessing time is 14 s at dr $= 50$ cm). The homogeneous blocks are the larger multi resolution nodes made of one material only (see Figure 11.8.a). The time to compute a source propagation was 30 s. An horizontal slice of the computed coverage area is represented on Figure 11.8(b).

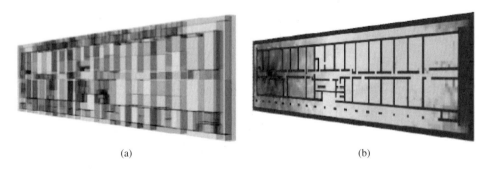

(a)   (b)

Figure 11.8 3D Propagation in one building based on MR-FDPF method. The homogeneous multi resolution nodes are computed during the preprocessing phase (a). The Horizontal slice of the coverage area, from -40 dBm to -100 dBm, led to an accuracy of 4.5 dB after calibration.

After calibration a RMSE of 4.5 dB was obtained between measurements and simulation, which is similar to the performance of the 2.5D model in Section 11.3.5. However, there has still been no attempt to use 3D FD models in large scenarios. Having the full wave propagation in 3D for scenarios like dense urban could be very interesting in very dense scenarios where the number of reflections would be too high for ray tracing and where a high accuracy would be expected. Moreover, simulating in 3D would be helpful to study fading as well and see the impact of height where receivers/emitters are located.

We expect that with increasing performance of hardware, full 3D FD models will be used in the near future for simulating large scenarios [17]. In the mean time, it would be helpful to propose simplified methods. One approach could be to use the irregular spatial grid. Indeed, FDTD using an irregular mesh is commonly used. With such an approach it would be possible to use a large spatial step in areas where the scenarios is simple (e.g., open space), whereas complex areas would use a fine spatial step (e.g., buildings with irregular shape).

Nevertheless, nowadays full 3D FD models for outdoor simulations are not ready yet. That is why a possible technique to overcome this drawback is to combine them with ray tracing as will be discussed next.

11.5.2   Combination with Ray Tracing Models

Geometric based models like ray tracing and ray launching are described in the previous chapter. Finite difference models are very different and both of them have advantages and drawbacks. Comparisons between them are given in [8]. In the following, the main criteria are compared:

- Complexity: For FDTD it depends mainly on the size of the scenario, whereas for RO it depends mainly on the number of walls.
- Accuracy: FDTD is in general more accurate because the number of reflections is not limited unlike RO.
- 3D extension: RO is in general less computationaly demanding than FDTD, that is why a 3D version of the model is easier to implement.

In the literature, combined models, also referred to as hybrid models, have been proposed [18–20], where RO and FDTD models are combined to take advantage of the properties of each model. Thus, in our chapter, taking into consideration the previous properties, it appears as a good choice to combine two models and choose between them depending on the scenarios:

- *Indoors*: The scenario is not very large, and made of numerous walls, that is why the number of reflections is very high. Moreover, in multi floor buildings, the scenario at each floor is quite flat that is, a 2D approximation of the propagation is not a bad assumption. Hence in this case the 2D FDTD model is a good option.
- *Outdoors*: The environment is large and propagation can not be easily approximated with a 2D model, in particular in scenarios with high buildings and antennas located on the roofs. Furthermore, there are more open space areas and the number of reflections to compute is smaller than indoors. In such a scenario 3D RT is preferred.

IRLA (Intelligent Ray Launching) is described in [21]. It is a full 3D ray launching especially developped for urban network planning. In this model, the buildings are approximated with a 2.5D database (representing the shape of the buildings and their heights). IRLA is based on a discretization of the environment into cubes, in order to reduce the number of reflections and diffractions to compute. Optimizations to avoid missing rays are also implemented in this model [22].

The new model we propose combines *IRLA* for the outdoor signal prediction with *MR-FDPF* for the indoor part. A great advantage of the models we use is that they are both based on a discrete resolution of the environment for the following reasons:

- *MR-FDPF*, as a FDTD-like model, solves the Maxwell's equations on a 2D grid.
- *IRLA* divides the environment into cubes for complexity reduction.

Hence, the main idea of the combined approach is to find how to link the two models, that is, how to use the *IRLA* 3D outdoor radio coverage as an input for the 2D indoor *MR-FDPF* simulation. The method is illustrated in Figure 11.9 and can be divided into the following steps:

- Run the *IRLA* prediction (Outdoor ray launching) of the emitter.
- Compute equivalent *MR-FDPF* source flows on the borders of the building, by summing the rays arriving at each cube on the borders of the indoor floor.
- Run the indoor *MR-FDPF* using the new equivalent sources as incoming flows of the bottom-up-down approach [6].
- Combine IRLA/MR-FDPF maps to plot both the outdoor and indoor coverages.

The scenario for the initial evaluation of the model is the *INSA* university campus in Lyon, France (see Figure 11.10). The directive antenna (E on Figure 11.10) was placed

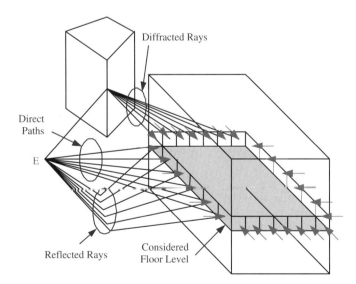

Figure 11.9 Schematic representation of the combined approach. First the outdoor part is simulated, then the incoming indoor flows are computed and used for the indoor simulation.

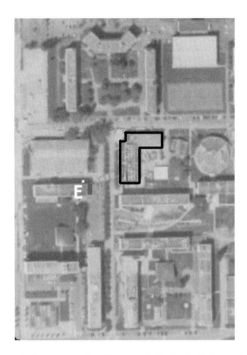

Figure 11.10 Outdoor to Indoor scenario. Surrounded: the building where the indoor measurements were performed. E: represents the position of the emitter.

on a window in one building and was pointing in the direction of the *CITI* building (see on Figure 11.10) where the indoor measurements have been performed.

The equipment for the measurements is detailed below:

- The emitter was an Agilent signal generator, with ouput power equal to 0 dBm and sending a 3.5 GHz frequency signal (frequency of WiMAX in Europe). The generator was located in the street and its directive antenna was at 3 m from street level, focusing its main beam in the direction of the building.
- The measurements were performed both indoor and outdoors (32 indoors and 72 outdoors) using a Signal Analyser equipped with an omnidirectional antenna. Measurements were performed at 1.5 m from floor level.

In order to avoid fading effects, for each point the mean value after a 20 second time average was recorded.

After the indoor and outdoor measurements were loaded, a calibration of the tool has been performed. First *IRLA* has been calibrated using a simulated annealing approach. As an illustration, the rays of the outdoor IRLA computation are plotted in Figure 11.11.

Using the rays on the borders of the *CITI* building (see Figure 11.10), the incoming virtual sources are computed using the combined approach and the combined approach was run. Finally the Indoor part of the signal was also been calibrated, in particular because the properties of the walls and the windows were not known.

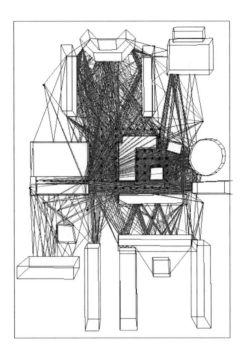

Figure 11.11 Outdoor reflections and diffractions rays computed with IRLA model.

In Figure 11.12 the simulated signal inside the *CITI* building is plotted. It is verified in this figure that the effect of the windows are very well taken into account.

In order to evaluate the accuracy of the model, the *RMSE* (Root Mean Square Error) was used.

The performance of the model has been summarized in Table 11.5, where the results concerning respectively the outdoor measurements, the indoor measurements, and all the measurements are given.

It is important to notice that, without combining MR-FDPF with Intelligent Ray Launching (IRLA), it would not have been possible to compute the whole scenario with MR-FDPF only, due to high memory requirements during the preprocessing step. However, by supposing that this amount of memory is large enough, it is then possible to interpolate the simulation time duration it would take for simulating the whole scenario with MR-FDPF. Indeed, and as detailed in [6], the complexity of the propagation phase of MR-FDPF varies in $O(log_2(N) \cdot N^2)$, where N is the smallest dimension of the scenario in pixels. Thus a simulation of the full environment (560 meters large) at the same resolution would be $log_2(560/100) \cdot (560/100)^2$, that is, 78 times slower, that is, it would take approximately 2.5 hours instead of less than 2 minutes with the proposed combined model. Furthermore, such simulation would only show a 2D cut, where the height of the outdoor emitters would not be properly taken into account, hence it would provide a low accuracy, compared to the approach we use where the outdoor signal effects are simulated in 3D. Consequently, the new model proposed in this chapter is advantageous both in terms of speed and accuracy.

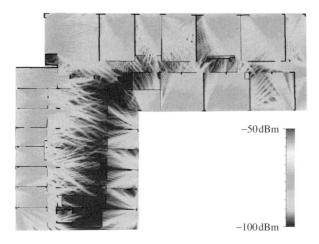

−50 dBm

−100 dBm

Figure 11.12 The final indoor radio coverage. The effects of signal penetration through windows are easily seen.

Table 11.5 Performance of the combined model

X	Outdoor points	Indoor points	All the points
Number of points	72	32	104
Pre-processing	0 s	41 s	41 s
Simulation	58 s	57 s	115 s
RMSE	7.9 dB	2.4 dB	6.2 dB

Measurements were also performed at other locations, and the details can be found in [23]. It seems that combining FD models with ray tracing is an interesting approach and more studies at other frequencies and scenarios would be interesting. Moreover, using such an approach can also be performed for indoor and outdoor. For instance, in the work of Umansky [24] similar models are used for indoor to outdoor predictions. In this case it is a bit more complicated because a technique needs to be implemented which can extract from the FD simulations the directions of the geometrical rays to compute. Therefore in this model it is proposed to use Space-Alternating Generalized Expectation Maximization (SAGE) for this purpose: after simulating the indoor radio coverage, the SAGE algorithm was applied to the fields on the borders of this building [24], and the directions and powers of most significant rays can be computed from their phases. Then, in a second stage the outdoor rays were computed.

It is to be noted that the previous technique, which combines ray tracing and FDTD does not consider the signal that goes from outside to inside multiple times. For instance we could imagine a scenario where the indoor emitter sends a signal outside, which is reflected on another building and comes back inside the building where the emitter is located. In such cases the outdoor-to-indoor or outdoor-to-indoor should be applied multiple times, which would increase the complexity. Hence we believe there is still some research to be done on how to combine the different approaches.

11.5.3   Application to Wideband Channel Modeling

As explained in Chapter 1, future wireless networks tend to use larger frequency bands (like Carrier Aggregation or Wideband techniques). FDTD-based attempts at characterizing wide-band channels have already been done in the past. The objective of such simulations is to predict the channel impulse response in a given propagation scenario. This can be done by exciting the grid with a source that numerically generates a wideband pulse (e.g., a Gaussian pulse). As this pulse interacts with the environment through reflections and other phenomena, the received numerical fields can be recorded at different locations. The recent work from Valcarce [25] focuses on these aspects, that is, how to design sources considering the fact that the FDTD grid has a dispersive nature which leads to errors in the simulation of wide-band pulses. Indeed, the dissimilarities between the simulation and physical reality increase with lower spatial grid resolution. Hence, reliable wide-band FDTD simulations require a large spatial resolution that gives rise to enormous computational requirements (memory and computational power). Such errors have been studied [25, 26], but there is still some research to be done in order to be able to simulate wideband channels at low computational cost.

11.6   Summary and Perspectives

Due to the increasing performance of computers, FD models appear as a good alternative to the popular ray tracing models. In particular we have demonstrated in this chapter that after proper calibration they can lead to a good accuracy for both path loss estimation and fading estimation. There is still a lot of research to be done in this area. For instance, it will be helpful to propose new simplified models for 3D and wideband scenarios. Furthermore, due to the advantages they offer, we believe FD will play a role in the range of possible solutions for simulating future generation networks.

Acknowledgements

Most of the results presented in this chapter were part of my PhD work as well as the IPLAN project at the University of Bedfordshire. Special thanks to all the colleagues at CITI/INSA and CWIND for their support and more especially: Jean-Marie Gorce, Jie Zhang, Guillaume Villemaud, Zhihua Lai, Dmitry Umansky and Meiling Luo. A web page with more details about the tool based on ParFlow can be found on this link [27].

References

1. A. Valcarce, G. De La Roche, A. Juttner, D. Lopez-Perez, and J. Zhang, "Applying FDTD to the coverage prediction of wimax femtocells," *EURASIP Journal of Wireless Communications and Networking*, vol. ID 308606, 2009.

2. J. Wagen, "Indoor service coverage predictions: How good is good enough?," in *2010 Proceedings of the Fourth European Conference on Antennas and Propagation (EuCAP)*, April 2010.

3. P. Zakharov, E. Mikhailov, A. Potapov, A. Korolev, and A. Sukhorukov, "Comparative Analysis of Ray tracing, Finite Integration Technique and Empirical Models using Ultra-Detailed Indoor Environment Model and Measurements," in *3rd IEEE Microwave, Antenna, Propagation and EMC Technologies for Wireless Communications*, October 2009.

4. K. Yee, "Numerical solution of inital boundary value problems involving Maxwell146s equations in isotropic media," *IEEE Transactions on Antennas and Propagation*, vol. 14, pp. 302–307, 1966.

5. B. Chopard, P. O. Lüthi, and J.-F. Wagen, "Multi-cell coverage predictions: a massively parallel approach based on the ParFlow method," in *IEEE International Symposium on Personal, Indoor and Mobile Radio Communications*, vol. 1, Boston, MA, USA, Sep. 1998–, pp. 60–64.

6. J.-M. Gorce, K. Jaffres-Runser, and G. de la Roche, "Deterministic approach for fast simulations of indoor radio wave propagation," *IEEE Transactions on Antennas and Propagation*, vol. 55, pp. 938–942, March 2007.

7. G. de la Roche, J.-F. Wagen, G. Villemaud, J.-M. Gorce, and J. Zhang, "Comparison between Two Implementations of ParFlow for Simulating Femtocell Networks," in *20th International Conference on Computer Communications and Networks (ICCCN)*, August 2011.

8. L. Nagy, R. Dady, and A. Farkasvolgyi, "Algorithmic complexity of FDTD and ray tracing method for indoor propagation modelling," in *The Third European Conference On Antennas and Propagation*, Berlin, Germany, 3 2009.

9. A. Valcarce, G. D. L. Roche, and J. Zhang, "On the use of a lower frequency in finite difference simulations for urban radio coverage," in *IEEE Vehicular Technology Conference*, Marina Bay, Singapore, 5 2008, pp. 270–274.

10. G. Villemaud, G. D. la Roche, and J. M. Gorce, "Accuracy Enhancement of a Multi-Resolution Indoor Propagation Simulation Tool by Radiation Pattern Synthesis," in *IEEE AP-S International Symposium*, Albuquerque, New Mexico, July 2006.

11. Y. Corre and Y. Lostanlen, "3D urban propagation model for large ray-tracing computation," in *International Conference On Electromagnetics In Advanced Applications*, Torino, Italy, September 2007.

12. G. de la Roche, J.-M. Gorce, and J. Zhang, "Optimized implementation of the 3D MR-FDPF method for Indoor radio propagation predictions," in *European Conference on Antennas and Propagation (EuCAP 2009)*, Berlin, Germany, March 2009.

13. A. Austin, M. Neve, and G. Rowe, "Modelling propagation in multi-floor buildings using the fdtd method," *IEEE Transactions on Antennas and Propagation*, vol. 59, no. 11, pp. 4239–4246, November 2011.

14. G. de la Roche, X. Gallon, J.-M. Gorce, and G. Villemaud, "2.5D extensions of the Frequency Domain ParFlow Algorithm for Simulating 802.11b/g Radio Coverage in multifloored buildings," in *Vehicular Technology Conference Fall (VTC-Fall 2006)*, Montreal, Canada, September 2006.

15. G. D. L. Roche, G. Villemaud, and J.-M. Gorce, "Efficient Finite Difference Method for Simulating Radio Propagation in dense urban environments," in *European Conference on Antennas and Propagation (EuCAP 2007)*, Edinburgh, UK, November 2007.

16. A. Abdi, C. Tepedelenlioglu, M. Kaveh, and G. Giannakis, "On the estimation of the k parameter for the rice fading distribution," *Proc. IEEE*, vol. 5, no. 3, pp. 92–94, March 2001.

17. A. Valcarce, G. de la Roche, L. Nagy, J.-F. Wagen, and J.-M. Gorce, "A New Trend in Propagation Prediction," *IEEE Vehicular Technology Magazine*, vol. 6, pp. 73–81, 2011.

18. Y. Wang, S. Safavi-Naeini, and S. K. Chaudhuri, "A hybrid technique based on combining ray tracing and fdtd methods for site-specific modeling of indoor radio wave propagation," *IEEE Transaction on Antennas and Propagation*, vol. 48, no. 5, pp. 743–754, May 2000.

19. S. Reynaud, A. Reineix, C. Guiffaut, and R. Vauzelle, "Modeling indoor propagation using an indirect hybrid method combining the utd and the fdtd methods." in *7th European Conference on Wireless Technology*, Amsterdam, October 2004.

20. M. Thiel and K. Sarabandi, "3D-Wave Propagation Analysis of Indoor Wireless Channels Utilizing Hybrid Methods," *IEEE Transactions on Antennas and Propagation*, vol. 57, pp. 1539–1546, 2009.

21. Z. Lai, N. Bessis, G. de la Roche, H. Song, J. Zhang, and G. Clapworthy, "An intelligent ray launching for urban propagation prediction," in *The Third European Conference On Antennas and Propagation (EuCAP)*, Berlin, Germany, 3 2009.

22. Z. Lai, N. Bessis, G. de la Roche, H. Song, J. Zhang, and G. Clapworthy, "A new approach to solve angular dispersion of discrete ray launching for urban scenarios," in *Loughborough Antennas and Propagation Conference (LAPC)*, Loughborough, UK, November 2009.

23. G. de la Roche, P. Flipo, Z. Lai, G. Villemaud, J. Zhang, and J.-M. Gorce, "Implementation and Validation of a New Combined Model for Outdoor to Indoor Radio Coverage Predictions," *EURASIP Journal of Wireless Communications and Networking*, vol. ID 215352, 2010.

24. D. Umansky, G. de la Roche, Z. Lai, G. Villemaud, J.-M. Gorce, and J. Zhang, "A New Deterministic Hybrid Model for Indoor-to-Outdoor Radio Coverage Prediction," in *European Conference on Antennas and Propagation (EuCAP 2011)*, Rome, Italy, April 2011.

25. A. Valcarce, H. Song, and J. Zhang, " On the design of pulsed sources and spread compensation in ?nite-difference time-domain electromagnetic simulations," *IEEE Transactions on Microwave Theory and Techniques*, vol. 58, pp. 2838–2849, 2010.

26. A. Alighanbari and C. Sarris, "Parallel time-domain full-wave analysis and system-level modeling of ultrawideband indoor communication systems," *IEEE Transactions on Antennas and Propagation*, vol. 57, no. 1, pp. 4239–4246, January 2009.

27. "Wiplan propagation tool," in *INRIA ARES/CITI Laboratory, INSA Lyon, France*, http://wiplan.citi.insa-lyon.fr.

12

Propagation Models for Wireless Network Planning

Thomas Kürner[1] and Yves Lostanlen[2]
[1] *Technische Universität Braunschweig, Germany*
[2] *University of Toronto, Canada*

The initial set-up, the extension and the optimization of a cellular mobile radio network still has to rely to a large extent on simulations. Based on the results of these simulations, decisions on the number and locations of sites have been taken. These decisions are triggering investments partly in the order of hundreds of millions of Euros. In subsequent planning steps the network configurations have to be determined. The way these tasks are done has a strong influence on the network quality. Although with the introduction of self-organizing network features in Long Term Evolution (LTE) [1] these configurations can be determined partly by the network itself, the need for supporting off-line simulations will still exist for a couple of years. Radio wave propagation models are the indispensable components in radio network simulations and the quality of the simulation's output critically depends on the accuracy of radio wave propagation models. This chapter provides a brief introduction into propagation models used for Radio Network Planning (RNP) and optimization, discusses the required input data, the accuracy and limits of the models and shows how propagation models can be applied to determine cell ranges and coverage probabilities.

12.1 Geographic Data for RNP

Radio Planning Tool (RPT) makes use of digital descriptions of the geographical environment. This information allows the transmitters to be placed at desired locations on the digital map, and to visualize the areas in which the signal reception can be observed. Those geographical data also contain crucial information to predict the radio-wave propagation in the area. Following a definition of usual terms, a brief overview of the production

LTE-Advanced and Next Generation Wireless Networks: Channel Modelling and Propagation, First Edition.
Edited by Guillaume de la Roche, Andrés Alayón Glazunov and Ben Allen.
© 2013 John Wiley & Sons, Ltd. Published 2013 by John Wiley & Sons, Ltd.

methods is given, then various types of data are described, and finally advanced aspects such as the multi-resolution and related transactions are detailed.

12.1.1 Terminology

Let us define some common terms in terrain modeling.

- Geo data layer: A geographical map data is composed of a set of distinct layers; each layer contains the description of one terrain characteristic such as a geographical obstacle (also called clutter or land usage) type, clutter height above ground, and ground altitude (or terrain elevation).
- Raster layer or Pixel layer: A pixel matrix (aka raster) layer is a collection of geo-localized rectangular pixel matrices; the value at each pixel may represent at one precise location a clutter type in a Digital Land Usage (DLU), a clutter height in a Digital Height Model (DHM), or an altitude in a Digital Terrain Model (DTM) or Digital Elevation Model (DEM).
- Mono-resolution: When a raster layer is composed of matrices with a unique pixel resolution.
- Multi-resolution: When a raster layer contains matrices with distinct pixel resolutions.
- High Resolution (HR): A raster matrix with a resolution lower than 10 meters is generally referred to as HR matrix providing clear and precise contours of each single building, rows of trees or of small urban woods.
- Low Resolution (LR): A raster matrix with a resolution greater than 20 meters is generally referred to as LR matrix, for which only the contours of building blocks or large vegetation blocks are represented; Figure 12.1 illustrates the example of raster layers.
- Vector layer: A clutter type may be represented by closed horizontal polygons that are vector data; sometimes the vector layer contains only building data but actually all clutter types can be represented (building, vegetation, bridge, and water).
- 3D vector layer: A 3D vector layer contains attributes that give the clutter height associated to each above-mentioned vector polygon. Figure 12.2 gives an example of such data.

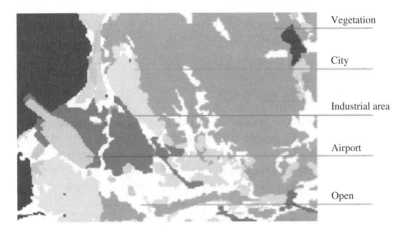

Figure 12.1   LR clutter type (or land usage).

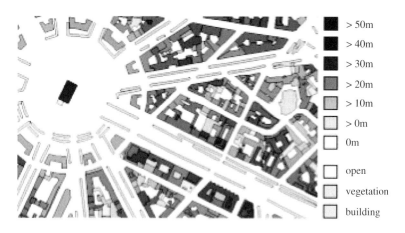

> 50m
> 40m
> 30m
> 20m
> 10m
> 0m
0m

open
vegetation
building

Figure 12.2   3D vector data.

12.1.2   Production Techniques

The production technique and source data depend on the required final resolution and accuracy, the considered environment and the type of data (raster, vector). The low resolution map data can be produced at a low cost thanks to satellite data (Landsat, Quickbird, Ikonos), and could be well suited for general purposes and in rural areas. However, the most challenging wireless network designs occur in suburban and urban areas, where the environment is more complex and the clutter heterogeneity larger. High resolution map data are the only possible input to achieve reliable engineering work in those zones. The progresses of the capturing sensors such as digital camera, LIght Detection And Ranging (LIDAR): a laser technique, and high resolution satellites have opened more ambitious objectives in terms of 3D urban digital mapping. The LIDAR (data still suffer from scattering effects of the lasers on the obstacle that blurs the final image, when a high precision is targeted. Today aerial imagery, and to some extent emerging satellite data, remain the best option to achieve optimal accuracies, and we shall only focus on those sources data in the remainder of the text. Combined with efficient algorithms, an automatic production of urban DTM, including buildings and vegetation, is possible. Yet, the automatic computation of 3D vector data is much more difficult because of low-contrasted building contours, hidden areas and complex-shaped buildings. Semi-automatic solutions have arisen consisting in using external 2D vector information and refined algorithms. However, for most applications requiring high quality vector data, a manual or a semi-automated process is necessary based on a photogrammetric approach.

The production of high quality digital 3D map data usually follows three main steps. First, pairs of overlapping stereo aerial (or satellite) pictures need to be well oriented thanks to analytical stereo plotters. The scale is usually between 1:10000 and 1:25000, the pixel size between 20 cm and 40 cm and a well-chosen overlapping of the stereo aerial pictures(enabling the 3D information). Then a photogrammetric analysis determines the contours observed in the images representing the various "objects". Different types of vector data are considered at this stage. Planimetric contours are 3D closed polygons representing the buildings, bridges, hydro and vegetation, whereas altimetry points and break-lines consist of 3D points and lines describing the terrain morphology (with increasing details when elevation varies) including also roads and railway tracks connected at the crossroads. The break-lines refine the object description and provide more robust and reliable digital descriptions of the environment. However, their capture is time-consuming and costly as it is done manually. A semi-automated approach partially releases this constraint as it requires a smaller number of break-lines to yield an equivalent accuracy. A set of heuristic manufacturing rules conditions the optimal production of the geographical data. The rules concern for example the minimum area to capture (e.g., 25 m), the minimum height difference to create an independent new object (e.g., 2 m). This essential photogrammetric operation yields the 3D vectors. Finally, raster data are computed by triangulating the altimetry vectors including the main roads, the railway tracks, and the hydro contours. A semi-automated algorithm makes use of both source images and vectors to create the raster data. To get reliable digital 3D map data, it is important to perform automatic analyses to guarantee the quality such as the consistence of the planimetry objects (closed contours, no self-intersection, well connected adjacent polygons). The final accuracy (actual 3D geographical location of the object) of the DTM is between 0.5 m and 1 m by using such a photogrammetric capture. The data set is then post-processed to be available at the required resolution (pixel size) that is generally 5 m in urban areas, 10 m – 25 m in extended urban areas and above 25 m in rural areas.

12.1.3 Specific Details Required for the Propagation Modeling

When using digital map data in the wireless network planning industry, it is important to consider specific land usage details and to pay attention to the transitions between heterogeneous data sets.

As transmitter antennas are located on building rooftops or podiums in urban areas (terrace or large balconies of a high building), the description of those must be detailed enough. If those details are not present, optimistic signal coverage will be predicted, because the shadowing and path loss will be under-estimated. It is useful to classify the buildings into different categories with distinct properties such as material type or construction type (modern glass building, historical stone wall building, and industrial building). Beside the propagation interaction properties that may differ according to the building type, the building penetration losses may also be distinguished. This allows a better accuracy of the outdoor-to-indoor (and vice-versa) prediction accuracy. The vegetation description may also play an important role in urban areas. Rows of trees will scatter and attenuate the signals. It is therefore wise to include 3D vegetation in addition to the 3D buildings in the raster and vector data sets. A seasonal effect may also be integrated by assigning different values for vegetation in winter and in summer accounting for the presence or not of leaves. In large cities, the bridges can have a different impact on the wave propagation modeling, depending on the signal reception location. In some cases, it is useful to consider the bridges as obstacles obstructing the signal (reception below the bridges or at street level), when in other situations, the bridges may be seen as road (reception above the bridge). This behavior may be taken into account either in the propagation model, or in the geographical map data by the assignment of different bridge properties.

In order to carry out efficient simulations in operational industrial contexts, trade-offs must be found between the level of details (nature and accuracy of the objects, number of useless diffracting edges, internal courts, and vertical details) describing the environment, the resolution of the raster matrices and the prediction computation time. This kind of accurate geographical map data may be expensive.

An efficient way to gather the relevant details for each environment consists in using a raster data set with various resolutions leading to a multi-resolution map data set. In order to ensure a smooth transition between the environments (to avoid the coverage ruptures), attention must be paid to the homogeneity of the data set (different land usage descriptions by different vendors, different production methods for LR and HR layers). The more advanced propagation models make the best use of multi-resolution geographical data set and will even complement the smoothness of the transition between two resolutions by ad hoc modeling techniques.

12.1.4   Raster Multi-Resolution

Most HR raster layers for clutter type or clutter height in urban environments are not available outside the city border. Thus, the HR layers are generally complemented by LR data that cover the whole radio deployment region. In Figure 12.3, the HR and LR clutter layers around the city are superimposed. Inside the city border, the clutter description is available in both resolutions, although the deterministic clutter height is only available in the HR layer. Outside the city border, only the LR clutter description is available. The clutter height is approximately known thanks to the mean height attributed to each clutter type.

A propagation model must be able to carry out calculations spread out over both HR and LR representations. The information along a single vertical profile must be extracted from HR layers inside the city border and from the LR layer outside. In areas where different resolutions are available, different strategies may be given to the prediction tool user depending on the wanted trade-off between computation time (use of LR data) and accuracy (use of HR data).

Figure 12.3   High-resolution clutter superimposed to the low-resolution clutter.

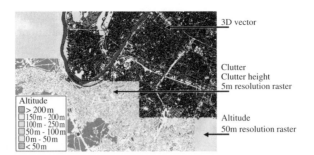

Figure 12.4 High-resolution land usage partly modeled by a 3D vector layer and partly by 5-meter raster layers; the ground altitude is available in a large 50-meter resolution layer.

12.1.5 Raster-Vector Multi-Resolution

Another situation commonly met in an operational terrain database is illustrated in Figure 12.4. The dense metropolitan area is modeled in a 3D vector layer, whereas the suburban region is represented in raster layers with 5-meter resolution. In this example, the clutter data is not available in the entire region, neither in the 3D vector layer nor in the raster layers. The ground altitude is given in a wide raster layer with 50-meter resolution.

To perform calculations on the whole area, the propagation model must have facilities to extract the vertical profile either from the raster layers or from the 3D vector layer. And the profiles must be composed of both types of data when the coverage is spread out over both representations.

In the region where raster and vector layers are both available, the extraction may be dependent on the user preferences. The user gives priority for clutter extraction (to have shorter computation times) or for vector extraction (to have better accuracy).

12.2 Categorization of Propagation Models

In principle investigations of the propagation phenomena can be done either experimentally based on measurements or derived from theory, see Figure 12.5. In the pre-GSM era mainly empirical propagation models have been derived based on measurements [2–5]. With the availability of digital terrain data, see Section 12.1, in the 1990s deterministic models based on theoretical approaches became more popular, see for example [6–11]. These developments enabled the integration of realistic propagation characteristics into simulation tools. Generally speaking, propagation models can be subdivided into the categories site-general and site-specific models, see Figure 12.1. In both categories a further sub-division into path loss models and channel models [12] is possible. Site-general channel models like the COST 207 or the WINNER model [13, 14] are more relevant in link-level simulations used for the development of systems and in the corresponding standardization process [15] and less important for network planning and optimization. Both site-general and site-specific path loss models are used in different phases in radio network planning. Therefore, this chapter focuses mainly on path loss models and includes a brief discussion on site-specific channel models in Section 12.5.

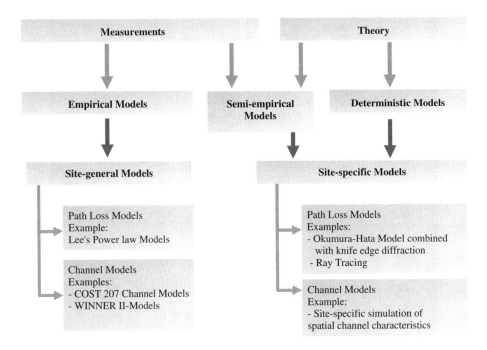

Figure 12.5 Categories of propagation models.

12.2.1 Site-General Path Loss Models

Site-general path loss models are useful to study the principle behavior of system-level concepts or to enable a rough estimation of the number of required sites in a larger area for example in greenfield planning during a license bidding. These models do not require site-specific terrain information. Instead input parameters are for example path loss decay exponents, effective antenna heights or average clutter loss factors characterizing the average propagation environment. Prominent models of this type are the Okumura-Hata-Model [3, 4], the ITU-R370 prediction model [5], Lee's model [2] and Erceg's suburban path loss model [16]. Details of the latter two models are presented in Section 12.3.

12.2.2 Site-Specific Path Loss and Channel Models

The widespread availability of digital terrain data, which include terrain height, land use information and building data, triggered the development of site-specific propagation models and their integration into radio planning tools. Site-specific propagation models are based on the detailed terrain characteristics extracted along the individual propagation paths between transmitter and receiver. The first site-specific models applicable to practicable cellular planning tasks are semi-empirical models (Figure 12.6), where the path loss calculation is based on a combination of deterministic approaches and empirical models, see Section 12.4. These models use LR geographic data, see Section 12.1. They give reasonably good results for the coverage prediction of large macro cell sites even deployed

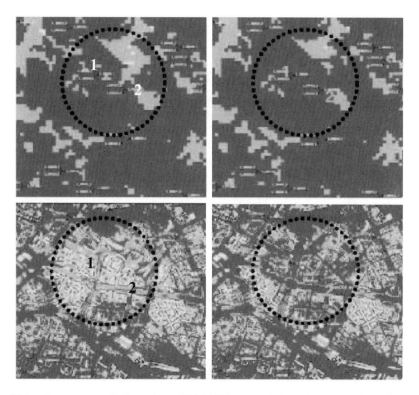

Figure 12.6 Coverage predictions for a GSM1800 network based on a semi-empirical macro cell model (top) and a ray tracing model (bottom) with (right) and without (left) the two base stations at position 1 and 2; dark colors mean excellent coverage; light colors indicate problem areas (from [12]).

in urban areas [17, 18], but due to their limited resolution these models have their limits in predicting for example indoor coverage problems in dense urban areas. With the availability of more powerful computers and detailed building data for the larger cities, ray tracing techniques (see Chapter 10) have been introduced as a deterministic approach to the standard planning process for dense urban areas with many cellular operators. The difference in predicting coverage in a dense urban area applying the semi-empirical model from [18] using LR geographic data with a resolution of 50 m and the ray tracing model from [19] with a resolution of 5 m (including a distinction between indoor and outdoor pixels) to the same area, is shown in Figure 12.2 [12]. Whereas the ray tracing model clearly shows coverage problems within the dotted circle, if base stations at sites 1 and 2 are not installed, the semi-empirical model is not able to detect these problem areas. Furthermore, ray-tracing models enable the determination of site-specific channel characteristics, for example delay spread or angular spread maps. With the introduction of multi-antenna systems, the prediction of these parameters may also become relevant.

12.3   Empirical Models

This section provides the details of two empirical path loss models, which can be used as site-general path loss models.

12.3.1   Lee's Model

Lee's power law model has been derived from measurement campaigns at 900 MHz [2, 20]. The path loss L_{Lee} in dB is expressed as

$$L_{Lee} = 10n \log R - 20 \log h_{eff} - D - 10 \log h_{mobile} + 29 \qquad (12.1)$$

where n is the path loss decay exponent, h_{eff} is the effective base station antenna height, D is a clutter loss correction factor derived from measurements and h_{mobile} is the height of the mobile station.

The parameter values n and D vary with the environment. Table 12.1 summarizes exemplary values of these parameters from [20]. Note that these values are valid for a carrier frequency of 900 MHz only. For other carrier frequencies and/or environments the model has been calibrated by tuning these parameters (see also Section 12.6.2). In case digital terrain data is available, h_{eff} can be determined freely for each specific terrain profile using one of the methods described in [2].

12.3.2   Erceg's Model

The suburban path loss model proposed by Erceg [16] has been derived from measurements of 95 existing macro cells at 1.9 GHz in three common terrain categories found across the United States. The three terrain categories are:

- Type A: Hilly terrain with moderate-to-heavy tree densities.
- Type B: Hilly terrain with light tree densities or flat terrain with moderate-to-heavy tree densities.
- Type C: Mostly flat terrain with light tree densities.

Table 12.1   Parameter settings for Lee's model from [20]

Environment	n	D
free space	2	45
open	4.35	49
suburban	3.84	61.7
urban (Philadelphia)	3.68	70
urban (Newark)	4.31	64
urban (Tokio)	3.05	84
urban (New York City)	4.8	77

Table 12.2 Parameter settings for Erceg's model from [16]

Model parameters	Terrain Type A	Terrain Type B	Terrain Type C
a	4.6	4	3
b	0.0075	0.0065	0.005
c	12.6	17.1	20

The median path loss L is given by

$$L_{Erceg} = 20\log\frac{4\pi d_0}{\lambda} - 10\gamma\log\frac{d}{d_0} \tag{12.2}$$

where λ is the wave length in meters, d is the distance between the base station and the mobile station in meters, $d_0 = 100\,\mathrm{m}$ is the reference distance and γ is the path loss exponent given as

$$\gamma = a - bh_b + \frac{c}{h_b} \tag{12.3}$$

and h_b is the height of the base station in meters. In Table 12.2 the values of a, b and c for the three environments are given. In [16] an additional term is included to characterize the random variation about the median path loss l. The Erceg model was originally proposed for frequencies close to 2 GHz with antenna heights of the mobile station close to 2 m, $10\,\mathrm{m} \le h_b \le 80\,\mathrm{m}$ and $100\,\mathrm{m} \le d \le 8000\,\mathrm{m}$. In [21] an extension of this model to mobile station antenna heights h_{mobile} in the range from 2 m to 10 m and to frequencies other than 2 GHz is given. The path loss of the extended Erceg model is given by

$$L_{Erceg,extended} = L_{Erceg} + L_f + L_h \tag{12.4}$$

$$L_f = 6\log\frac{f}{2000} \tag{12.5}$$

$$L_h = \begin{cases} -10.8\log\frac{h_{mobile}}{2} & \text{for categories A and B} \\ -20\log\frac{h_{mobile}}{2} & \text{for categories A and B} \end{cases} \tag{12.6}$$

with f as the frequency in MHz.

In the literature, see for example [22], the extended Erceg model is also known as the Stanford University Interim (SUI) model.

12.4 Semi-Empirical Models for Macro Cells

In this section, semi-empirical models for macro cells are described. A detailed description of a model applicable for both rural and urban terrain is described in Section 12.4.1 and the COST231-Walfisch-Ikegami model suitable for urban areas is presented in Section 12.4.2 (Figure 12.7). A brief overview of other existing models is given in Section 12.4.3.

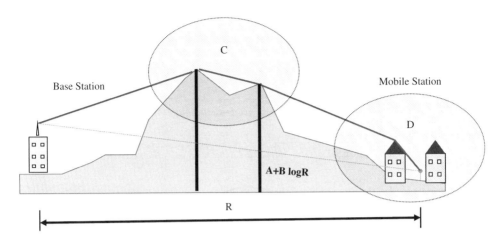

Figure 12.7   Propagation scenario for the semi-empirical models for macro cells.

12.4.1   A General Formula for Semi-Empirical Models for Macro Cells

In this section, we assume that LR geographic data consists of a DLU and DEM, see Section 12.1. The path loss L in dB observed in a macro cell at an arbitrary terrain, see Figure 12.7, can be calculated by the following equation:

$$L = A + B \log r + C + D \tag{12.7}$$

where A is the loss at a distance of 1 km, B is the propagation coefficient, r is the distance between the base station and the mobile station in km, C is the diffraction loss through topographical obstacles and D is the clutter loss correction factor. The parameters A and B are determined by the basic path loss obtained from an empirical model and the diffraction loss C is determined by a knife edge diffraction model (Figure 12.8). D has to be determined for each land use class based on measurements, see Section 12.5.6.

Combinations of Knife-Edge diffraction models to consider shadowing by terrain deterministically with the empirical Okumura-Hata model [4, 3] and its extensions [23, 24] are still the most frequently used semi-empirical macro cell path loss models applied in commercial radio planning tools. Implementation details for one exemplary combination are given in the following subsection.

12.4.1.1   Extended Hata Model

This model is based on a large measurement campaign published by Okumura et al. [4] and equations to calculate A and B published by Hata [3]. This model was originally valid for a frequency range between 150 MHz and 1000 MHz. COST 231 [23] extended this model to frequencies betweeen 1500 MHz and 2000 MHz. A further extension to a frequency range up to 3 GHz is available in the SEAMCAT tool [24]. The equation to calculate parameter A depends on the frequency range:

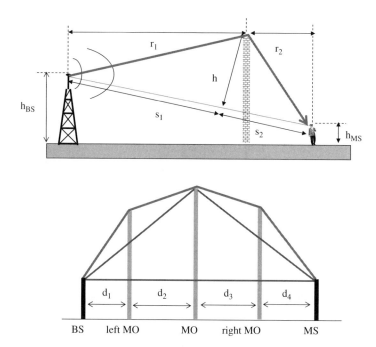

Figure 12.8 Parameters and geometry for knife edge modeling: diffraction on a single knife edge (top) and multiple knife edge diffraction using Deygout's method (bottom).

for $150 \, \text{MHz} \le f \le 1500 \, \text{MHz}$:

$$A = 69.55 + 26.26 \log f - 13.82 \log h_{eff} - a(h_{mobile}) \tag{12.8}$$

for $1500 \, \text{MHz} \le f \le 2000 \, \text{MHz}$:

$$A = 46.3 + 33.9 \log f - 13.82 \log h_{eff} - a(h_{mobile}) \tag{12.9}$$

for $2000 \, \text{MHz} \le f \le 3000 \, \text{MHz}$:

$$A = 46.3 + 33.9 \log 2000 + 10 \log(f/2000) - 13.82 \log h_{eff} - a(h_{mobile}) \tag{12.10}$$

where

$$a(h_{mobile}) = (1.1 \log f - 0.7)h_{mobile} - (1.56 \log f - 0.8) \tag{12.11}$$

f, is the carrier frequency in MHz, h_{mobile} is the height of the mobile station in m and h_{eff} is the effective base station antenna height. Note that the results achieved by the Hata model depend on the algorithm selected to calculate h_{eff}. For more details on algorithms to calculate h_{eff}, see for example [2, 18].

Parameter B is independent of the frequency and is calculated as:

$$B = 44.9 - 6.55 \log h_{eff} \tag{12.12}$$

The model is valid for base station antenna heights (h_{eff}) between $30\,\text{m}$ and $200\,\text{m}$, h_{mobile} between 1 and $10\,\text{m}$ and d between 1 and $20\,\text{km}$. In [24] the validity range for d is extended up to $100\,\text{km}$.

12.4.1.2   Knife Edge Models

For the calculation of the diffraction loss caused by topographical obstacles, various knife edge models have been proposed in the literature [25–28]. In all these models arbitrarily shaped natural obstacles are replaced by one or more knife edges. In this section, Deygout's method [25] is explained in detail.

In the case of only one obstacle and the geometry for the single knife edge case the additional diffraction loss C can be calculated by the following approximation [29]:

$$C_{KE} = 6.9 + 20 \log(\sqrt{(v - 0.1)^2 + 1} + v - 0.1) \qquad (12.13)$$

$$v = h\sqrt{\frac{2}{\lambda} \frac{s_1 + s_2}{r_1 r_2}} \qquad (12.14)$$

Note that in practical implementations, the diffraction loss C_{KE} is also multiplied by an additional empirical correction factor k. Typical values for k are in the range of 0.2 to 0.5.

In cases where there is more than one obstacle, the Deygout method describes the following algorithm:

- Step 1: The main obstacle between Base Station (BS) and Mobile Station (MS) is determined. The obstacle with the largest value of v in Equation 12.14 considering only a single knife edge between BS and MS is regarded as the Main Obstacle (MO). The diffraction loss for this obstacle is $C_2(v_2)$.
- Step 2: Repetition of step 1 for the sub-paths BS-MO ($\Rightarrow C_1(v_{1,left})$) and MO-MS ($\Rightarrow C_3(v_{3,right})$).
- Step 3: In cases where further obstacles exist in the sub-paths, repetition of the 2nd step were required.

If three obstacles exist and the main obstacle is in the middle, the total diffraction loss is calculated as follows:

$$C_D = C_2(v_2) + C_1(v_{1,left}) + C_3(v_{3,right}) \qquad (12.15)$$

The Deygout method can also be extended to the case of more than three obstacles. However, in practical applications often no more than three obstacles are considered. With the Deygout method the diffraction loss is overestimated, especially in cases of short distances between diffractive edges. In order to reduce this problem, Causebrook [27] established a method for the correction of the diffraction loss.

$$C_D = C_2(v_2) + C_1(v_{1,left}) + C_3(v_{3,right}) - \hat{C}_1 - \hat{C}_2 \qquad (12.16)$$

$$\hat{C}_1 = (6 - C_2(v_2) + C_1 v_1)) \cos \alpha_1 \qquad (12.17)$$

$$\hat{C}_2 = (6 - C_2(v_2) + C_3 v_3)) \cos \alpha_2 \qquad (12.18)$$

$$\cos \alpha_1 = \sqrt{\frac{d_1(d_3 + d_4)}{(d_1 + d_1)(d_2 + d_3 + d_4)}} \qquad (12.19)$$

$$\cos \alpha_2 = \sqrt{\frac{d_4(d_1 + d_2)}{(d_1 + d_1)(d_2 + d_3 + d_4)}} \qquad (12.20)$$

The meaning of d_i can be extracted from Figure 12.7.

12.4.2 COST231-Walfisch-Ikegami-Model

Propagation in urban macro cells, see Figure 12.5 is dominated by diffraction effects on the roofs of the building. In cases of Non Line Of Sight (NLOS) two main propagation effects can be identified, which mainly determine the path loss between the BS and the MS. First, multiple screen diffraction is observed along the path between the BS and the roof-top close to the street canyon, where the mobile is located. This effect has been modeled by Walfisch and Bertoni [30]. The second effect is the roof-top-to-street-diffraction from the last roof-top to the MS. This effect combined with a reflection on the building wall of the opposite building in the street canyon has been modelled by Ikegami [32]. COST Action 231 [23] has proposed a combination of the Walfisch-Bertoni model and the Ikegami model, the so-called COST-231-Walfisch-Ikegami model. The path loss in this model depends on the following parameters describing the urban environment (see also Figure 12.9):

- h: heights of the buildings;
- w: street width;
- b: building separation;
- φ: street orientation with respect to the direct radio path.

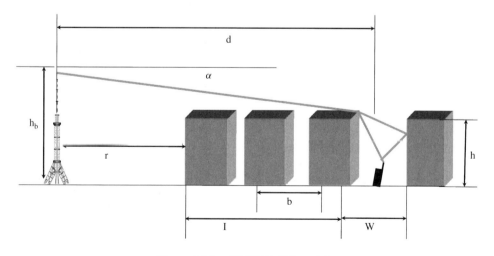

Figure 12.9 COST231-WI model.

Although these parameters can be extracted from detailed digital building data, it is not a deterministic model since it assumes that the parameters are not varying along the terrain profile. Therefore, only average values can be used as input data.

For the NLOS case depicted in Figure 12.5 the path loss $L_{WI,NLOS}$ is given by

$$L_{WI,NLOS} = \begin{cases} L_0 + L_{rts} + L_{msd} & \text{for } L_{rts} + L_{msd} > 0 \\ L_0 & \text{for } L_{rts} + L_{msd} \leq 0 \end{cases} \tag{12.21}$$

where L_0 is the free space loss

$$L_0 = 32.4 + 20 \log d + 20 \log f \tag{12.22}$$

L_{rts} is the path loss caused by roof-top-to-street-diffraction:

$$L_{rts} = -16.9 - 10 \log w + 10 \log f + 20 \log \Delta h_{mobile} + L_{ori} \tag{12.23}$$

$$L_{ori} = \begin{cases} -10 + 0.354\varphi & \text{for } 0 \leq \varphi \leq 35 \\ 2.5 + 0.075(\varphi - 35) & \text{for } 35 \leq \varphi \leq 55 \\ 4 - 0.11(\varphi - 55) & \text{for } 35 \leq \varphi \leq 55 \end{cases} \tag{12.24}$$

$$\Delta h_{mobile} = h - h_{mobile} \tag{12.25}$$

$$\Delta h_{base} = h_b - h \tag{12.26}$$

L_{msd} is the path loss caused by muliple-screen-diffraction:

$$L_{msd} = L_{bsh} + k_a + k_d \log d + k_f \log f - 9 \log b \tag{12.27}$$

where

$$L_{bsh} = \begin{cases} -10 \log(1 + \Delta h_{base}) & \text{for } h_{base} > h \\ 0 & \text{for } h_{base} \leq h \end{cases} \tag{12.28}$$

$$k_a = \begin{cases} 54 & \text{for } h_{base} > h \\ 54 - 0.8\Delta h_{base} & \text{for } d \geq 0.5 \text{ km} \text{ and } h_{base} \leq h \\ 54 - 1.6\Delta h_{base} d & \text{for } d < 0.5 \text{ km} \text{ and } h_{base} \leq h \end{cases} \tag{12.29}$$

$$k_d = \begin{cases} 18 & \text{for } h_{base} > h \\ 18 - 15\frac{\Delta h_{base}}{h} & \text{for } h_{base} \leq h \end{cases} \tag{12.30}$$

$$k_f = -4 + \begin{cases} 0.7(\frac{f}{925} - 1) & \text{for medium sized cities and suburban centers} \\ & \text{with medium tree density} \\ 1.5(\frac{f}{925} - 1) & \text{for metropolitan centers} \end{cases} \tag{12.31}$$

d is the distance in km, f is the frequency in MHz, h_b is the height of the base station in m and h_{mobile} is the height of the mobile station in m.

For the case of Line Of Sight (LOS) propagation within the street canyon COST 231 has derived an empirical path loss model from measurements carried out in Stockholm:

$$L_{WI,LOS} = 42.6 + 26 \log d + 20 \log f \tag{12.32}$$

for $d \geq 20$ m

The COST-231-Walfisch-Ikegami model is restricted to the frequency range between 800 and 2000 MHz, a base station antenna height between 4 m and 50 m, mobile station antenna heights between 1 m and 3 m and a distance d between 20 m and 5 km.

Although the model gives reasonable results, when compared to measurements [23] and [33] has found that the expression accounting for the diffraction loss from the last rooftop to the street was erroneously obtained.

12.4.3 Other Models

Saunders and Bonar [34] as well as Maciel, Bertoni and Xia [35, 36] have published closed-form solutions for similar propagation envionments as depicted in Figure 12.5. In [37] an extension of the Walfisch-Bertoni model is proposed that takes into account additional attenuation by vegetation. In [22] Ericsson's implementation of the Hata model is described, which introduces additional parameters to allow a better fitting to the propagation environment. ITU-R recommendation P1411-4 [38] covers propagation over short paths below 1 km affected primarily by buildings and trees. In [18] a hybrid propagation model for macro cells is presented, which combines many of the above mentioned propagation models. Originally developed for 1800 MHz macro cells the model has been extended to ultra-high sites [39] and shows reasonable accuracy also at 900 MHz [17].

12.5 Deterministic Models for Urban Areas

The complex air interfaces involved in LTE and Long Term Evolution Advanced (LTE-A) wireless systems require advanced simulation tools for the planning and optimization of radio access networks. In particular, sophisticated techniques like beam-forming and Multiple Input Multiple Output (MIMO) allow the air interfaces to fully take benefit from the multipath radio channels. Based on OFDM, the access techniques offer adaptive modulation and coding. To finely predict the capacity and the coverage of the deployed systems, an accurate propagation model is mandatory. To get a reliable coverage for the whole network, the propagation prediction techniques should provide seamless coverage, excluding signal prediction ruptures, from the high towers acting as umbrella cells down to street-level transmitters. A propagation prediction with a high dynamic range of reception levels is required to determine the useful service coverage areas for various transmitter antenna heights and over large areas.

12.5.1 Waveguiding in Urban Areas

In dense urban areas, the large buildings along the streets create an environment that guides the propagation of the waves. Also called urban canyoning, this phenomenon results from multiple successive interactions on the building facades (reflection) and building corners (diffraction). These physical effects need to be taken into account in reliable propagation predictions for two main reasons. First, the canyoning yields signal strengths beyond the classical free space distance. Second, the multiple interactions will dominantly contribute to NLOS areas coverage. The ray-based methods inherently provide the multiple interactions along the various ray paths and are therefore the preponderant methods in urban areas.

12.5.2   Transitions between Heterogeneous Environments

A wireless network operator will have to deploy its network in different environments: rural, suburban and urban areas. As discussed previously, the geographical representation of those areas will differ in terms of accuracy, resolution and type of the obstacles. Yet a seamless coverage must be ideally determined. A key feature of operational models is to ensure a smooth transition between the various heterogeneous areas. A common vertical plane technique for all those areas helps to achieve this goal.

12.5.3   Penetration Inside Buildings

Although mobile phone calls are realized on the street or in cars, it has been observed that in urban areas 75 percent of voice and data communications occur inside buildings: home, office, train station, airport, and so on. Consequently deterministic models should efficiently predict the signal coverage inside buildings, at several levels and ultimately with a sense of direction of arrival of the dominant multipath occurring in the streets outside the building and penetrating with various angles and strengths.

12.5.4   Main Principles of Operational Deterministic Models

Many propagation prediction techniques have been proposed over the last decade [23, 40–43]. Ray-based models generate a high interest, as they succeed in predicting with high accuracy the field strength around low transmitters in urban environments. For very low transmitter heights, the propagation is confined between the buildings and thus the first approaches have mainly been in two-dimensions (horizontal plane) [44, 45]. 3D ray-tracing models (or similar techniques) are well suited to estimate the site-specific space-time characteristics of the narrowband or wideband urban propagation channel, which is essential for the simulation of the performance of new radio systems. Many works form the recent years present comparisons of the multipath predictions to wide-band channel characteristics, such as the power delay profile, the delay spread or the angular spread [10, 11, 46]. These techniques are consequently useful and reliable to assess deterministically the wideband radio channel. The concept of Artificial Neural Networks (ANNs) associated with the Walfish-Ikegami model or street graphs [47, 48] were introduced a couple of years ago. The training of the artificial neural network is a complex technique, but the method proved to be efficient in terms of computation times. However, the reliability strongly depends on a large quantity of measurements of high quality. As there is no possibility to get space-time information by the ANNs contrary to the ray-based techniques, we will not consider them further for real 3D space-time urban propagation predictions, as the time and angular dispersion of the received fields are crucial for multiple antenna techniques that are at the heart of urban LTE deployments.

12.5.4.1   3D Ray-Tracing

Yet the main restrictive aspect of the 3D ray-tracing (image method) for a large amount of radio-planning remains generally the computation time. Some techniques have been elaborated recently and are subject to continuous work to speed up the computation, especially for coverage computation. When the 3D ray-tracing is based on computer image theory, the acceleration techniques generally involve a pre-processing of the vector database and the extraction of the main geometrically and physically meaningful elements leading to a computation of the field strength. Often the simplifications are efficient in terms of computation time, but they result in the prediction of only dominant paths that may be dependent on transmitter site characteristics. For areas usually corresponding to small 3G cell radius, these propagation models give satisfying coverage results.

12.5.4.2 Mixing Ray-Tracing and Ray-Launching

Another way to optimize the computation times consists in mixing ray-launching and ray-tracing [11, 49] separated in two planes (vertical and horizontal). The approach of considering the propagation first in the horizontal plane and then in the vertical plane creates a small error on the trajectory of rays diffracted by the horizontal edge of buildings, but comparisons of the predictions to actual measurements match very well.

12.5.4.3 3-Step Model

More recently, within in the framework of the European Project MOMENTUM, [19] and [50] suggested a progressive propagation "city model" decomposed into three main steps: vertical plane (knife-edge diffraction over building), multipath (single scattering processes considering Lambertian transmitters) and vegetation models (for all paths).

12.5.4.4 Advanced Operational Model

The 3D urban model presented in [10] is an efficient alternative to most recent models. As the model gathers the main advantages of each previous solution, we detail hereafter the main principles. The focus is set on the operational use of these propagation models for radio network planning and optimization over large areas. The method consists also of separating the vertical effects and the lateral effects. In a first step, the direct path is computed based on an extraction of a vertical cut from the geographical data set – possibly combining LR and HR data – containing a list of possible obstacles (building, vegetation). A multiple-knife edge technique (e.g., a modified Deygout) determines the diffraction loss on the vertical edges of those obstacles. In a second step, the lateral effects – providing the canyoning – are taken into account using a ray-launching technique. Specular reflections, diffractions and indoor penetrations are computed in 2D after launched ray beams interact with the building contours found in the area.

The concept of Visibility Masks (VMs) is advantageously introduced in [10] to identify the reception areas of the ray beams – offering an alternative to the reception spheres used in most ray-launching techniques. Indeed, a VM consists of a constructed area containing the whole 2D area that can be reached by the ray beam after each interaction. As an example, the VM for the first order reflection of the ray beam in the left part of Figure 12.10 is limited by the following reflection wall on the upper part, and by the obstructing portions of another building wall in the lower part. Similar VM boundaries are observed for a ray-beam diffraction in the right part of Figure 12.10. In the case

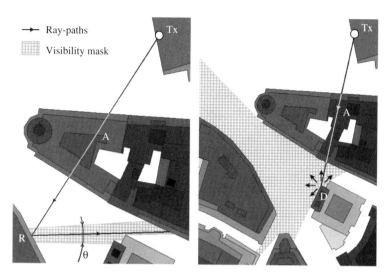

Figure 12.10   Specular reflection and visibility mask.

of multiple successive interactions, multiple masks are built until the maximum allowed higher order interaction is reached. At the end of this ray-launching phase, the multiple ray-beam trajectory is obtained in 2D, see Figure 12.12. Note that the method is able to compute the reflected rays on complex building facades presenting a podium (such as a terrace on a building, see pointed arrow in Figure 12.11). Indoor penetration is computed to simulate the coverage inside buildings. All the outdoor ray-paths are prolonged through the building walls that compose the incident ray-path VM, and then along a straight line inside the building.

A third step of the method unfolds the lateral ray paths computed in 3D. The 3D trajectory is determined by putting together the vertical terrain profiles of each 2D ray path into an "unfolded profile" that gathers the variations of the ground altitude, clutter type and clutter height for each radio link (transmitter to receivers), as illustrated in Figure 12.12. Then, the same multiple knife-edge technique as for direct rays is used to detect the main obstacles (building, vegetation) of the "unfolded profile" and to compute a diffraction loss. Other mentioned methods make use of the Uniform Theory of Diffraction (UTD) or Walfish-Bertoni methods for these unfolded profiles. These methods, however, cannot be efficiently applied to rural and suburban areas. Therefore, for the sake of homogeneity in the methods in various environments leading to smooth transitions, using similar multiple-knife edge techniques for the direct paths in rural, suburban and dense urban areas is certainly the best option. At this stage, the algorithm calculates a kind of polyline that represents the shortest path from the transmitter to the receivers passing over the main obstacles detected in the "unfolded profile". This polyline represents the ray trajectory in the vertical plane. When the trajectory goes below the ground level or above the interacting buildings (interaction found in the 2D ray-path) then the ray is rejected. Finally, the field strengths of the ray contributions are calculated thanks to a combination of the UTD and the multiple knife-edge methods. Generally, transmit and receive antenna gains are extrapolated from the 2D-plane antenna patterns given in one

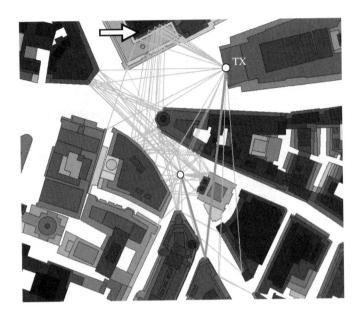

Figure 12.11   Ray paths after the ray-launching phase.

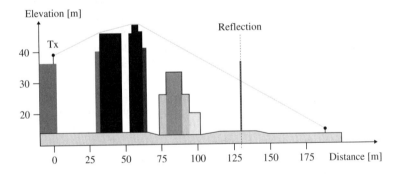

Figure 12.12   Unfolded 3D ray trajectory.

specific polarization by the manufacturer. The extrapolated 3D antenna gain is given from a sophisticated technique [51]. Usually, the operational ray-based models do not calculate the complex field of the ray contribution but only the norm. The reason behind this is that the phase error is likely to be larger than 2π due to inaccuracies in the geographical map data for wireless network wavelengths. Moreover, the complex antenna gain is usually not known. Thus, the phase of the ray contribution is not calculated and must be considered as a random variable uniformly distributed over the interval $[0, 2\pi]$ radian. At the end of the process, when all ray beams have been taken into account, the ray-spectrum of the predicted receiver is obtained expressing each individual ray contribution characterized by its norm, delay, angle of departure and angle of arrival. As a result,

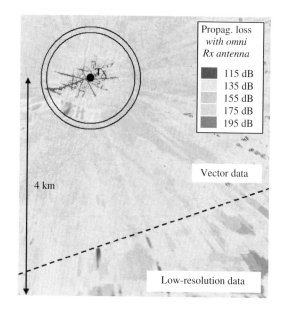

Figure 12.13 Coverage prediction of a 3D urban area.

this type of site-specific technique is very well suited to the prediction of the space-time propagation channel response for small and large urban coverage areas. An operational deterministic model should positively answer to operational constraints such as efficient computation times (distribution and Graphical Processor Unit implementations may be helpful), the indoor signal coverage assessment (including a sense of 3D dominant arrival direction), the interference estimation over large distances and for weak signals (high dynamic range), and by the heterogeneity of the digitized geographical representation of the environments (with smooth transitions). For the latter purpose, the authors of [10] have implemented the notion of predicting regions over a large distance, where the full scene is subdivided into a near area hosting a full multipath, a far area experiencing only the direct path, and a transition area where a weighted combination of the multipath and direct path on heterogeneous LR and HR geographical data is computed. Figure 12.13 presents a 5m-resolution coverage map predicted around a 25m-high transmitter. The limits of the near-reception and transition-reception regions are represented by black circles. The geographical data set is heterogeneous, as the South-East part contains LR raster data only. Thus, a mix of low- and high-resolution data is used in the construction of the vertical terrain profiles. Wideband channel parameters are predicted in the near-reception region, like the horizontal arrival angle-spread or the delay-spread shown in Figures 12.14 and 12.15. The computation times on modern IT systems are less than a minute for an 8 km by 8 km area. This shows that this kind of deterministic ray-based predictions can be introduced into the operational radio-planning and optimization process with very acceptable computation times.

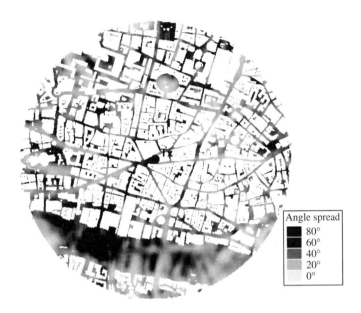

Figure 12.14 Coverage of angular spread in an urban area.

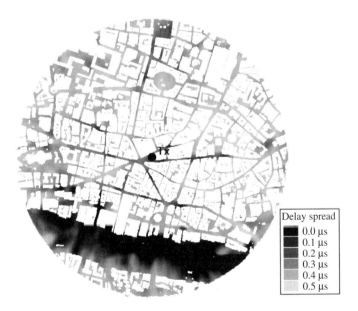

Figure 12.15 Coverage of delay spread in an urban area.

12.5.5   Outdoor-to-Indoor Techniques

In urban areas most communications take place inside buildings. A propagation prediction model must reliably estimate the indoor coverage obtained by penetration of outdoor transmitters. A simple version consists in applying a simple additional loss per building that could potentially depend on the reception height (at different floors of the building) [19]. As mentioned in the previous detailed description of the urban model, a more refined version consists in prolonging the ray paths inside the buildings [10]. The coverage is then obtained thanks to a penetration loss at the building interface that may depend on the angle of arrival. An additional linear loss is added to further attenuate the signal for distances deeper away from the exterior walls and windows [42, *pp 137–138*].

12.5.6   Calibration of Parameters

Site-specific propagation models usually predict the signal strength for LOS and NLOS locations with a better accuracy than semi-empirical models. This comes from the many deterministic details from the digital geographical models taken into account and the more sophisticated algorithmic approaches. However, for a better accuracy it is highly recommended to calibrate some model parameters to achieve even higher performances. The model tuning usually adds offsets and weighting to free space losses, diffraction and reflection coefficients. A measurement data set of high quality and a good statistical representation of common situations are required to yield better performance. A well-tuned model should indeed exhibit optimal performances in the areas where the measurements were collected, but the model should also be applicable to other similar areas.

12.6   Accuracy of Propagation Models for RNP

Radio propagation predictions like any other prediction of complex physical phenomena are subject to approximations leading to uncertainties. A radio engineer must first adapt his propagation models to the final application and to constraints brought by the new technologies (like using the space-time channel modeling for 3G LTE systems). The main factors impacting the accuracy are the input data (geographical map data, transmitter parameters, receiver sensitivity), the modeling technique (statistical, semi-empirical or site-specific), and finally the goodwill of the radio engineer (interpretation of predictions and their compliance with the required performance). Radio-frequency received a power measurements permit to adjust the parameters of the propagation prediction models to get optimal accuracy. The calibration is mandatory for statistical and semi-empirical models, and recommended to achieve optimal performance and applicability from site-specific propagation prediction models. Generally a new calibration is realized when a new site topology is introduced: the propagation behavior and the propagation channel characteristics are different for high transmitters (macro cells) and lower transmitters (small cells). The availability of a new spectrum frequency usually leads to a recalibration of the models, as the propagation is also different at 800–900 MHz and at 2.1–2.6 GHz. Specific adjustments may be carried out to diversify the set of propagation models, for example, combining morphological or topological characteristics: flat and hilly terrain, high or low density of building and vegetation, strong presence of hydro land usage or not.

In the following, we elaborate on good practices to get optimal operational calibrated propagation models, starting with the measurement campaign specification and post-processing, then the parameter tuning methods and finally we try to introduce some insights into the interpretation and use of the calibrated models.

12.6.1 Measurement Campaign

In order to characterize correctly the radio environment without interferences and signal reception impairments due to the other transmitting devices, it is common practice to transmit a single frequency Continuous Wave (CW) in an available spectrum band, and to collect the Radio Frequency (RF) signal, at the same frequency, in the area under investigation. A temporary transmitter must be deployed on a pumped mast or on a rooftop, with possible building access issues by the landlords. The reception part consists of a receiver (spectrum analyzer or professional dedicated equipment), a GPS receiver linked to the receiver to time and position stamp the measured data. For urban areas, a dead-reckoning system will prevent erroneous location estimation due to the GPS tracking loss caused by the urban multipath and shadowing by the high-rise buildings. When a system has already been deployed, it is possible to collect, as a complement, scanned RF data with decoding receivers, the characteristics of which are quite similar to actual mobile handsets. The advantages of this type of campaign are that the system is already on air, with operational antennas and radio system parameters, thus only the receiving part generates some work. The main drawback is that, when users communicate, the live network adds some interferences and changing reception conditions. This prevents a reliable environment characterization. In both cases, it is important to obtain reliable information on the transmitting and receiving technical characteristics (antenna azimuth and tilt, transmitted power, actual location of the antennas). A professional measurement team will also describe, with the help of pictures, all the impacting elements such as a rooftop description, drive testing conditions (dense traffic, detour), surrounding obstacles and building types. The roadmap designed for the drive testing should take into account the geographical environment description (low and high resolution data). In rural and suburban areas (see Figure 12.16), the vehicle should go through all preponderant clutter types. In dense urban areas, a spider web scheme should be followed: radial from the transmitter (center of the web), and concentric circles around the transmitter at various distances to capture the main shadowing effects (Figure 12.17). When characterizing radio environments of small cells, at street level, the channel dynamics (pedestrian and vehicular traffic) are a constraint, and could disturb the interpretation of the collected data set. It is better to carry out the measurement at night and to drive along the same measurement routes a couple of times. Finally, to ensure a reliable measurement data set, the points should be collected with a balance representation in each clutter type or urban environment (large open areas, Grand Boulevard, narrow streets) and at intervals permitting to prevent disturbance by the fast fading (Lee Criterion [51]).

The raw data set must be analyzed (to keep only the well-located points, that is, with no obvious position error), and then post-processed, for example, by applying some filtering (noise level). Attention must be paid to the threshold effect that could bias the data set or the interpretation of the calibration metrics.

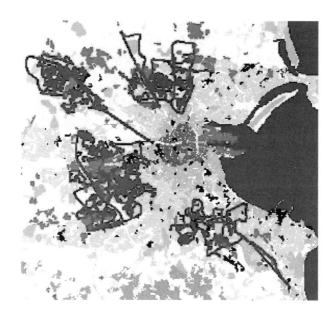

Figure 12.16   Route pattern across various clutter types on LR map data.

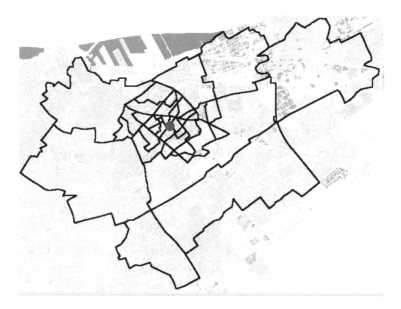

Figure 12.17   Spider-web pattern on HR map data.

12.6.2   Tuning (aka Calibration) Process

The idea is to compare the simulated prediction with the actual measured data, and to analyze some statistical metrics. Usually a portion of the measured data is used for calibration purposes and the remaining portion for the validation. The most commonly analyzed

Figure 12.18   LOS and LOS regression lines for a received power vs. distance.

statistical metrics are the mean, standard deviation, the correlation and dispersion of the errors between the prediction and the measurement. The measured data set is used to calibrate the propagation model parameters. Then, calibration algorithms adjust the propagation parameters and optimize their value to reach targeted values. Several methods have been used in the industry such as the determinant-based method, iterative techniques with regression analyses, and basically any statistical method that could converge quickly enough to yield optimum parameters. The average error according to the various distances is also key for a stable model. It is quite common to calibrate the model parameters for LOS and NLOS areas. Figure 12.18 shows the received power level versus the distance for LOS and NLOS collected points. A couple of propagation models introduce an additional slope at a breakpoint distance from the transmitter. This introduces additional model parameters that could be conveniently tuned. Yet we do not recommend this approach, especially for small cells, as dual slope models could potentially exhibit some discontinuities in practice that would completely imbalance the model and provide wrong results which are difficult to solve. Separating the LOS and NLOS collected points has proven to be a far more reliable technique. For statistical propagation models used with low resolution geographical data, the calibrated parameters are usually the attenuations (land use classes), the heights (2D clutter, effective heights), the LOS and NLOS offsets (intercept of the curves), and the slopes. The vertical diffraction attenuation method (Deygout, Epstein-Peterson, Millington, Giovanelli) could also be calibrated. For site-specific models associated with high resolution geographical data, the goal is to calibrate coefficients for vertical and horizontal diffraction, the reflection coefficients, possibly the roughness and material properties. Some advanced models also tune some antenna-related parameters. In practice, the operational calibration is made thanks to automated tuning modules. For advanced and reliable site-specific propagation models, the calibration must only be a slight parameter refining, and the final parameter values should not be highly different from the initial values.

12.6.3 Model Accuracy

After the calibration phase, a different data set (the remaining portion not used for calibration) is used to validate the performance of the tuned parameters. The values of the metrics will differ according to the configuration of the transmitter (macro or small cells), the quality and relevance of the transmitter information and the type of measurement campaigns (CW or scanned data). A calibrated propagation model is usually considered as reliable if the model is well centered (mean error < 2 dB), with a small variation around the mean values (error standard deviation <8 dB), a small dispersion of the mean error per transmitter (<5 dB), and a correlation factor which is quite high (>0.8). For dense urban areas and small cell configurations, it is also important to observe the street and indoor behavior: canyoning or waveguiding effects, indoor penetration close to windows and deeper indoor, and so on. Experience is key in the analysis of those effects for a reliable and reusable propagation prediction model. It is important to search for objectivity through derivations of statistical metrics. One should not stick to pure statistical values though. Even if accuracy is definitely important, at least as important characteristics are the relevance, reliability, reusability in different environments, and homogeneity of the calibrated propagation model. To this end, a combination of statistical metrics and engineering experience is the key to success. Figure 12.19 illustrates the good coherence between the predicted and measured values in a dense urban area using the 3D ray-launching propagation model presented in Section 12.5. One may observe the simulations below noise-level of the prediction, when the measured data set experiences a low threshold (−93 dB in this example). It is important to handle correctly those filtering and threshold in the calibration process to avoid biased calibrated parameters that could lead to better performance in appearance (i.e., with better statistical values), but that actually experience some biases in reality.

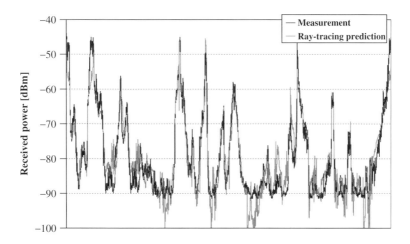

Figure 12.19 A well-tuned advanced propagation prediction model compared to measurements.

However, the interpretation is important and accuracy depends on available inputs. An accuracy of 9 dB in highly dense areas with a dense traffic, low height antennas might not be a bad prediction. On the other hand, 7 dB in open areas with line of sights might not be a good prediction.

The application and above all the phase in the radio planning process drive the required performance targets. Although it is optimal to use a well calibrated propagation model at any stage of the process, given time and budget constraints, heuristically tuned models are usually used at a bidding stage, when the user mainly focuses on a rough estimation of the transmitter number, location, and overall initial coverage of a targeted deployment area. More refined (and calibrated) propagation models are usually used in the subsequent detailed planning and optimization phases of the mobile network deployment process.

12.7   Coverage Probability

The propagation models presented in this chapter provide predictions for a median path loss, based on which a median received power $P_{Rx,pred}$ can be derived. However, the measured power level is subject to slow fading, which can be described by a lognormal distribution. If depicted in logarithmic scale, this transforms into a Gaussian distribution characterized by a mean value and a standard deviation σ, see Figure 12.20. Coverage at a certain location is only possible, if the received power P_{Rx} is larger than the minimum required power level $P_{Rx,min}$, which corresponds to the receiver sensitivity. To assure that the location is covered with a certain coverage probability P, the following condition has to be fulfilled:

$$P_{Rx,pred} \geq P_{Rx,min} + A \tag{12.33}$$

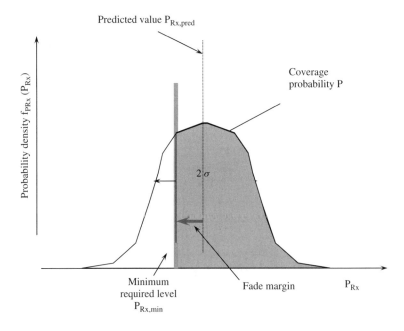

Figure 12.20   Probability density distribution for the received power level.

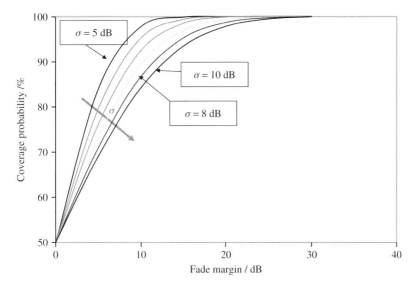

Figure 12.21 Coverage probability as a function of the standard deviation of the shadowing process.

The coverage probability P is then given by

$$P = 0.5 - 0.5 erfc\left(\frac{A}{\sqrt{2}\sigma}\right) \qquad (12.34)$$

where $erfc$ is the error function. Figure 12.21 shows coverage probabilities as a function of the shadowing standard deviation σ and the fading margin A.

References

1. T. Kürner (Ed.) et. al., "Final report on self-organisation and its implications in wireless access networks," in *FP7-ICT-SOCRATES Deliverable D5.9*, www.fp7-socrates.eu, 2011.
2. W. C. Y. Lee, *Mobile Communications Design Fundamentals*. Wiley, 1993.
3. M. Hata, "Empirical formula for propagation loss in land mobile radio services," *Vehicular Technology, IEEE Transactions on*, vol. 29, no. 3, pp. 317–325, Aug. 1980.
4. Y. Okumura, E. Ohmori, T. Kawano, and K. Fukuda, "Fieldstrength and its variability in vhf and uhf land-mobile service," *Review of the Electrical Communication Laboratory*, vol. 16, no. 9–10, pp. 825–873, 1968.
5. *ITU-R Recommendation P370-7: VHF and UHF Propgation Curves for the freqeuncy range form 30 MHz to 100 MHz*, 1997.
6. J. Hviid, J. Andersen, J. Toftgard, and J. Bojer, "Terrain-based propagation model for rural area-an integral equation approach," *Antennas and Propagation, IEEE Transactions on*, vol. 43, no. 1, pp. 41–46, Jan. 1995.
7. M. Lebherz, W. Wiesbeck, and W. Krank, "A versatile wave propagation model for the vhf/uhf range considering three-dimensional terrain," *Antennas and Propagation, IEEE Transactions on*, vol. 40, no. 10, pp. 1121–1131, Oct. 1992.
8. M. Lawton and J. McGeehan, "The application of a deterministic ray launching algorithm for the prediction of radio channel characteristics in small-cell environments," *Vehicular Technology, IEEE Transactions on*, vol. 43, no. 4, pp. 955–969, Nov. 1994.

9. H. Bertoni, W. Honcharenko, L. Macel, and H. Xia, "Uhf propagation prediction for wireless personal communications," *Proceedings of the IEEE*, vol. 82, no. 9, pp. 1333–1359, Sep. 1994.

10. Y. Corre and Y. Lostanlen, "Three-dimensional urban em wave propagation model for radio network planning and optimization over large areas," *Vehicular Technology, IEEE Transactions on*, vol. 58, no. 7, pp. 3112–3123, Sept. 2009.

11. J.-P. Rossi and Y. Gabillet, "A mixed ray launching/tracing method for full 3-d uhf propagation modeling and comparison with wide-band measurements," *Antennas and Propagation, IEEE Transactions on*, vol. 50, no. 4, pp. 517–523, Apr. 2002.

12. T. Kürner, "The role of propagation models for the development and deployment of future wireless communication systems," in *Electromagnetics in Advanced Applications, 2007. ICEAA 2007. International Conference on*, Sept. 2005, 4 pages (electronically).

13. M. Failli (Chairman), *COST Action 207: Digital Land Mobile Radio Communications-Final Report*. Office for the Official Publications of the European Communities, 1989, http://www.lx.it.pt/cost231.

14. P. Kyösti and et. al., *WINNER II Channel Models*, http://www.ist-winner.org/deliverables.html, 2007.

15. *3GPP Spatial Chanel Model Extended (SCME)*, http://www.ist-winner.org/3gpp_scme.html.

16. V. Erceg, L. Greenstein, S. Tjandra, S. Parkoff, A. Gupta, B. Kulic, A. Julius, and R. Bianchi, "An empirically based path loss model for wireless channels in suburban environments," *Selected Areas in Communications, IEEE Journal on*, vol. 17, no. 7, pp. 1205–1211, Jul. 1999.

17. T. Kürner and M. Neuland, "Application of bertoni's work to propagation models used for the planning of real 2g and 3g cellular networks," in *Antennas and Propagation, 2009. EuCAP 2009. 3rd European Conference on*, March 2009, pp. 1686–1690.

18. T. Kürner, R. Fauss, and A. Wäsch, "A hybrid propagation modelling approach for dcs 1800 macrocells," in *Vehicular Technology Conference*, 1996. 'Mobile Technology for the Human Race'., IEEE 46th, vol. 3, Apr-1 May 1996, pp. 1628–1632 vol. 3.

19. T. Kürner and A. Meier, "Prediction of outdoor and outdoor-to-indoor coverage in urban areas at 1.8 ghz," *Selected Areas in Communications, IEEE Journal on*, vol. 20, no. 3, pp. 496–506, Apr. 2002.

20. S. Saunders, *Antennas and Propagation for Wireless Communication Systems*. Wiley, 1999.

21. *Doc. IEEE 802.16a-03/01: Channel Models for Fixed Wireless Applications*, http://www.ieee802.org/16/tga/docs/80216a-03_s01.pdf, 2003.

22. J. Milanovic, S. Rimac-Drlje, and K. Bejuk, "Comparison of propagation models accuracy for wimax on 3.5 ghz," in *Electronics, Circuits and Systems, 2007. ICECS 2007. 14th IEEE International Conference on*, Dec. 2007, pp. 111–114.

23. E. Damosso and L. M. Correia (Eds.), *COST Action 231: Digital mobile radio towards future generation systems-Final Report*. Office for the Official Publications of the European Communities, 1999, http://www.lx.it.pt/cost231.

24. *Information document for SEAMCAT-3 Wiki Help database-SEAMCAT implementation of Extended Hata and Extended Hatam]SRD model*, http://tractool.seamcat.org/wiki/Manual/PropagationModels/ExtendedHata.

25. J. Deygout, "Multiple knife-edge diffraction of microwaves," *Antennas and Propagation, IEEE Transactions on*, vol. 14, no. 4, pp. 480–489, Jul. 1966.

26. J. Epstein and D. Peterson, "An experimental study of wave propagation at 850 mc," *Proceedings of the IRE*, vol. 41, no. 5, pp. 595–611, May 1953.

27. J. H. Causebrook and B. Davis, "Tropospheric radio wave propagation over irregular terrain: the computation of field strength for uhf broadcasting," BBC Res. Dept., Report no. 1971/43, Tech. Rep., 1971.

28. C. Giovaneli, "An analysis of simplified solutions for multiple knife-edge diffraction," *Antennas and Propagation, IEEE Transactions on*, vol. 32, no. 3, pp. 297–301, Mar. 1984.

29. *ITU-R Recommendation P526-6: Propagation by Diffraction*, 1997.

30. J. Walfisch and H. Bertoni, "A theoretical model of uhf propagation in urban environments," *Antennas and Propagation, IEEE Transactions on*, vol. 36, no. 12, pp. 1788–1796, Dec. 1988.

31. F. Ikegami, S. Yoshida, T. Takeuchi, and M. Umehira, "Propagation factors controlling mean field strength on urban streets," *Antennas and Propagation, IEEE Transactions on*, vol. 32, no. 8, pp. 822–829, Aug. 1984.

32. D. Har, A. Watson, and A. Chadney, "Comment on diffraction loss of rooftop-to-street in cost 231-walfisch-ikegami model," *Vehicular Technology, IEEE Transactions on*, vol. 48, no. 5, pp. 1451–1452, Sep. 1999.

33. S. Saunders and F. Bonar, "Prediction of mobile radio wave propagation over buildings of irregular heights and spacings," *Antennas and Propagation, IEEE Transactions on*, vol. 42, no. 2, pp. 137–144, Feb. 1994.

34. L. Maciel, H. Bertoni, and H. Xia, "Unified approach to prediction of propagation over buildings for all ranges of base station antenna height," *Vehicular Technology, IEEE Transactions on*, vol. 42, no. 1, pp. 41–45, Feb. 1993.

35. L. Maciel, H. Bertoni, and H. Xia, "Propagation over buildings for paths oblique to the street grid," in *Personal, Indoor and Mobile Radio Communications, 1992. Proceedings, PIMRC '92., Third IEEE International Symposium on*, oct 1992, pp. 75–79.

36. S. Torrico, K. L. Chee, and T. Kürner, "A propagation prediction model in vegetated residential environments-a simplified analytical approach," in *Antennas and Propagation (EUCAP), Proceedings of the 5th European Conference on*, April 2011, pp. 3279–3283.

37. *ITU-R Recommendation P1411-4: Propagation Data and prediction Methods for the Planning of Short-Range Outdoor Radi Communication Systems and Radio Local Area Networks in the Frequency Range 30 MHz to 100 GHz*, 2007.

38. A. Hecker and T. Kürner, "Analysis of propagation models for umts ultra high sites in urban areas," in *Personal, Indoor and Mobile Radio Communications, 2005. PIMRC 2005. IEEE 16th International Symposium on*, vol. 4, Sept. 2005, pp. 2337–2341 vol. 4.

39. T. Sarkar, Z. Ji, K. Kim, A. Medouri, and M. Salazar-Palma, "A survey of various propagation models for mobile communication," *Antennas and Propagation Magazine, IEEE*, vol. 45, no. 3, pp. 51–82, June 2003.

40. L. M. Correia (Ed.), *Wireless Flexible Peronalised Communicatons (COST 259 Final Report)*. Wiley, 2001.

41. L. M. Correia (Ed.), *Mobile Broadband Multimedia Networks (COST 273 Final Report)*. academic Press, 2006.

42. R. Verdone and A. Zanella (Eds.), *Pervasive Mobile and AmbientWireless Communications: COST Action 2100*, 2012.

43. K. Rizk, R. Valenzuela, S. Fortune, D. Chizhik, and F. Gardiol, "Lateral, full-3d and vertical plane propagation in microcells and small cells," in *Vehicular Technology Conference, 1998. VTC 98. 48th IEEE*, vol. 2, May 1998, pp. 998–1003 vol. 2.

44. S. Tan and H. Tan, "A microcellular communications propagation model based on the uniform theory of diffraction and multiple image theory," *Antennas and Propagation, IEEE Transactions on*, vol. 44, no. 10, pp. 1317–1326, Oct. 1996.

45. T. Fugen, J. Maurer, T. Kayser, and W. Wiesbeck, "Capability of 3-d ray tracing for defining parameter sets for the specification of future mobile communications systems," *Antennas and Propagation, IEEE Transactions on*, vol. 54, no. 11, pp. 3125–3137, Nov. 2006.

46. T. Binzer and F. Landstorfer, "Radio network planning with neural networks," in *Vehicular Technology Conference, 2000. IEEE VTS-Fall VTC 2000. 52nd*, vol. 2, 2000, pp. 811–817 vol. 2.

47. G. R. Cerri and P., "Application of an automatic tool for the planning of a cellular network in a real town," *Antennas and Propagation, IEEE Transactions on*, vol. 54, no. 10, pp. 2890–2901, Oct. 2006.

48. G. Liang and H. Bertoni, "A new approach to 3d ray tracing for site specific propagation modeling," in *Vehicular Technology Conference, 1997, IEEE 47th*, vol. 2, May 1997, pp. 1113–1117 vol. 2.

49. T. Kürner (Ed.) and et al, "Final report on automatic planning and optimisation," *IST-2000-28088 MOMENTUM*, Oct. 2003.

50. Y. Corre and Y. Lostanlen, "An enhanced method to extrapolate in 3d antenna radiation patterns in the context of the radio network planning," in *The European Conference on Antennas and Propagation: EuCAP 2006 (ESA SP-626)*, Nov. 2006, pp. 441.1.

51. W.-Y. Lee, "Effects on correlation between two mobile radio base-station antennas," *Vehicular Technology, IEEE Transactions on*, vol. 22, no. 4, pp. 130–140, Nov. 1973.

13

System-Level Simulations with the IMT-Advanced Channel Model

Jan Ellenbeck

Technische Universität München, Germany

13.1 Introduction

One of the main applications of channel models is to enable the performance evaluation of wireless systems by means of computer simulations. Simulations of wireless systems like LTE-Advanced are essential in the research and standardization phase when the technology is still being developed. At this stage, prototypes are often not available and testing different candidate features in the field would be too expensive and time consuming. In comparison, simulations can be set up easily and deliver quick results. For the simulation results to be reliable, however, both the channel as well as the investigated wireless system have to be modeled accurately enough. The accuracy of the channel model is important because the performance gains over legacy systems delivered by LTE-Advanced systems mainly stem from a better exploitation of the radio channel's selectivity in the time, frequency, and spatial domain. Consequently, the International Telecommunication Union (ITU) adopted a well-established spatial channel model developed within the WINNER project when it defined detailed guidelines [1] for the evaluation of LTE-Advanced and IEEE 802.16m-2011 WirelessMAN-Advanced as so-called International Mobile Telecommunication (IMT)-Advanced compliant technologies.

In Sections 13.2 and 13.3 of this chapter, we will present the ITU's evaluation guidelines and the channel model. They serve as de facto standards which were followed by both the 3GPP and IEEE standardization bodies with their individual members during their self-evaluation as well as by the independent IMT-Advanced evaluation groups. During their evaluation process, the individual members of the standardization and evaluation groups calibrated their simulation tools. In Section 13.4 we will present the metrics used to calibrate the channel model implementations and show example calibration results. To model a whole wireless system for system-level simulations, a so-called link-to-system

LTE-Advanced and Next Generation Wireless Networks: Channel Modelling and Propagation, First Edition.
Edited by Guillaume de la Roche, Andrés Alayón Glazunov and Ben Allen.
© 2013 John Wiley & Sons, Ltd. Published 2013 by John Wiley & Sons, Ltd.

model is typically employed to reduce simulation complexity while preserving the main performance characteristics of the wireless links. In Section 13.5 we will introduce some commonly used link-to-system mapping techniques. Building on that, in Section 13.6 we will show how to use the channel model and the link-to-system modeling techniques for the simulation of an LTE system. Finally, in Section 13.6.3 we will discuss the results of the IMT-Advanced system-level simulator calibration campaign conducted by the 3GPP Radio Access Network (RAN) 1 working group.

Most likely, all companies and organizations actively taking part in the standardization process have their own proprietary implementations of the channel model and abstraction techniques presented in this chapter. They use their simulators to contribute simulation results to standardization meetings but they do not share simulation tools and there is no publicly available system-level simulator reference implementation. MATLAB® implementations of the IMT-Advanced channel model are available from the ITU's website [2] and independent MATLAB® link-level simulators have been made available as well [3–5]. System-level simulators in MATLAB® [6] and C++ [7] allow system-level simulations of LTE systems to a certain extent. An efficient C++ implementation of the ITU channel model, the presented link-to-system abstraction techniques as well as a Multiple Input Multiple Output (MIMO)-capable LTE protocol stack have been released by the author as an open-source project called IMTAphy under the GNU Public License on the Internet [8, 9].

13.2 IMT-Advanced Simulation Guidelines

In its "Guidelines for evaluation of radio interface technologies for IMT-Advanced", first published in 2008 in report ITU Radiocommunication Sector M.2135 [10] with subsequent corrections [1], the ITU has defined the general evaluation methodology, the relevant figures of merit, test environments, scenarios, and corresponding channel models, which we will present in the following. The aim of that report was to define test environments that are representative of expected use cases for IMT-Advanced systems and to allow comparable evaluation results by providing well-defined evaluation guidelines. In this chapter we will focus on system-level simulations in accordance with report M.2135 because the ITU requires their use for evaluating the system capacity. Other (simpler) evaluation methods can be used to assess characteristics like peak rates or latency bounds. In essence, system-level simulations consider the performance across multiple layers of the protocol stack for a whole cellular system consisting of many communication links in multiple cells. In contrast to link-level simulations, which usually focus on a single communication link only, system-level simulations model all communication links in a system. Modeling all links allows the mutual interference between multiple links to be considered which is an important factor limiting the system's performance.

13.2.1 General System-Level Evaluation Methodology

For system-level simulations, Base Stations (BSs) and Mobile Stations (MSs)[1] are placed on a two-dimensional scenario that is generic, that is, not specific to a certain location.

---

[1] In this chapter, we stick to the ITU-R nomenclature and use the words *Base Station* (BS) and *Mobile Station* (MS) instead of *Evolved NodeB* (eNB) and *User Equipment* (UE) as the 3GPP does in the context of LTE.

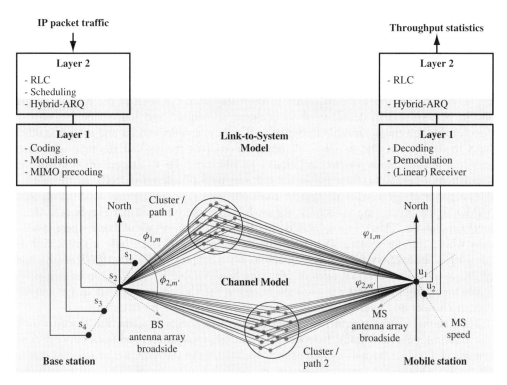

Figure 13.1 Downlink transmission through protocol stack and over MIMO channel with multi-path propagation (only two paths shown) between a BS and an MS.

The users (MSs) are dropped randomly into the scenario and for the duration of this *drop* they exchange packet-based data traffic in the DownLink (DL) and UpLink (UL) direction with their serving base station. Figure 13.1 shows IP packets traveling through a simplified protocol stack and over the wireless channel from one of the base stations to one of its associated mobile stations in the downlink direction. In a system-level simulation multiple base stations at different sites each serve a number of mobile stations so that hundreds or thousands of such links are modeled. Depending on the underlying channel conditions of each link and depending on how the radio technology handles them, a smaller or bigger number of packets correctly arrives at the receiving end of the link. The carried traffic (goodput) for each link together with other metrics of interest like packet delays or Signal to Interference plus Noise Ratios (SINRs), is included in the statistical evaluation of the system-level simulation.

Figure 13.1 also shows the typical scope of a system-level simulation and of the models that are involved: IP packets are created by a traffic generator that models the behavior of higher layers. When determining spectral efficiencies, an inexhaustible traffic source keeps the senders' buffers filled at all times (full buffer assumption). Other evaluations, for example, for determining the Voice over IP (VoIP) capacity, make use of more sophisticated traffic models.

The investigated radio interface technology (e.g., LTE) is modeled from the point where the outgoing packets reach the transmitter's *data link layer* (layer 2). Typical layer

2 functions are Radio Link Control (RLC) and scheduling. In a base station the scheduler makes resource allocation decisions. To perform channel-dependent scheduling and link-adaptation, the BS scheduler needs Channel State Information (CSI), which is provided by the *physical layer* (layer 1). In the uplink, the BS layer 1 can directly measure the CSI but in the downlink it has to rely on MS feedback. From a system-level perspective, the scheduling and resource allocation functionality is at the heart of the evaluation and should be modeled in detail: Channel-adaptive scheduling and link adaptation have a direct impact on the achievable throughput. But they also depend on and influence other links in the system due to inter-cell interference. This means that resource allocation and *interference management* techniques should only be evaluated in system-level simulations where not only a single cell but a number of interfering cells are modeled.

The physical layer with its typical functions like channel coding, modulation, and MIMO signal processing is usually not modeled on the bit or symbol-level in system-level simulations. This level of detail is only found in *link-level simulations* where instead of a whole cellular system with hundreds or thousands of links only a single link is considered. Instead, system-level simulations often employ a so-called *PHY abstraction* or *link-to-system model* that aims at capturing how reliable the physical layer transports a data block from the sender to the receiver given the chosen resource allocation and the current channel and interference conditions (see Section 13.5 for details).

The channel model together with the deployment scenario forms the basis for link and system-level simulations. It models the general propagation conditions between base stations and mobile stations as well as their fluctuations in the time, frequency, and spatial dimension. Due to the temporal fluctuations of the channel and because traffic and thus interference conditions in each cell are potentially dynamic, system-level simulations are performed over a certain simulation time frame. During that time (typically a couple of seconds), statistics are gathered to compute the metrics of interest. The channel conditions are not uniform over the scenario area so that each user's conditions depend on its randomly assigned position. As users do not move but are at a fixed position during one drop, each user only samples the channel at its position. To achieve a statistically sound result that is representative for the whole scenario area, it is thus common to re-drop users to different positions during the course of the simulation. Usually, statistics are not gathered during the transient phase after a new drop to allow, for example, control loops to settle.

13.2.2   System-Level Performance Metrics

In its report M.2134 [11] the ITU-R outlines the requirements that IMT-Advanced systems have to meet and defines the following metrics of interest, amongst others: the *normalized user throughput*, the *cell spectral efficiency*, the *cell edge user spectral efficiency*, and the *user plane latency*.

The *normalized user throughput* $\gamma_i^{(\mathrm{Dir})} = \chi_i^{(\mathrm{Dir})}/(T_i \omega_{\mathrm{Dir}})$ measured in bit/s/Hz is defined as the ratio of correctly delivered bits $\chi_i^{(\mathrm{Dir})}$ for a user i over its active session time T_i and over the bandwidth $\omega^{(\mathrm{Dir})}$ available in the whole system. Here, *Dir* serves as a placeholder for DL and UL because the amount of delivered bits and the available bandwidth can be different for downlink and uplink. The throughput is measured separately for the DL and UL direction at the interface between layers 2 and 3, for example, between the IP and LTE data link layer. For $\omega^{(\mathrm{Dir})}$ the total system bandwidth (including guard bands) available for the DL or UL direction is considered even if the actual used bandwidth per cell is less due to, for example, a frequency reuse pattern. For example, $\omega^{(DL)} = \omega^{(UL)} = 10\,\text{MHz}$ if an operator has $10\,\text{MHz} + 10\,\text{MHz}$ of paired Frequency Division Duplexing (FDD) spectrum or if a $20\,\text{MHz}$ Time Division Duplexing (TDD) spectrum is equally shared between DL and UL. For a given number of users per cell, the Cumulative Distribution Function (CDF) of the normalized user throughput $\gamma_i^{(Dir)}$ shows which throughputs the users can expect and how fairly the throughputs are distributed among the users. Based on $\gamma_i^{(Dir)}$ two important scalar performance metrics are defined. The *cell spectral efficiency* $\eta^{(Dir)} = \sum_{i=1}^{N_{MS}} \gamma_i^{(Dir)}/N_{BS}$ in bit/s/Hz/cell is the sum of the normalized user throughputs over all N_{MS} users in the system divided by the number of cells N_{BS}. It shows an operator how much capacity to expect per cell. The *cell edge user spectral efficiency* as the 5th percentile of the normalized user throughput CDF shall indicate how much throughput a "cell edge" user can expect. The users at the 5th percentile level of the CDF are not necessarily at the cell edge or in an area with weak signal conditions though. They could also be disadvantaged by the scheduling process. Thus, a low metric value can also indicate an unfair user scheduling process. Finally, the *user plane latency* is defined as the time that passes from the moment an IP packet enters the layer 2 on the transmitter side until it leaves the layer 2 on the receiver side of the link, see Figure 13.1.

13.2.3   Test Environment and Deployment Scenario Configurations

The ITU-R report M.2135 [1] distinguishes between so-called *test environments*, the corresponding *deployment scenarios*, and the associated *channel models*. Table 13.1 lists the deployment scenarios that have been selected by the ITU for the IMT-Advanced evaluation: the *indoor hotspot scenario*, the *urban micro-cell scenario*, the *urban macro-cell scenario*, and the *rural macro-cell scenario*. They belong to the *indoor*, the *micro-cellular*, the *base coverage urban*, and the *high speed* test environment, respectively. For

Table 13.1   Overview of deployment scenarios and simulation parameters [1]

Parameter/ deployment scenario	Indoor hotspot	Urban micro-cell	Urban macro-cell	Rural macro-cell
Test environment	indoor	microcellular	base coverage urban	high speed
Channel model	InH	UMi	UMa	RMa
Total BS transmit power	21 dBm	44 dBm	49 dBm	49 dBm
Max. MS transmit power	21 dBm	24 dBm	24 dBm	24 dBm
Center frequency	3.4 GHz	2.5 GHz	2 GHz	800 MHz
BS height	6 m	10 m	25 m	35 m
BS antenna downtilt [12]	0°	12°	12°	6°
Inter-site distance	60 m	200 m	500 m	1732 m
Min. BS−MS distance	3 m	10 m	25 m	35 m
User locations	indoor	50% indoor, 50% outdoor	in vehicle	in vehicle
User speed	3 km/h	3 km/h	30 km/h	120 km/h

each deployment scenario there is a channel model of the same name. In this text we will refer to the different deployment scenarios and channel models by using the acronyms (InH, UMi, UMa, and RMa) for the channel models as introduced by the ITU [1]. The aforementioned scenarios are considered mandatory by the ITU. In addition, there is an optional Suburban Macro (SMa) scenario and channel model that belongs to the base coverage urban test environment as well. It is fully specified in report M.2135 [1] but we will not discuss it here.

Most of the deployment scenario assumptions presented in Tables 13.1 and 13.2 and in the following sections are based on the original ITU guidelines [1] and subsequent "Guidelines for using IMT-Advanced channel models" [13]. These were issued to help the evaluation groups interpret the original guidelines. Some other parameters like the BS antenna downtilt have not been explicitly specified by the ITU. For these we will use the 3GPP's assumptions for system-level simulator calibration as described in the annex of TR 36.814 [12].

All deployment scenarios except the InH scenario use a conventional cellular deployment model with 19 base station sites. Three base stations, each equipped with a sectorized antenna array serving a hexagonal cell, are co-located per site as depicted in Figure 13.2(a). Note that, despite the highly regular deployment geometry, the various propagation effects explained in Section 13.3 lead to non-homogeneous perturbations of the signal strength as shown in Figure 13.2(b). The different hexagonal scenario types all consist of $19 \times 3 = 57$ cells in total. But the cell sizes and thus the covered area vary greatly due to different Inter-Site Distances (ISDs) that range from 200 m (for UMi) to 1732 m (for RMa). The InH scenario, in contrast, is a very small scenario with just two base stations equipped with omnidirectional, that is, non-sectorized antennas. The base stations are located at 60 m ISD in the middle of a small 120 m × 50 m indoor area. Table 13.1 lists further differences between the deployments like different BS and MS transmit power budgets for 20 MHz of spectrum. Some parameters are common to all deployments, see Table 13.2.

For each simulation drop, $10 \times 57 = 570$ users are randomly and uniformly distributed over the whole area of the hexagonal scenarios and $10 \times 2 = 20$ users are dropped onto the rectangular indoor hotspot area. The user positions are kept fixed during the drop duration. But to model the fast-fading channel, the users are imagined to move at a scenario-specific speed into randomly assigned directions. A BS serves 10 mobiles on average. Each MS is always only associated to the base station of one cell which is called its *serving cell*.

Table 13.2 Common deployment scenario parameters [1]

Parameter	Value
BS noise figure	5 dB
MS noise figure	7 dB
BS antenna gain	17 dBi (0 dBi for Indoor Hotspot (InH))
MS antenna gain	0 dBi
Avg. users per cell	10
Feeder loss at BS	2 dB for large-scale calibrations
	0 dB otherwise (see Tab-A.2.2-1 [12])
Handover margin	1 dB [12]

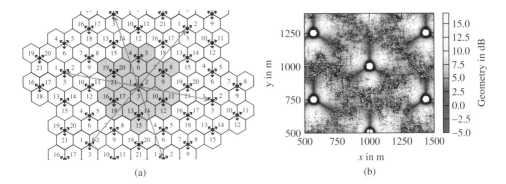

(a)   (b)

Figure 13.2 Hexagonal cell layout as used in the UMa, UMi, and RMa deployment scenarios. (a) Wraparound configuration surrounding a 21 cells (7 sites) scenario with 6 shifted copies of itself; arrows mark the offset of each copy; (b) Geometry distribution over the central area of a 19 sites UMa scenario.

The exact amount of users per BS depends on the random spatial distribution of users, on the random shadowing (see below), and on the association and handover process of the radio interface technology. The handover process itself usually cannot be simulated because the MS positions are fixed and only the fast-fading varies during the drop. But to account for the handover hysteresis which keeps a mobile associated to its previous BS while another is already stronger, the 3GPP assumes [12] that the mobiles are associated to one of the base stations whose Reference Signal Received Power (RSRP) power level is within a 1 dB *handover margin* from that of the BS with the strongest received signal.

Many of the channel model properties presented in Section 13.3 depend on distances and angles between base station sites and mobiles. All distances and angles are measured in the xy-plane in two dimensions only without taking BS or MS heights into account. The only exception are the antenna field patterns, which are modeled in three dimensions (see Section 13.2.4) so that the elevation angles between mobiles and base stations are computed from their relative heights and distances. To assure a minimum pathloss and coverage by the BS antenna elevation pattern, users that are too close to a base station site are re-dropped to a different position in the system area.

Real-world cellular systems cover large areas with thousands of potentially interfering cells. They are typically inter-cell interference-limited. For modeling cellular systems, a good trade-off between a low complexity and an accurate interference level is to surround the three cells of a site under consideration by two rings of interfering sites ($6 + 12 = 18$ sites with $3 \times 18 = 54$ interfering cells). The interference from cells farther away can be neglected due to the high pathloss. However, from the 57 cells in total, only the performance of the three inner cells could be evaluated because the remaining cells are more or less on the edge of the scenario area where they do not receive equally strong interference from all directions. To avoid this, a so-called *wraparound* technique can be used. It creates a torus-like deployment topology where all cells appear to be surrounded by two rings of interfering sites. This way, all cells face a realistic and uniform inter-cell interference situation and can all be included in the performance evaluation. For IMT-Advanced system-level simulations wraparound modeling is mandatory [1]. In IMTAphy we realize this on a per BS–MS link basis by shifting an MS to that position among seven possible positions that yields the smallest Euclidean distance in the xy-plane to the BS. Note that for each station we model a link to come from the direction of the shortest distance and not from all possible directions at once. Figure 13.2(a) shows an exemplary 21 cells scenario (highlighted cells in the center) surrounded by six virtual copies, which are shifted from the scenario's center according to the plotted shift vectors. Of course, a wraparound is not applied in the InH scenario because it models two isolated base stations in an indoor setting.

13.2.4 Antenna Modeling

The double-directional IMT-Advanced channel model introduced in the following section separates the modeling of BS and MS antennas from the modeling of the spatial propagation channel itself because there is a fundamental difference: the propagation channel, on the one hand, is caused by the environment and can only be accurately modeled, but it cannot be changed. The antennas, on the other hand, can actually be designed by a system engineer so that it is important to support arbitrary far-field antenna patterns. The channel model distinguishes between horizontal and vertical polarization components of the antenna field pattern. However, the IMT-Advanced evaluation guidelines [1] only specify simple antenna patterns with purely vertical polarization. The 3GPP describes the modeling of polarized antennas with arbitrary polarization slant angles in the annex of TR 36.814 but only uses purely vertically polarized antennas for their calibration setups [12]. The antennas of the mobiles as well as the indoor hotspot BS antennas have an omnidirectional pattern with 0 dBi gain. The BS antennas for the hexagonal scenarios should only radiate into a 120° wide cell sector and also focus the radiated power downwards into their own cell. The ITU thus specifies a three-dimensional (3D) pattern whose azimuth (in the xy-plane) and elevation (upward from the xy-plane) power gain patterns are shown as polar plots in Figure 13.3.

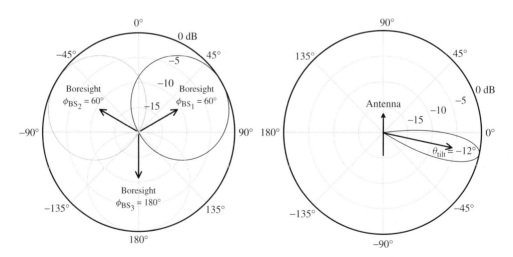

Figure 13.3 Polar plots of antenna gain patterns. (a) Top view with 3 azimuth sector patterns; (b) Side view of elevation pattern at $\theta_{\text{tilt}} = -12°$.

The 3D antenna pattern is defined as having a maximum gain $G(\phi, \theta) = 0\,\text{dB}$ in addition to a 17 dBi gain in the boresight direction and a forward-backward attenuation of $G_{\text{Min}} = 20\,\text{dB}$ resulting in a maximum attenuation of $G(\phi, \theta) = -20\,\text{dB}$. The pattern is defined in two components with respect to the azimuth angle ϕ as $G_{\text{Azimuth}}(\phi)$ in (13.1) and with respect to the elevation angle θ as $G_{\text{Elevation}}(\theta)$ in (13.2). As can be seen in Figure 13.3, the elevation pattern with a 3 dB beamwidth $\theta_{3\,\text{dB}} = 15°$ is much narrower than the azimuth pattern with a 3 dB beamwidth $\phi_{3\,\text{dB}} = 70°$. The final 3D pattern is computed as the sum of the attenuations $G_{\text{Azimuth}}(\phi)$ and $G_{\text{Elevation}}(\theta)$ by applying (13.3) in a way that assures that the maximum attenuation of $G_{\text{Min}} = 20\,\text{dB}$ is not exceeded. The patterns are defined relative to the BS antenna array broadside direction azimuth angle ϕ_{BS} and the downtilt angle θ_{tilt}. The downtilt is not specified by the ITU so that Table 13.1 lists the 3GPP default values.

$$G_{\text{Azimuth}}(\phi) = -\min\left[12\left(\frac{\phi - \phi_{\text{BS}}}{\phi_{3\,\text{dB}}}\right)^2, G_{\text{Min}}\right] \text{ in dB}, \quad -180° \leq \phi - \phi_{\text{BS}} \leq 180°$$

(13.1)

$$G_{\text{Elevation}}(\theta) = -\min\left[12\left(\frac{\theta - \theta_{\text{tilt}}}{\theta_{3\,\text{dB}}}\right)^2, G_{\text{Min}}\right] \text{ in dB}, \quad -90° \leq \theta - \theta_{\text{tilt}} \leq 90° \quad (13.2)$$

$$G(\phi, \theta) = -\min\left[-\left(G_{\text{Azimuth}}(\phi) + G_{\text{Elevation}}(\theta)\right), G_{\text{Min}}\right] \text{ in dB} \quad (13.3)$$

13.3   The IMT-Advanced Channel Models

For the evaluation of IMT-Advanced systems the ITU has defined [1] a channel model consisting of a *primary module* and an *extension module*. The primary module consists of pathloss models, scenario-specific parameter sets for the InH, UMa, UMi, RMa, and SMa scenarios, and a so-called *generic model*. The generic model specifies the mathematical model and the algorithms used for channel modeling that apply to all scenarios. It is based on the WINNER 2 channel model [14] which, in turn, is similar to the 3GPP/3GPP2 SCM channel model [15]. The scenario-specific parameters are used to instantiate scenario-specific spatial channel models. The optional extension module allows alternate parameter sets for the UMa, RMa, and SMa scenarios to be generated. Here, we will focus on the primary module with its default parameter sets because this was actually used for the IMT-Advanced evaluations conducted by the 3GPP and the independent evaluation groups.

The channel model, as shown in Figure 13.1, models the propagation channel between each BS antenna s and each MS antenna u for a link between a base station and mobile station. In a system level simulation, the channel model has to cover all links, both serving and interfering. In fact, being able to model the different (but potentially correlated) channel conditions on different links and between different antenna pairs of the same link, is what makes the channel model suitable for MIMO system-level simulations.

The generic model employs a geometry-based stochastic approach to represent the multipath propagation channel between base stations and mobile stations. As an end result, the model provides the time-variant Channel Impulse Response (CIR) $h_{u,s}(t)$ for an antenna pair (u, s):

$$h_{u,s}(t) = \sum_{n=1}^{N} h_{u,s,n}(t)\delta(t - \tau_n). \tag{13.4}$$

For each of the N paths, a complex path coefficient $h_{u,s,n}(t)$ describes the change in amplitude and phase a complex baseband signal perceives propagating along the considered path that causes a delay of τ_n relative to the fastest path. Besides these multipath effects, the signal also suffers from large-scale fading effects such as pathloss and shadowing that directly scale the norm of the path coefficients $h_{u,s,n}$. By applying the Fourier transform, the CIR can be transformed into the frequency domain resulting in the Channel Transfer Function (CTF) $H_{u,s}(t, f)$ at a given frequency f. Note that path coefficients, delays, and all related values generally differ between different BS–MS links.

The ITU generic model is not site-specific and does not model actual objects in the environment of a station like buildings that serve as scatterers and reflectors for the multipath components. Instead, it generates a number N of scattering clusters. Each cluster reflects the signal on its way from the transmitter to the receiver and contributes one path to the multipath propagation. The geometric location of the n-th cluster is implicitly defined by its propagation delay τ_n and the angles under which cluster n is seen from the BS and MS. The nomenclature follows a downlink perspective so that the angles ϕ_n at the BS are called *angles of departure* (AODs) while the angles φ_n at the MS are called *angles of arrival* (AOAs). The model is stochastic because the actual angles and delays are generated randomly from a properly configured probability distribution. With the AOAs, the AODs, and the delays, the model captures the properties of multipath propagation that determine the fading nature of the MIMO channel: In the frequency domain the individual delays τ_n of the paths cause the *frequency-selective* fading. Wavefronts of multipath components arriving with distinct AOAs at the receive array are seen with different relative phase offsets at distinct receive antenna array elements $u_i \neq u_j$. This leads to a *spatially-selective* fading behavior.

Figure 13.1 shows two exemplary paths between a base station and a mobile station antenna array. Depending on the chosen deployment scenario and on the propagation condition, between 10 and 20 clusters per link have to be modeled in the IMT-Advanced channel model. In addition, the clusters are assumed to have a substructure of $M = 20$ individual scatterers, as the figure shows. Each of these scatterers reflects a ray m under slightly different angles of departure $\phi_{n,m}$ and arrival $\varphi_{n,m}$ as seen from the BS and MS, respectively. Introducing such a per-cluster angular spread increases the angular resolution of the channel model which is important for accurate MIMO evaluations. In (13.16) on page 365 we will show how a single path coefficient $h_{u,s,n}(t)$ is computed from the superposition of these 20 rays per cluster. Before we get there, however, we will first introduce how the necessary large-scale and small-scale parameters are generated.

13.3.1 Large-Scale Link Properties

As a first step, a set of so-called [1] *large-scale parameters* is determined for each link in the system. There are two broad classes of large-scale parameters. The first class of large-scale parameters, including pathloss and shadowing, describes the large-scale fading a link experiences. The other class, including the delay and angular spread of a link, serves to parameterize the probability distributions from which the individual delays and angles are drawn (the *small-scale parameters*).

13.3.1.1 Determining Line-of-Sight Propagation Conditions

After the users have been dropped into the selected deployment scenario, the first step to initialize the channel model is to randomly classify each BS–MS link to exhibit Line Of Sight (LOS) or Non Line Of Sight (NLOS) propagation. As Table 13.3 shows, depending on the two-dimensional (2D) distance d, a LOS probability $\Pr(d)$ is applied, reflecting that LOS propagation is more likely the closer a user is to a base station site. Note that the propagation condition is decided on a per-site basis so that a user always experiences identical propagation conditions to all base stations co-located at one site. Table 13.3 also lists the percentage of LOS propagation observed in system-level simulations. Except for InH, NLOS propagation dominates when all (serving + interfering) links are considered. Most serving links are LOS though because users are close to their serving base station and having a line-of-sight to a certain BS increases the chances that this BS is seen as the strongest BS and thus gets selected as the serving BS.

Depending on the propagation condition, different parameters are used for the pathloss and spatial channel model as explained in the following sections. The UMi scenario is a special case because in UMi half of the users are outdoors and half of them are indoors. This classification is done randomly with a 50 percent probability when a UMi drop is initialized. For the outdoor users, the propagation can be LOS or NLOS and according pathloss models and channel model parameter sets are used. For the indoor users, there is a special UMi Outdoor to Indoor (O2I) channel model parameter set. While the UMi O2I propagation does not exhibit LOS Rician fading, both NLOS and LOS pathloss models can be applied to the outdoor portion d_{out} of the distance between the MS and the BS [13]. The outdoor distance $d_{out} = d - d_{in}$ is computed by subtracting a uniformly randomly chosen indoor distance $d_{in} \sim \mathcal{U}(0, \min[d, 25])$ from the total distance d in meters.

Table 13.3 Probability of line-of-sight propagation

Scenario	Probability as a function of distance [1]	Per link probability	
		All links	Serving links
InH	$\Pr(d) = \begin{cases} 1 & d \leq 18 \\ \exp((18 - d)/27) & 18 < d < 37 \\ 0.5 & d \geq 37 \end{cases}$	67%	92%
RMa	$\Pr(d) = \begin{cases} 1 & d \leq 10 \\ \exp((10 - d)/1000) & d > 10 \end{cases}$	12%	82%
UMa	$\Pr(d) = \min\left(\dfrac{18}{d}, 1\right)\left[1 - \exp\left(\dfrac{-d}{63}\right)\right] + \exp\left(\dfrac{-d}{63}\right)$	4%	37%
UMi	$\Pr(d_{out}) = \min\left(\dfrac{18}{d_{out}}, 1\right)\left[1 - \exp\left(\dfrac{-d_{out}}{36}\right)\right] + \exp\left(\dfrac{-d_{out}}{36}\right)$	9%	75%

13.3.1.2 Pathloss Models

The user locations are also relevant when computing the pathloss on a link because for UMi indoor users a wall penetration loss $PL_{tw} = 20\,dB$ and an additional indoor pathloss $PL_{in}(d_{in}) = d_{in} \times 0.5\,dB/m$ are added to the basic outdoor pathloss PL_{basic}:

$$PL^{(UMi,O2I)}(d) = PL_{basic}(d) + PL_{tw} + PL_{in}(d_{in}) \text{ in dB.} \qquad (13.5)$$

In the UMa and RMa scenarios all users are in vehicles (see Table 13.1) so that the basic outdoor pathloss $PL_{basic}(d)$ is adjusted by a log-normal distributed vehicle penetration loss PL_{Veh} with a mean of $9\,dB$ and a standard deviation of $5\,dB$:

$$PL(d) = PL_{basic}(d) + PL_{Veh} \text{ in dB} \quad \text{with} \quad PL_{Veh} \sim \mathcal{N}(9, 5^2) \text{ in dB.} \qquad (13.6)$$

Note that the vehicle penetration loss PL_{Veh} is specific to each vehicle so that the same loss from one user to all base stations is assumed. The presented pathloss adjustments were already mentioned in the initial 2008 ITU guidelines [10] but were subsequently clarified in 2009 [13]. In the InH scenario no adjustments to the basic pathloss are necessary because no outdoor walls are penetrated and the loss due to indoor walls is already considered in the basic pathloss formulas. Some simulations account for a feeder loss of, for example, $2\,dB$ which can simply be added to the overall pathloss.

The ITU adopted the basic pathloss $PL_{basic}(d)$ formulas from the WINNER channel model, which is based on measurement campaigns and literature results [14]. There are special pathloss formulas for each deployment scenario (InH, UMa, UMi, RMa, and SMa) and for both LOS and NLOS propagation types. For LOS propagation in the hexagonal scenarios a dual slope pathloss model is used that assumes a higher pathloss exponent for distances above a certain break point distance d_{BP}. The pathloss models are valid for system center frequencies in the range of $2\,GHz$–$6\,GHz$ ($450\,MHz$–$6\,GHz$ for RMa) and allow certain parameters like BS and MS antenna heights (all scenarios) as well as street widths and average building heights (UMa, RMa, and SMa) to be specified. Due to this configurability, the pathloss formulas given by the ITU [1] are lengthy. For that reason, we choose to only present the UMa pathloss model in a simplified form assuming standard values as given in Table 13.1 and in report M.2135 [1]. The basic pathloss $PL_{basic}(d)$ for a given distance d in meters and NLOS propagation is computed according to (13.7) with a pathloss exponent of 3.9. For LOS propagation (13.8) distinguishes between two different pathloss exponents based on a break point distance threshold of $d_{BP} = 4(h_{BS} - 1)(h_{MS} - 1)f_{center}/c$ in meters, where c is the speed of light. Here, $h_{BS} = 25\,m$, $h_{MS} = 1.5\,m$, and $f_{center} = 2\,GHz$ so that $d_{BP} = 320\,m$:

$$PL_{basic}^{(NLoS)}(d) = 9.63 + 39\log_{10}(d) \text{ in dB}, \qquad (13.7)$$

$$PL_{basic}^{(LoS)}(d) = \begin{cases} 34.02 + 22\log_{10}(d) \text{ in dB} & \text{if } 10\,m < d < d_{BP} \\ -11.02 + 40\log_{10}(d) \text{ in dB} & \text{if } d_{BP} < d < 5000\,m. \end{cases} \qquad (13.8)$$

13.3.1.3 Correlated Large-Scale Parameters

As a last step before small-scale parameters and channel coefficients can be generated, a set of five correlated large-scale parameters is generated for each link in the scenario.

Table 13.4   UMa large-scale parameter distribution parameters [1]

Large-scale parameter	Mean μ		Std. dev. σ		Correlation distance d_{corr}	
	LoS	NLoS	LoS	NLoS	LoS	NLoS
Delay spread (DS) in $\log_{10}$ s	−7.03	−6.44	0.66	0.39	30 m	40 m
Angular spread in $\log_{10}$ (degree) of:						
– departure angles (ASD)	1.15	1.41	0.28	0.28	18 m	50 m
– arrival angles (ASA)	1.81	1.87	0.20	0.11	15 m	50 m
Shadow fading (SF)	0 dB	0 dB	4 dB	6 dB	37 m	50 m
Rician K factor (K)	9 dB	n/a	3.5 dB	n/a	12 m	n/a

The five large-scale parameters are (1) the delay spread σ_{delay} between the paths of one link, (2 and 3) the angular spreads σ_{AOD} and σ_{AOA} for the departure and arrival angles of all $N \times M$ rays, (4) the shadow fading, and (5) the Rician K factor specifying the power ratio between the direct LOS ray and the sum power of all other rays. The shadow fading is added to the pathloss to obtain the total large-scale fading of the link. The K factor is used to scale the ray powers in the case of LOS propagation. The other large-scale parameters (σ_{delay}, σ_{AOD}, and σ_{AOA}) are used to parameterize the probability distributions from which the small-scale parameters are obtained. In the following we present how the large-scale parameters are generated for the example of the UMa scenario with NLOS and LOS propagation. Other scenarios use different parameter sets than those shown in Table 13.4. Note that for NLOS propagation there are only four large-scale parameters because the Rician K factor is not needed.

The large-scale parameters are correlated in two ways: First, for each of the five large-scale parameters, a geographic correlation between different links that go to the same base station site is established depending on the geographic proximity of the involved mobiles. This is done separately for each of the five large-scale parameters. For example, the shadowing is identical for links going from one user position to the same site (full intra-site correlation). It is somewhat correlated for links of geographically close users if the links go to the same site and it is independent for links to different sites (no inter-site correlation)[2]. Second, for each link a cross-correlation between the five large-scale parameters is established by means of a correlation matrix.

The ITU has adopted [1] the correlation procedure proposed by the WINNER project [14]. It establishes the correlations in a normal Gaussian domain before transforming the large-scale parameters to their desired distribution. For each of the 19 sites and separately for LOS and NLOS propagation links, five matrices G_{DS}, G_{ASD}, G_{ASA}, G_{SF}, and $G_K \in \mathbb{R}^{\dim_x \times \dim_y}$ containing i.i.d. standard normal random variables for each $1\,m \times 1\,m$ grid point of the scenario area are created. Note that the dimensions $\dim_x$ and $\dim_y$ of the hexagonal scenarios can be over 1 km so that each matrix contains millions of entries. The geographic correlations inside these matrices are then established

---

[2] This means that interfering channels to different sites are not correlated in the IMT-Advanced model. The WINNER project states [14] that inter-site correlations were high in one drive test they conducted and low in another one. Because their data was limited, they assumed no correlations for links to different sites although their measurements showed such correlations could exist.

by a two-step convolution (e.g., first in the x-dimension and then in the y-dimension) with a one-dimensional Finite Impulse Response (FIR) filter $g(d) = \exp(-d/d_{\mathrm{corr}})$ depending on the distance d between two x or y coordinates as plotted in Figure 13.4(a). The parameter d_{corr} specifies the correlation distance, which differs for each combination of scenario and propagation condition, as listed in Table 13.4. Figure 13.4(b) shows a part of a filtered shadow fading grid G_{SF} belonging to the central site of the scenario already pictured in Figure 13.2(b). After introducing the geographic correlations, for each link a vector

$$\xi^{(x,y)} = \left([G_{\mathrm{DS}}]_{x,y} \quad [G_{\mathrm{ASD}}]_{x,y} \quad [G_{\mathrm{ASA}}]_{x,y} \quad [G_{\mathrm{SF}}]_{x,y} \quad [G_{\mathrm{K}}]_{x,y}\right)^{\mathrm{T}} \qquad (13.9)$$

is created by taking the values corresponding to the (x,y) position of the respective MS from the grids. Now, cross-correlations between the large-scale parameters of each link are introduced by a multiplication $s = \sqrt{R} \times \xi^{(x,y)}$ with a scenario and propagation-specific correlation matrix with $\sqrt{R} \times \sqrt{R} = R$. Equation (13.10), as an example, shows the cross-correlation coefficients for UMa LOS links as specified in report M.2135 [1]:

$$
\begin{array}{c}
 \begin{array}{ccccc} \mathrm{DS} & \mathrm{ASD} & \mathrm{ASA} & \mathrm{SF} & \mathrm{K} \end{array} \\
\begin{array}{c} \mathrm{DS} \\ \mathrm{ASD} \\ \mathrm{ASA} \\ \mathrm{SF} \\ \mathrm{K} \end{array}
\begin{pmatrix}
1 & 0.4 & 0.8 & -0.4 & -0.4 \\
0.4 & 1 & 0 & -0.5 & 0 \\
0.8 & 0 & 1 & -0.5 & -0.2 \\
-0.4 & -0.5 & -0.5 & 1 & 0 \\
-0.4 & 0 & -0.2 & 0 & 1
\end{pmatrix} = R^{(\mathrm{UMa,LoS})}
\end{array}
\qquad (13.10)
$$

After element-wise normalization, the vector $s = \left(s_{\mathrm{DS}} \quad s_{\mathrm{ASD}} \quad s_{\mathrm{ASA}} \quad s_{\mathrm{SF}} \quad s_{\mathrm{K}}\right)^{\mathrm{T}}$ of properly correlated standard normal random variables can now be used to obtain the final large-scale parameters per link. Note that for NLOS propagation, there is no Rician K large scale parameter. The grid G_{K} is thus not needed and R and $\xi^{(x,y)}$ have less entries, accordingly. All large-scale parameters are log-normal distributed. For a variable x whose logarithm $\log_{10}(x)$ follows a normal distribution $\mathcal{N}(\mu, \sigma^2)$ with mean μ and standard deviation

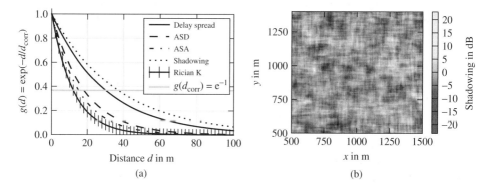

Figure 13.4 Large-scale parameter filtering for the UMa scenario with LOS propagation. (a) Correlation filter functions; (b) Filtered shadowing grid G_{SF}.

σ we write $x \sim \log_{10} \mathcal{N}(\mu, \sigma^2)$. The transformations from the normal Gaussian to the log-normal domain are shown in (13.11) to (13.15) with the respective mean and standard deviations given in report M.2135 [1] (see Table 13.4 for UMa NLOS and LOS values):

$$\sigma_{\text{delay}} = 10^{(\mu_{\text{DS}} + s_{\text{DS}}\sigma_{\text{DS}})} \sim \log_{10} \mathcal{N}(\mu_{\text{DS}}, \sigma_{\text{DS}}^2) \qquad \text{in s} \qquad (13.11)$$

$$\sigma_{\text{AOD}} = 10^{(\mu_{\text{ASD}} + s_{\text{ASD}}\sigma_{\text{ASD}})} \sim \log_{10} \mathcal{N}(\mu_{\text{ASD}}, \sigma_{\text{ASD}}^2) \qquad \text{in degree} \qquad (13.12)$$

$$\sigma_{\text{AOA}} = 10^{(\mu_{\text{ASA}} + s_{\text{ASA}}\sigma_{\text{ASA}})} \sim \log_{10} \mathcal{N}(\mu_{\text{ASA}}, \sigma_{\text{ASA}}^2) \qquad \text{in degree} \qquad (13.13)$$

$$\text{Shadowing} = s_{\text{SF}}\sigma_{\text{SF}} \sim \mathcal{N}(0, \sigma_{\text{SF}}^2) \qquad \text{in dB} \qquad (13.14)$$

$$\text{Rician K} = \mu_{\text{K}} + s_{\text{K}}\sigma_{\text{K}} \sim \mathcal{N}(\mu_{\text{K}}, \sigma_{\text{K}}^2) \qquad \text{in dB} \qquad (13.15)$$

13.3.2 Initialization of Small-Scale Parameters

After generating the correlated large-scale parameters for each link, the actual small-scale parameters can be determined. The exact process is specified in the ITU report [1]. Here we will only give a rough description: First the individual path delays τ_n are generated following an exponential distribution according to the delay spread σ_{DS}. The path delays τ_n are normalized by subtracting the smallest delay from all others and then sorted so that $\tau_1 = 0 \le \tau_2 \le \ldots \le \tau_N$. Afterwards, the power contribution P_n of each path is derived from the delay distribution. The path powers are normalized to $\sum_{n=1}^{N} P_n = 1$. Note that at this stage, the path powers are only fractions denoting how the power is distributed among paths. They have to be scaled by the transmit power and the large-scale attenuation effects to get the actual received power in watt. Within each path, the power is distributed uniformly over the 20 rays reflected by the corresponding cluster. Then, arrival and departure angles $\varphi_{n,m}$ and $\phi_{n,m}$ of each ray m of each cluster n are computed given the angular spread large-scale parameters σ_{AOA} and σ_{AOD}. A wrapped Gaussian distribution (Laplacian for InH) is used to generate arrival and departure angles for the cluster centers. The individual ray angles are then deterministically distributed around the cluster center angle according to a scenario-specific per-cluster angular spread. At this point, arrival and departure rays of the same cluster are randomly coupled to each other. Random initial phases $\Phi_{n,m} \sim \mathcal{U}(-\pi, \pi)$ are finally drawn (separately for all vertical (V) and horizontal (H) polarization combinations: VV, VH, HV, and HH) to model different ray path lengths and phase changes induced by the reflecting material. Before the path coefficients can be computed from these small-scale parameters, the 20 rays of the two strongest clusters (according to P_n) are split into three subclusters each (with 10, 6, and 4 rays each). The delays of the two strongest clusters are spread by fixed offsets of 0 ns, 5 ns, and 10 ns leading to more delay taps in the CIR as a mean to support bandwidths of up to 100 MHz [14].

The IMT-Advanced channel model is said to exhibit three levels of randomness: First, the correlated large-scale parameters are determined from random numbers. Second, the small-scale parameters are drawn randomly from distributions that are parameterized with the correlated large-scale parameters. Third, the initial phases are randomly selected from a uniform distribution. Once all of these parameters have been set up, the channel model behaves deterministically with respect to the model parameters.

13.3.3 Coefficient Generation

The computation of the complex path coefficients $h_{u,s,n}$ follows a sum of sinusoids approach that sums over all rays $m = 1 \ldots M$ that belong to a cluster n. Equation (13.16) shows that the summands for each ray consider the horizontal and vertical polarization components of the MS and BS antenna field patterns, the coupling between departure and arrival rays, the phase offsets between the antenna elements in the MS and BS array, and the time-dependent phase shift resulting from the ray's Doppler component $v_{n,m}$:

$$
h_{u,s,n}(t) = \sqrt{P_n} \sum_{m=1}^{M} \left[\overbrace{\begin{pmatrix} F_{MS,u}^{(V)}(\varphi_{n,m}) \\ F_{MS,u}^{(H)}(\varphi_{n,m}) \end{pmatrix}^{T}}^{\text{MS antenna pattern}} \overbrace{\begin{pmatrix} e^{j\Phi_{n,m}^{(VV)}} & \frac{e^{j\Phi_{n,m}^{(VH)}}}{\sqrt{\kappa}} \\ \frac{e^{j\Phi_{n,m}^{(HV)}}}{\sqrt{\kappa}} & e^{j\Phi_{n,m}^{(HH)}} \end{pmatrix}}^{\substack{\text{coupling between po-} \\ \text{larization components} \\ \text{with random phases}}} \overbrace{\begin{pmatrix} F_{BS,s}^{(V)}(\phi_{n,m}) \\ F_{BS,s}^{(H)}(\phi_{n,m}) \end{pmatrix}}^{\text{BS antenna pattern}} \right.
$$

$$
\left. \times \underbrace{\exp(2\pi j d_u \lambda_0^{-1} \sin(\varphi_{n,m}))}_{\text{MS array phase offset element } u} \underbrace{\exp(2\pi j d_s \lambda_0^{-1} \sin(\phi_{n,m}))}_{\text{BS array phase offset element } s} \underbrace{\exp(2\pi j v_{n,m} t)}_{\text{Doppler phase shift}} \right] \qquad (13.16)
$$

The computation in (13.16) depends on a number of parameters, which are summarized in Table 13.5. Besides the previously computed small-scale parameters (P_n, $\phi_{n,m}$, $\varphi_{n,m}$, and $\Phi_{n,m}$), a scenario-specific [1] cross-polarization power ratio κ is considered as well as a number of link-specific parameters: The (electrical) field patterns of the BS and MS antennas are derived, for example, from the 3D antenna power gain pattern introduced in (13.3) as $F_{BS,s}^{(V)}(\phi_{n,m}) = \sqrt{G(\phi_{n,m}, \theta)}$. When only vertically polarized antennas are assumed, the horizontal field pattern is set to 0 and only the vertical-to-vertical coupling with initial phase $\Phi_{n,m}^{(VV)}$ remains in (13.16). Note that the spatial channel model is

Table 13.5 Spatial channel model parameters

Parameter	Description
$h_{u,s,n}(t)$	complex path coefficient of CIR
s	BS antenna index $s = 1 \ldots S$
u	MS antenna index $u = 1 \ldots U$
n	path (cluster) index
m	ray index $m = 1 \ldots 20$
τ_n	delay of path n
P_n	power fraction of path n
$F_{MS,u}^{(V)}(\varphi_{n,m})$, $F_{MS,u}^{(H)}(\varphi_{n,m})$	electrical field pattern at MS antenna u and BS antenna s
$F_{BS,s}^{(V)}(\phi_{n,m})$, $F_{BS,s}^{(H)}(\phi_{n,m})$	with vertical (V) and horizontal (H) polarization
$\Phi_{n,m}^{(VV)}$, $\Phi_{n,m}^{(VH)}$, $\Phi_{n,m}^{(HV)}$, $\Phi_{n,m}^{(HH)}$	random initial phases
κ	cross-polarization power ratio
d_s, d_u	distances of antennas s and u to reference antenna element
$\phi_{n,m}$	angle of departure (at the BS) of ray m in cluster n
$\varphi_{n,m}$	angle of arrival (at the MS) of ray m in cluster n
λ_0	wavelength at center frequency
$v_{n,m}$	Doppler frequency component of ray m in cluster n
$\boldsymbol{v}$ and φ_v	mobile's speed vector and direction of travel in the xy-plane

two-dimensional only (clusters have no elevation) so that the elevation angle θ for the antenna pattern has to be derived from, for example, the line-of-sight elevation between the BS and MS. For uniform linear arrays at the BS and MS, d_s and d_u denote the distances between the reference antenna element (i.e., s = u = 1) and the considered elements s and u, respectively. For arbitrary antenna array layouts, effective distances d_s and d_u have to be computed depending on the arrays' geometries [1]. Finally, the Doppler frequency component $v_{n,m} = \|v\| \cos(\varphi_{n,m} - \varphi_v)/\lambda_0$ expresses the mobile's speed component along the ray direction with respect to the system center wavelength λ_0 for a mobile traveling in the azimuth direction φ_v.

For each link, the path coefficients for all paths $n = 1 \ldots N$ in the CIR are computed according to (13.16). If a link has LOS propagation, a single line-of-sight ray with power $P^{(\text{LOS})} = K/(K+1)$ is added to the zero-delay coefficient of the first path $h_{u,s,n=1}$ where K is the Rician power ratio large-scale parameter introduced in (13.15). Before adding the LOS ray, all path coefficients are scaled down as $h'_{u,s,n} = \sqrt{1/(K+1)} \times h_{u,s,n}$ so that the path powers still sum to 1.

The antennas are assumed to be in the far field so that each ray is seen as a planar wavefront arriving (and departing) with identical angles $\varphi_{n,m}$ at different antenna array elements resulting in spatially correlated fading. Different rays have different Doppler components so that the superposition of rays will cause temporal fading on each path. If an FDD system with different center frequencies for DL and UL is modeled, all parameters except λ_0 and the initial phases can be shared between the DL and UL channel model. For large frequency offsets, the DL and UL fading are uncorrelated though.

13.3.4 Computationally Efficient Time Evolution of CIRs and CTFs

As we have just seen, computing the CIR for a time instant t involves an elaborate channel model initialization procedure and then the summation over 20 rays in (13.16) for each of up to 24 (sub)clusters and for each antenna pair – and that is only for a single BS–MS link! Considering that a typical system-level simulation scenario consists of 57 cells with 10 mobiles per cell, we need to compute CIRs and CTFs for $57 \times 57 \times 10 = 32490$ links since we need to model serving as well as interfering links. Thus, for a 4×4 MIMO configuration up to $32490 \times 4 \times 4 \times 24 \times 20 \approx 250$ million complex ray coefficients have to be computed and summed for each time instant t. Clearly, an efficient implementation is needed to deploy the IMT-Advanced channel model in computer simulations given that system-level simulations should cover a significant time span.

The important insight here is that the specification of the channel model allows to shift the main computation effort from the time evolution phase to the initialization phase. Equation (13.16) can be rewritten to show that only one term depends on the current time t:

$$h_{u,s,n}(t) = \sum_{m=1}^{M} \left[a_{u,s,n,m} \exp(jb_{n,m}t) \right] \tag{13.17}$$

This means that at the beginning of a drop, after the large-scale correlation and small-scale initialization processes have been performed, we can compute the constants $a_{u,s,n,m}$ and $b_{n,m}$. For the above example, 250 million complex-valued constants $a_{u,s,n,m}$ are still needed though. But now an efficient computer implementation like, for example, in

`IMTAphy` [8, 9] is possible: We can stack the constants $a_{u,s,n,m}$ and $b_{n,m}$ for all links, antenna pairs, paths, and rays into huge vectors $\boldsymbol{a}$ and $\boldsymbol{b}$. Then for each new time evolution step t, we scale the vector $\boldsymbol{b}$ by t. Then we feed each element of the scaled vector $\boldsymbol{b}$ into the cis function defined as $\text{cis}(x) = \exp(jx)$. Both the scaling and the $\text{cis}(x)$ function are available in vector math libraries that distribute the workload over different processors and utilize the vector extensions of modern processors. The element-wise multiplication of the $\text{cis}(x)$ output with the vector $\boldsymbol{a}$, followed by a summation over $M = 20$ rays, can then be executed as multiple dot products of two length-M vectors for each link, antenna pair, and path.

With a suitable array layout that avoids copying intermediate results in the computer's memory, the resulting path coefficients can be regarded as a sequence of $\boldsymbol{Z} \in \mathbb{C}^{(U \cdot S) \times N}$ CIR matrices for all links. By matrix multiplications, the frequency domain CTF channel matrices $\boldsymbol{H}' = \boldsymbol{Z} \times \boldsymbol{F}$ with $\boldsymbol{H}' \in \mathbb{C}^{(U \cdot S) \times F}$ can be computed from the Fourier coefficients $\boldsymbol{F} \in \mathbb{C}^{N \times F}$ with $[\boldsymbol{F}]_{n, f_{\text{idx}}} = \exp(-2\pi j f_{\text{center}}(f_{\text{idx}}) \tau_n)$, where $f_{\text{center}}(f_{\text{idx}})$ is the center frequency of sub-band $f_{\text{idx}} = 1 \ldots F$. Using a different memory indexing, the complex coefficients of the CTF channel matrices $\boldsymbol{H}' \in \mathbb{C}^{(U \cdot S) \times F}$ can be accessed as a series of frequency domain channel matrices $\boldsymbol{H} \in \mathbb{C}^{U \times S}$ organized in a conventional Rx-by-Tx layout (here for the DL).

High-efficiency implementations of all of the discussed vector and matrix operations are included in standard vector math libraries. As such libraries also exploit the vector extensions of modern Central Processing Units (CPUs), using single-precision floating point numbers results in twice the computational throughput with half the memory consumption compared with double-precision numbers. The slightly reduced numeric precision can be tolerated because the channels are random anyway. On a 2012 off-the-shelf desktop computer with a single processor (four cores), the `IMTAphy` [8, 9] implementation of the IMT-Advanced channel model is able to perform a time evolution step for a 32490 links UMa scenario with a 4×4 MIMO configuration in less than one second. About 10 percent of that time is spent on the Fourier transform to derive the CTF on 100 frequency bins. Considering that during this one second about 400 MByte of CTF matrices are updated in the computer's memory, an on-the-fly computation approach like the one taken by `IMTAphy` probably outperforms loading pre-computed channel data from an external memory. When both downlink and uplink are simulated simultaneously in an FDD system, almost twice the CPU time is needed because (13.16) and (13.17) depend on the center frequency.

13.4 Channel Model Calibration

Different companies and organizations involved in the 3GPP standardization or in the IMT-Advanced evaluation process have implemented the IMT-Advanced evaluation guidelines together with the IMT-Advanced channel model in their own simulation tools. Due to a myriad of parameters and because of the complexity of system-level simulators and channel model implementations, simulation results published by different sources usually differ, and sometimes significantly so. There are multiple reasons for the divergence of simulation results even for a fully standardized system like LTE Rel-8. First, system specifications have many parameters that can be freely chosen and different organizations might choose different settings. Second, certain algorithms like the scheduling algorithm are not standardized and some organizations might have better solutions than others. Third, even detailed models and guidelines sometimes leave room for interpretation. For example, the ITU-R guidelines [1] published in 2008 had to be clarified in 2009 [13] to avoid different interpretations. Fourth and finally, simulation software – like all software – is usually not free from implementation errors. To avoid the two latter error sources, the 3GPP as well as some of the independent evaluation groups like the WINNER+ project conducted calibration campaigns to align the simulation results of their member organizations. They aimed at excluding the first two mentioned reasons by a detailed specification of simulation parameters and algorithm assumptions. During the course of the calibration campaigns, a common understanding of how to interpret the guidelines was reached as well. A system-level simulation relies on a number of random inputs. Thus, not specific results but rather statistical evaluations like the distribution of certain metrics can be used for the calibration process.

In this section we will focus on the calibration efforts by the 3GPP and some IMT-Advanced evaluation groups regarding the IMT-Advanced channel model itself. In addition, we will introduce some further metrics that can be used to verify the generated channel matrices. In Section 13.6.3 we will present the aspects of the 3GPP's calibration campaign that are specific to LTE.

13.4.1 Large-Scale Calibration Metrics

The two large-scale metrics presented here are the *coupling loss* CDF and the *geometry* CDF that are evaluated over all links. The coupling loss is the sum of all the losses between an MS and its serving BS without considering the fast-fading: the feeder loss, the influence of BS and MS antenna patterns, the pathloss on the link, potential O2I or Outdoor to Vehicle (O2V) penetration losses, and the shadowing. Besides testing the implementation of the mentioned components, the CDF of the coupling loss shown in Figure 13.5(a) also helps to identify problems with the LOS classification, the user placement (i.e., wrap-around and minimum distances), and the association process between BSs and MSs (handover hysteresis). The geometry, which is the per-user wideband and long-term downlink SINR linear average in a full reuse scenario with Round Robin scheduling,

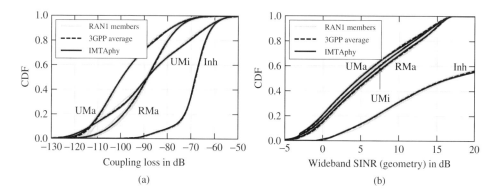

Figure 13.5 Large-scale calibration comparing IMTAphy against 3GPP results [16]. (a) Coupling loss; (b) DL wideband long-time average SINR (geometry).

is plotted in Figure 13.5(b). It can hide modeling problems with one of the mentioned coupling loss factors if the losses on the serving and interfering links deviate from the correct value by the same factor. In contrast, the geometry CDF helps to reveal correlation issues between links, for example, if different propagation conditions are applied to MS–BS links going to the same site. Figure 13.5 shows the results obtained with IMTA-phy (averaged over multiple random drops) compared to the results obtained by more than a dozen 3GPP RAN1 members who participated in this "step 1A" of the 3GPP calibration campaign [16]. As can be seen, there are only small deviations between the curves of the 3GPP RAN1 members. From the RAN1 members' individual results an average curve is computed (see also TR 36.814 [12]), which is almost identical to and thus concealed by the IMTAphy curve. A technical document [17] presents CDFs for LOS and NLOS propagation conditions separately, which can be very helpful for debugging purposes.

13.4.2 Small-Scale Calibration Metrics

To test the small-scale parameters that are used to generate the channel coefficients, two metrics defined in TR 25.996 [15] are commonly used: The CDF (again over all links) of the weighted *Root Mean Square* (RMS) delay spread $\sigma_\tau^{(RMS)}$ as defined in (13.18) and the CDF of the weighted Root Mean Square (RMS) circular angular spread. For each link the weighted RMS delay spread $\sigma_\tau^{(RMS)}$ is computed from the delay of each path τ_n weighted by that path's power P_n:

$$
\sigma_\tau^{(RMS)} = \sqrt{\frac{\sum_{n=1}^{N}(\tau_n - \mu_\tau)^2 P_n}{\sum_{n=1}^{N} P_n}} \quad \text{with} \quad \mu_\tau = \frac{\sum_{n=1}^{N}\tau_n P_n}{\sum_{n=1}^{N} P_n}. \tag{13.18}
$$

Note that for LOS propagation links, the first cluster's power also contains the power of the line-of-sight ray. Figure 13.6(a) shows the CDFs for the delay spreads in the UMa scenario in comparison to reference results provided by WINNER+ project members [18] and two members of the Chinese IMT-Advanced evaluation group [19]. In addition,

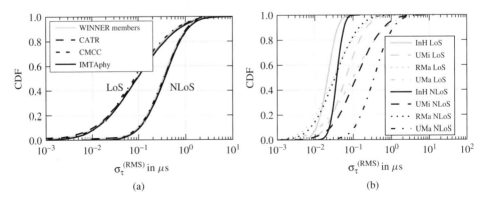

Figure 13.6 CDF of the RMS delay spreads computed according to TR 25.996 [15]. (a) IMTAphy curves compared to reference curves from the WINNER+ [18] and Chinese evaluation groups [19] in the UMa scenario; (b) IMTAphy delay spread scenario comparison.

Figure 13.6(b) illustrates the different delay spreads for NLOS and LOS propagation conditions in the four IMT-Advanced deployment scenarios as derived with IMTAphy.

The RMS angular spread $\sigma_{\text{ASA}}^{(\text{RMS})}$ of arrival angles is computed in a similar way from the weighted squared differences between each ray's angle $\varphi_{n,m}$ and the weighted mean μ_φ. However, there is a subtle difference: As already pointed out in 2003 [20], a straight-forward RMS angular spread computation would depend on the orientation of the coordinate system. Therefore, the $\sigma_{\text{ASA}}^{(\text{RMS})}$ is defined as the minimum RMS spread for arbitrary rotations of the coordinate system's origin by the angle Δ: $\sigma_{\text{ASA}}^{(\text{RMS})} = \min_\Delta \left[\sigma_{\text{ASA}}^{(\text{RMS})}(\Delta) \right]$. Taking the minimum over all possible rotations, the RMS angular spread is always smaller than $104°$ with the maximum value reached for a uniform distribution of power in the azimuth plane [20]. In addition to taking the minimum over all shifts Δ, the angles have to be mapped to the $[-\pi, \pi]$ range during the computation to achieve comparable results as pointed out by the WINNER+ project [18]. To weight each ray's angle, a per-ray power $P_{n,m} = P_n/20$ is used which is computed by dividing the cluster's power P_n equally among all 20 rays. The only exception is the LOS case where the power of all 20 rays is scaled down so that the single LOS ray carries the whole line-of-sight power $P^{(\text{LOS})} = K/(K+1)$. The computation of the RMS angular spread of departure angles $\sigma_{\text{ASD}}^{(\text{RMS})}$ from the departure angles $\phi_{n,m}$ is done in a completely analogous way.

Figure 13.7 shows the CDFs of the RMS angular spreads for arrival and departure angles in the UMa scenario generated by IMTAphy compared with the results by members of the WINNER+ and Chinese IMT-Advanced evaluation groups. It can be seen that the rooftop BS antennas experience a much smaller angular spread than the mobile users who are typically surrounded by scatterers. Both delay and angular spreads for LOS links are significantly smaller because most power is concentrated in the LOS ray of the first cluster. Again, the IMTAphy calibration curves and the curves provided by most organizations match very well.

Of course, the delay and angle small-scale parameters generated by an implementation of the channel model can be tested by calibrating against RMS delay and angular spread reference CDFs. But also the cluster power small-scale parameters as well as the Rician

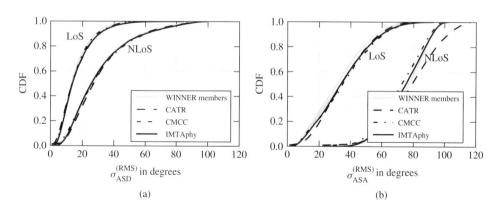

Figure 13.7 CDFs of the RMS angular spreads computed according to TR 25.996 [15] for the UMa scenario compared with WINNER+ [18] and Chinese [19] reference curves. (a) At the base station (AOD); (b) At the mobile station (AOA).

K factor distributions can be verified because they influence the delay and angular spread computations.

13.4.3 CIR and CTF Calibrations

The small-scale calibrations presented above only cover the inputs to the channel matrix computation in the generic model. In order to validate the resulting outputs, further tests are desirable. One way to check if the time evolution and the speeds are handled correctly is to evaluate the temporal autocorrelation properties of a single path of the generated CIR [21]. Figure 13.8(a) shows that the IMTAphy and the WINNER implementations [2] of the IMT-Advanced channel model, as well as the older 3GPP SCM channel model [22], exhibit a very similar temporal autocorrelation. The generated fading also matches the expected theoretic Rayleigh fading autocorrelation function which is a 0-th order Bessel function of the first kind.

A common way [21, 23] to characterize the correlation properties of the frequency domain channel matrices $H \in \mathbb{C}^{N_{Rx} \times N_{Tx}}$ produced by a channel model for a link with N_{Rx} receive and N_{Tx} transmit antennas, is to consider the distribution of the resulting MIMO capacity C according to (13.19). Here, $I_{N_{Rx}}$ denotes the identity matrix of size $N_{Rx} \times N_{Rx}$ and $(\cdot)^H$ the Hermitian transpose operation:

$$C = \log_2 \det\left(I_{Rx} + \frac{\text{SINR}}{N_{Tx}} H H^H\right). \tag{13.19}$$

Figure 13.8(b) shows the capacity CCDFs comparing IMTAphy, the WINNER implementation [2], and the SCM channel model [22] against the capacity distribution of purely random Gaussian i.i.d. distributed channel matrices. For all models, the values are generated assuming a 4×4 Uniform Linear Array (ULA) MIMO configuration with 10λ element spacing at the BS and 0.5λ at the MS for an SINR of $14\,\text{dB}$ in the UMa scenario. In addition, tabulated values for the 5%, 50%, and 95% CDF levels from Chong et al. [23] are plotted as a further reference. Obviously, the correlations considered by the different channel models yield lower capacities than in the Gaussian case, especially

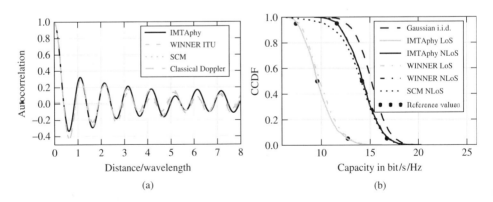

Figure 13.8 Comparison of different channel model implementations in the UMa scenario. (a) Temporal autocorrelation of path coefficients; (b) 4×4 ergodic open-loop MIMO capacity CCDFs at a fixed SINR of 14 dB (reference values [23]).

for LOS conditions. It can be seen that the `IMTAphy` and WINNER implementations of the IMT-Advanced channel model match very well. The SCM model, for which only NLOS propagation is available, shows a slightly different performance which is to be expected because the modeling assumptions are not identical to the IMT-Advanced channel model.

13.5 Link-to-System Modeling for LTE-Advanced

Now we turn our attention from the channel model in the lower part of Figure 13.1 to the link-to-system model as the next highest modeling block. Even though link-to-system modeling techniques are a standard feature of system-level simulations, they are not specified in detail for IMT-Advanced evaluations. In its guidelines [1] the ITU-R merely mentions their use and requires them to be "sound in principle". While our focus here is on LTE and LTE-Advanced, we often draw on the IEEE 802.16m Evaluation Methodology Document (EMD) [24] published by the IEEE 802.16 Task Group m (TGm). The EMD is one of the most comprehensive references for system-level simulation in general and link-to-system modeling in particular.

We begin the presentation of link-to-system modeling techniques by discussing how link-to-system modeling, or PHY abstraction techniques, as they are sometimes called, allow to extend the scope from single-link link-level simulations to modeling system aspects in a larger multi-cellular setting. Afterwards, we will discuss specific aspects in detail.

13.5.1 System-Level Simulations vs. Link-Level Simulations

To understand the differences between link and system-level simulations, we first give a very brief description of the LTE protocol stack. An extensive description can be found, for example, in Dahlman et al. [26] and of course in the standard documents [25, 27–29]. The LTE protocol stack transmits data that arrives in the form of Internet Protocol (IP) packets over a time-frequency grid of physical resources using Orthogonal Frequency Division Multiple Access (OFDMA) in the downlink and Single-Carrier Frequency Division Multiple Access (SC-FDMA) in the uplink. The medium access is controlled by a scheduler that for each 1 ms Transmission Time Interval (TTI) decides which user is allocated to each of the 180 kHz Physical Resource Blocks (PRBs) available in the system. From the IP packet payload bits of a scheduled user a so-called *transport block* is formed. Its size (in bits) is adapted to the number of PRBs scheduled to the user and to the modulation scheme and effective code rate (the Modulation and Coding Scheme (MCS)) decided by the *link adaptation* mechanism. For a transmission with a single spatial data stream (called *single-layer* in LTE) a single transport block, and for a transmission using spatial multiplexing (*multi-layer*) a maximum of two transport blocks per user and TTI are transmitted. After the scheduler has made its resource allocation decisions, the transport blocks for each user are formed and each transport block travels down to the layer 1 of the protocol stack. Depending on its size, the transport block is split into multiple code blocks which are encoded by a fixed rate-1/3 Turbo coder. The coded bits are mapped by the LTE rate matcher to the available Resource Elements (REs) which are single Quadrature Amplitude Modulation (QAM) symbols per 15 kHz subcarrier. Each RE can hold 2, 4, or 6 bits for Quadrature Phase Shift Keying (QPSK), 16-QAM, or 64-QAM

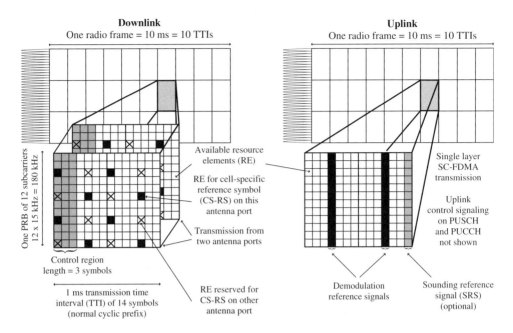

Figure 13.9 Overhead in the LTE downlink and uplink resource grids [25].

modulation, respectively. As shown in Figure 13.9, out of the 12×14 total resource elements per PRB, TTI, and spatial layer, the first 3 out of 14 symbols are occupied by the control region in the downlink. Other REs are occupied by *reference symbols* and the physical broadcast channel in the downlink (not shown). The exact effective code rate thus depends on how many REs are available to hold the rate-1/3 coded bits. For each of the 14 symbol durations during the TTI, the bits mapped to each resource element are translated to the I/Q constellation for each Orthogonal Frequency Division Multiplexing (OFDM) subcarrier. Finally, the complex baseband signal $s(t)$ is obtained after IFFT-based OFDM modulation and the insertion of a Cyclic Prefix (CP). For multi-antenna transmissions, there can be multiple spatial layers $s_m(t)$ which are mapped to the transmit antennas by a precoding matrix P as shown in Figure 13.10. At the receiver, the above steps are reversed. With multiple antennas at the receiver, for example, a linear filter W first recovers the spatial layers as $x_m(t)$. Then in each layer, the CP is removed and the OFDM signal is demodulated by means of a Fast Fourier Transform (FFT). The coded bits are recovered by detecting the constellation on each subcarrier for each symbol duration during the TTI. Each recovered code block is Turbo-decoded and the transport block is successfully received if each of its code blocks is successfully decoded. The receiver acknowledges a successful transport block reception by sending an Acknowledgement (ACK) message. Otherwise it sends a Hybrid ARQ (HARQ) Negative Acknowledgement (NACK) message triggering a retransmission.

Both link and system-level simulations cover the presented functions of the protocol stack but differ in the simulation fidelity and complexity. In link-level simulations, the exact complex baseband transmit signal $s(t)$, as it would result from the transmitter, is

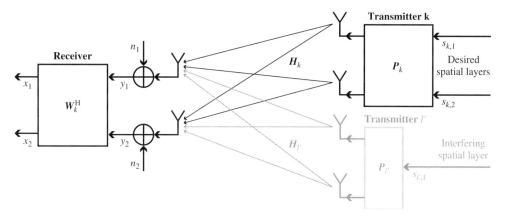

Figure 13.10   Receiver employing a filter matrix W_k to receive two spatial layers $s_{k,1}$ and $s_{k,2}$ over a 2×2 MIMO channel H_k from a desired transmitter k using a precoding matrix P_k with interference coming from an undesired transmitter l' using a precoding matrix $P_{l'}$.

generated. This means that the transmission chain is modeled on the bit and resource element level. The transmit signal $s(t)$ is convolved with the channel impulse response $h(t)$ to obtain the received signal $y(t) = s(t) \star h(t) + n(t)$ with $\star$ denoting the convolution operator. Here, only noise $n(t)$ but no interfering transmissions are considered (single cell). This received signal $y(t)$ is fed into the receiver implementation which performs demodulation, detection, and decoding to obtain the transmitted data block. The transport block is received correctly, if the decoded bits match the transmitted ones.

Conducting a system-level simulation at this level of detail would be too complex. The transmit signal would have to be created for all serving links in all cells, and it would have to be convolved with all (e.g., 32490) links if interference from all neighboring cells are considered. In system-level simulations, individual bits and resource elements are thus not considered and no signal $s(t)$ is generated. Instead of modeling the channel in a 12×14 (subcarriers $\times$ OFDM symbols) resolution, usually only one frequency domain channel coefficient is generated per PRB and TTI assuming that the channel's coherence time and bandwidth are large enough. In order to model the performance of the system with respect to the resource allocation decisions given the current channel and interference conditions, a suitable link-to-system model is now used. The SINR, which is the ratio of the signal to interference and noise power a transmission encounters on a certain PRB, plays a central role. Based on pre-computed BLock Error Rate (BLER) performance results that relate the probability of successfully receiving a code block to the SINR incurred by the code block during transmission, link-to-system mapping techniques are able to provide the probability of successful reception of a transport block. A system-level simulator with a valid link-to-system model is thus able to deliver "correct" average performance results by randomly dropping each incoming transport block with its designated error probability without having to model the transmission on the bit level [30].

In Section 13.5.2 we will see how to compute the SINR values per PRB considering the channel gains between a receiver, its desired transmitter, and all interferers while also taking the transmit precoding and receive filtering matrices into account. Transport blocks are usually transmitted over multiple PRBs, which can encounter different fading or inter-ference conditions, so that the SINR values of the different PRBs within a transport block might differ. Using the techniques to compute *effective SINRs* presented in Section 13.5.3, we are able to condense multiple SINR values into a single value that is representative for the whole transport block when used to obtain a BLER from the pre-computed curves as shown in Section 13.5.4.

13.5.2 Modeling of MIMO Linear Receiver and Precoder Performance

Here we consider a MIMO system as depicted in Figure 13.10. We take the perspective of one fixed receiver (pictured on the left) for which we compute the SINRs on all spatial layers $m = 1 \ldots M_k$ it receives from its desired transmitter $k \in \mathcal{K}$ where $\mathcal{K}$ is the set of all transmitters in the system. For the sake of simplicity, we omit indices indicating which receiver we mean. However, we do note that all receiver-related quantities can be different for different receivers. Also, most variables like, for example, the channel matrices $\boldsymbol{H}(t, f)$, are time and frequency-dependent and the SINR computations discussed below would be repeated, for example, for each TTI and each PRB in a typical system-level simulation. In the presentation we omit the dependency on t and f and consider the system at some fixed time instant and at some fixed PRB. The desired transmitter k transmits M_k spatial layers with complex baseband modulation symbols $s_{k,1}, \ldots, s_{k,M_k}$ by applying a precoding $\boldsymbol{P}_k \in \mathbb{C}^{N_{\text{Tx},k} \times M_k}$ over an $N_{\text{Rx}} \times N_{\text{Tx},k}$ MIMO channel[3] $\boldsymbol{H}_k$. The receiver employs a receive filter $\boldsymbol{W}_k^H \in \mathbb{C}^{M_k \times N_{\text{Rx}}}$ to recover the spatial layers as $x_1, \ldots, x_{M_k}$ from transmitter k. In Figure 13.10 transmitter k transmits two spatial layers over a 2×2 channel by applying a suitable precoding matrix $\boldsymbol{P}_k \in \mathbb{C}^{2 \times 2}$. Another transmitter l' transmits just a single spatial layer precoded by a precoding vector $\boldsymbol{P}_l' \in \mathbb{C}^2$ causing undesired interference to the receiver.

In a multi-cell scenario not only the signal layers $\boldsymbol{s}_k \in \mathbb{C}^{M_k}$ from the desired transmitter k but also a number of interfering signals from other transmitters $l \in \mathcal{K}$ with $k \neq l$ arrive over the effective channels $\boldsymbol{H}_l \boldsymbol{P}_l$ at the receiver. The antennas receive the superposition $\boldsymbol{y} \in \mathbb{C}^{N_{\text{Rx}}}$ of all transmitted signals and noise $\boldsymbol{n} \in \mathbb{C}^{N_{\text{Rx}}}$ with $[\boldsymbol{n}]_i \sim \mathcal{CN}(0, \sigma^2)$ as:

$$y = \sum_{l \in \mathcal{K}} H_l P_l s_l + n. \tag{13.20}$$

Equation (13.21) shows that the signal layers $\boldsymbol{x}$ recovered by the receiver from the received signal $\boldsymbol{y}$ by means of the decoding linear filter $\boldsymbol{W}_k^H \in \mathbb{C}^{M_k \times N_{\text{Rx}}}$ contain not only the desired

---

[3] In Section 13.3.4 we assumed the channel matrices $\boldsymbol{H} \in \mathbb{C}^{U \times S}$ to have as many rows and columns as the MS and BS, respectively, have antennas. Here we will always assume them to be organized as $\boldsymbol{H} \in \mathbb{C}^{N_{Rx} \times N_{Tx}}$, that is, with as many rows as Rx antennas and as many columns as Tx antennas. That way, the presentation will hold for the downlink and uplink direction. Further, we assume the amplitudes of the channel matrix coefficients to be scaled by (the square root of) the corresponding pathloss and shadowing terms and the transmit power. Both pathloss and shadowing are constant during a drop, so the scaling can be precomputed and included either in the constant factor $a_{u,s,n,m}$ used for CIR computation (see Equation (13.17)) or in the Fourier coefficients $\boldsymbol{F}$).

signal but also the filtered inter-cell interference and noise:

$$x = W_k^H y = W_k^H \left(\sum_{l \in \mathcal{K}} H_l P_l s_l + n \right)$$

$$= \underbrace{W_k^H H_k P_k s_k}_{\substack{\text{desired signal and} \\ \text{inter-stream inter-} \\ \text{ference}}} + \underbrace{\sum_{\substack{l \in \mathcal{K} \\ l \neq k}} W_k^H H_l P_l s_l}_{\text{inter-cell interference}} + \underbrace{W_k^H n}_{\substack{\text{filtered} \\ \text{noise}}}. \tag{13.21}$$

If we sort the contributions to x as in (13.21) above, we can derive the following expression for the SINR of each spatial layer m received from transmitter k:

$$\text{SINR}_{k,m} = \frac{\overbrace{\left| \left[W_k^H H_k P_k \right]_{m,m} \right|^2}^{\text{signal power}}}{\underbrace{\sum_{\substack{i=1 \\ i \neq m}}^{M_k} \left| \left[W_k^H H_k P_k \right]_{m,i} \right|^2}_{\text{inter-stream interference}} + \underbrace{\sigma^2 \sum_{i=1}^{N_{\text{Rx}}} \left| \left[W_k^H \right]_{m,i} \right|^2}_{\text{filtered noise}} + \underbrace{\left[W_k^H \sum_{\substack{l \in \mathcal{K} \\ l \neq k}} C_l W_k \right]_{m,m}}_{\text{inter-cell interference}}}, \tag{13.22}$$

where $C_l = (H_l P_l)(H_l P_l)^H$ denotes the covariance matrix $C_l \in \mathbb{C}^{N_{\text{Rx}} \times N_{\text{Rx}}}$ of the interference from transmitter l as seen at the receiver. The modulation symbols of different layers in s_k and s_l are assumed to be uncorrelated and to have unit power, that is, their covariance matrix is the identity matrix: $\mathbb{E}[s_k s_k^H] = I_{M_k}$.

We can understand (13.22) by considering the $M_k \times M_l$ matrix $W_k^H H_l P_l$ which maps the spatial layers in s_l to the layers x that we want to receive. The m-th component of the desired signal s_k is mapped to x by matrix-multiplication with the m-th column of $W_k^H H_k P_k$. Of course, the signal $[s_k]_m$ of layer m in s_k should only contribute to the m-th received signal layer $[x]_m$. Hence, the gain $|[W_k^H H_k P_k]_{m,m}|^2$ as the only desired component of the signal is to be found in the nominator of (13.22). All entries in row m and column $i \neq m$ of $W_k^H H_k P_k$ cause inter-stream interference and are thus to be found in the denominator. The other terms in the denominator are the filtered noise and the filtered inter-cell interference.

The SINR model in (13.22) allows for arbitrary linear precoding matrices P and filter matrices W. In LTE, the precoders could be taken from the codebooks defined in the standard [25]. At the receiver, a linear Minimum Mean Square Error (MMSE) filter is usually assumed for mobiles supporting the reception of multiple spatial streams:

$$W_k^{(\text{MMSE})} = \left(\hat{H}_k P_k P_k^H \hat{H}_k^H + \hat{C}_{\text{IIPN}} \right)^{-1} \hat{H}_k P_k. \tag{13.23}$$

A Maximum Ratio Combining (MRC) receiver that combines the signals from multiple antennas when receiving a single stream can be realized with:

$$W_k^{(\text{MRC})} = \hat{H}_k P_k. \tag{13.24}$$

Note that an actual receiver does not have perfect channel knowledge. It has to perform some kind of channel estimation to obtain estimates of its serving channel $\hat{\boldsymbol{H}}_k$ and of the inter-cell interference plus noise covariance matrix $\hat{\boldsymbol{C}}_{\text{IIPN}}$. As there is no standard way to model for estimation errors, perfect serving channel knowledge (i.e., $\hat{\boldsymbol{H}}_k = \boldsymbol{H}_k$) is often assumed for system-level simulations. A baseline MMSE receiver performance can be realized by modeling the interference covariance matrix as a diagonal matrix $\hat{\boldsymbol{C}}_{\text{IIPN}} = \text{diag}(\sigma_i^2)$ with σ_i^2 being the total interference plus noise power at receive antenna i.

13.5.3   Effective SINR Values

There are two main methods used in the literature to combine multiple SINRs that are encountered by a code block into a single effective SINR. Both the *Exponential Effective SINR Mapping* (EESM) and the *Mutual Information Effective SINR Mapping* (MIESM) transform the individual SINR values using a suitable mapping $g(\cdot)$ to a domain where the transformed values can be arithmetically averaged. This average is then transformed back using the inverse mapping $g^{-1}(\cdot)$, see Brueninghaus et al. [24, 30]. Schematically, for N_{PRB} individual values SINR_n, the effective SINR is computed as:

$$\text{SINR}^{(\text{eff})} = g^{-1}\left(\frac{1}{N_{\text{PRB}}} \sum_{n=1}^{N_{\text{PRB}}} g(\text{SINR}_n)\right). \tag{13.25}$$

Here, we will focus on an MIESM variant that maps SINRs to the *Mean Mutual Information per Bit* (MMIB) domain [24, 31]. Figure 13.11 exemplifies how the MMIB-based effective SINR computation process works. In the example two SINR values (0 dB and 20 dB) are mapped to the corresponding MMIB values (0.486 and 1) indicated on the right figure axis. In that domain the arithmetic mean $(0.486 + 1)/2 = 0.743$ is computed and then mapped back to an effective SINR value of 3.33 dB.

The rationale for mutual information-based averaging approaches is easy to understand. Each individual SINR_n value represents the conditions some of the modulation symbols in the code block encountered during the transmission. For example, each SINR_n could stand for one PRB in an LTE transport block or for one of the layers over which the transport block is distributed. Looking at a single symbol, the mutual information conveyed by that symbol is constrained by its modulation order. Even for arbitrary high SINRs, a QPSK

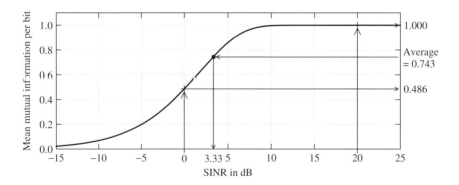

Figure 13.11   QPSK MMIB-based effective SINR computation example [24].

symbol can carry a maximum of 2 bits, a 16-QAM symbol a maximum of 4 and so on. For QPSK modulation the maximum MMIB value of 1 is already reached at about 10 dB. Thus, increasing the SINR value of the QPSK symbol beyond 10 dB does not contribute additional mutual information. If no effective SINR technique was used and if, for example, the SINR values of 0 dB and 20 dB in the example above were to be averaged directly in the SINR domain, the 20 dB value would be overrepresented and lead to a too high effective SINR value.

Numerical approximations of the MMIB mapping function for different modulations are given in the EMD [24, 31]. The mapping and its inverse are often implemented by means of a lookup table. Note that the reverse mapping for MMIB values of 1 is not well-defined so that effective SINRs are constrained to a certain range.

For the link-to-system model in the SC-FDMA uplink in LTE a slightly different aggregation approach has to be used. Although uplink transmission is scheduled onto PRBs as in the downlink, those PRBs have to be adjacent because uplink transmissions actually happen on a wider single carrier. In contrast to the OFDMA downlink, the need for more sophisticated frequency domain equalization arises in the uplink [26]. Usually, a model for MMSE-based equalization in the frequency domain is used that yields a single SINR value per symbol duration. Conventional effective SINR computation approaches must be used in addition if multiple SINRs of individual symbol durations during the TTI or multiple streams are modeled. According to 3GPP assumptions [32, 33] the post-equalization $\text{SINR}^{(\text{MMSE-FDE})}$ can be computed from the SINR_n values on the N_{PRB} subcarriers as:

$$\text{SINR}^{(\text{MMSE-FDE})} = \frac{\Gamma^2}{N_{\text{PRB}}\Gamma - \Gamma^2} \quad \text{with} \quad \Gamma = \sum_{n=1}^{N_{\text{PRB}}} \frac{\text{SINR}_n}{\text{SINR}_n + 1}. \tag{13.26}$$

13.5.4 Block Error Modeling

Once an effective SINR is computed for a code block, the expected BLER can be looked up in pre-computed BLER tables like the ones plotted in Figure 13.12. Of course, the error probability depends on the code rate, that is, the ratio of actual information bits versus the number of coded bits. A smaller code rate implies more redundancy bits and thus allows the Turbo decoder to successfully decode the code block in the presence of more (coded) bit errors.

Figure 13.12 shows BLER curves that were generated by the author during a detailed link-level-simulation campaign with a MATLAB®-based link-level simulator [4]. These and some additional curves are stored as lookup tables in IMTAphy. The link-level simulations were performed as described in Section 13.5.1 assuming a single BS–MS link over a Single Input Single Output (SISO) Additive White Gaussian Noise (AWGN) channel. During the simulations, the SINR was varied over the whole range and for each SINR value, one transport block per TTI was transmitted for 5000 TTIs in a row. After these 5000 trials, the empirical BLER was computed from the ratio of erroneously received transport blocks. The link-level simulator [4] was slightly modified to vary the size of the transport block from 50 to 6144 information bits, which is the size of the largest possible code block in LTE. The modulation scheme and code rates for the transmission of each transport block were chosen according to one of the 15 possible CQI values in LTE, see Table 13.6.

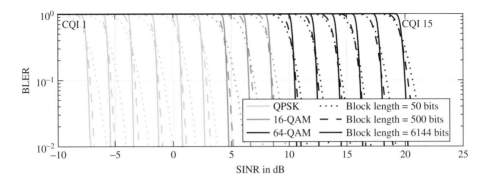

Figure 13.12   Block error rates for different MCSs and block lengths based on LTE CQIs.

Table 13.6   LTE CQI definition [29]

CQI index	Modulation scheme	Code rate
0	out of range	
1	QPSK	0.076
2	QPSK	0.117
3	QPSK	0.188
4	QPSK	0.301
5	QPSK	0.438
6	QPSK	0.588
7	16-QAM	0.369
8	16-QAM	0.479
9	16-QAM	0.602
10	64-QAM	0.455
11	64-QAM	0.554
12	64-QAM	0.650
13	64-QAM	0.754
14	64-QAM	0.853
15	64-QAM	0.926

As can be seen in Figure 13.12, different modulations and code rates (cf. Table 13.6) lead to different threshold SINRs at which the BLERs start to fall from 1 towards 0. Different block lengths also influence the decoding chances because Turbo coding works better for larger code blocks. However, small block lengths already show BLERs < 1 at lower SINRs because chances are that a small block encounters a period of very good reception conditions. In LTE the goal of the link adaptation mechanism is to estimate the channel conditions (i.e., the SINR that a transport block will encounter) in order to select the highest modulation and coding scheme that can still be received successfully. Figure 13.13(a) shows that with a proper link adaptation, it is possible to closely follow the Shannon capacity by selecting the most suitable modulation and coding scheme. However, the spectral efficiencies (computed as the product of modulation bits, BLER, and

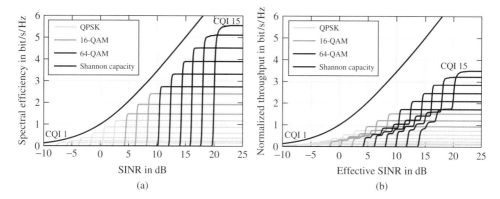

Figure 13.13 Performance of different MCSs based on the 15 LTE CQIs. (a) Gross spectral efficiency and Shannon bound; (b) Normalized net throughput with Chase combining HARQ.

code rates) shown in Figure 13.13(a) are not achievable in practice due to the overhead mentioned in Section 13.5.1. Taking a typical overhead situation into account, the normalized throughputs as depicted in Figure 13.13(b) can be achieved in an LTE system. The figure also shows that with the help of HARQ retransmissions and *Chase combining* (CC) the SINR threshold of each MCS can be lowered by 3 dB per retransmission. The availability of HARQ allows the link adaption process to be tuned to aim at a residual BLER of about 10% because at this point the gains from using better MCSs outweigh the losses incurred by a retransmission.

In an actual system-level simulation, many different code block sizes between 50 and 6144 bits are possible. Also, arbitrary code rates can result from the rate matching process that depends on the number of resource elements within the allocated PRBs that are not used for other purposes. Therefore, the link-to-system interface in `IMTAphy` performs an interpolation if the actual block size or the code rate lie in between two of the precomputed curves. To look up an arbitrary code rate, the curves for the next higher and lower code rate are shifted in the SINR direction. To look up an arbitrary block length, a linear interpolation between the BLERs of two available block lengths is performed. The LTE rate matcher achieves code rates below the mother code rate of the Turbo coder by repeating some bits. For low code rates the rate-1/3 curves are thus used and the SINRs are scaled to reflect the power gain from repetition coding. Similarly, the effect of a Chase combining HARQ retransmission can be modeled by adding the SINRs of the initial transmission and of each retransmission on a PRB to PRB basis before computing the effective SINR. Modeling an *Incremental Redundancy* (IR) retransmission involves a mixture between adjusting the code rate and a fractional scaling of the SINR depending on how many bits are retransmitted and how many new bits are provided by the IR retransmission.

13.6   3GPP LTE-Advanced System-Level Simulator Calibration

Building on the channel and link-to-system model introduced in the preceding sections, we are now able to look at the system-level simulation of a complete protocol stack of an IMT-Advanced radio interface technology. As an example, we consider the simulation assumptions and the results of the simulator calibration campaign the 3GPP conducted to prepare for the submission of LTE-Advanced as an IMT-Advanced technology in 2009. The calibration campaign does not reflect the performance that can be expected from an actual LTE-Advanced system because only a very basic setup was chosen. The setup is well-defined and does not leave much room for vendor-specific optimizations like sophisticated schedulers or receivers, cf. Section 13.4. What it does allow, is to capture the effects the fast-fading channel and the link-to-system modeling assumptions have on the performance metrics. For an independent implementation like IMTAphy it can thus serve as a benchmark that allows comparison with the simulation tools used by 3GPP members.

The simulation assumptions are described in the annex of TR 36.814 [12]. The four mandatory IMT-Advanced scenarios InH, UMa, UMi, and RMa are considered with all the parameters like user speeds and so on, as introduced in Section 13.2.3. Inter-cell interference is explicitly modeled by performing the simulation with the same level of detail in each cell. The most important LTE system parameters are summarized in Table 13.7. We will briefly discuss the simulation assumptions separately for downlink and uplink in the following. Finally, in Section 13.6.3 we will present the results of the calibration simulations.

Table 13.7 3GPP simulation assumptions for IMT-Advanced simulator calibration [12]

Parameter	Value
Common:	
Bandwidth	50 PRBs (10 MHz) in UL and DL each (FDD) for UMa, UMi, and RMa and 100 PRBs for InH
Receiver type	MRC receiver
Channel estimation	perfect channel estimation
Feeder loss at the BS	0 dB, in contrast to the calibration in Section 13.4.1
HARQ	with up to three retransmissions
Antenna configuration	1 Tx and 2 Rx vertically polarized antennas $\lambda/2$ separation at MS and 10λ separation at BS
Downlink:	
Control region overhead	3 out of 14 symbols occupied each 1 ms TTI over the whole bandwidth
CQI feedback	5 ms periodicity, error-free, estimate from TTI n can be used in TTI $n+6$
Uplink:	
Frequency domain equalization	MMSE assuming perfect channel estimates
Power control	open-loop only with $P_0 = -106$ dBm, $\alpha = 1$
SRS overhead	1 out of 14 symbols occupied each 1 ms TTI over the whole bandwidth
PUCCH overhead	4 PRBs deducted from the system bandwidth
Link adaptation	ideal estimate from TTI n can be used for TTI $n+7$
Scheduling	see text

13.6.1   Downlink Simulation Assumptions

The downlink simulation setup assumes a Single Input Multiple Output (SIMO) system with a single BS transmit antenna and two receive antennas at the mobile stations. The 10 users per cell (on average) are associated to their serving cells taking a 1 dB handover margin into account. The mobiles are configured to provide wideband CQI reports to their serving base station every 5 ms. As the BSs cannot perform any precoding with their single antennas, no rank or precoding matrix indicators are signaled by the mobiles. The base stations employ a very simple Round Robin scheduler that assigns an equal share of resources to all users. In a strictly sequential order, one user after the other is assigned all PRBs during one TTI. The users get different data rates though because the scheduler performs link adaptation based on the CQI status reports which are expected to be available 6 ms after they are measured. For the reception the mobiles use their two antennas to perform Maximum Ratio Combining (MRC). A fixed control region size of three symbol durations is modeled. In addition, six resource elements are occupied by single-port cell-specific reference symbols in each PRB and TTI. Every 10-th TTI the primary broadcast channel occupies some of the six center PRBs which are then also not available for data transmission.

13.6.2   Uplink Simulation Assumptions

The uplink simulations also use a SIMO model, this time with a single transmit antenna at the mobile and two antennas at the BS, which are again used for MRC combining. Compared to the downlink, there are three major differences in the uplink:

First, the overhead situation in the uplink is different from the downlink as shown in the right half of Figure 13.9. There is no control region in the first symbols of the subframe and no pattern of cell-specific reference symbols, like in the downlink. Instead, all subcarriers for 2 out of 14 symbol durations per TTI are used for transmitting demodulation reference symbols and one symbol duration is assumed to be used for an uplink Sounding Reference Signal (SRS) in this setup. In addition, 4 out of 50 uplink PRBs are not available for transmissions because they are permanently used for the Physical Uplink Control Channel (PUCCH).

Second, the Round Robin scheduler assumed for calibration works differently: In each cell, the $N_{PRB} = 50 - 4 = 46$ available PRBs are divided equally among all associated users in each TTI. Each TTI, every user is served with always the same PRB allocation regardless of the transmission being a new transmission or a HARQ retransmission. If the $N_{PRB} = 46$ available PRBs cannot be divided evenly among the MSs associated to the BS, a number of MSs will be allocated one PRB more than the rest so that all PRBs are always utilized. This special variant of Round Robin scheduling leads to a quite unrealistic interference situation in the uplink. The resulting SINR and throughput distributions do not exhibit the typical uplink interference effect. Under realistic conditions, a base station suffers from a highly dynamic level of inter-cell interference which fluctuates depending on the pathlosses of the interfering mobiles that happen to be co-scheduled in neighboring cells.

Third, the standard LTE open-loop power control with full pathloss[4] (PL) compensation $\alpha = 1$ and a target power of $P_0 = -106\,\text{dBm}$ is used. The transmit power for MS i in the uplink is computed based on its maximum allowed Tx power P_{Max} and on the number $N_{\text{PRB},i}$ of PRBs associated to it:

$$P_i^{(\text{UL})} = \min\left(P_{\text{Max}}, \ P_0 + \alpha \times \text{PL}_i + 10 \times \log_{10}(N_{\text{PRB},i})\right) \text{ in dBm} \qquad (13.27)$$

13.6.3 Simulator Calibration Results

The first step ("step 1A") of the 3GPP calibration campaign focusing on the large-scale fading was already presented in Section 13.4.1. As part of the second step ("step 1C"), we now look at the CDFs of the post receiver-combining SINR for the downlink and uplink as shown in Figure 13.14 for the four mandatory scenarios.

The SINR is averaged per-user in the linear scale over all PRBs used in the transmissions to and from a user. For the `IMTAphy` results, each drop was simulated for 2 seconds of simulation time and a total of 15 drops were generated for the cellular scenarios. For the InH scenario 350 drops were used. That many InH drops are necessary because only 20 users are placed on the scenario per drop. In addition, the area is so small that due to the correlated shadowing no reliable average can be derived from a single drop even if there were more users.

At first glance, the downlink SINR CDFs plotted in Figure 13.14(a) look similar to the geometry ones shown in Figure 13.5(b). But in contrast to the previous results, two receive antennas with MRC-combining are now used with a fast-fading channel. The resulting MRC gain of up to 3 dB is visible when carefully comparing the non-fading SISO geometry CDF in Figure 13.5(b) to the CDF in Figure 13.14(a). For example, the curves top out at about 20 dB instead of 17 dB as before.

Another effect of working with fast-fading serving and interfering links is visible when closely examining the improvement in the InH CDF between Figures 13.5(b) and 13.14(a).

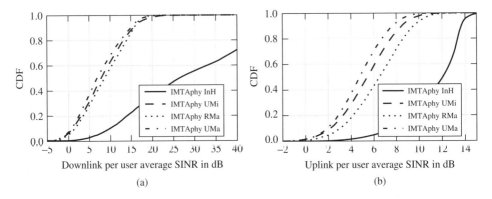

Figure 13.14 Average per-user post-receiver SINRs for the IMT-Advanced scenarios. (a) Downlink; (b) Uplink.

---

[4] Not to be confused with the pure pathloss described in Section 13.3.1. What the MS sees is the coupling loss, cf. Figure 13.5(a).

For example, at the 20% level of the CDFs, a gain of more than the 3 dB attributable to MRC-combining can be seen. The reason for this effect is the unique characteristic of the InH scenario where only one interferer exists. It occurs when averaging the SINR, which in the InH case is the ratio of two χ^2-distributed random variables: the fast-fading gains of the desired and of the interfering signal. In the InH scenario the single interferer can exhibit deep fades leading to high SINR values that shift the average. In the hexagonal scenarios with 56 interferers, it is highly unlikely that all interferers are faded at the same time so that the fast-fading of the signal power averages out over time and no such effect is visible.

The distribution of uplink SINRs shown in Figure 13.14(b) is heavily influenced by the special power control settings and the Round Robin variant used for calibration purposes. The visible SINR variance in the CDFs does not stem from pathloss effects because they are fully compensated by the power control mechanism. Neither is the fast-fading responsible for the SINR variance because the CDF shows the distribution of long-time average values. The main reason why the MSs achieve different average SINRs is the arbitrary distribution of favorable and unfavorable static interference constellations caused by the scheduler used for the calibration.

Figure 13.15 compares the DL and UL post-combining SINR distributions obtained with `IMTAphy` in the UMi scenario to the 3GPP average results as published in TR 36.814 [12]. In addition, the CDFs obtained by the individual contributors of the 3GPP RAN1 working group are also plotted. Compared to the non-fading SISO geometry SINR distribution in Figure 13.5(b), the results obtained from the fading channels with the respective link-to-system models exhibit a much higher spread with significant deviations from the average curve published in TR 36.814 [12]. One effect contributing to these deviations, which is especially visible in the RMa scenario [9], seems to be that some companies model the fading to co-located base stations in the LOS case as uncorrelated. Most of the high geometry links exhibit LOS propagation with the two co-located BSs limiting the geometry to 17 dB due to the 20 dB forward-backward attenuation of the antenna pattern. For LOS propagation, most power is then concentrated in the single LOS rays that arrive at identical angles from all co-located BSs. Thus, the channels should

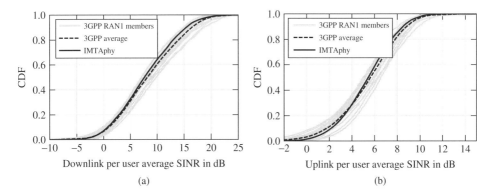

Figure 13.15 Average per-user post-receiver SINRs for the UMi scenario compared to results obtained by 3GPP RAN WG1 members [34]. (a) Downlink; (b) Uplink.

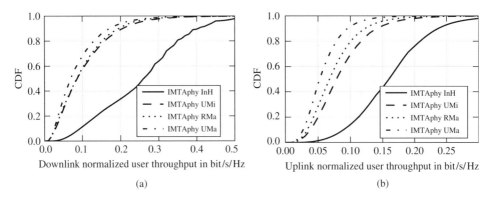

Figure 13.16 CDFs of the normalized per-user throughput for the IMT-Advanced scenarios. (a) Downlink $\gamma_i^{(DL)}$; (b) Uplink $\gamma_i^{(UL)}$.

be fully correlated thereby preventing the MRC combining from realizing a 3 dB gain achievable with uncorrelated interference.

Finally, Figure 13.16 shows the per-user throughput CDFs for the four scenarios and Figure 13.17 compares the throughput achieved by IMTAphy to 3GPP results. As the throughput performance directly depends on the achievable SINRs, the picture is similar to the SINR case. The IMTAphy curve exhibits slightly lower throughput than most of the 3GPP RAN1 members which might be due to a more conservative link-to-system model. For example, some 3GPP members use IR HARQ, which can offer slightly better performance than the CC HARQ employed by IMTAphy. Again, a high variation in the results provided by different organizations is visible because all aspects of the link-to-system model and the protocol implementation contribute to the performance result.

We should note again that these results do not represent the LTE performance that can be expected in a real-world deployment. In the uplink, the full pathloss compensation power control will certainly not be used by operators. In addition, the problems of the UL

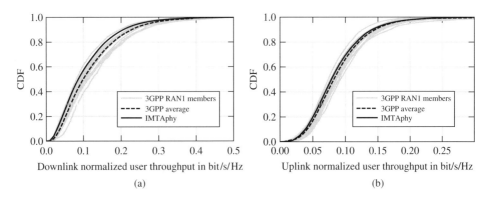

Figure 13.17 CDFs of the normalized per-user throughput for the UMi scenario compared to results obtained by 3GPP RAN WG1 members [34]. (a) Downlink $\gamma_i^{(DL)}$; (b) Uplink $\gamma_i^{(UL)}$.

scheduler used for calibration have been discussed above. In the downlink, an LTE BS can choose from a large number of transmission modes (e.g., diversity transmission, spatial multiplexing, or MU-MIMO) and could use channel-adaptive scheduling algorithms like a frequency-selective *proportional-fair* scheduler to increase the system performance. After calibrating their simulators, the 3GPP demonstrated LTE's IMT-Advanced target fulfillment with simulation results generated using more realistic simulation assumptions, cf. TR 36.814 [12].

13.7 Summary and Outlook

In this chapter we have presented the ITU-R's guidelines and channel models to be used in system-level simulations of IMT-Advanced systems like LTE-Advanced. Although the IMT-Advanced evaluation process has been completed, these guidelines will serve as the de facto standard for the performance evaluation of cellular systems in the foreseeable future.

The presentation has shown that the IMT-Advanced guidelines and channel models are specified in detail and are thus complex to implement in system-level simulation tools. This has motivated the 3GPP, amongst others, to conduct extensive calibration campaigns to align the tools used by member organizations. Simple large-scale and pure channel model calibrations show a very good alignment between project members. But as small-scale channel effects and link-to-system modeling influences are also considered in system-level simulator calibrations, bigger deviations between the results provided by different organizations become visible.

Even though the 3GPP conducted the simulator calibration with LTE-Advanced in mind, two prominent LTE-Advanced features were not yet considered at the time: First, carrier aggregation using bandwidths up to 100 MHz will be part of LTE-Advanced. But the 3GPP's calibrations were only conducted for 10 MHz (20 MHz for InH) of spectrum. However, the IMT-Advanced channel model is designed [1] to support bandwidths of up to 100 MHz and the WINNER project actually conducted measurements with an appropriate bandwidth to provide suitable parameters for the channel model [35]. The impact of extending the bandwidth on the simulation complexity of the channel model will be marginal because the CIR is not affected and performing the Fourier transform accounts only for a small part of the time. However, more PRBs and more traffic in general will directly scale the simulation complexity for the link-to-system model and the layer 2.

The second feature of LTE-Advanced that has not been considered and which will become more important in the future is Coordinated Multipoint transmission or reception (CoMP) operation where transmissions to or from one MS are coordinated by multiple BSs. The IMT-Advanced channel model already supports *intra-site* CoMP by modeling the correlations between channels from one MS to all BSs co-located at one site as described in Section 13.3.1. To support *inter-site* CoMP, the model needs to be extended to also account for correlations between channels to different sites. Extending the channel model in such a way would not add much complexity but it would require suitable parameter sets derived from measurements in the field.

All results presented in this chapter are generated with the open-source system-level simulator IMTAphy developed at the Institute for Communication Networks (LKN) of the Technische Universität München in Germany. The simulator follows the ITU-R guidelines and features an efficient C++ implementation of the IMT-Advanced channel model. In addition, all the presented link-to-system modeling techniques and an LTE layer 2 are implemented. The source code and documentation [9] is freely available under the GNU Public License at http://launchpad.net/imtaphy/ allowing the reader to reproduce all results presented in this chapter.

References

1. ITU-R, "Guidelines for evaluation of radio interface technologies for IMT-Advanced," ITU Radiocommunication Sector, Rep. ITU-R M.2135-1, Dec. 2009.
2. Finland, "Software implementation of IMT.Eval channel model," Tech. Rep. Document 5D/478-E.
3. J. Blumenstein, J. C. Ikuno, J. Prokopec, and M. Rupp, "Simulating the long term evolution uplink physical layer," in *Proc. of the 53rd International Symposium ELMAR-2011*, Zadar, Croatia, 2011.
4. C. Mehlführer, M. Wrulich, J. C. Ikuno, D. Bosanska, and M. Rupp, "Simulating the long term evolution physical layer," in *Proc. of the 17th European Signal Processing Conference (EUSIPCO 2009)*, Glasgow, Scotland, Aug. 2009.
5. C. Mehlführer, J. C. Ikuno, M. Šimko, S. Schwarz, M. Wrulich, and M. Rupp, "The Vienna LTE simulators–enabling reproducibility in wireless communications research," *EURASIP Journal on Advances in Signal Processing*, vol. 2011: 29, 2011.
6. J. C. Ikuno, M. Wrulich, and M. Rupp, "System level simulation of LTE networks," in *Proc. IEEE Vehicular Technology Conference (VTC-Spring 2010)*, Taipei, Taiwan, May 2010.
7. D. Bültmann, M. Mühleisen, K. Klagges, and M. Schinnenburg, "openWNS–open wireless network simulator," in *Proc. European Wireless Conference (EW 2009)*, May 2009, pp. 211–215.
8. J. Ellenbeck, "IMTAphy source code hosted on launchpad.net", http://launchpad.net/imtaphy.
9. J. Ellenbeck, "IMTAphy channel model and LTE simulator documentation website", http://www.lkn.ei.tum.de/personen/jan/imtaphy/index.html.
10. ITU-R, "Guidelines for evaluation of radio interface technologies for IMT-Advanced," ITU Radiocommunication Sector, Rep. ITU-R M.2135, Nov. 2008.
11. ITU-R, "Requirements related to technical performance for IMT-Advanced radio interface(s)," ITU Radiocommunication Sector, Rep. ITU-R M.2134, Nov. 2008.
12. 3GPP, "Evolved Universal Terrestrial Radio Access (E-UTRA); Further advancements for E-UTRA physical layer aspects," 3rd Generation Partnership Project (3GPP), TR 36.814 version 9.0.0 Release 9, Mar. 2010.
13. Finland, "Guidelines for using IMT-Advanced channel models," Tech. Rep.
14. P. Kyösti, J. Meinilä, L. Hentilä, X. Zhao, T. Jämsä, C. Schneider, M. Narandžić, M. Milojević, A. Hong, J. Ylitalo, V.-M. Holappa, M. Alatossava, R. Bultitude, Y. de Jong, and T. Rautiainen, "WINNER II channel models; Part I channel models," IST-4-027756 WINNER II, Tech. Rep. Deliverable D1.1.2V1.2, Feb. 2008.
15. 3GPP, "Spatial channel model for Multiple Input Multiple Output (MIMO) simulations," 3rd Generation Partnership Project (3GPP), TR 25.996 version 6.1.0 Release 6, Sep. 2003.
16. Ericsson, "Summary from email discussion on calibration step 1+2," 3GPP TSG-RAN WG1, TDoc R1-092019, May 2009.
17. NTT DOCOMO, "Evaluation of ITU test environments," 3GPP TSG-RAN WG1, TDoc R1-091482, Mar. 2009.
18. WINNER+, "Calibration for IMT-Advanced evaluations," Tech. Rep.
19. Chinese Evaluation Group, "Calibration activities in Chinese evaluation group," Chinese Evaluation Group, Tech. Rep.
20. Nortel Networks, "A discussion on angle spread calculation," 3GPP-3GPP2 SCM AHG, Tech. Rep. SCM-119-Angle Spread, Mar. 2003.
21. M. Narandžić, C. Schneider, R. S. Thomä, T. Jämsä, P. Kyösti, and X. Zhao, "Comparison of SCM, SCME, and WINNER channel models," in *Proc. IEEE Vehicular Technology Conference (VTC-Spring 2007)*, Dublin, Ireland, Apr. 2007, pp. 413–417.
22. J. Salo, G. Del Galdo, J. Salmi, P. Kyösti, M. Milojević, D. Laselva, and C. Schneider, "MATLAB implementation of the 3GPP Spatial Channel Model (3GPP TR 25.996)," Jan. 2005.

23. C.-C. Chong, F. Watanabe, K. Kitao, T. Imai, and H. Inamura, "Evolution trends of wireless MIMO channel modeling towards IMT-Advanced," *IEICE Transactions on Communications*, vol. 92, no. 9, pp. 2773–2788, Sep. 2009.

24. R. Srinivasan, J. Zhuang, L. Jalloul, R. Novak, and J. Park, "IEEE 802.16m evaluation methodology document (EMD)," IEEE 802.16 Task Group m (TGm), Tech. Rep. IEEE 802.16m-08/004r5, Jan. 2009.

25. 3GPP, "Evolved Universal Terrestrial Radio Access (E-UTRA); Physical channels and modulation," 3rd Generation Partnership Project (3GPP), TS 36.211 version 10.0.0 Release 10, Jan. 2011.

26. E. Dahlman, S. Parkvall, and J. Sköld, *4G LTE/LTE-Advanced for mobile broadband*. Academic Press, 2011.

27. 3GPP, "Evolved Universal Terrestrial Radio Access (E-UTRA) and Evolved Universal Terrestrial Radio Access Network (E-UTRAN); Overall description; Stage 2," 3rd Generation Partnership Project (3GPP), TS 36.300 version 10.2.0 Release 10, Jan. 2011.

28. 3GPP, "Evolved Universal Terrestrial Radio Access (E-UTRA); Multiplexing and channel coding," 3rd Generation Partnership Project (3GPP), TS 36.212 version 10.0.0 Release 10, Jan. 2011.

29. 3GPP, "Evolved Universal Terrestrial Radio Access (E-UTRA); Physical layer procedures," 3rd Generation Partnership Project (3GPP), TS 36.213 version 10.2.0 Release 10, Jun. 2010.

30. K. Brueninghaus, D. Astély, T. Sälzer, S. Visuri, A. Alexiou, S. Karger, and G. Seraji, "Link performance models for system level simulations of broadband radio access systems," in *Proc. IEEE Personal, Indoor and Mobile Radio Communications Conference (PIMRC 2005)*, Berlin, Germany, Sep. 2005, pp. 2306–2311.

31. K. Sayana, J. Zhuang, and K. Stewart, "Link performance abstraction based on Mean Mutual Information per Bit (MMIB) of the LLR channel," IEEE 802.16 Broadband Wireless Access Working Group, Tech. Rep. IEEE C802.16m-07/097, May 2007.

32. Samsung, "Simulation methodology for EUTRA uplink: SC-FDMA and OFDMA," 3GPP TSG-RAN WG1, TDoc R1-051352, Nov. 2005.

33. Motorola, "Simulation methodology for EUTRA UL: IFDMA and DFT-spread-OFDMA," 3GPP TSG-RAN WG1, TDoc R1-051335, Nov. 2005.

34. Ericsson, "Email discussion summary on calibration step 1c," 3GPP TSG-RAN WG1, TDoc R1-092742, Jun. 2009.

35. P. Kyösti, J. Meinilä, L. Hentilä, X. Zhao, T. Jämsä, M. Narandžić, M. Milojević, C. Schneider, A. Hong, J. Ylitalo, V.-M. Holappa, M. Alatossava, R. Bultitude, Y. de Jong, and T. Rautiainen, "WINNER II channel models; Part II radio channel measurement and analysis results," IST-4-027756 WINNER II, Tech. Rep. Deliverable D1.1.2V1.0, Sep. 2007.

14

Channel Emulators for Emerging Communication Systems

Julian Webber

Hokkaido University, Japan

14.1 Introduction

The aim of a hardware channel emulator is to present accurate and sufficiently varied representations of the required channel environment (e.g., indoor, urban or rural and so on) to the hardware receiver system. The channel emulator presents a physical interface to measure the performance under repeatable conditions for a transceiver-under-test. The emulator assists in understanding the effects of key Multiple Input Multiple Output (MIMO) channel parameters, and enables accurate Bit Error Rate (BER) results to be obtained, for a number of channel scenarios and receiver parameters, in a much shorter time than is possible by simulation on a desktop computer.

Channel fading depends on a considerable number of physical processes and parameters (i.e., many reflections, refractions, diffraction and so on) all of which cannot be simulated in real-time on a single Field Programmable Gate Array (FPGA) device without simplifying assumptions. Furthermore, in recent years the complexity has increased not only with multiple antennas elements, but also in modeling the number of resolved paths and sub-paths. Therefore the objective is to accurately model the key fading processes under the constraint of limited hardware resources. Fortunately, the size of FPGAs in terms of the number of gates has rapidly increased from just 64 logic cells in 1985 (Xilinx XC2064) up to 2 million (Xilinx XC7V2000T) currently, and have become large enough to accommodate a complete baseband channel emulator system.

In the development of a MIMO channel emulator, a suitable hardware transceiver interface is useful for functional testing and also for system evaluation. The LTE advanced system employs MIMO-OFDMA in the downlink and Single Carrier Frequency Division Multiple Access (SC-FDMA) in the uplink. MIMO-OFDM systems can achieve high data-rates, are robust to the effects of multipath fading and have low complexity equalizer

LTE-Advanced and Next Generation Wireless Networks: Channel Modelling and Propagation, First Edition.
Edited by Guillaume de la Roche, Andrés Alayón Glazunov and Ben Allen.
© 2013 John Wiley & Sons, Ltd. Published 2013 by John Wiley & Sons, Ltd.

implementations. SC-FDMA on the other hand exhibits a lower Peak-to-Average Power Ratio (PAPR) and is deployed in the power-constrained mobile transmitter for the uplink.

The contents of this chapter are as follows. The top-level design features of a channel emulator are first presented. We then look in more detail at the individual building blocks. Uniform and Gaussian number generation techniques used in both the noise and fading generator modules are then described. Four different fading models are introduced including a method to create correlation between channels. The convolution process both in the time and frequency domains is covered. An example transceiver system that can be used in conjunction with the emulator for evaluating advanced generation systems is finally outlined.

14.2 Emulator Systems

The function of the MIMO channel emulator is to convolve the input data with a randomly generated fading channel as described by the equation

$$y = Hx + z, \tag{14.1}$$

where y is the $M \times 1$ received signal vector, H is an $M \times N$ channel matrix, x is the $N \times 1$ input data vector, and z is a $1 \times N$ vector of Additive White Gaussian Noise (AWGN) samples. The concept of the 2×2 MIMO time domain system is shown in Figure 14.1, where there are $N \times M$ fading and convolution processes and $N = 2$ noise generation units. Here, $x_0(t)$ represents the signal at input port-0 at time t; $h_{00}(t)$ is the single-input single-output (SISO) channel between input port-0 and output port-0; $z_0(t)$ is the noise sample at port-0 and $y_0(t)$ is the received signal at port-0, time t.

The MIMO channel emulator system comprises four principle system blocks: fading channel generator; convolution; noise generator; and signal conditioning processes (i.e., analogue-to-digital converters and digital-to-analogue converters) as shown in Figure 14.2. Common parameters required in the set-up configuration are: number of transmit / receive antenna elements, Ricean K-factor, correlation between elements, channel tap gains, number of multipaths, relative delays of each path. These features are covered in more detail later in this chapter.

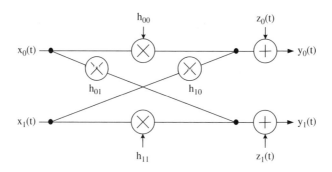

Figure 14.1   MIMO 2×2 channel emulator conceptual diagram.

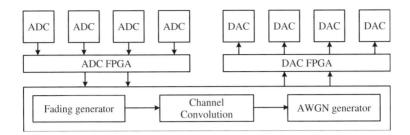

Figure 14.2 Channel emulator block diagram.

The channel emulation processes can be implemented in either the time or frequency domain. When the number of delayed paths is limited the time domain implementation using linear convolution is more efficient. However, as the number of channels, paths, and in recent designs, sub-paths increases, frequency domain convolution becomes favorable [1]. The frequency domain design consists of fading channel generation through filtering a Gaussian noise source and multiplication with the frequency domain input data. There is a fixed cost for the IFFT/FFT unit but the complexity is independent of the number of path delays. For time domain emulators the number of channels and paths is limited by the number of multiplication operations and is a key parameter in testbed design.

Examples of commercially available emulators include : Anritsu MF6900A [2]; Agilent N5106A [3]; Azimuth Systems [4]; and Spirent SR5500 [5], and selected specifications are listed in Table 14.1. Examples of testbeds developed at universities and research institutes include [6–12]. Parts of these systems will be discussed in more detail and referenced later in this chapter. Commercial systems usually have many features with a variety of optional packages for emulating specific channel environments such as for the Long Term Evolution (LTE) MIMO channels (e.g., MX690010A [2]).

14.3 Random Number Generation

White Gaussian Noise Generated (WGNG) samples are naturally produced by amplifying the thermal noise of a resistive element. However, the components are expensive and the generated data is not repeatable, with repeatability being an important feature when debugging and in understanding problematic channels for a particular system. Random sequences with long repeat times can be implemented using a Pseudo Random Noise Generator (PRNG).

Table 14.1 Features of recent channel emulators

Name	Max Doppler	Channels	Paths
Agilent N5106A [3]	1.6 kHz	2–8	24
Propsim F8 [13]	≡ 1150 km/h	2–32	48
Spirent SR5500 [5]	2 kHz	2–16	24
Anritsu MF6900A [2]	20 kHz	2–8	12
JRC NJZ-1600D [14]	2 kHz	2	12

The pseudo-random generator is an essential component of both the noise and fading generator modules. A Uniform Random Variable (URV) is generated from the output of a feedback shift register with specific feedback taps to generate a maximal length sequence [15]. The feedback delay-line requires very few resources and so long registers are used (e.g., 32–56 bits) to ensure each uniform sequence has repeat times far longer than a typical session. At a frequency of 100 MHz a 32-bit sequence would repeat after about 43 seconds. By initializing the register with a unique seed for sufficiently long registers, a different starting position can be selected.

14.3.1 Pseudo Random Noise Generator (PRNG)

A PRNG is created from a Linear Feedback Shift Register (LFSR). The LFSR is a delay line consisting of D-type flip-flops with feedback taps connected to an exclusive-or (XOR) gate. The taps are at specific positions to generate a maximal length sequence. An example generator consisting of a 4-bit LFSR is shown in Figure 14.3.

14.3.2 Gaussian Look-Up-Table

One method to generate Gaussian sequences is to read pre-computed samples from a Look Up Table (LUT). A 256K-word LUT was used in the design in [16]. This method requires a very large memory in order to accurately represent the distribution in the tail-region of the normal distribution. As a single-user $N \times M$ MIMO channel requires $N \times M$ Gaussian samples per delay path, this method becomes impractical for large MIMO systems particularly when the samples are held in on-chip FPGA memory.

14.3.3 Sum of Uniform (SoU) Distribution

The Central Limit Theorem states that the sum of N random samples will follow a Normal distribution as $N \rightarrow \infty$. The noise sample, z_G, approaches the Gaussian distribution as the number of summed uniform samples increases, and is given by

$$z_G = \frac{\sqrt{2}}{\sigma} \left(\frac{1}{N} \sum_{n=0}^{N-1} X_N - \mu \right) \tag{14.2}$$

where, μ, σ^2 are the mean and variance of X respectively. For practical reasons a value of $8 < N \leq 20$ is common, with $N \geq 20$ recommended in [17]. The normalized auto-correlation calculated over 13000 samples using $N = 12$ is shown in

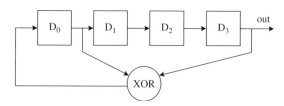

Figure 14.3   Pseudo random noise generator design using a 4-bit LFSR.

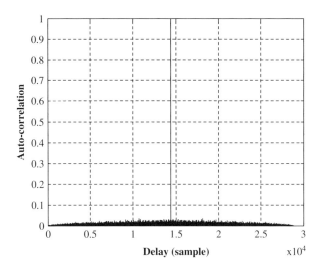

Figure 14.4   AWGN autocorrelation from summing $N = 12$ uniform random numbers.

Figure 14.4. Here, each generator used a different combination of seed, seed-offset and M-sequence length.

14.3.4   Box-Muller

The Box-Muller algorithm converts two uniform random variables into two Gaussian random variables using an inverse transformation. It is the commonly used method in Matlab and C-language function calls to create Gaussian noise. In this algorithm, two uniform random numbers, x_1 and x_2, distributed over [0,1) are first generated. These samples are then input into two functions, $f(x_1)$ and $g(x_2)$, and multiplied together to create a normally distributed sample $\mathcal{N}(0,1)$. The formula is given as:

$$f(x_1) = \sqrt{-2 \ln(x_1)} \tag{14.3}$$

$$g(x_2) = \cos(2\pi x_2) \tag{14.4}$$

$$v = f(x_1)g(x_2) \tag{14.5}$$

The Probability Density Function (PDF) of the Box-Muller and Sum of Uniform (SoU) algorithms have very similar distributions up to sample standard deviations of $\sigma = \pm 4$. Beyond this point the SoU method has less samples than the actual Gaussian distribution and consequently the BER is under-estimated at very low values.

The challenge when designing the Box-Muller block is to efficiently implement the square root of the logarithmic function $f(x_1)$ on the FPGA. To generate the required number of samples for the tail region requires a large look-up-table in order to store samples for the asymptotic regions near $x_1 = 0$ and 1. On the other hand, much fewer samples would suffice for the linear region around $x_1 = 0.5$. A non-uniform piecewise linear approximation for $f(x_1)$ and $f(x_2)$ was proposed in [18]. The logarithmic term

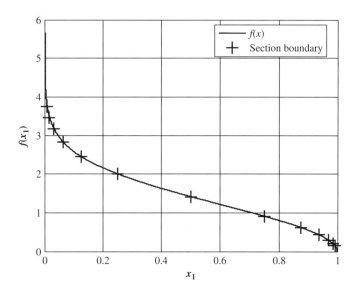

Figure 14.5 Logarithmic component of the Box-Muller equation showing the region boundary positions marked by '+'.

$f(x_1)$ of the Box-Muller is shown by a solid line in Figure 14.5. The whole function is shown here divided into 14 regions and the '+' markers indicate a boundary position. Starting at the center of the linear region ($x_1 = 0.5$) the size of regions in either direction is successively halved and this allows for a simplified address generation method. An improved system that extended the range of σ from 6.7 to 8.2 using fewer logic resources was later proposed by the same authors in [19].

An iterative approximation technique to compute $\ln(x_1)$ was proposed in [20]. The technique produced samples in the range $\pm 6.55\sigma$ and used approximately 3% of the configurable slices of a Virtex XC2V4000 FPGA. The logarithmic function and sinusoidal terms were stored in five 4-bit and one 8-bit look-up-tables respectively in [6]. In [21] several draws from a quantized Box-Muller distribution are added and averaged through the central limit theorem to form a smooth distribution. Xilinx and Altera both produce IP cores for AWGN generators. Cores are also available from FPGA vendors (e.g., Xilinx [22] produces a core yielding a PDF that deviates less than 0.2 percent from the Gaussian PDF for $|x| < 4.8\sigma$). The third-party Ukalta ultra-compact core generates samples up to 3.1σ with 1 percent relative error and uses 46 slices [23].

14.4   Fading Generators

Clarke first produced a reference model for Rayleigh fading which is detailed in [24]. This classical model describes a rich isotropic scattering environment surrounding a mobile receiver. The process is described by a sum-of-sinusoids where each sinusoid represents a scattered signal. This model is characterized by its auto-correlation function that is given by the first order Bessel function, $C(\tau) = J_0(2\pi f_d \tau)$, where τ is the delay lag and f_d is the maximum Doppler shift [24]. Sum-of-sinusoid simulators can be grouped into those that are statistical or deterministic. For statistical models, at least one of either the Doppler frequencies, gain or phase can vary with each trial, and a sufficient number of independent trials should be run so that ensemble averaging can be applied and for the sample statistics to converge with that of the classical model. Care has to be taken at the trial boundaries when a new channel is loaded for wave continuity reasons. For deterministic simulations, both the phase, gain and Doppler frequencies are kept constant throughout a simulation trial [25]. With these parameters set, a time-varying channel cannot be modeled and typically twice the number of oscillators are required for each fading channel compared to the statistical models.

A simple model for a moving transmitter surrounded by a ring of scatterers communicating with a fixed base-station receiver is depicted in Figure 14.6. The Doppler shift is proportional to the vehicle speed, v, and the cosine of the angle of arrival (AoA) between the scatterer and horizontal plane between Tx and Rx. The angular spread of scattered rays at the receiver is represented by α. This model was originally used to calculate the correlation at the base-station as a function of the angle of arrivals at the receiver [26]. There are several parameters or features that can be added to improve either the fading stability (e.g., [27]) or for modeling specific channel scenarios (e.g., [28]).

Jakes proposed a popular sum of sinusoids model [29] to characterize wideband Rayleigh fading, based on the flat-fading Clarke model. It was later shown that the in-phase and quadrature components of the model are slightly correlated, and that for a fixed number of oscillators, the components do not follow a perfect Gaussian distribution [30]. Therefore various models have since been proposed that improve certain characteristics of the Jakes model, for example, [31, 32]. We describe four different models that have been popularly implemented in hardware:

a) Gaussian Independent Identically Distributed also referred to as I.I.D;
b) Modified Jakes';
c) Zheng; and
d) Random walk models.

14.4.1 *Gaussian I.I.D.*

To create fading with a desired Doppler spectra, complex Gaussian random samples are passed through a filter having the appropriate response. For a line-of-sight channel where

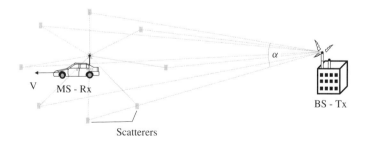

Figure 14.6 Model showing ring of scatterers evenly distributed around a mobile transmitter and fixed receiver.

a monopole antenna is used, the power density spectrum $\Phi(\omega)$ is given by [33]

$$\Phi(\omega) = \frac{a}{\sqrt{1 - (\frac{\omega}{\omega_D})^2}} : |\omega| < \omega_D, 0 \text{ otherwise.} \tag{14.6}$$

where a is the amplitude, and ω_D is the Doppler shift at velocity v. The symmetrical filter response can be implemented with a band-pass or low-pass filter followed by an interpolator [34].

A single channel tap of a block fading I.I.D. channel is created by a Gaussian generator such as the aforementioned Box-Muller or Sum of Uniform methods, for example [35]. Several multipath delays are then independently generated with an average Tapped-Delay Line (TDL) profile following a standard model such as those defined within the IEEE Task Group TGn [36].

14.4.2 Modified Jakes' Model

The modified Jakes or Dent model [31] has been a popular algorithm for fading channel realization in hardware. This algorithm redefines the scatterer angles of the original Jakes' model and uses orthogonal Walsh codes to remove correlation between sequences. The algorithm is expressed as

$$h_J(t, k) = \sqrt{\frac{2}{N_0}} \sum_{n=1}^{N_0} C_{kn}\{[\cos(\beta_n) + j\sin(\beta_n)]\cos(\omega_n t + \theta_n)\} \tag{14.7}$$

where, t is the sample time, n is the ray, k is the channel, β_n is a constant phase offset, N_0 is the number of oscillators/4, C_{kn} is an orthogonal Walsh-Hadamard code bit that is, +1 or -1, θ_n is an initial phase and ω_n is the Doppler shift.

Each scattered ray is modeled by a sinusoidal signal, and can be created from either a dedicated direct digital synthesis core, Coordinate Rotational Digital Computer (CORDIC) or by storing a quantized version in a Read Only Memory (ROM) LUT. A single quadrant only needs to be held in ROM due to its symmetry. The read address step-size determines the frequency and hence is used to control the Doppler frequency. The Doppler frequencies generated from the output of a LUT are shown in Figure 14.7. Typically a small number (e.g., between 8 and 32) of the scattered rays are then summed. For each channel, an independent URV is created for β_n and θ_n.

14.4.3 Zheng Model

The Zheng model produces samples with statistics converging to the Clarke model with only a small number of oscillators, but requires a relatively large number of simulation trials. The fading on the in-phase and quadrature components at sample index m is expressed by [37]

$$h_I(m) = \sqrt{\frac{2}{N}} \sum_{n=1}^{N}\{\cos(2\pi f_D T_s m\cos(\alpha_n + \phi_n))\} \tag{14.8}$$

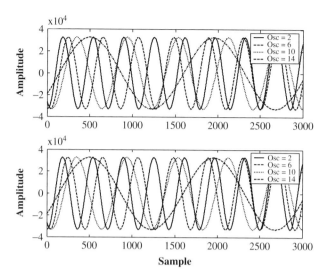

Figure 14.7 Scattered ray components from the Modified Jakes' model, each with a different Doppler frequency (four shown for clarity).

$$h_Q(m) = \sqrt{\frac{2}{N}} \sum_{n=1}^{N} \{\cos(2\pi f_D T_s m \sin(\alpha_n + \psi_n)\} \tag{14.9}$$

and the arrival angle of the n-th ray is

$$\alpha_n = \frac{2\pi n - \pi + \theta}{4N} \tag{14.10}$$

where f_D is the maximum Doppler frequency, T_s is the sample period, n is the sinusoid number, and ϕ_n and θ are phases. Both ϕ_n and θ are initialized at the start of each trial with a different uniform random number distributed on $[-\pi, \pi)$. The autocorrelation function of the Zheng model with $N = 16$ averaged over four trials is compared with $N = 32$ averaged over 16 trials in Figure 14.8. Although both models perform close to the Clarke reference model the increase in accuracy at larger lag delays can be seen with the latter configuration and 50–100 trials are typically required for convergence. As the auto-correlation between in-phase and quadrature components does not quite match the reference model over a single trial, a faster deterministic model, with a lower envelope correlation between I-Q channels and a statistical model that requires about 30 trials was proposed in [25]. A good discussion of the merits of various schemes is found in [38].

14.4.4 Random Walk Model

A random walk model was proposed by Alimohammad in [39] that produced very accurate fading statistics over a single simulation trial. The fading signal is given by

$$h_I(m) = \sqrt{\frac{1}{N}} \sum_{n=1}^{N} \{\cos(2\pi f_D T_s m \cos(\alpha_n(m) + \phi_n(m)))\} \tag{14.11}$$

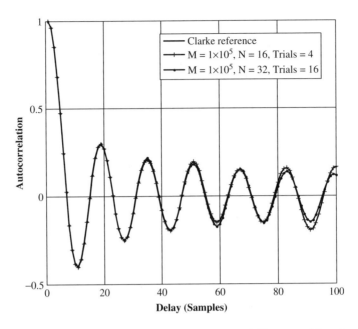

Figure 14.8 Autocorrelation function for Clarke reference and Zheng models using 10^5 samples with variable number of oscillators and number of ensemble trials.

$$h_Q(m) = \sqrt{\frac{1}{N} \sum_{n=1}^{N} \{\cos(2\pi f_D T_s m \sin(\alpha_n(m)) + \psi_n(m))\}} \tag{14.12}$$

$$\alpha_n(m) = \frac{2\pi n - \pi + \theta(m)}{4N}, \tag{14.13}$$

where $\phi_n(m)$, $\psi_n(m)$ and $\alpha_n(m)$ are distributed uniformly over $[-\pi, \pi)$ and now are functions of m. The remaining variables have the same definition as those in eqn 14.10.

The random variable $\theta(m)$ in eqn 14.13 is a slowly changing quantity, and is calculated from the current value plus a very small random update and thus appears to vary in a *random walk* manner with a high temporal correlation as shown in Figure 14.9. The update rate, δ_0, is controlled by the normalized Doppler frequency and the new variable is given by [38]

$$\theta(m) = \theta(m-1) + \delta_0 \times u(m), \tag{14.14}$$

where δ_0 is a small constant and u is a random variable distributed on $[0,1)$. If the value of θ exceeds the limits $[-\pi, \pi)$, they are held and δ_0 becomes equal to $-\delta_0$.

14.4.5 Ricean K-Factor

For the dense urban environment, it is well known that the fading process can be accurately approximated with the Rayleigh profile (such as in downtown Manhattan [40]). In other

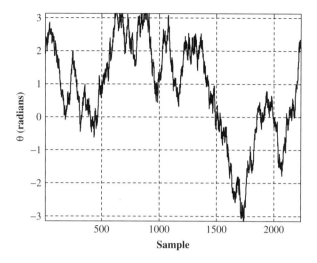

Figure 14.9 Random walk phase trajectory with $\delta_0 = 1 \times 10^{-6}$.

environments (i.e., rural) a single line-of-sight ray may be dominant while the remaining scattered rays sum to form a Rayleigh distribution. The Ricean K-factor describes the ratio of the magnitudes of the direct to non-direct components. The channel can then be expressed as

$$H = \sqrt{\frac{K}{1+K}} H_{\mathrm{d}} + \sqrt{\frac{1}{1+K}} H_{\mathrm{n}}, \tag{14.15}$$

where H_{d} and H_{n} are the coherent and incoherent channel components respectively and K is the Ricean K-factor.

The channel suffers deep fades when K is zero, and as K increases the direct component becomes more dominant. The above gain coefficients are multiplied, together with the path gain, in the tapped delay line. To avoid computing the square root and division operations, a small 128-word Random Access Memory (RAM) can store pre-computed values from -32 to +32, at 0.5 unit intervals for example.

14.4.6 Correlation

Correlation between channels is caused by insufficient spacing of the antenna elements at both the transmitter and receiver to obtain independent (decorrelated) signals on each branch [41]. As the correlation between MIMO sub-channels increases, the system performance degrades [42]. The Kronecker model estimates the MIMO channel covariance matrix R, from the Kronecker product of the transmit R_{TX}, and receive R_{RX}, correlation matrices [43]:

$$R = R_{\mathrm{TX}} \otimes R_{\mathrm{RX}}. \tag{14.16}$$

The model allows for a low complexity technique to create correlated channels and has been shown to have good accuracy on the assumption of three criteria:

1. the antenna elements are sufficiently separated for there to be no mutual coupling,
2. there is low correlation at one end of the link, and
3. the array is small for example typically 3×3 [44].

Through Cholesky factorization of the covariance matrix, $R = LL^T$, a lower diagonal matrix L is obtained. A vector of correlated samples Z, can then be calculated by multiplying L with a column vector of I.I.D. samples V, as shown by

$$
\begin{pmatrix} Z_1 \\ Z_2 \\ Z_3 \\ \vdots \\ Z_m \end{pmatrix} = \begin{pmatrix} 1 & 0 & 0 & \cdots & 0 \\ l_{21} & l_{22} & 0 & \cdots & 0 \\ l_{31} & l_{32} & l_{33} & \ddots & \vdots \\ \vdots & \vdots & \vdots & \ddots & 0 \\ l_{m1} & l_{m2} & l_{m3} & \cdots & l_{mm} \end{pmatrix} \begin{pmatrix} V_1 \\ V_2 \\ V_3 \\ \vdots \\ V_m \end{pmatrix}. \tag{14.17}
$$

The factorization can be computed in real-time using a Cholesky factorization IP core such as by AccelDSP [45]. Alternatively, the lower diagonal matrix can be computed off-line for a variety of different element correlations and stored in a ROM look-up-table. The fading channel data for a highly correlated system with $\rho_{12} = 0.85$, $\rho_{13} = 0.8$, $\rho_{14} = 0.85$, $\rho_{23} = 0.85$, $\rho_{24} = 0.9$ and $\rho_{34} = 0.95$ is plotted in Figure 14.10.

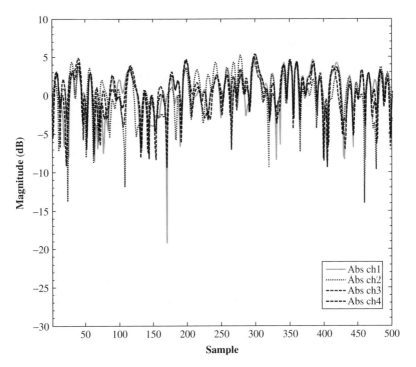

Figure 14.10 Plot of correlated fading envelope on four channels implemented using Kronecker model and Cholesky factorization.

The correlation coefficients depend on the array characteristics and can be computed off-line or computed by the host computer at the start of the simulation trial once the required parameters of the channel environment have been entered for example, antenna height, element spacing and so on. The correlation, ρ, [46] is given by

$$\rho = \int_{-\pi}^{\pi} \exp(-j\sin(\theta))G(\theta)P(\theta)d\theta \qquad (14.18)$$

where $d = 2\pi D/\lambda$, D is the distance between array elements, $G(\theta)$ is the antenna radiation pattern, $P(\theta)$ is the power azimuth spectrum antenna element.

A generalized algorithm for creating an arbitrary number of spectrally or spatially correlated samples after Doppler filtering, with equal or unequal powers, was described in [47]. The variance of the random variables are changed as a result of passing through a Doppler filter, and so extra processing steps are required when using this method.

14.5 Channel Convolution

The wideband MIMO channel is modeled by tapped delay lines, one for each channel delay profile between Tx and Rx antenna pairs. The received signal at element n is

$$y_n(t) = \sum_{m \neq n} \sum_k h_{m,n,k} x_m(t - \tau) + z_n \qquad (14.19)$$

where x_m is the transmitted signal from element m, $h_{m,n,k}$ is the kth path gain from element m to n, τ_k is the kth path delay and z_n is an AWGN sample.

Time domain designs using a Finite Impulse Response (FIR) filter have a trade-off between the number of channel taps and MIMO sub-channels available, due to the limited number of hardware multiplier units available. The delayed paths can be stored in First In First Out (FIFO) or block RAM memory. The maximum channel delay depends then on the available memory and, 10k words, for example, will support $100\,\mu s$ when using a 10 ns sample period. Alternatively, the multipath delays can be implemented using a standard TDL model. A programmable TDL architecture is shown in Figure 14.11. Here, specific tap delays can be set at run-time using a Xilinx SRL16E programmable shift register [15]. Each register has a maximum size of 16 delay elements, and five can be cascaded together, for example, to provide delays of up to 80 samples. The n-th path delay is donated by D_n. The specific number of sample delays is set at run-time according to the specific channel profile using the INIT register. The multipath gains at time sample m are donated by $g_n(m)$ and Ricean coefficients $r_n(m)$. The channel resolution R_{max} is derived from

$$D = 20b\log_{10}(2) + 20\log_{10}\left(10^{\frac{R_{max}}{20}} - 1\right) \qquad (14.20)$$

where D is the dynamic range and b is the number of bits [48]. For example, if 40 dB dynamic range is required, with $R_{max} = 0.1\,dB$ resolution, then 14 bits are required. A shift operation after the data-tap multiplication can increase the dynamic range without reducing the resolution for a given number of bits.

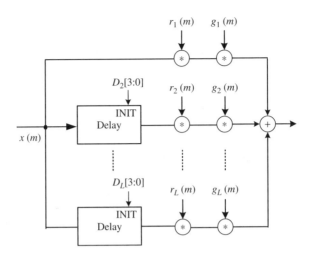

Figure 14.11   Tapped delay line filter architecture.

The linear convolution in the time domain can also be efficiently computed in the frequency domain using the circular convolution theorem [49]. To apply circular convolution, the size of the Fast Fourier Transform (FFT) needs to be greater or equal to the length of the data block size plus the length of the delay spread [50]. The emulator then multiplies the frequency domain input and channel signals together and applies a final Inverse Fast Fourier Transform (IFFT) operation to return back to the time domain.

$$y(n) = \text{IFFT}\left[\text{FFT}(x(n)\left[\text{FFT}(h(n))\right]\right] \tag{14.21}$$

When computing the convolution in the frequency domain, there is an overhead involved in computing the FFT/IFFT operations once per channel profile. The multiplication complexity is a function of the MIMO array size, but unlike its time domain counterpart, it is independent of the number of channel taps [51]. To reduce complexity, the CORDIC operator can be used to further convert "multiply" units into simpler "adder" units. A comparison between the two methods was made noting the relative complexity, C, of a multiplier compared to an adder and for a 3×3 MIMO system with FFT size 1024 and assuming $C = 16$, the frequency domain system required 67 percent less resources [51].

Clusters around a main path can be modeled using a tapped delay line with additional taps and finer resolution delays. The model still does not include directional information which affects the MIMO correlation. The Spatial Channel Model (SCM) was based on measurements in the 2 GHz band and with a 5 MHz bandwidth. There were six clusters with up to 20 sub-paths per cluster [52]. The Spatial Channel Model Extended (SCME) expanded the work of the SCM project with 100 MHz measurement bandwidth, 5 GHz frequency band and with 3-4 mid-paths per cluster [53]. The processing of additional sub-paths clearly adds complexity to the emulator, and hence there is a trade-off between the number of paths/sub-paths and MIMO channels supported.

14.6   Emulator Development

The hardware logic can be described at various levels of abstraction. Rapid prototyping software such as Xilinx System Generator for Digital Signal Processor (DSP) [54] enables high productivity, flexibility and ease of maintenance by removing the need to program at Register Transfer Level (RTL) or gate level. This is a graphical block-based approach using Matlab Simulink and a suite of DSP functions that are compiled into a gate netlist. An example of a channel emulator built using this method is detailed in [7]. Development software such as SystemC [55] that enables the design to be described in C-language has also been growing in popularity. National Instruments produce a 6.6 GHz MIMO test system for beamforming applications [56]. The system supports up to 4×4 and has ready-to-use code written in the high-level LABview [57] language. The two most widely used languages for gate-level description and implementation are still Verilog and Very High speed integrated circuits hardware Description Language (VHDL). Although, programming with fixed-point RTL provides control over the design of each module, the code development, testing and debugging times become much longer.

To reduce development time, IP cores can be used. The cores have efficient implementations in terms of gate sizes, are rigorously tested and well documented. In some cases, the core may offer extra functionality that is not required. Cores from the Xilinx IP core generator used in a typical emulator are FFT/IFFT, CORDIC and FIR compilers. The universal CORDIC core can compute sine, cosine, hyperbolic sine/cosine, square roots, phase rotations and translations.

The fading and noise generation processes are implemented in real-time on the vast majority of channel emulators. An off-line computation of the channel taps was demonstrated in [8] where they are loaded into RAM for 2 minutes run-time simulation. To compute very low BER accurately, much longer run-times would be required in the order of hours or days.

14.7   Example Transceiver Applications for Emerging Systems

In this section two example transceiver applications are described for use with a channel emulator [58]. The first application is MIMO-OFDM and the second application under development is frequency domain equalization for single carrier uplink such as that specified for the LTE uplink. The transmitter and receiver units each consist of a Koden E-1071 base unit [59] as shown in Figure 14.12. Each system contains a digital signal processing card with four TigerSharc TS201 DSPs and three Xilinx Virtex-4 FPGA including an LX100 device where the MIMO signal processing operations are computed. The circuit architecture description was written at RTL gate level using VHDL language.

14.7.1   MIMO-OFDM

Eigenbeam Space Division Multiplexing (E-SDM) is used to optimize the capacity of a MIMO-OFDM system. On each substream, orthogonal beams are directed between the transmit and receive antennas [60]. The transmit and receive beamforming weights are

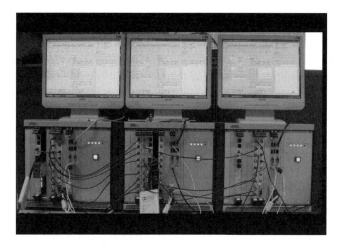

Figure 14.12 Testbed set-up consisting of transmitter, channel emulator and receiver systems.

represented by the weight matrices W_{TX} and W_{RX} respectively. The optimal weights are given by equations

$$W_{Tx} = U\sqrt{P_T}, \tag{14.22}$$

$$W_{Rx} = U^H H^H, \tag{14.23}$$

where $P_T = \text{diag}(P_1, \cdots, P_K)$ is the transmit power matrix, H^H is the frequency domain channel transpose complex conjugate and U is obtained from the eigenvalue decomposition (EVD) as (14.24):

$$H^H H = U\Lambda U^H, \tag{14.24}$$

$$\Sigma = \text{diag}(\sigma_1, \cdots, \sigma_K). \tag{14.25}$$

where, $\sigma_1, ..., \sigma_K$ are the K eigenvalues of $H^H H$ contained in the vector Σ, and the columns of U are the corresponding eigenvectors. A top-level block diagram of the MIMO-OFDM receiver system [61] is displayed in Figure 14.13. At the transmitter, the data is scrambled and encoded for Forward Error Correction (FEC). An interleaver multiplexes the data bits onto non-adjacent sub-carriers. The data is then punctured, mapped and a time domain signal generated by the IFFT operation. To transmit one data packet, an additional sounding packet is first transmitted allowing the receiver to estimate the MIMO channel and channel beamforming is then applied to the subsequent packet.

The beamforming bit-error performance was evaluated using Modelsim^{TM} software. This simulates the hardware logic response using the same VHDL code as that used in the final logic gate synthesis compilation. The number of bits per substream were as follows {4,4}, {4,4,4}, {2,2,2,2}, {6,2} and refer to the draft IEEE 802.11n standard [62] modulation coding scheme (MCS) values 11, 19, 25 and 34 respectively. The Spatial Division Multiplexing (SDM) results are computed in an AWGN uncorrelated Gaussian channel and shown in Figure 14.14. The packets with E-SDM beamforming were of length

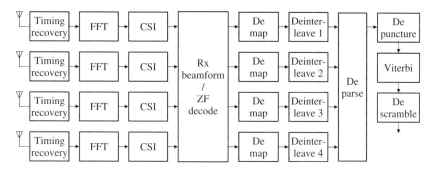

Figure 14.13 MIMO-OFDM receiver block diagram.

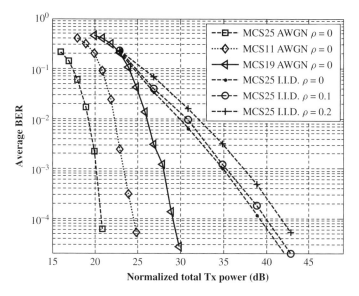

Figure 14.14 BER performance of SDM in AWGN and E-SDM systems in I.I.D. channels with increasing elemental correlation.

N_T times 40 OFDM symbols and were transmitted over random Gaussian 2×2, 3×3 and 4×4 single-tap MIMO channels with increasing ρ. By increasing the correlation from $\rho = 0.0$ to 0.2 an additional 2.6 dB transmit power is required to maintain the BER at 10^{-3}.

14.7.2 Single Carrier Systems

Single carrier systems provide a number of advantages over multi-carrier systems such as OFDM. The system is less reliant on coding and interleaving as the information on each symbol is modulated across the transmission bandwidth. The PAPR is lower compared to an OFDM system where the individual sub-carriers can add up with the same size and phase. The IFFT operator in the OFDM transmitter is moved to the receiver in a Single Carrier Frequency Domain Equalization (SC-FDE) system (Figure 14.15), and

Figure 14.15 SC-FDE receiver block diagram showing frequency domain channel equalization blocks.

hence the complexity is shifted from the power-limited mobile transmitter to the base-station receiver [63].

Figure 14.16 depicts a doubly-selective channel where fading occurs both at individual frequencies during the data slot-time, and also at each sample instant across the signal bandwidth. In this situation, Frequency Domain Equalization (FDE) is a low complexity technique to equalize the inter-symbol interference inflicted channel. To maintain cyclicity of the channel a Cyclic Prefix (CP) longer than the channel excess delay is appended to an FDE data block. The received signal after CP removal is given by

$$r = Hs + n, \tag{14.26}$$

where H is the $N \times N$ channel matrix and n is Gaussian noise. An $N \times N$ Discrete Fourier Transform (DFT) is computed at the receiver giving

$$Fr = FHs + Fn \tag{14.27}$$

$$= FHF^H Fs + Fn. \tag{14.28}$$

Assuming the channel is static within a block, H can be diagonalized by the DFT operation F as $FHF^H = D = \mathrm{diag}(d_1, \ldots, d_N)$. The received signal is then given as

$$r_f = Ds_f + n_f, \tag{14.29}$$

where $s_f = Fs$, $r_f = Fr$, and $n_f = Fn$ are the frequency domain transmitted, received and noise vectors respectively. The equalized signal is obtained by

$$y_f = W_r^H r_f, \tag{14.30}$$

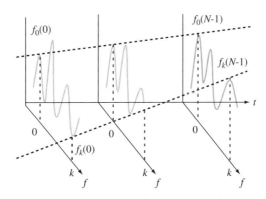

Figure 14.16 Doubly-selective channel profile change in high mobile environments.

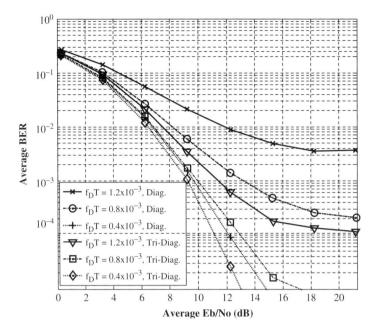

Figure 14.17 Error performance of SC-FDE in fading channel with increasing $f_D T$ product.

where W_r is the minimum mean square error (MMSE) weight. The computation of the weight matrix has low complexity ($\mathcal{O}(N)$) as $DD^H + \sigma^2 I$ is a diagonal matrix.

The channel is no longer cyclic when it changes within an FDE time slot, and FHF^H becomes a non-diagonal square matrix T. The signal components at each delay and frequency can be interpolated after considering the relationship between T and the channel variation [64]. By estimating the components either side of the diagonal (termed tri-diagonal) a significant improvement in performance can be obtained. Further, the complexity of calculating T is reduced from $\mathcal{O}(N^2 \log N)$ to $\mathcal{O}(N \log N)$.

The single carrier uplink was simulated for a system with block size 256 samples and Binary Phase-Shift Keying (BPSK) modulation. The standard Jakes' fading model was used with 16 delay taps and 1 dB attenuation per tap. As the normalized Doppler frequency $f_D T$ increases, the performance degrades. Figure 14.17 compares the performances of the diagonalization and tri-diagonalization equalization techniques, with the latter providing a 3.0 dB improvement at a target BER of 10^{-3} when $f_D T = 0.8 \times 10^{-3}$.

14.8   Summary

This chapter has reviewed the design of recent channel emulators. It is not possible to describe in detail all of the available emulators in a single chapter and the reader is encouraged to make use of the supplied references. The design and implementation aspects of a channel emulator for investigating the performance of modern communications systems such as LTE in fading channels has been described. The description of four implementation efficient and low-complexity fading generators were described.

Methods to create uniform and Gaussian samples from a linear feedback shift register were discussed. Finally, a testbed transceiver system was outlined for use with the fading channel emulator.

References

1. H. Eslami and A. Eltawil, "A scalable wireless channel emulator for broadband MIMO systems," in *IEEE International Conf. on Communications (ICC 2007)*, June 2007, pp. 2592–2597.
2. Anritsu, "MF6900A fading simulator product introduction," June 2009.
3. Agilent, "N5106A PXB baseband generator and channel emulator datasheet," September 2011.
4. Azimuth, "ACE 400WB MIMO channel emulator," 2008.
5. S. Communications, "Spirent SR5500 wireless channel emulator," September 2010.
6. E. Fung, K. Leung, N. Parimi, M. Purnaprajna, and V. Gaudet, "ASIC implementation of a high speed WGNG for communication channel emulation," in *IEEE Signal Processing Systems (SIPS), proc.*, October 2004, pp. 304–309.
7. J. Hwang, K. Lin, J. Li, and J. Deng, "Fast FPGA prototyping of a multipath fading channel emulator via high-level design," in *IEEE International Symposium on Communic. Inform. Tech. (ISCIT), proc.*, October 2007, pp. 168–171.
8. A. Dassatti, G. Masera, M. Nicola, A. Concil, and A. Poloni, "High performance channel model hardware emulator for 802.11n," in *IEEE Field-Programmable Tech. 2005, proc.*, December 2005, pp. 303–304.
9. C. Wang, D. Yuan, H. Chen, and W. Xu, "An improved deterministic SoS channel simulator for multiple uncorrelated Rayleigh fading channels," *IEEE Trans. Wireless Commun.*, vol. 7, no. 9, pp. 3307–3311, September 2008.
10. K. Sakaguchi, J. Takada, and K. Araki, "A novel architecture for MIMO spatio-temporal channel sounder," *IEICE Trans. Electron.*, vol. E85-C, no. 3, pp. 436–441, March 2002.
11. F. Ren and Y. Zheng, "Hardware implementation of triply selective Rayleigh fading channel simulators," in *IEEE ICASSP'10*, June 2010, pp. 1498–1501.
12. X. Qian, S. Hong, G. Liang-Cai, and J. Sen, "Design and implementation of a wideband HF channel sounder," *Wuhan University Journal of Natural Sciences*, vol. 9, no. 2, pp. 220–224, February 2004.
13. Elektrobit, "Scalable tool for radio channel emulation: EB Propsim F8," 2010.
14. JRC, "Multi-path fading simulator: NJZ-1600D," November 2005.
15. M. George and P. Alfke, "Linear feedback shift registers in Virtex devices," *Xilinx Application Report XAPP210*, April 2007.
16. M. Jezequel, C. Berrou, J. Inisan, and Y. Sichez, "Test of a turbo-encoder/decoder," in *Turbo coding seminar, Lund, Sweden*, no. 9, August 1996, pp. 35–41.
17. M. Jeruchim, P. Balaban, and S. K, *Simulation of communications systems*, first edition ed. Springer, 1992.
18. D. Lee, W. Luk, J. Villasnor, and P. Leong, "A Gaussian noise generator hardware-based simulations," *IEEE Trans. Computers*, vol. 53, no. 12, pp. 1523–1533, December 2004.
19. D. Lee, J. Villasnor, W. Luk, and P. Leong, "A hardware Gaussian noise generator using the Box-Muller method and its error analysis," *IEEE Trans. Computers*, vol. 55, no. 6, pp. 659–671, June 2006.
20. A. Alimohammad, S. Fard, B. Cockburn, and C. Schlegel, "An iterative hardware Gaussian noise generator," in *IEEE Pacific Rim Conf. Commun., Comp. Sig. Process., proceed.*, August 2005, pp. 649–652.
21. A. Ghazal, E. Boutillon, J. Danger, and G. Gulak, "Design and performance analysis of a high speed AWGN communication channel emulator," in *IEEE PACRIM'01*, August 2001.
22. Xilinx, "Additive white Gaussian noise (AWGN) core v1.0," *Xilinx*, October 2001.
23. UKalta, "3.1 sigma uncorrelated Gaussian noise generator IP core," *UKalta Engineering Product Brief*, 2009.
24. R. Clarke, "A statistical theory of mobile-radio reception," *Bell System Technical Journal*, vol. 47, pp. 957–1000, 1968.
25. A. Zajic and G. Stuber, "Efficient simulation of Rayleigh fading with enhanced de-correlation properties," vol. 5, no. 7, July 2006, pp. 1866–1875.
26. W. Lee, *Mobile Communications Engineering*, first edition ed. McGraw Hill, 1982.
27. H. Nishimoto, T. Nishimura, T. Ohgane, and Y. Ogawa, "Arrangement of scattering points in Jakes' model for i.i.d. time-varying MIMO fading," vol. E90-B, no. 11, November 2007, pp. 3311–3314.

28. A. Chelli and M. Patzold, "A MIMO mobile-to-mobile channel model derived from a geometric street scattering model," in *IEEE ISWCS'07*, October 2007, pp. 792–797.

29. W. Jakes, *Microwave Mobile Communications*, first edition ed. John Wiley and Sons, 1974.

30. M. Pätzold and F. Laue, "Statistical properties of Jakes' fading channel simulator," in *IEEE Vehicular Technology Conference (VTC 1998)*, May 1998, pp. 712–718.

31. P. Dent, G. Bottomley, and T. Croft, "Jakes fading model revisited," *Electronics Letters*, vol. 29, pp. 1162–1163, June 1993.

32. M. Pop and N. Beaulieu, "Design of wide-sense stationary sum-of-sinusoids fading channel simulators," in *IEEE ICC'02*, May 2002, pp. 709–716.

33. J. Parsons and A. Bajwa, "Wideband characterization of fading mobile radio channels," in *Proc. Inst. Elect. Eng.-Part F*, vol. 129, no. 2, April 1982, pp. 95–101.

34. M. Kars and C. Zimmer, "Digital signal processing in a real-time propagation simulator," *IEEE Trans. Instrument. Measure.*, vol. 55, no. 1, pp. 197–205, February 2006.

35. S. Orfanidis, *Optimum signal processing*, first edition ed. MacMillan, 1988.

36. T. Paul and T. Ogunfunmi, "Wireless LAN comes of age: IEEE 802.11n amendment," *IEEE Circuits and Systems Magazine*, pp. 28–54, 2008.

37. Y. Zheng and C. Xiao, "Improved models for the generation of multiple uncorrelated Rayleigh fading waveforms," *IEEE Commun. Lett.*, vol. 6, no. 6, pp. 256–258, June 2002.

38. A. Alimohammad, S. Fard, B. Cockburn, and C. Schlegel, "Compact Rayleigh and Rician fading simulator based on random walk processes," *IET Communications Journal*, vol. 3, no. 8, pp. 1333–1342, 2009.

39. A. Alimohammad, S. Fard, B. Cockburn, and C. Schlegel, "An accurate and compact Rayleigh and Rician fading channel simulator," in *VTC2008-Spring*, May 2008, pp. 409–413.

40. D. Chizhik, J. Ling, P. Wolniansky, R. Valenzuela, N. Costa, and K. Huber, "Multiple-input multiple-output measurements and modeling in Manhattan," *IEEE Journ. Sel. Areas Commun.*, vol. 21, no. 3, pp. 321–331, April 2003.

41. B. Allen, R. Brito, M. Dohler, and H. Aghvami, "Performance comparison of spatial diversity array topologies in an OFDM based wireless LAN," *IEEE Trans. on Consumer Elec.*, vol. 50, no. 2, pp. 420–428, May 2004.

42. V. Tarokh, N. Seshadri, and A. Calderbank, "Space-time codes for high data rate wireless communications: performance and codes construction," *IEEE Trans. Inform. Theory*, vol. 44, no. 6, pp. 744–765, March 1998.

43. D. Shiu, G. Foschini, M. Gans, and J. Kahn, "Fading correlation and its effect on the capacity of multi-element antenna systems," *IEEE Trans. Commun.*, vol. 48, no. 3, pp. 502–513, March 2000.

44. C. Oestges and B. Clerkx, *MIMO wireless communications: From real-world propagation to space-time code design*, first edition ed. Academic Press, 2007.

45. Xilinx, "AccelDSP synthesis tool v10.1.1," *Xilinx*, April 2008.

46. L. Shumacher, K. Pedersen, and P. Morgensen, "From antenna spacing to theoretical capacities-guidances for simulating MIMO systems," in *IEEE PIMRC'02*, September 2002, pp. 587–592.

47. L. Tran, T. Wysocki, A. Mertins, and J. Seberry, "Algorithm for generating correlated Rayleigh envelopes," in *EURASIP Journal on Wireless Communications and Networking*, May 2005, pp. 801–815.

48. K. Borries, G. Judd, D. Stancil, and P. Steenkiste, "FPGA-based channel simulator for a wireless network emulator," in *Proc. IEEE VTC*, May 2009.

49. J. Proakis and D. Manolakis, *Digital signal processing*, fourth edition ed. Prentice-Hall, 2006.

50. S. Stearns, *Digital signal processing examples in Matlab*, first edition ed. CRC Press, 2003.

51. H. Eslami and A. Eltawil, "Design and implementation of a scalable channel emulator for wideband MIMO systems," *IEEE Trans. on Vehic. Tech.*, vol. 58, no. 9, pp. 4698–4709, November 2009.

52. "Spatial channel model for multiple input multiple output MIMO simulations (release 10)," in *3GPP TR 25.996 V10.0.0*, March 2011.

53. Elektrobit, "Beyond conformance testing in 3GPP LTE: white paper," June 2009.

54. Xilinx, "System generator for DSP-user guide v13.3," October 2011.

55. D. Black, J. Donovan, B. Bunton, and A. Keist, *SystemC: From the ground up*, second edition ed. Springer, 2009.

56. N. Instruments, "Multichannel RF signal generation and acquisition," *National Instruments Data Sheet*, 2010.

57. N. Instruments, "Creating custom hardware with labview," *National Instruments Data Sheet*, 2005.

58. J. Webber, T. Nishimura, T. Ohgane, and Y. Ogawa, "Experimental MIMO Pseudo E-SDM and channel emulation system," in *IEEE PIMRC'09, proc.*, no. 9, September 2009.

59. "E-1072 signal processing platform," *Koden Electronics*, September 2008.

60. T. Ohgane, T. Nishimura, and Y. Ogawa, "Applications of space division multiplexing and those performance in a MIMO channel," *IEICE Trans. Commun.*, vol. E88-B, no. 5, pp. 1843–1851, May 2005.

61. Y. Cho, J. Kim, W. Yang, and C. Kang, *MIMO-OFDM wireless communications with Matlab*, first edition ed. Wiley, 2010.

62. "IEEE P802.11 wireless LANs joint proposal: High throughput extension to the 802.11 standard: PHY," *IEEE 802.11-05/1102r4*, January 2006.

63. T. Tavangaran, A. Wilzeck, and T. Kaiser, "MIMO SC-FDMA system performance for space time/frequency coding and spatial multiplexing," in *IEEE WSA'08*, 2 2008.

64. K. Saito, J. Webber, T. Nishimura, T. Ohgane, and Y. Ogawa, "Tri-diagonalizing approach on frequency domain equalization in a doubly-selective channel," in *IEEE Globecom '09, Hawaii, proc.*, no. 9, December 2009.

15

MIMO Over-the-Air Testing

Andrés Alayón Glazunov[1], Veli-Matti Kolmonen[2] and Tommi Laitinen[2]

[1]*KTH Royal Institute of Technology, Sweden*
[2]*Aalto University, Finland*

15.1 Introduction

A marked trend in the wireless communication industry is to increase the bandwidths of services offered to customers. For this to become reality, highly optimized and complex systems have been developed. It has been predicted that in new-coming systems, such as the Long Term Evolution (LTE) and Long Term Evolution Advanced (LTE-A), the connections Over The Air (OTA), that is, over the wireless link, will reach above 100 Mbit/s in data throughput. These requirements combined with the increasingly complex usage environments pose challenging system design problems. For example, today's tough requirements on high speed connections both indoors and in highly mobile outdoor environments as well as in city centres need to be met simultaneously by numerous wireless devices sharing the same system resources. This inevitably creates a challenging operating environment which is highly dynamic and unpredictable.

In this context, the effects of the user's head, body and hands as well as the variety of usage positions of wireless devices pose further challenges to antenna design engineers. Indeed, a modern mobile phone can be used in various ways, for example, for gaming, browsing or talking. In the first two cases, the user may hold the phone with one or two hands in front of his body. On the other hand, the user can hold the phone beside his head while talking, or he can just put it in his pocket and use a hands free connection. All these positions rotate the device differently in space and alter its antenna properties due to contact with the user's bodily tissue resulting in antenna mismatching and radiation pattern distortion. On top of this, in order to cope adaptively with the varying radio channel conditions, the wireless systems are becoming much more complex. For example, there are multiple modulation and coding schemes in LTE that can be applied depending on the channel and network conditions, see for example, [1].

The challenges mentioned above have previously been dealt with in SISO OTA antenna testing, [2]. However, in LTE and LTE-A systems, multiple antennas will be used both

LTE-Advanced and Next Generation Wireless Networks: Channel Modelling and Propagation, First Edition.
Edited by Guillaume de la Roche, Andrés Alayón Glazunov and Ben Allen.
© 2013 John Wiley & Sons, Ltd. Published 2013 by John Wiley & Sons, Ltd.

on the base station and on the terminal sides. This not only complicates designing the device but also testing becomes rather challenging due to increased complexity. Since the introduction of Multiple Input Multiple Output (MIMO) techniques, great interest has been shown to the utilization of the spatial domain of the channel, [3]. These multiple antenna techniques have been seen as the enabling technology for the high data speed requirements. The frequency, the time and the code domains as well as the spatial domain at either end of the communication link has been all exploited in previous wireless communication systems. However, after the introduction of MIMO techniques, the focus has shifted towards the utilization of the spatial domain *at both ends* of the wireless link. For traditional base stations this is not problematic; on the other hand, for wireless devices, this creates a true challenge since accommodating multiple antennas that operate efficiently on a small volume is not a trivial engineering task. Indeed, the customers require small and light wireless devices, but if antennas are closely spaced, the antenna coupling effects increase which may lead to operation degradation of the antennas.

In LTE, the network can benefit from multiple antennas in different ways, [1]: 1) diversity gain, 2) array gain, or 3) spatial multiplexing gain. The evaluation of the performance of wireless MIMO devices in realistic environments and their compliance to the standards are crucial in ensuring a reliable operation of the whole wireless network. Different methods have been devised for evaluating different parts of a wireless system, ranging from computer simulations and laboratory tests to actual field tests. In simulations, the different parts of the system, such as, the antennas and the receiver architecture, can be simulated individually. This way, the performance of the whole system can then be evaluated by emulating the most relevant features of the propagation channel. Although these simulations have ideally statistical properties similar to real environments, they lack several important features relevant to real devices, such as, manufacturing errors, realistic antennas, and so on. Another option would be to test these devices in real environments using existing networks. However, this is rather time consuming and in the worst case, the networks may have not even been deployed yet, which renders this option impractical. Hence, testing is preferred in laboratory conditions with full control over the whole emulation environment.

Different aspects of the MIMO OTA testing have been extensively discussed in the technical literature, for example, see [4–6]. Here, we focus on the general MIMO OTA testing problem and provide a review of up to date advances in this field of research.

15.1.1 Problem Statement

The main goal of MIMO OTA testing is to provide an accurate and reliable performance evaluation of wireless MIMO devices with good repeatability of results while keeping costs low. This generic statement can be further explained as follows:

Accuracy here refers to the question of how accurately a realistic wireless environment can be replicated in laboratory conditions. Obviously, this depends on various aspects, and among them, the choice of channel model is of paramount relevance. Therein, the methods and techniques used play an important role in the emulation process, which is the main topic of this chapter.

Reliability is concerned with the capability of the test methodology to fully reflect the complexity of the different aspects of the system. Here, it is important to evaluate whether the test methodology grasps all the important features of the propagation channel, the user effects, and all the relevant system aspects. Furthermore, the uncertainty of the measurement should be sufficiently small in order to readily detect unwanted performance.

Performance deals with the evaluation of appropriate figures of merit that are sensitive to the different aspects of a wireless system that enable reliable and accurate evaluation.

Repeatability is essential to assure that the same testing conditions can be created for all the devices tested.

Cost should be reduced as much as possible. In practice, this means that the testing system should be as simple as possible, which puts direct limitations on the complexity of the test system.

15.1.2 General Description of OTA Testing

Several candidate solutions for MIMO OTA testing have been studied in [6] as well as in the scientific community. The proposed methods can be categorized in various ways. Here, we follow a classification similar to that presented in [6], where the MIMO OTA techniques are divided into three main approaches:

- multi-probe systems;
- reverberation chamber;
- two-stage method.

The multi-probe systems refer here to methods that use an anechoic chamber, where multiple antennas are distributed in some configuration around the Device Under Test (DUT). Examples of OTA antenna configurations include, in addition to those proposed in the 3GPP standards, [6], the circular, [7], the single-cluster, [8], and the two-path, [9] configurations. In the multi-probe approach, the spatial characteristics of a propagation channel are fully or partially reproduced. For example, in the circular OTA configuration the waves can arrive, in principle, from any azimuthal direction and hence be capable of reproducing any channel conditions in the azimuthal plane. In the single-cluster and the two-path configurations the incoming waves are restricted to one or two cluster directions, respectively. The multi-probe method does not utilize the DUT radiation pattern during reproduction of the field. With this method, in theory, any field conditions can be recreated; however, this approach is somewhat costly due to space and equipment requirements.

In the Reverberation Chamber (RC) system the electromagnetic fields are generated within a metallic cavity, where metallic stirrers produce a random variation of the field, [10]. In this way, a uniform angular distribution over all directions is created which is used as a reference channel. This method is more cost efficient than the multi-probe system as no large anechoic room is required.

In the two-stage method, the DUT radiation pattern is measured first in an anechoic chamber. Then, the DUT antennas are bypassed and the Radio Frequency (RF) signal is fed directly into the device. However, before feeding the signal into the DUT the signal is weighted by coefficients obtained from a channel model, for example, according to (15.1). This means that the DUT has to provide special connectors for the two-stage method to become applicable in practice, whereas the multi-probe and RC methods do not need to meet such a requirement. This method is very cost efficient and both geometry- and correlation-based models can be used.

Figure 15.1 shows the different parts of a wireless system and the concept of OTA testing. In Figure 15.1(a), schematic representation of the real world usage scenario of a wireless device is illustrated. The main feature is the multipath propagation, that is, the device receives multiple copies of the signal transmitted by, for example, the base station. These multipath components interact with the surrounding environment and may arrive at the receiver from different angles and at different time instants. In Figure 15.1(b) the same scenario is illustrated using a block diagram. In the figure, block "1." denotes the transmitter electronics circuitry (digital and analog), block "2." illustrates the transmit antenna configuration, block "3." denotes the propagation channel (no effect of antennas), block "4." shows the receive antenna configuration, and block "5." denotes the receiver electronics circuitry (digital and analog). This division is adopted from the double directional channel model concept, where the propagating waves are separated from the antenna systems, [11]. Finally, Figure 15.1(c) shows the general concept of MIMO OTA testing. In the figure, the electronics represent the base station emulator used to replicate the performance of the real world counterpart. From the base station emulator the signal is fed to a testing facility which may be comprised of a metallic cavity (see Section 15.3), an anechoic chamber along with fading emulator (see Section 15.2), or the facility can be created using purely simulations (see two-stage method in 15.3). For the multi-probe and the RC methods, the field created within the testing facility is represented by blocks "1."-"3." of Figure 15.1(b), whereas the DUT is represented by blocks "4."-"5.". This way, a real world scenario is transformed into a controlled test environment where wireless devices can be objectively tested.

The OTA testing of Single Input Single Output (SISO) antenna systems has been extensively studied which has resulted in a standardized methodology, [2]. On the other hand, the OTA testing of MIMO capable devices is much more complex and to this end, a study item in 3GPPP [6] has been set up to investigate different approaches. The high level requirements for the study of the MIMO devices are, [6]

1. The measurement of the radiated performance of MIMO and multi-antenna reception for High Speed Packet Access (HSPA) and LTE terminals must be performed over-the-air, that is, without RF cable connections to the DUT;
2. The MIMO OTA method(s) must be able to differentiate between a good terminal and a bad terminal in terms of MIMO OTA performance;
3. The desired primary Figure Of Merit (FOM) is throughput.

The MIMO OTA system used for standardized device testing should naturally fulfil these requirements.

15.2 Channel Modelling Concepts

As described above, it is a fundamental requirement to test the performance of wireless devices in realistic environments by recreating the appropriate propagation environment in laboratory conditions. In turn, this demands realistic channel models that are capable of reproducing the fundamental characteristics of the propagation environment. The radio channel modelling focuses on the properties of the channel impulse response that characterizes the channel behavior. The impulse response describes the effect of the channel on

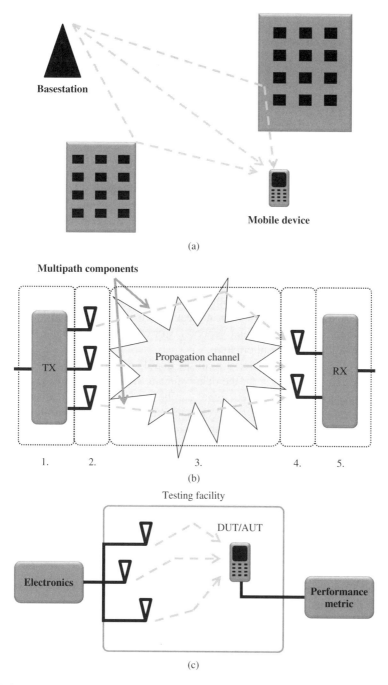

Figure 15.1 MIMO OTA concept. (a) Schematic representation of the propagation channel; (b) Schematic representation of the usage environment; (c) Schematic representation of laboratory testing environment.

the input (transmitted) signal. For MIMO systems this is better illustrated using a channel matrix $\mathbf{H}(t)$, which describes the radio channel including each transmit and receive antenna element as well as the propagation channel. The elements $H_{u,s}$ of the matrix are the impulse responses from the uth receive antenna to the sth transmit antenna at the interface of the blocks "1." and "2.", and "4." and "5." in Figure 15.1(b). The MIMO channel matrix is the fundamental measure that is being reproduced within the testing facility, for example, as shown Figure 15.1(c).

15.2.1 Geometry-Based Modelling

According to the double-directional channel concept under the plane wave assumption for the electromagnetic waves impinging at the receive antenna, [11], the channel matrix can be factored into various terms representing the propagation channel and antennas separately. The DUT antennas interact with the incoming wavefronts and map the incoming fields into electric currents. Therefore, as indicated in Chapter 2 and in (15.1), the antenna configuration of the device plays a crucial role in determining the performance of the system. For example, if the channel contains a large number of multipath components and has a large AoA/AoD spread, then spatial multiplexing will benefit from antennas with omnidirectional patterns. On the other hand, if the antenna patterns are highly directive, that is, it is directive with a narrow beam, then the system cannot utilize all the available dimensions that the channel supports. Conversely, if the angle spread is small, beamforming is a much better alternative.

In addition to the spatial characteristics of the double directional channel, other channel metrics, such as, the fading pattern, the delay spread, the Doppler spread, and the MIMO correlation properties of the received signal can be reproduced in a realistic way.

Under the plane wave assumption the channel matrix describing the input-output relation of the channel can be written for nth cluster in 2D case as (e.g. [12, 13])

$$H_{s,u,n}(t,\tau) = \sum_{m=1}^{M} e^{jd_s 2\pi \lambda_0^{-1} \sin(\varphi_{n,m})} \underbrace{\begin{bmatrix} F_{tx,s,V}(\varphi_{n,m}) \\ F_{tx,s,H}(\varphi_{n,m}) \end{bmatrix}^{T}}_{2.} \underbrace{\begin{bmatrix} \alpha_{n,m,VV} & \alpha_{n,m,HV} \\ \alpha_{n,m,VH} & \alpha_{n,m,HH} \end{bmatrix}}_{3.}$$

$$\underbrace{\begin{bmatrix} F_{rx,u,V}(\phi_{n,m}) \\ F_{rx,u,H}(\phi_{n,m}) \end{bmatrix}}_{4.} e^{jd_u 2\pi \lambda_0^{-1} \sin(\phi_{n,m})} e^{j2\pi v_{n,m}t} \delta(\tau - \tau_{n,m}), \quad (15.1)$$

where "2."–"4" refer to the different blocks in Figure 15.1(b). The u and s represent the receiver and transmitter indices; n is the cluster index; m is the multipath index; ϕ and φ are the angle of arrival and departure; d_u and d_s are the distances of the uth and sth receiver and transmit antenna elements from the reference phase centre; F_{tx} and F_{rx} are the radiation patterns of the transmit and receive antennas, respectively; $\alpha_{n,m}$ are the polarimetric path gains; λ_0 is the wavelength; and $v_{n,m}$ is the Doppler frequency component dependent on the $\varphi_{n,m}$ and the mobile wireless device speed and the traveling direction. The α's and φ's and ϕ's define the Angular Power Spectrum (APS) of the propagation given by the selected channel model. To this end, the channel parameters represent the propagation channel that the DUT operates in.

The exploitation of the angular domain is one of the most important advances of LTE as compared to previous systems. Hence, it is crucial to accurately reproduce the spatial characteristics of the channel in the testing facility. Equation (15.1) can easily be separated into two terms corresponding to the different parts shown in Figure 15.1(c). For a single-cluster and single path channel the field generated by the OTA system can be represented as follows

$$E_{\text{OTA}}(t) = \begin{bmatrix} F_{\text{tx,V}}(\varphi) \\ F_{\text{tx,H}}(\varphi) \end{bmatrix}^{\text{T}} \begin{bmatrix} \alpha_{\text{VV}} & \alpha_{\text{HV}} \\ \alpha_{\text{VH}} & \alpha_{\text{HH}} \end{bmatrix} e^{jd_s 2\pi\lambda_0^{-1}\sin(\varphi)} e^{j2\pi vt}, \tag{15.2}$$

while the part corresponding to the DUT is then given by

$$G_{\text{DUT}} = \begin{bmatrix} F_{\text{rx,V}}(\phi) \\ F_{\text{rx,H}}(\phi) \end{bmatrix} e^{jd_u 2\pi\lambda_0^{-1}\sin(\phi)}. \tag{15.3}$$

The multi-probe and the RC OTA testing systems strive to reproduce the field described in (15.2) as accurately as possible (by completely different approaches) for each single path following predefined statistical distribution laws. In both cases the radiation patterns of the DUT given by (15.3) is not measured, but enters (15.1) implicitly. On the other hand, in the two-stage method, the radiation pattern of the DUT (15.3) is measured first and then entered into (15.1) to produce the signal fed to the DUT without the antennas.

In the classical Clarke's model, [14], the Multipath Components (MPCs) arriving at the mobile are uniformly distributed in azimuth. This model produces a Rayleigh fading pattern of the signal as the wireless device travels across the environment. However, it has been observed from measurements that this is not always valid, since MPCs are rather concentrated to groups, or clusters, with some finite angular spread, [15]. For this reason a considerable effort has been made to develop realistic channel models that incorporate the clustering behaviour. As a consequence, several models have been proposed during recent years to reproduce the channel parameters for the single link case in realistic scenarios. These geometry-based stochastic channel models are based on the approach where the MPCs are grouped together to form clusters. They are often homonymous with the projects where these were developed, for example, COST273 [16], COST2100 [17], Spatial Channel Model (SCM) [18], Spatial Channel Model Extended (SCME) [19], WINNER [12], and IMT-Advanced [13]. The clusters' properties are defined by various random distributions whose parameters have been extracted from extensive measurement campaigns. All these models provide plane wave parameters that a MIMO OTA system should be able to reproduce in laboratory conditions.

The RC approach produces a 3D uniform angular distribution, [20]; however, recent work has been conducted to modify the angular characteristics of the RC [21]. On the other hand, the multi-probe approach can reproduce (if a sufficiently large number of OTA antennas is used) any APS defined by the aforementioned models, along with the polarization and delay characteristics. Clearly, this also applies to the two-stage method. However, since a full emulation of all the relevant channel characteristics is rather involved, a simplified single-cluster model has also been proposed for MIMO OTA testing purposes.

For standardized LTE OTA testing, the considered MIMO channel models are the following, [6]

- SCME Urban micro-cell;
- Modified SCME Urban micro-cell;

- SCME Urban macro-cell;
- WINNER II Outdoor-to-indoor.

15.2.2 Correlation-Based Modelling

In the correlation-based modelling approach, the channel transfer matrix is recreated based on the correlation statistics at the TX and the RX sides. The Kronecker model is a popular correlation-based model due to its simplicity, [22]. Indeed, it assumes that the full-correlation matrix of the MIMO channel can be separated as follows

$$\mathbf{R} = \mathbf{R}_{rx} \otimes \mathbf{R}_{tx}, \tag{15.4}$$

where $\mathbf{R}_{tx}$ and $\mathbf{R}_{rx}$ are the transmit and receive correlation matrices, respectively. Hence, the MIMO channel matrix is then given by

$$\mathbf{H} = \mathbf{R}_{rx}^{\frac{1}{2}} \mathbf{H}_G \mathbf{R}_{tx}^{\frac{1}{2}}, \tag{15.5}$$

where $\mathbf{H}_G$ is a matrix of i.i.d. complex Gaussian distributed random variables. The correlation matrices $\mathbf{R}_{tx}$ and $\mathbf{R}_{rx}$ depend on both the antenna configuration (e.g. radiation patterns and antenna spacing) and the propagation channel. Hence, the channel matrix will depend on the radiation pattern of the DUT. For this reason, the correlation-based modelling approach cannot be directly[1] utilized in combination with multi-probe or RC systems. On the other hand, since (15.5) can replace (15.1), the correlation-based models are especially suited for the two-stage testing method.

In addition to the geometry-based channel models listed above, correlation models are also given in, [6]

- Extended Pedestrian A;
- Exponential decay.

The correlation matrices used for OTA testing purposes are given in [23]. It is worthwhile mentioning that the geometrical models can be used to calculate the transmit and receive correlation matrices required in the correlation-based modelling approach; however, the converse is not applicable.

15.3 DUTs and Usage Definition

In principle, there are no general constraints on the wireless devices that can be tested for OTA performance. However, the two-stage method requires the possibility to bypass the device antennas during the second stage. Furthermore, the multi-probe and the RC methods can be used to test any given wireless device; the only requirement is that the device has to fit physically inside the *test zone* of the testing facility. Therefore, a DUT can be a mobile phone, USB dongle, laptop, and tablet PC. Furthermore, almost any channel condition can be recreated in multi-probe systems. This is true if a sufficiently large

---

[1] It is worthwhile recalling that there is a clear relationship between the transmit/receive correlation matrix and the APS of the AoD/AoA.

Table 15.1 UE category requirements for LTE [1, 25]

	1	2	3	4	5
Supported downlink data rate [Mbps]	10	50	100	150	300
Supported uplink data rate [Mbps]	5	25	50	50	75
Number of receive antennas required	2	2	2	2	4

number of OTA antenna-probes are arranged in a 3D configuration, but also if the size of the DUT is sufficiently small as compared to the distance to the OTA antenna-probes. This means that the spatial operation of the DUT can be tested too. The RC methods can be used to generate a wide range of channel conditions if used in combination with a channel emulator. Research in RC chambers for OTA testing offers many interesting problems.

As mentioned above, the user may affect the performance of a wireless device in an unpredictable way. However, the effect of the user can be taken into account by using different types of phantoms that model the user's head, hands, torso or even the whole body. Clearly, the phantoms can be used during testing if they fit within the test zone. Phantoms have been used since the first SISO antenna testing methods were devised, [24], and are being developed continuously.

The channels can be recreated for arbitrary systems and environments using different methods. However, a comprehensive communication system testing requires the appropriate network functionalities to be created in laboratory conditions. Hence, base station emulators have to be used to fulfil this requirement. Testing arbitrarily many features is not feasible in practice. Therefore, different user equipment (UE)[2] categories are used to indicate the mandatory features specified for LTE and LTE-A, [25]. For LTE (Rel. 8) and LTE-A (Rel. 10) there are 5 and 8 UE requirement categories, respectively. An example of the UE category requirements for LTE (Rel. 8) are shown in Table 15.1, [1, 25].

15.4   Figures-of-Merit for OTA

The Total Radiated Power (TRP) and the Total Radiated Sensitivity (TRS) are the two main FOMs of a mobile device tested with SISO OTA methods, [2]. However, these metrics cannot fully characterize the spatial behaviour of a DUT with multiple antennas. Therefore, additional MIMO OTA FOMs have been introduced into the standards. For the evaluation of the OTA system itself, that is, its capability to produce the desired test environment, a calibration of the emulated propagation environment is required. This calibration is based on the emulated electromagnetic field. It can be done relative to the performance of reference antennas satisfying (15.2); or by verifying the desired channel characteristics obtained from the reproduced channel matrix (15.1), again by using calibrated reference antennas. The field domain FOMs include accuracy of the reproduced electromagnetic field, [26], and the spatial correlation. The reproduced channel matrices

---

[2] User equipment, or UE for short, is the common denomination of a wireless device according to the 3GPP standardization organ.

Table 15.2 Excerpt of the figure of merits listed in [6]

Category	1	2	3	4
FOM	MIMO throughput CQI	TRP TRS	Gain imbalance Spatial correlation MIMO capacity	Antenna efficiency MEG

are evaluated using different metrics, such as, diversity gain, MIMO capacity, spatial correlation, and throughput.

For the standardized OTA testing, however, the aim is to evaluate the performance of the whole DUT including the full receiver (and/or transmitter) chain. Therefore, the selection of FOMs is critical to be able to distinguish a good device from a bad device. Thus, in order to perform reliable measurements, the chosen FOM should be sensitive to the OTA performance of the DUT and, at the same time, the test method itself should provide accurate enough results of the FOM, [5]. In [6, 27] the FOMs currently considered in standardized OTA testing are listed in Table 15.2.

As can be seen from Table 15.2, the *MIMO throughput* is the most significant FOM for MIMO-capable UEs. Indeed, in only one parameter, several important aspects are encompassed at once, that is, both the usage environment and the user impact on the system performance are tested by measuring throughout, [6, 28]. In LTE, the *channel quality indicator* (CQI) is used to feedback the channel conditions to the transmitter. The UE can report the CQI information to the base station so that the base station can select appropriate modulation and coding scheme (MCS). For example, the UE informs the base station of the highest MCS so that the block error rate (BLER) probability is not exceeding 10 percent. Thus, in addition to the channel conditions, the CQI takes into account the performances of antenna and the UE receiver too, [1]. The *total radiated power* (TRP) is defined as the power radiated by the UE, while the *total radiated sensitivity* (TRS) is the lowest receiver sensitivity level of the UE's receiver, for example, required to attain certain BLER, [2]. These metrics depend on the system and hence they require a system level simulator or emulator.

The *MIMO capacity* provides the theoretical capacity that a system with multiple antennas at both ends of the link can achieve in a given propagation environment, [29]. The *spatial correlation* measures the similarity of the received signals between different antennas under certain channel conditions, [14]. The capacity is directly related to the rank of the channel matrix and therefore, any correlation between the different antenna branches may result in capacity degradation. The gain imbalance contains possible uneven losses in the antenna elements and also this value directly affects the channel matrix and the received signal power, and hence, it directly influences the MIMO capacity. The *antenna efficiency* measures the losses in the antenna elements as well as missmatch losses and as such decreases the received power, [30]. The *mean effective gain* (MEG) is a useful metric to quantify the operation of an antenna in a multipath environment, [31, 32]. The MEG has been widely adopted as one essential parameter in antenna design for multipath channel applications as shown in Chapter 17. These metrics can be evaluated using measured DUT radiation patterns. Once the radiation patterns are known, the capacity and the spatial correlation can be evaluated using (15.1) or (15.5) for given channel conditions. The gain imbalance and the antenna efficiency can be evaluated directly from the radiation pattern measurement of the DUT antennas.

15.5   Multi-Probe MIMO OTA Testing Methods

In this section, multi-probe MIMO OTA testing systems are discussed. As opposed to basically any other technique discussed later in Section 15.3, the multi-probe techniques do not pose any major theoretical limitations on the emulation of different propagation channel conditions. For this reason, multi-probe systems have become one of the main candidates for MIMO OTA testing, [6]. In line with the perspective of the book, the focus here is on the electromagnetic field analysis, but less attention is paid to hardware realizations and communications aspects of the MIMO OTA testing.

This section begins with the introduction of a multi-probe MIMO OTA test set-up. This is followed by a discussion on propagation channels modelling issues relevant to the MIMO OTA testing, such as, the probe configurations, the number of required probes and the field synthesis principle. Then, the MIMO OTA test set-ups and probe configurations for single-cluster channel models with narrow angular spreads are considered. An example illustrating the validity of the 2D probe configurations is presented. The section is ended with a brief discussion on a technique for compensating near-field effects and range reflections for practical multi-probe systems.

15.5.1   Multi-Probe Systems

A schematic representation of the set-up of a typical multi-probe MIMO OTA testing system is shown in Figure 15.2. The main equipment of the set-up consists of a communications tester, a fading emulator and the probe antennas. The test zone depicted in the middle of the chamber is defined as the volume inside which the targeted radio channel environment is created and the volume occupied by the DUT.

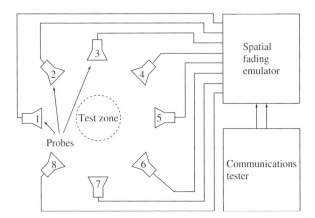

Figure 15.2   A schematic representation of the set-up of a typical multi-probe MIMO OTA testing system consisting of eight probes, a fading emulator and a communications tester. From [33].

The basic idea is to use a fading emulator and a communications tester to emulate the multi-path propagation channel and the base station, respectively. The desired field is synthesized in the test zone by means of probes appropriately located at some distance around the test zone. Typically, multi-probe MIMO OTA testing is performed in an anechoic chamber, where the probe antennas are placed in the far field of the DUT and vice versa. It is further assumed for the sake of simplicity that the mutual scattering between the neighbouring probes and between the probes and the DUT are negligible. Later in this section, we provide the examples of two techniques for compensating near-field and scattering effects.

15.5.2   Channel Synthesis

The MIMO channel matrix can be expressed by (15.1), where each element is generated by combining the field radiated by the probe antennas. The amplitude and phase of each field component impinging at the DUT shall describe the contribution of real scattering clusters to the total impinging field. For this purpose, different models are considered that emulate single- or multiple-cluster scenarios. Furthermore, the impinging field to the DUT is assumed to have a certain angular power spectrum. The so-called Laplacian APS, which has been derived from extensive propagation channel sounding experiments, is often assumed to be a suitable model. In multipath channels, it is assumed that the statistical distributions of the AoA/AoD corresponding to the vertical and horizontal polarizations are decoupled in elevation and azimuth. In addition, the behaviour of the co-polarized channel matrix component is similar for vertical-to-vertical and horizontal-to-horizontal channel links in (15.1). Therefore, a Power Azimuth Spectrum (PAS) and a Power Elevation Spectrum (PES) are defined separately. The possible combinations of clustering and angle spread are illustrated in Figure 15.3.

The SCM, SCME, WINNER and IMT-Advanced propagation channel models assume that the PAS is Laplacian distributed [12, 13, 18, 19]. On the other hand the PES is often assumed to be a delta function with the peak at $\theta = 90°$, that is, at the horizontal plane. Hence, the impinging field to the DUT is assumed to arrive from the horizontal plane at

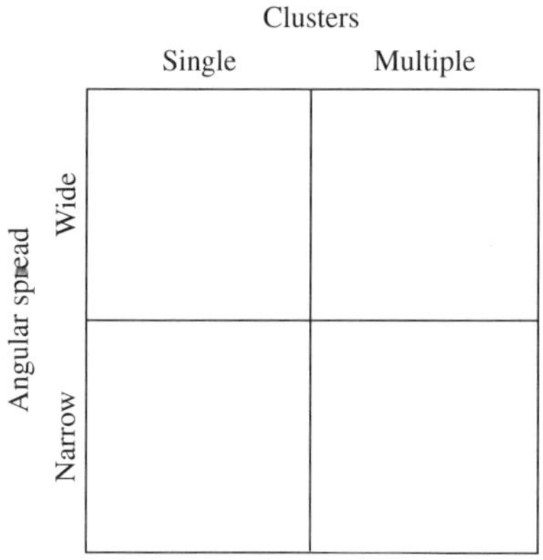

Figure 15.3   The power angular spread dimension against the number of clusters dimension.

different azimuthal directions. In some cases the Laplacian distribution is applied for the PES too, [34].

The equations for the Laplacian distributed PAS and PES can be written as

$$P_\phi(\phi) = \begin{cases} e^{-\sqrt{2}\frac{|\phi|}{\sigma_\phi}} & \phi \in [-\pi \ldots \pi) \\ 0 & \text{otherwise} \end{cases} , \tag{15.6}$$

$$P_\theta(\theta) = \begin{cases} e^{-\sqrt{2}\frac{|\theta-\frac{\pi}{2}|}{\sigma_\theta}} & \theta \in [0 \ldots \pi] \\ 0 & \text{otherwise} \end{cases} , \tag{15.7}$$

respectively. Here, σ_ϕ and σ_θ are the angular spreads of the PAS and PES, respectively.

It is worthwhile mentioning here that the field synthesis method presented in the following can be used in combination of a newly developed formalism that expands the propagation channel into spherical vector wave multimodes, [35–37]. This method is especially appealing since it relies on the fact that the channel matrix (15.1) can be directly expressed through the multimode expansion coefficients of the antennas and those corresponding to the AoA/AoD distributions (15.6) and (15.7).

15.5.3   Field Synthesis

15.5.3.1   Probe Configurations

We now consider the field synthesis problem in MIMO OTA testing based on an electromagnetic field approach. Clearly, choosing the proper probe configuration is here a fundamental issue. For example, the number of required probes largely dictates the overall system cost. In this context, the theoretically optimum probe configuration from the field synthesis point of view should allow generating any given field distributions inside the test zone. This can be achieved if the following conditions are met: 1) the probes are located on a surface of a sphere and pointed towards the centre of the test zone, 2) the number of probes is sufficiently large compared to the dimensions of the test zone, and 3) both vertically (θ) and horizontally (ϕ) polarized probes are employed. This case is referred to as the 3D case; hence, the test zone is also a spherical volume with radius r_0.

However, the vast majority of currently available channel models are so-called 2D models. The scattering clusters of these models are confined to the horizontal plane, that is, ($\theta = 90°$). Hence, since the impinging field to the antenna arrives at the horizontal plane, the required probe configuration can be confined to the horizontal plane too. This case is referred to as the 2D case. In a general 2D case, both vertically and horizontally polarized probes are required. For this case, the test zone becomes a circular cylindrical volume with radius r_c.

In many propagation channels, the elevation power spectrum of the impinging fields is such that most, if not all, of the power arrives at the wireless device from a narrow elevation range around $\theta = 90°$. Then, optimum probe locations could be chosen to cover only a small solid angle region at around $\theta = 90°$ defined by the angular spreads of the PES. This is referred to as the 2.5D case here. An illustration of such a narrow solid angle region is shown in Figure 15.4. Studies on elevation power angular spreads of channels have been reported, for example, in [38].

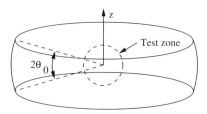

Figure 15.4   Illustration of the solid angle range for the possible probe locations in the 2.5D case. The ϕ-angle range for the probe locations is from 0 to 2π and the θ-angle range is from $90° - \theta_0$ to $90° + \theta_0$.

15.5.3.2   The Rules for the Number of Probes

In this section we discuss the number of probes required in multi-probe MIMO OTA testing based on the spherical vector wave multimode expansion of the electromagnetic field, [39]. We provide a relationship between the minimum number of spherical vector wave modes required to describe the impinging fields inside the test zone given its radius, for example, r_0 or r_c.

We first consider 2D configurations with probes equally spaced in azimuth. Then, we consider the 3D case with probes equally spaced in both azimuth and elevation with a uniform distribution over the 4π solid angle. The 3D probe configuration is clearly the primary candidate to emulate propagation channels with multiple clusters and/or with wide APS. For now, we ignore the configurations with unequally spaced probes and the 2.5D probe configuration.

We consider single-port probe antennas to synthesize a propagation channel with both θ- and ϕ-polarized impinging fields. Hence, in this case, there are two equally large sets of probes with orthogonal polarizations. Clearly, if only one polarization is of interest, then only a half of the total number of probes is required.

In the 2D case, the general rule for the required number of equally spaced probes in azimuth is given by, [39]

$$K_2 = 4M + 2, \tag{15.8}$$

where

$$M = [kr_c] + m_1, \tag{15.9}$$

is the number of spherical vector wave modes, $k = \frac{2\pi}{\lambda}$ is the wavenumber, λ is the wavelength, r_c is the above-mentioned radius of the minimum circular cylindrical test zone in the 2D case. The square brackets round up the number inside the brackets to the closest integer. The constant m_1 is an integer that determines the accuracy of the field synthesis. The resulting probe configuration assigns equally spaced points in a circle with $\phi \in [0, 2\pi - \delta\phi]$, where $\delta\phi$ is the azimuthal separation of consecutive probes.

In the 3D case, the general rule for the number of required probes in the full 4π solid angle range is

$$K_3 = 2N(N + 2), \tag{15.10}$$

where

$$N = [kr_0] + n_1, \tag{15.11}$$

is the number of spherical vector wave modes, r_0 is the above-mentioned radius of the spherical test zone in the 3D case. The constant n_1 is an integer that determines the accuracy of the field synthesis.

As we can see from (15.8)–(15.11), for large horizontal radii of the test zone, that is, large r_c and r_0, such that $r_c = r_0$, the number of required probes in the 2D case is smaller than in the 3D case. Here, we have assumed that the number of channels of the fading emulator is the same as the number of probes. Later in this section we will briefly discuss the case where this assumption is not necessarily satisfied.

15.5.3.3   Excitations for the Probes

The probe configurations discussed in the previous subsection form the basis for the field synthesis. The synthesis is based on exciting the probes with proper excitation voltages (coefficients), such that the sum field from the probes equals the desired impinging field to the test zone. Clearly, an incorrect probe configuration will lead to erroneous synthesis of the desired field, which in turn will result in a channel emulation with insufficient accuracy. Hence, for a given probe configuration, one should aim at finding an optimum set of excitations for the probes to minimize the field synthesis uncertainty.

From the theory of spherical antenna measurements we know that the electric and magnetic fields tangential to the surface of the test volume determine the field inside the test zone, [39]. Hence, the field synthesis can be realized by satisfying the condition that the tangential components of the electric and magnetic fields radiated by the probes and measured on the surface of the test zone are equal to those of the desired field. For the 2D case, it is sufficient to ensure that the tangential electric and magnetic fields radiated by the probes are equal to those of the desired field on the line where the horizontal plane intersects with the surface of the circular cylindrical test zone; henceforth referred to as the circle surrounding the test zone. The field synthesis also takes into account the known radiation patterns of the probes. Then, the location and orientation of the probes with respect to the test zone allows the radiated electric and magnetic fields on the surface of the test zone to be determined. If the probe configuration and the number of probes have been chosen appropriately, we can write the following expression

$$\bar{F}_t(\theta, \phi, r_\alpha) \approx \sum_{k'=1}^{K} c_{k'} \bar{F}_{k'}(\theta, \phi, r_\alpha) \tag{15.12}$$

where $\bar{F}_t(\theta, \phi, r_\alpha)$ is the target tangential electromagnetic field (electric or magnetic fields) on the surface of the spherical test zone (3D case) or on the circle surrounding the cylindrical test zone (2D case, $\theta = \frac{\pi}{2}$), $\bar{F}_{k'}(\theta, \phi, r_\alpha)$ is the tangential electromagnetic field of k'th probe on the same surface or circle, and $c_{k'}$ is the excitation coefficient for the k'th probe. The α refers to either 0 or c, and K is the total number of probes. Since the functions $\bar{F}_{k'}(\theta, \phi, r_\alpha)$ are known, the excitation coefficients $c_{k'}$ can be found by minimizing the difference between the left and right-hand sides of the equation, for example, in a minimum least-squares sense.

15.5.4   Two Examples of Field Synthesis Methods

This subsection reviews two recently published studies of the influence of the probe configuration on the field synthesis accuracy. The first one concerns the single-cluster case with narrow angular spread. Here we also consider the influence of the APS of the impinging fields on the adequate MIMO OTA set-up and the required number of probes for synthesizing the desired field in the 2D case. The second study concerns the applicability of the 2D probe configuration for 2.5D channels.

15.5.4.1   Single-Cluster Case with Narrow Angular Spread

The application of the rules in (15.8) and (15.10) directly to the single-cluster case with a narrow angular spread may lead to a situation where a large number of probes remains unused during testing. In this case, we could consider choosing an adequate subset of probes for the field synthesis. An illustration of the MIMO OTA test set-up with eight probes but only four channels of the fading emulator is shown in Figure 15.5. A switch matrix is used to choose a subset of probes to produce the required field.

The benefits of using this system are, on one hand, improved channel synthesis accuracy and a lower number of required channels of the fading emulator on the other hand. In [26], an investigation on this topic has been reported. The study addressed the accuracy of the field synthesis in the 2D case for various PAS of the impinging field. One of the aims of the study was to find the minimum number of probes required for synthesizing a plane-wave field in the test zone with the maximum relative error of $-15\,\mathrm{dB}$ for different azimuthal spreads.

The obtained results are shown in Figure 15.6 for $r_c \approx 0.5\lambda$, [40]. The continuous line represents results for equally spaced probes over the complete 2π azimuth region, while the dashed-line curve corresponds to the case where probes are located in the optimum azimuthal sector obtained following the method discussed above. As we can see from

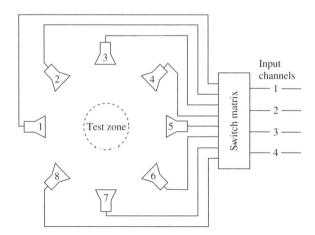

Figure 15.5   Illustration of a MIMO OTA set-up with the number of probes exceeding the number of channels of the fading emulator.

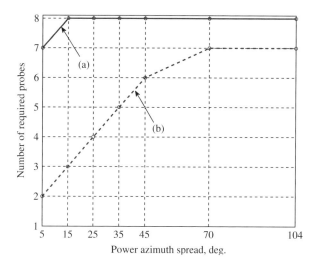

Figure 15.6   The number of probes required for synthesizing a plane-wave field in the test zone with the maximum relative error of $-15\,\mathrm{dB}$ as a function of the power azimuthal spread in the example case with $\frac{r_0}{\lambda} \approx 0.5$ for the case with (a) equal spacing of the probes in ϕ over the 2π angular region and (b) probes in the optimum angular sector in ϕ. From [40].

the figure, for large angular spreads the benefits of placing the probes in the optimal azimuthal sector are comparable to having the probes uniformly placed over the complete 2π azimuthal range. On the other hand, for narrow power PAS, that is, small azimuthal spreads, the benefits may be substantial since the number of probes could be reduced considerably.

15.5.4.2   Applicability of the 2D Probe Configuration for 2.5D Channels

From (15.8) and (15.10) we can see that in the 2D case the number of required probes, even for a laptop type of a wireless device operating, for example, at 6 GHz remains reasonably low whereas the number of probes increases rapidly by increasing the ratio $\frac{r_0}{\lambda}$ in the 3D case. Although many channel models are 2D models, and hence the field synthesis based on these models would not require a 3D probe configuration, it can well be argued what the validity of the 2D model ultimately is.

In [38], the applicability of the 2D probe configuration to the 2.5D case has been investigated. In that study, the impinging field to the test zone is assumed to have a fixed azimuth angle spread of $104°$, that is, fixed PAS. On the other hand, the main focus has been on the influence of the PES of the impinging field on the field synthesis accuracy. An example result of this study is shown in Figure 15.7, [38]. The number of probes in azimuth has been chosen to minimize the field synthesis error corresponding to the elevation angular spread of $0°$, that is, the PES is a δ-function. It is clear from Figure 15.7 that the size of the test zone influences the accuracy. For example, according to this result, if the synthesis accuracy criterion is $-15\,\mathrm{dB}$ and for $\frac{r_0}{\lambda} = 0.4$, the 2D probe configuration could be applied for channels with elevation angle spread equal $7°$ or less.

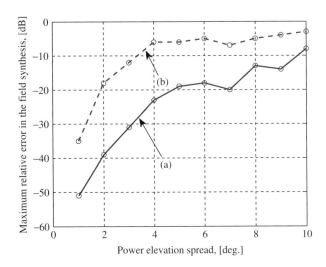

Figure 15.7 The maximum relative error of the synthesized field as a function of the power elevation spread. The case with (a) $\frac{r_0}{\lambda} = 0.4$ and (b) $\frac{r_0}{\lambda} = 3.2$. From [38].

However, assuming the same accuracy, if $\frac{r_0}{\lambda} = 3.2$ then the applicability of the 2D probe configuration is reduced to elevation angle spread equal $2°$ or less. The result shows that both the size of test zone in wavelengths and the elevation angle spread both affect the field synthesis accuracy in this case.

15.5.5 Compensation of Near-Field Effects of Probes and Range Reflections

We made above the assumption that the probes are located in the far field of the test zone and vice versa, and that there are no reflections from the neighbouring probes in the range. We further assumed that the multiple reflections between the DUT and the probes are negligible. The validity of these assumptions may be, of course, argued. In particular, if the distance of the probes from the centre of the test zone is desired to be made small such that, for example, the chamber dimensions could be reduced, or if the number of probes is large, the near-field effects and range reflections may become potentially a significant error source if they are not compensated for. Different methods for fully or partially compensating the near-field effects and range reflections have been presented in the literature [41–44].

For the general 3D case, a technique for compensating the range reflections from the neighbouring probes in a multi-probe range has been presented in [43]. It is noted that the multiple reflections between the probes and the DUT are not compensated for by this technique. The idea is to scan the test zone for a sufficient number of locations for both polarizations with a known calibrating probe. In this way, the actual impinging field to the test zone are found separately for each excited probe and the fields tangential to the surface of test zone can be obtained. It is then possible to write an equation similar to (15.11) by simply replacing the calculated tangential fields from the probes, $\bar{F}_{k'}(\theta, \phi, r_\alpha)$, on the surface of the test zone with the measured ones. In this way, a new set of excitation coefficients is obtained that includes the compensation of the near-field effects and range reflections. A study, analogous to that in [43], has been performed for a 2D case in [44]. The difference between the two techniques is that in the 2D case a full compensation of the near-field and range reflections can generally not be achieved.

15.6   Other MIMO OTA Testing Methods

15.6.1   Reverberation Chambers

Electromagnetic RCs (also known as mode-stirred chambers (MSC)) have become standard tools for Electromagnetic Compatibility (EMC) measurements on electronic equipment in the past three decades, [45]. EMC deals with measurements of the generation, propagation and reception of electromagnetic energy, and the associated unwanted effects, such as, Electromagnetic Interference (EMI).

As mentioned above, the RC technology has found its way into the OTA performance testing of wireless devices too. Figure 15.8(a) shows a schematic representation of a RC used in the OTA testing of a wireless device or just an antenna in so-called *passive* measurements. Passive measurements are used to test the performance of an antenna design not necessarily implemented in an actual device. The method relies on accurate measurements of the $S-$parameters of the system involving the DUT, the measurement antennas and the chamber. Hence, the contributions to the $S-$parameters will come from both the antennas and the chamber. While the contribution from the antennas is deterministic, the contribution from the chamber is stochastic with zero mean due to the mode stirring effect. The desired antenna performance parameter is then obtained by averaging over the randomly distributed modes of the RC. However, the ultimate goal is to test the performance of a wireless device in conditions similar to that encountered in a real wireless network, that is, the performance of the DUT shall be tested while connected to a base station emulator as shown in Figure 15.8(b). The range of propagation channels emulated in a RC can be further broadened if it is used in combination with a channel emulator as shown in Figure 15.8(c).

In the following we provide insights into the current developments in this field of research. A recent review of OTA RC technology can be found in [21]. The focus here is on the channel emulation capability of RC since, as well-known, all other things being equal, the device performance is directly connected to the interplay between the device and the propagation channel.

15.6.1.1   Working Principles of a RC

The RC has been proven to be an exceptional mean for recreating the desired high field strength levels in a confined test environment with moderate input power. A RC is basically an electrically large, highly conductive closed cavity or room of commonly rectangular measures operating in the overmoded regime [46]. The energy entering the cavity will be reflected from its walls building up a 3D standing wave pattern, which intensity reinforces as power is delivered to the cavity until the losses in the cavity equal the input power. The standing wave pattern is, at high frequencies, a spatially rapidly changing

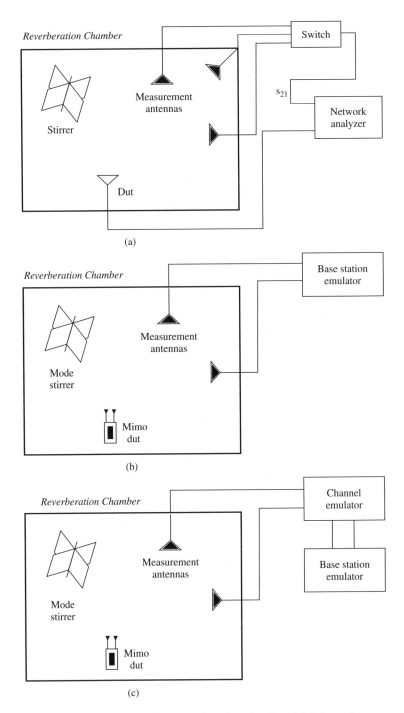

Figure 15.8 MIMO OTA concept with a reverberation chamber. (a) Schematic representation of the use of a RC in (passive) radiated or received power measurements; (b) Schematic representation of the use of a RC in (active) radiated or received power measurements; (c) Schematic representation of the use of a RC in (active) radiated or received power measurements with channel emulator.

field pattern with static positions of minima and maxima. Those can be dynamically distorted or moved by displacement of mode stirrers or paddles which change the boundary conditions for the bouncing waves inside the cavity. The stirrers' dimensions must be comparable to the wavelength corresponding to the lowest usable frequency. The produced field variation inside the mode-stirred cavity has the same statistical distribution over time at each position except at the boundaries of the cavity and the objects present inside the cavity. Moreover, in an ideal RC, each field component has an in-phase component and a quadrature component that are i.i.d. zero-mean complex Gaussian variables. Then, in turn, the distribution of the magnitude of any field component is Rayleigh distributed while the phase is uniformly distributed. Hence, the magnitude of the received voltage of a test antenna inside the RC will also be Rayleigh distributed and its phase uniformly distributed.

15.6.1.2 Why are the RC Used in MIMO OTA Testing?

The main reason why the RC has become an attractive tool for OTA testing, in addition to cost efficiency, is its capability to reproduce a reference wireless channel with the following characteristics, [46]

1. Isotropic AoA distribution;
2. Rayleigh distributed fading pattern;
3. Exponential decay of the power delay profile;
4. Balanced powers of orthogonal polarizations.

Extensive industrial and academic research has been undertaken during the past decade to adapt, miniaturize and enhance the RC technology to produce a cost efficient, time efficient, portable, accurate OTA testing technology with excellent repeatability. This has already resulted in the development of a second-generation RCs capable of performing MIMO OTA testing of devices operating according to the LTE/LTE-A cellular standards. However, many challenges are still ahead before a complete understanding is achieved so as to fulfill all the requirements dictated by the latest cellular radio communication standards, [47]. First-generation RC for OTA testing focused on SISO systems and wireless devices with a single antenna. For these systems, the emulated propagation environment in a RC came as a bonus since the focus had been on the isotropic radiated performance. The highest priority was therefore to measure standardized figures of merit such as the TRP and the Total Isotropic Sensitivity (TIS) that are essential for the adequate performance of 3G radio communication systems (e.g. GSM and UMTS)operating over SISO channels, [48]. The introduction of more advanced radio communication standards has resulted in the urge for new types of OTA testing equipment and techniques capable of providing the required performance evaluation of advanced wireless devices, [47]. Channel emulators have now become indispensable to test demanding performance specifications that support OFDM/MIMO techniques and higher-order modulation schemes. The TRP and the TIS alone are not enough to provide a comprehensive performance characterization of MIMO-based devices with multiple antennas. More relevant figures of merit have been introduced (see above) to perform MIMO OTA testing of LTE/LTE-A wireless devices operating over multipath fading channels. The channel provided by a standard single-cavity RC is a good reference. However, more realistic channels following the standard specifications need to be incorporated into the second-generation RC. These channels are not necessarily Rayleigh fading, or isotropic in terms of AoA and AoD distributions with balanced power of orthogonal polarizations and, normally, they have several clusters in the power-delay-profile.

15.6.1.3 Rayleigh Fading

In an ideal RC, the Rayleigh fading channel is naturally obtained by perfect mode stirring, [49]. However, a practical RC that uses *mechanical stirring* does not automatically provide such a perfect scenario. Therefore, in addition to the stirred component there might be an unstirred component too. The statistical distribution of the modes excited in the RC depends upon the efficiency of operation of the mode stirrers defined by their sizes, shapes and moving patterns. For example, it has been shown in [50] that the larger the stirrers and the more pronounced their asymmetry, the better the stirring since they can cover a larger volume and perturb the boundary conditions more efficiently. In addition to mechanical stirring, three other methods are being widely used: *platform stirring*, *polarization stirring* and *frequency stirring*. In platform stirring, the DUT is placed on a moving platform, where the platform itself acts as a stirrer and an averaging effect is obtained due to the DUT movement [51]. In polarization stirring [52], three orthogonal wall-mounted antennas are used to measure the transmitted/received power by the DUT, which is especially suited for TRP/TIS measurements. Frequency stirring is achieved by performing averaging over the frequency bandwidth of interest and therefore is appropriate for narrow bandwidth systems [53]. Improved accuracy in terms of Rayleigh fading statistics can also be obtained if there is no direct Line Of Sight (LOS) between the DUT and the measurement antenna and therefore no unstirred field component is present. It should be noted that in a realistic RC there will be some irreducible errors which lead to Ricean fading channel behaviour with a non-zero K-factor ($K = 0$ corresponds to the theoretical Rayleigh distribution).

15.6.1.4 Ricean Fading

Realistic wireless channels are more likely to follow the Ricean fading pattern as compared to the Rayleigh fading pattern, which will depend on the propagation scenario. So in order to model the desired fading behaviour, we need to recreate this channel by controlling the K-factor in the chamber. The Ricean K-factor can be defined in terms of the RC fields as the ratio between the power in the unstirred component to the power of the stirred component of the total field. It has been shown in [54] that the K-factor can be controlled in several ways that are related to the geometry and the physical properties of the channel in the reverberation chamber. If one measurement antenna is used, then the following expression summarizes the main result

$$K = \frac{V}{\lambda Q r^2} G_m G_d \eta_{md} \qquad (15.13)$$

where λ is the wavelength, V is the volume of the RC, Q is the quality factor of the RC, that is, the ratio between the stored and dissipated power in the chamber, r is the distance between the measurement antenna and the DUT and η_{md} is the polarization matching factor between the DUT and the measurement antenna. As we can see from

(15.13) the K-factor depends upon the interaction between the DUT, the measurement antenna and the RC. To obtain a reference channel, the DUT shall be replaced with a reference antenna. Now, we have several alternatives to obtain the desired K-factor, that is, it can be tuned by changing the loading of the RC which will impact Q, by changing the measures of the RC so as to impact its volume V, by changing the distance between the DUT and the measurement antenna, by redirecting the measurement antenna (if it is directive) or just by choosing a measurement antenna with suitable gain and polarization. An extra degree of freedom can be obtained by introducing an additional measurement antenna and controlling its input power.

Several investigations have been carried out to study the effects of these parameters on the K-factor. For example, in [55] the K-factor was found to be dependent both on the number and position of absorbers placed within the RC. In [56] a stationary-LOS K-factor is introduced to predict the measurement uncertainty as a function of frequency for both frequency and polarization stirring.

A technique that relies on software design rather than on hardware design employs a post-processing approach to emulate Ricean fading channels with desired K-factor, [57–59]. The method requires that the unstirred field component is present in the RC so that its magnitude is changed off-line by an off-set to obtain the desired K-factor with preserved phase of the unstirred (deterministic) component and preserved power of the stirred (stochastic) component.

The correct estimation of the K-factor from measurement data is an important problem related to Ricean fading emulation in RC, for example, it can be used to verify that the required K-factor has been emulated correctly or it can be used to derive other important parameters. In [60], the focus is on the estimation of the Ricean K-factor in a RC that uses mechanical and frequency stirring. The obtained results provide estimates of K-factor within the bandwidth of the antennas. Additional results on K-factor estimation in a RC can be found in [61] where an analysis of the performance and applicability of various estimators has been performed.

A method for estimating the gain of an antenna in a mode-stirred RC based on the estimation of the Ricean K-factor which provides the relative transmitting power level of the line-of-sight path is provided in [62]. In [63] a method is proposed to minimize the impact of the unstirred field component from a statistical analysis of the K-factor. They show that looking for the smallest Ricean K-factor gives a better result than applying some goodness-of-fit tests.

15.6.1.5 Other Fading Statistics

Many different statistical distributions can be used to describe the fading behaviour of the wireless channel depending on the specific application and on the propagation channel. For example, in vehicle-to-vehicle systems, the fading variation may exceed that of the Rayleigh fading pattern. This means that the likelihood of observing a low signal level is much higher as compared to the corresponding signal level of a Rayleigh fading pattern with equal average power. A modified RC has been used in [64] to model this channel. There, an electrically switched multi-element antenna array was added to a RC with reduced measures as compared to conventional RC for the same tested frequency range.

15.6.1.6 AoA and AoD Distributions

The spatial behaviour of the propagation channel has gained more relevance in the evaluation of LTE and LTE-A standards due to the introduction of MIMO technologies. As it has been explained in the previous chapters, the distributions of the AoA and the AoD play a fundamental role in the correlation properties of the channel transfer matrix of MIMO systems. In addition, it has a fundamental impact on the shape of the Doppler spectrum and therefore on the speed of the fading variation, that is, the autocorrelation function. First-generation RC focused on the isotropic channel, which was required to perform accurate TRP and TIS measurements. Second-generation RCs have continued focusing on the isotropic channel to some extent. However, methods have been devised for creating non-isotropic fading environments in RC. In [65] it was shown that the APS could be changed by placing absorbers of different type and shape in the RC. In [66], a so called Scattered Field Chamber (SFC) RC is used to reduce the angle-spread of the AoA; the method uses a shielded anechoic box placed inside the chamber in combination with highly directive antennas. A method to emulate multipath fading using a random time-variable phase for every direction of arrival has been devised in [67]. In [68], the APS was controlled by opening the door of the RC resulting in non-isotropic AoA distributions of the MPC as well as lower power levels due to power escaping from the chamber. Further research is required to recreate the non-isotropic environments required by the standards.

15.6.1.7 Polarization Power Balance

The average power of the vertically polarized component normally differs from the average power of the horizontally polarized component in real propagation environments. The polarization power imbalance is characterized by the ratio of the former to the latter and is known as the Cross Polarization Ratio (XPR) (also known as cross-polar discrimination (XPD)). The XPR is usually larger than 1 (linear scale) due to the vertical polarization of the transmitting base station antennas. However, the research on the application of RC to OTA testing has focused on the emulation of channels with balanced power of the orthogonal polarizations since the beginning, [52]. This can be explained by the main use of first-generation RC for OTA focusing on isotropic figures of merit such as TRP and TIS. Consequently, there is lacking a proper treatment of the emulation of polarization power imbalance in RC and its impact on OTA testing. It's worthwhile to note that the channel XPR is defined with respect to the ideal isotropic antennas. However, in practice, realistic reference antennas such as the half-wavelength dipole antennas shall be used to estimate the channel XPR inside the RC.

15.6.1.8 Doppler Spectrum and Fading Autocorrelation

The Ricean/Rayleigh fading pattern usually refers to the probability distribution of the signal envelope, that is, a first-order statistic of the channel. However, the rate at which the fading occurs over time is described by the auto-correlation function, which is a second-order statistic of the channel. In real environments, the Doppler spread is a consequence of the movement of the transmitting antenna, the receiving antenna, the environment or all of them at the same time. The fading rate depends on the speed of the wireless device and on the AoA distribution of the waves incoming at the receiver. The ideal isotropic AoA distribution implies a uniform Doppler power spectrum for a moving wireless device, that is, the power is the same for all observable Doppler frequencies. The corresponding autocorrelation function is the *sinc*-function, [69]. In practice, even if perfect mode stirring is achieved, the shape of the Doppler spectrum deviates from the ideal. Nevertheless, the required coherence times can be still emulated. In a RC, the Doppler spectrum or at least the coherence time (which is inversely proportional to the maximum Doppler frequency) can be emulated by changing the rotation speed of the mechanical stirrers and/or the platform stirrer. In [70], it has been demonstrated how the Doppler power spectrum can be measured in an RC during step-wise stationary stirring conditions, and how this can be used to determine the actual rms Doppler bandwidth achieved during continuous stirring. The latter is the desirable mode of operation in OTA testing of an active wireless device. In [71], different Doppler spreads in RC are generated by using different loading conditions using step-wise stationary stirring, and as a result the maximum Doppler shift decreased with increasing loading of the RC. In [72], an accurate control of the coherence time of the channel recreated in a RC is achieved by modulating the stirrer velocity. A closed-form expression for the autocorrelation function of the fading signal inside a RC is given in [73], which corresponds to the fixed wireless scenario rather than to the mobile wireless scenario. A Doppler spectrum close to the classical "bath-tube" shape, which is observed when the APS only depends on the azimuth and it is uniform, can be obtained by using a cascade connection of a channel emulator and reverberation chamber as demonstrated in [74].

15.6.1.9   Power Delay Profile and Frequency Correlation

The frequency selective behaviour of a wideband channel is modeled by the power delay profile, that is, it is the spectrum of the resolved multipath components arriving at the receiver position at different delays. The delay resolution improves by increasing the system bandwidth. Furthermore, the multipath nature of the wireless propagation channel is such that power delay profile consists of various clusters with different fading characteristics. The clustered behaviour of the channel is a result of scattering from the objects present in the path between the receiver and transmitter and those surrounding them.

A single-cavity RC supports a single-cluster power delay profile with exponential decay with rms delay spread given by, [75]

$$\tau_{rms} \simeq \frac{Q\lambda}{2\pi c}, \tag{15.14}$$

where strict equality is achieved for an ideal RC; λ is the wavelength and Q is the quality factor of the RC. Hence, any single-cluster with exponential power decay can be emulated by adding certain amounts of absorbers and averaging over many different paddle positions since the rms delay spread can be controlled by the chamber load as shown in [76, 77]. It is well-known that the rms delay spread is inversely proportional to the coherence bandwidth of the channel, the latter is defined as the maximum frequency range over which different frequency components are correlated. This level, usually 0.5 or 0.7, is defined at the complex frequency correlation function. In [78], it has been shown that the average mode bandwidth and the coherence bandwidth of the channel are highly correlated and they are almost frequency invariant for a given loading of the RC.

In order to emulate a more realistic channel, that is, with multiple clusters, a few other techniques have been devised. For example, in [74, 79] a fading or channel emulator was used to generate a signal with multiple clusters that was later passed into a RC (e.g. see Figure 15.8(c) for a schematic representation). The desired power delay profile was obtained by adjusting the amplitudes and delays of the discrete components of the fading emulator. Short RF pulses were fed into the fading emulator in order to monitor the impulse response of the system. The method made use of visual inspection to tune the profile into the desired profile shape. Good accuracy of emulation was obtained; however, the method requires further development to become fully automated. Another approach relies on the use of several metal cavities coupled by means of, for example, waveguides, [80], or slots, [57]. However, no comprehensive study of the impact of this method on the power delay profile has been performed and therefore further investigations are required. A single-cluster with dual-slope power delay profile was investigated in [81] employing multiply-connected reverberant spaces.

15.6.1.10 Radiation Pattern, MIMO Throughput, BER

We have mentioned above some of the most important parameters of the propagation channel and the methods used to recreate them in a RC. It is well-known that free-space FOM of the DUT, such as, the TRP, the TIS and the radiation efficiency can be accurately measured in a RC. However, it is just recently that a technique has become available allowing the measurement of the radiation pattern of a DUT in a RC, [82]. The proposed technique is based on the time-reversal electromagnetic chamber concept, which allows the number of probe antennas needed for the measurement to be reduced by a factor four as well as avoiding mechanical and platform stirring. With the arrival of LTE/LTE-A standards the focus has been shifted to include FOM that are relevant for MIMO OTA systems. For example, the BER performance of different systems was studied in [83], which included SISO system, MISO system (transmit beamforming), SIMO system (receive diversity), a combination of transmit beamforming with receive diversity and a full MIMO antenna system. BER measurements have also been conducted in [76], where it was shown that the BER changed significantly when the chamber loading or the paddle speed was changed. Measurements reported in [84] show the possibility of using OTA RC for discerning between a DUT with a good antenna design from that with a bad design for LTE MIMO at different bit rates. In [85], the authors proposed a simple model of the throughput data rate for wireless devices in LTE systems with MIMO and OFDM with good agreement between theory and experiments.

15.6.2 Two-Stage Method

The two-stage method was introduced in [86]. As illustrated in Figure 15.9 this method divides the MIMO OTA testing into two main stages: (1) the measurement of the DUT's antenna pattern inside an anechoic chamber and (2), the convolution of the measured antenna pattern in stage (1) with the desired channel model obtained from a channel emulator; the obtained signal is then used in a conducted throughput test on the DUT.

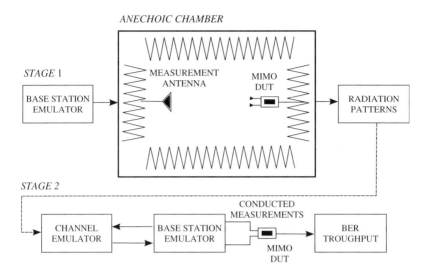

Figure 15.9   Schematic representation of the two-stage OTA testing.

Tests have shown that the obtained throughput and channel capacity results are comparable to similar results obtained with the multi-probe antenna based method for simplified SIMO HSDPA receive diversity performance. This method enables reuse of existing antenna pattern measurements and then provides accurate emulation of any required 2D or 3D channel model without further access to an anechoic chamber.

15.7   Future Trends

The OTA testing of future wireless devices will necessarily include more advanced technologies. As illustrated in Chapter 8, the channel modelling activities have evolved from single link modelling to geometrical multi-user MIMO (MU-MIMO) channel modelling. Currently, the knowledge on spatial interference characteristics in MU-MIMO environments and its effects on the overall system performance in realistic scenarios is limited. MIMO OTA testing systems need to be extended to comprise multi-link systems capable of evaluating the performance of DUTs in relevant usage and propagation scenarios. In this context, full 3D multi-probe systems may become an essential tool. Hence, in order to fully recreate the desired channel characteristics, accurate channel models of the spherical vector wave multimodes of the electromagnetic field will be required too.

References

1. S. Sesia, I. Toufik, and M. Baker, Eds., *LTE The UMTS Long Term Evolution, from theory to practice*, 2nd ed. John Wiley & Sons, UK, Sep. 2011.
2. 3GPP, "Technical Specification Group Radio Access Network; User Equipment (UE)/Mobile Station (MS) Over The Air (OTA) antenna performance; Conformance testing (Release 10)," Sep. 2011.
3. G. J. Foschini and M. J. Gans, "On limits of wireless communications in fading environments when using multiple antennas," *Wireless Personal Communications*, vol. 6, no. 3, pp. 311–335, Mar. 1998.

4. J. Krogerus, P. Mkikyro, and P. Vainikainen, "Towards an applicable OTA test method for multi-antenna terminals," in *COST 2100 6th Management Committee Meeting*, Oct. 6-8, 2008, Lille, France, TD(08)671.

5. M. Rumney, "Selecting figures of merit and developing performance requirements for MIMO OTA," in *COST 2100 10th Management Committee Meeting*, Feb. 3-5, 2010, Athens, Greece, TD(10)028.

6. 3GPP, "Technical Specification Group Radio Access Networks; Measurement of radiated performance for MIMO and multiantenna reception for HSPA and LTE terminals (Release 11)," May 2011.

7. P. Kyosti, J. Kolu, J.-P. Nuutinen, and M. Falck, "OTA Testing for Multiantenna Terminals," in *COST 2100 6th Management Committee Meeting*, Oct. 6-8, 2010, Lille, France, TD(08)670.

8. D. Kurita, Y. Okano, S. Nakamatsu, and T. Okada, "Experimental comparison of MIMO OTA testing methodologies," in *Proc. Fourth European Conf. Antennas and Propagation (EuCAP)*, 2010, pp. 1–5.

9. A. Tankelun, E. Böhler, and C. von Gagern, "Two-Channel Method for Evaluation of MIMO OTA performance of Wireless Devices," in *COST 2100 12th Management Committee Meeting*, Nov. 23–25, 2010, Bologna, Italy, TD(10)046.

10. P.-S. Kildal and K. Rosengren, "Correlation and capacity of mimo systems and mutual coupling, radiation efficiency, and diversity gain of their antennas: simulations and measurements in a reverberation chamber," vol. 42, no. 12, pp. 104–112, 2004.

11. M. Steinbauer, A. F. Molisch, and E. Bonek, "The double-directional radio channel," vol. 43, no. 4, pp. 51–63, Aug. 2001.

12. "The WINNER website," 2006. [Online]. Available: https://www.ist-winner.org/

13. 2008, Guidelines for evaluation of radio interface technologies for IMT-Advanced, Rep. ITU-R M.2135.

14. W. C. Jakes, *Microwave Mobile Communications*. The Institute of Electrical and Electronics Engineers, NY, USA, 1974.

15. H. Suzuki, "A statistical model for urban radio propogation," vol. 25, no. 7, pp. 673–680, 1977.

16. L. M. Correia, Ed., *Mobile Broadband Multimedia Networks*. Elsevier, UK, 2006.

17. "COST Action 2100-Pervasive Mobile & Ambient Wireless Communications," http://www.cost2100.org/.

18. 3GPP, "Technical Specification Group Radio Access Network; Spatial channel model for Multiple Input Multiple Output (MIMO) simulations (Release 10)," Mar. 2011.

19. D. S. Baum, J. Hansen, and J. Salo, "An interim channel model for beyond-3g systems: extending the 3gpp spatial channel model (scm)," in *Proc. VTC 2005-Spring Vehicular Technology Conf. 2005 IEEE 61st*, vol. 5, 2005, pp. 3132–3136.

20. K. Rosengren and P.-S. Kildal, "Study of distributions of modes and plane waves in reverberation chambers for the characterization of antennas in a multipath environment," *Microwave and Optical Technology Letters*, vol. 30, no. 6, pp. 386–391, 2001.

21. M. A. Garcia-Fernandez, J. D. Sanchez-Heredia, A. M. Martinez-Gonzalez, D. A. Sanchez-Hernandez, and J. F. Valenzuela-Valdes, "Advances in mode-stirred reverberation chambers for wireless communication performance evaluation," *Communications Magazine, IEEE*, vol. 49, no. 7, pp. 140–147, 2011.

22. J. P. Kermoal, L. Schumacher, K. I. Pedersen, P. E. Mogensen, and F. Frederiksen, "A stochastic mimo radio channel model with experimental validation," vol. 20, no. 6, pp. 1211–1226, 2002.

23. 3GPP, "LTE MIMO correlation matrices," Aug. 2007.

24. P. Degauque, A. Alayon Glazunov, A. Sibille, and J. Pamp, *Mobile Broadband Multimedia Networks: Techniques, Models and Tools for 4G.*, ser. ISBN 10: 0-12-369422-1. Academic Press, Elsevier, May 2006, ch. Antennas and diversity:from narrowband to ultra-wide band., pp. 218–276.

25. 3GPP, "Technical Specification Group Radio Access Network; Evolved Universal Terrestrial Radio Access (E-UTRA); User Equipment (UE) radio access capabilities (Release 10)," Dec. 2011.

26. T. Laitinen, P. Kyosti, T. Jamsa, and P. Vainikainen, "Generation of a field with a laplacian-distributed power azimuth spectrum scattered by a single cluster in a mimo-ota test system based on multiple probe antennas," in *2010 Asia Pasific Microwave Conference (APMC'10)*, Yokohama, Japan, Dec. 2010, pp. 2127–2130.

27. 3GPP, "Figure of Merits for MIMO OTA Measurements," May 2009.

28. 3GPP, "Technical Specification Group Radio Access Network; Evolved Universal Terrestrial Radio Access (E-UTRA); User Equipment (UE) conformance specification Radio transmission and reception Part 1: Conformance Testing (Release 10)," Dec. 2011.

29. E. Telatar, "Capacity of multi-antenna gaussian channels," *European Transactions on Telecommunications*, vol. 10, no. 6, pp. 585–595, 1999. [Online]. Available: http://dx.doi.org/10.1002/ett.4460100604.

30. C. A. Balanis, *Antenna Theory, Analysis and Design*, 3rd ed. John Wiley & Sons, NJ, USA, 2005.

31. T. Taga, "Analysis for mean effective gain of mobile antennas in land mobile radio environments," vol. 39, no. 2, pp. 117–131, 1990.

32. A. Alayon Glazunov, A. F. Molisch, and F. Tufvesson, "Mean effective gain of antennas in a wireless channel," *IET Microwaves, Antennas & Propagation*, vol. 3, no. 2, pp. 214–227, 2009.

33. R. Verdone and A. Zanella, Eds., *Pervasive Mobile and Ambient Wireless Communications: COST Action 2100 (Signals and Communication Technology)*, 1st ed. John Wiley & Sons, UK, Jan. 2012.

34. 2010, CELTIC-WINNER+ Deliverable 5.3 v.1.0. "WINNER+ Final Channel Models," CELTIC-WINNER+, Tech. Rep.

35. A. Alayon Glazunov, M. Gustafsson, A. F. Molisch, F. Tufvesson, and G. Kristensson, "Spherical Vector Wave Expansion of Gaussian Electromagnetic Fields for Antenna-Channel Interaction Analysis," *Antennas and Propagation, IEEE Transactions on*, vol. 57, no. 7, pp. 2055–2067, July 2009.

36. A. Alayon Glazunov, M. Gustafsson, A. Molisch, and F. Tufvesson, "Physical Modelling of Multiple-Input Multiple-Output Antennas and Channels by Means of the Spherical Vector Wave Expansion," *Microwaves, Antennas & Propagation, IET*, vol. 4, no. 6, pp. 778-791, June 2010.

37. A. Alayon Glazunov, "Expansion of the kronecker and keyhole channels into spherical vector wave modes," *Antennas and Wireless Propagation Letters, IEEE*, vol. 10, pp. 1112–1115, 2011.

38. T. Laitinen and P. Kyosti, "On appropriate probe configurations for practical mimo over-the-air testing of wirelss devices," in *Proc. European Conference on Antennas and Propagation*, 2012.

39. J. E. Hansen, *Spherical Near-Field Antenna Measurements*. London, UK: Peter Peregrinus, 1988.

40. T. Laitinen, J. Toivanen, P. Kyosti, J. Nuutinen, and P. Vainikainen, "On a mimo-ota testing based on multi-probe technology," in *Proc. URSI Int Electromagnetic Theory (EMTS) Symp*, 2010, pp. 227–230.

41. D. N. Black and E. B. Joy, "Test zone field compensation," *Antennas and Propagation, IEEE Transactions on*, vol. 43, pp. 362–368, Apr. 1995.

42. R. J. Pogorzelski, "Extended probe instrument calibration (EPIC) for accurate spherical near-field antenna measurements," *Antennas and Propagation, IEEE Transactions on*, vol. 57, pp. 3366–3371, Oct. 2009.

43. J. Toivanen, T. Laitinen, and P. Vainikainen, "Modified test zone field compensation for small-antenna measurements," *Antennas and Propagation, IEEE Transactions on*, vol. 58, pp. 3471–3479, Nov. 2010.

44. D. Parveg, T. Laitinen, A. Khatun, V.-M. Kolmonen, and P. Vainikainen, "Calibration procedure for 2-d mimo over-the-air multi-probe test system," in *Proc. European Conference on Antennas and Propagation*, 2012.

45. D. Hill, "Thirty Years in Electromagnetic Compatibility: Projects and Colleagues," *Electromagnetic Compatibility, IEEE Transactions on*, vol. 49, no. 2, pp. 219–223, May 2007.

46. D. Hill, *Electromagnetic Theory of Reverberation Chambers*. NIST TN, Dec. 1998, vol. 1506.

47. 3rd Generation Partnership Project; Technical Specification Group Radio Access Networks, "3GPP TR 37.976V1.5.0 (2011-05) Measurement of radiated performance for MIMO and multi-antenna reception for HSPA and LTE terminals (Release 11)," 3GPP and 3GPP2; download at http://www.3gpp.org, Tech. Rep., 2011.

48. 3rd Generation Partnership Project; Technical Specification Group Radio Access Networks, "3GPP TS 34.114V10.1.1 (2011-12) User Equipment (UE)/Mobile Station (MS) Over The Air (OTA) antenna performance; Conformance testing (Release 10)," 3GPP and 3GPP2; download at http://www.3gpp.org, Tech. Rep., 2011.

49. J. Kostas and B. Boverie, "Statistical model for a mode-stirred chamber," *Electromagnetic Compatibility, IEEE Transactions on*, vol. 33, no. 4, pp. 366–370, Nov. 1991.

50. J. Clegg, A. Marvin, J. Dawson, and S. Porter, "Optimization of stirrer designs in a reverberation chamber," *Electromagnetic Compatibility, IEEE Transactions on*, vol. 47, no. 4, pp. 824–832, Nov. 2005.

51. K. Rosengren, P.-S. Kildal, C. Carlsson, and J. Carlsson, "Characterization of antennas for mobile and wireless terminals by using reverberation chambers: improved accuracy by platform stirring," in *Proc. IEEE Antennas and Propagation Society Int. Symp*, vol. 3, 2001, pp. 350–353.

52. P.-S. Kildal and C. Carlsson, "Detection of a polarization imbalance in reverberation chambers and how to remove it by polarization stirring when measuring antenna efficiencies", *Microwave and Optical Technology Letters*, vol. 34, no. 2, pp. 145–149, 2002. [Online]. Available: http://dx.doi.org/10.1002/mop.10398

53. T. A. Loughry, "Frequency stirring: an alternate approach to mechanical mode-stining for the conduct of electromagnetic susceptibility testing," Phillips Laboratory, Kirtland Air Force Base, NM, Tech. Rep. Tech. Rep. PL-TR-91, 1036, 1991.

54. C. Holloway, D. Hill, J. Ladbury, P. Wilson, G. Koepke, and J. Coder, "On the Use of Reverberation Chambers to Simulate a Rician Radio Environment for the Testing of Wireless Devices," *Antennas and Propagation, IEEE Transactions on*, vol. 54, no. 11, pp. 3167–3177, Nov. 2006.

55. A. Sorrentino, G. Ferrara, and M. Migliaccio, "The Reverberating Chamber as a Line-of-Sight Wireless Channel Emulator," *Antennas and Propagation, IEEE Transactions on*, vol. 56, no. 6, pp. 1825–1830, 2008.

56. X. Chen, P.-S. Kildal, and S.-H. Lai, "Estimation of Average Rician K-Factor and Average Mode Bandwidth in Loaded Reverberation Chamber," *Antennas and Wireless Propagation Letters, IEEE*, vol. 10, pp. 1437–1440, 2011.

57. J. D. Sanchez-Heredia, M. Gruden, J. F. Valenzuela-Valdes, and D. A. Sanchez-Hernandez, "Sample-Selection Method for Arbitrary Fading Emulation Using Mode-Stirred Chambers," *Antennas and Wireless Propagation Letters, IEEE*, vol. 9, pp. 409–412, 2010.

58. J. D. Sanchez-Heredia, J. F. Valenzuela-Valdes, A. M. Martinez-Gonzalez, and D. A. Sanchez-Hernandez, "Emulation of MIMO Rician-Fading Environments With Mode-Stirred Reverberation Chambers," *Antennas and Propagation, IEEE Transactions on*, vol. 59, no. 2, pp. 654–660, 2011.

59. P. Hallbjorner, J. D. Sanchez-Heredia, E. de los Reyes, and D. A. Sanchez-Hernandez, "Limit for the proportion of remaining samples in the mode-stirred chamber sample selection technique," *Microwave and Optical Technology Letters*, vol. 53, no. 11, pp. 2608–2610, 2011. [Online]. Available: http://dx.doi.org/10.1002/mop.26336

60. C. Lemoine, E. Amador, and P. Besnier, "On the K-Factor Estimation for Rician Channel Simulated in Reverberation Chamber," *Antennas and Propagation, IEEE Transactions on*, vol. 59, no. 3, pp. 1003–1012, 2011.

61. M. I. Andries, P. Besnier, and C. Lemoine, "Rician channels in a RC: Statistical uncertainty of K estimations versus K fluctuations due to unstirred paths," in *Proc. 5th European Conf. Antennas and Propagation (EUCAP)*, 2011, pp. 1758–1762.

62. C. Lemoine, E. Amador, P. Besnier, J. Sol, J.-M. Floc'h, and A. Laisne, "Statistical estimation of antenna gain from measurements carried out in a mode-stirred reverberation chamber," in *Proc. XXXth URSI General Assembly and Scientific Symp*, 2011, pp. 1–4.

63. C. Lemoine, E. Amador, and P. Besnier, "Mode-stirring efficiency of reverberation chambers based on Rician K-factor," *Electronics Letters*, vol. 47, no. 20, pp. 1114–1115, 2011.

64. J. Frolik, T. M. Weller, S. DiStasi, and J. Cooper, "A Compact Reverberation Chamber for Hyper-Rayleigh Channel Emulation," *Antennas and Propagation, IEEE Transactions on*, vol. 57, no. 12, pp. 3962–3968, 2009.

65. M. Otterskog and K. Madsén, "On creating a nonisotropic propagation environment inside a scattered field chamber," *Microwave and Optical Technology Letters*, vol. 43, no. 3, pp. 192–195, 2004. [Online]. Available: http://dx.doi.org/10.1002/mop.20417.

66. M. Otterskog, "Modelling of propagation environments inside a scattered field chamber," in *Proc. VTC 2005-Spring Vehicular Technology Conf. 2005 IEEE 61st*, vol. 1, 2005, pp. 102–105.

67. A. Khaleghi, A. Azoulay, and J. C. Bolomey, "Evaluation of diversity antenna characteristics in narrow band fading channel using random phase generation process," in *Proc. VTC 2005-Spring Vehicular Technology Conf. 2005 IEEE 61st*, vol. 1, 2005, pp. 257–261.

68. J. F. Valenzuela-Valdes, A. M. Martinez-Gonzalez, and D. A. Sanchez-Hernandez, "Diversity Gain and MIMO Capacity for Nonisotropic Environments Using a Reverberation Chamber," *Antennas and Wireless Propagation Letters, IEEE*, vol. 8, pp. 112–115, 2009.

69. D. Hill, "Spatial correlation function for fields in a reverberation chamber," *Electromagnetic Compatibility, IEEE Transactions on*, vol. 37, no. 1, p. 138, Feb 1995.

70. K. Karlsson, X. Chen, P.-S. Kildal, and J. Carlsson, "Doppler Spread in Reverberation Chamber Predicted From Measurements During Step-Wise Stationary Stirring," *Antennas and Wireless Propagation Letters, IEEE*, vol. 9, pp. 497–500, 2010.

71. A. Sorrentino, G. Ferrara, and M. Migliaccio, "On the Coherence Time Control of a Continuous Mode Stirred Reverberating Chamber," *Antennas and Propagation, IEEE Transactions on*, vol. 57, no. 10, pp. 3372–3374, 2009.

72. J.-H. Choi, J.-H. Lee, and S.-O. Park, "Characterizing the Impact of Moving Mode-Stirrers on the Doppler Spread Spectrum in a Reverberation Chamber," *Antennas and Wireless Propagation Letters, IEEE*, vol. 9, pp. 375–378, 2010.

73. O. Delangre, P. De Doncker, M. Lienard, and P. Degauque, "Analytical angular correlation function in mode-stirred reverberation chamber," *Electronics Letters*, vol. 45, no. 2, pp. 90–91, 2009.

74. C. Wright and S. Basuki, "Utilizing a channel emulator with a reverberation chamber to create the optimal MIMO OTA test methodology," in *Proc. Global Mobile Congress (GMC)*, 2010, pp. 1–5.

75. D. Hill, M. Ma, A. Ondrejka, B. Riddle, M. Crawford, and R. Johnk, "Aperture excitation of electrically large, lossy cavities," *Electromagnetic Compatibility, IEEE Transactions on*, vol. 36, no. 3, pp. 169–178, Aug. 1994.

76. E. Genender, C. L. Holloway, K. A. Remley, J. M. Ladbury, G. Koepke, and H. Garbe, "Simulating the Multipath Channel With a Reverberation Chamber: Application to Bit Error Rate Measurements," *Electromagnetic Compatibility, IEEE Transactions on*, vol. 52, no. 4, pp. 766–777, 2010.

77. X. Chen and P.-S. Kildal, "Theoretical derivation and measurements of the relationship between coherence bandwidth and RMS delay spread in reverberation chamber," in *Proc. 3rd European Conf. Antennas and Propagation EuCAP 2009*, 2009, pp. 2687–2690.

78. X. Chen, P.-S. Kildal, C. Orlenius, and J. Carlsson, "Channel Sounding of Loaded Reverberation Chamber for Over-the-Air Testing of Wireless Devices: Coherence Bandwidth Versus Average Mode Bandwidth and Delay Spread," *Antennas and Wireless Propagation Letters, IEEE*, vol. 8, pp. 678–681, 2009.

79. H. Fielitz, K. A. Remley, C. L. Holloway, Q. Zhang, Q. Wu, and D. W. Matolak, "Reverberation-Chamber Test Environment for Outdoor Urban Wireless Propagation Studies," *Antennas and Wireless Propagation Letters, IEEE*, vol. 9, pp. 52–56, 2010.

80. M. Lienard and P. Degauque, "Simulation of dual array multipath channels using mode-stirred reverberation chambers," *Electronics Letters*, vol. 40, no. 10, pp. 578–580, May 2004.

81. J. S. Giuseppe, C. Hager, and G. B. Tait, "Wireless RF Energy Propagation in Multiply-Connected Reverberant Spaces," *Antennas and Wireless Propagation Letters, IEEE*, vol. 10, pp. 1251–1254, 2011.

82. A. Cozza and A. E.-B. A. El-Aileh, "Accurate Radiation-Pattern Measurements in a Time-Reversal Electromagnetic Chamber," *Antennas and Propagation Magazine, IEEE*, vol. 52, no. 2, pp. 186–193, 2010.

83. K. A. Remley, H. Fielitz, H. A. Shah, and C. L. Holloway, "Simulating MIMO techniques in a reverberation chamber," in *Proc. IEEE Int Electromagnetic Compatibility (EMC) Symp*, 2011, pp. 676–681.

84. C. S. L. Patane and C. Orlenius, "LTE MIMO throughput measurement method for characterization of multi-antenna terminal performance," in *Proc. IEEE Int Antennas and Propagation (APSURSI) Symp*, 2011, pp. 43–46.

85. P. Kildal, A. Hussain, X. Chen, C. Orlenius, A. Skarbratt, J. Asberg, T. Svensson, and T. Eriksson, "Threshold Receiver Model for Throughput of Wireless Devices With MIMO and Frequency Diversity Measured in Reverberation Chamber," *Antennas and Wireless Propagation Letters, IEEE*, vol. 10, pp. 1201–1204, 2011.

86. Y. Jing, Z. Wen, H. Kong, S. Duffy, and M. Rumney, "Two-stage over the air (OTA) test method for MIMO device performance evaluation," in *Proc. IEEE Int Antennas and Propagation (APSURSI) Symp*, 2011, pp. 71–74.

16

Cognitive Radio Networks: Sensing, Access, Security

Ghazanfar A. Safdar

University of Bedfordshire, UK

16.1 Introduction

Cognitive Radio is a topic which has attracted lots of interest from researchers recently; it is a promising technology that could play a strong role in the future communication systems, for example, next generation cellular systems such as Long Term Evolution (LTE), Wireless Interoperability for Microwave Access (WiMAX). The aim of this chapter is to briefly describe the concept of cognitive radio, what are cognitive radio networks and their types. Phenomenon and types of spectrum sensing, access mechanisms in cognitive radio networks, need and design of common control channel is also described before the chapter looks into the security of cognitive radio networks and a framework to realize cognitive radio networks security. The chapter concludes by describing several applications of the cognitive radio networks.

16.2 Cognitive Radio: A Definition

A *cognitive* radio is a radio that understands the context in which it finds itself and as a result can tailor the communication process in line with that understanding [1].

In simple words, a cognitive radio is a very smart radio. There are currently numerous communication systems making use of smart radios which can adapt their behaviour in many ways. For example, 3G systems to ensure power imbalances have the ability to dynamically alter their power output; mobile phones can intelligently process the incoming signals to mitigate different distortion effects; WiMAX has the ability to adapt the characteristics of the signals they transmit in order to maintain good throughput and link stability. It is important to note that the adaptations which occur in smart communications systems that we heavily employ in our everyday life are well defined and can be

LTE-Advanced and Next Generation Wireless Networks: Channel Modelling and Propagation, First Edition.
Edited by Guillaume de la Roche, Andrés Alayón Glazunov and Ben Allen.
© 2013 John Wiley & Sons, Ltd. Published 2013 by John Wiley & Sons, Ltd.

anticipated, where the adaptations are invoked by common and well known conditions. A cognitive radio takes this type of adaptive behavior much further, firstly by greatly increasing the level of adaptivity which applies to a wider range of parameters such as: operating frequency; power; modulation scheme; battery usage; processor usage and so on, and secondly the adaptation itself, unlike in other communication systems discussed above, can happen in both planned and unplanned ways where in the latter case the radio recognizes patterns of behaviour, learns from reoccurring situations and uses mechanisms to anticipate future events [2]. Thus a cognitive radio has the ability to learn from its actions and feed this learning to influence any future reactions. A cognitive radio is made up of software and hardware components to facilitate the wide variety of different configurations for its communication. In order to make any decisions on how to configure itself for the communication tasks, cognitive radio processes the inputs it receives; at this stage all the constraints and so on are taken into account to make a decision about configuration while attempts are made to match actions to requirements.

16.2.1 Cognitive Radio and Spectrum Management

The process of spectrum management describes how and by whom the spectrum would be used. The key purpose of spectrum management is to allow as many different users as possible while ensuring the least possible interference among the participants involved [3]. There are several approaches to spectrum management: the administrative spectrum management approach includes the usage of a particular range of frequencies (or frequency bands) and specifies the services delivered including the technologies which could be employed in the delivery of these particular services. In this approach the entire radio spectrum is formed of bands or blocks of frequencies established for a particular type of service by the process of frequency allocation, also known as Fixed Spectrum Assignment (FSA) policy [4]. While FSA has its own advantages, alongside it has some disadvantages too, for example, the spectrum planning involved and fixed frequency assignment to certain technologies is a slow process and could be less capable of keeping up with new innovations. The drawbacks of FSA has led to another approach to spectrum management called Dynamic Spectrum Access (DSA). DSA is supported by devices that have the capability of sensing their radio environment and gathering information about its usage. This very ability of scanning the spectrum for spectrum holes, empty bands or white spaces is called cognitive capability, and devices which are empowered with cognitive ability are called Cognitive Radios (CRs). CRs in DSA, rather than having static or fixed frequency bands to operate, could instead use whatever unused spectrum or white spaces are available, thus dynamic spectrum access enables what otherwise are wasted resources to be made use of. DSA centres on the concept of spectrum sharing between licensed and unlicensed users. In this case the licensed users of the spectrum, also known as Primary Users (PUs), have priority access to the spectrum, whereas the unlicensed or Secondary Users (SUs), also known as Cognitive Users can use the spectrum opportunistically when it is not being used by the PUs [1]. The secondary users should not cause any interference and cease to communicate as soon as the PUs reclaim or return to the licensed band. In DSA, no static frequency assignments are made, and users with static and non static assignment can coexist. In order to make this happen, SUs must be able to firstly detect the white space, perform configuration to continue transmission in that white space, scan

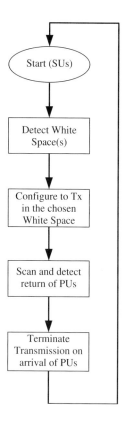

Figure 16.1   Cognitive User & DSA.

the white space to be able to detect the return of the PUs and then cease transmission immediately before it should start looking for another white space (Figure 16.1).

16.2.1.1   Overlay and Underlay DSA

As already mentioned, CRs perform DSA and this could be either overlay or underlay DSA. In the overlay case, which is also called Opportunistic Spectrum Access (OSA), Secondary or Cognitive users scan the spectrum for spectrum holes (white spaces) in the frequency bands and upon finding one/more of them, they then overlay their own transmission on the white space, and adjust their transmission parameters so that transmission could be carried out on them. However continuous monitoring of the selected white spaces needs to be performed because they must be vacated should the SUs return. It could be said that OSA involves sharing of spectrum between PUs and SUs with PUs having priority over the SUs. In Concurrent Spectrum Access (CSA), on the other hand, unlicensed cognitive users opportunistically operate in an underlay fashion. In this case, both the Primary User (PU) and the Secondary User (SU) share the spectrum on the condition that the SUs do not affect the PUs. Cognitive radio users thus may use spread spectrum or ultra-wideband techniques along with careful power control to ensure that no PUs receive a strong enough signal to cause any intolerable interference.

16.2.1.2 Need for Cognitive Radio

It has become a well known fact that the increase in the demand for wireless applications has rendered the radio spectrum a restricted resource. Studies reveal that the cause of spectrum scarcity is the under utilization of the licensed spectrum bands and it is proposed that alternative spectrum management techniques should be employed to improve spectrum usage, such as DSA, in other words resulting in the need of cognitive radios. As described above, a cognitive radio can change its transmission parameters as a result of its interaction with the environment in which it operates; however cognitive capability also encompasses tasks like sharing the sensed white spaces with other cognitive users, identifying any PUs or SUs in its vicinity and determining the kind of networks available for communication in its neighbourhood. Furthermore, continuous scanning of the environment for signals from other networks and reconfiguration of transmission parameters that minimize interference to nearby users is also a feature of the CR networks. CRs are becoming an increasingly important part of the wireless technology portfolio due to the mentioned scarcity of spectrum resources. They are seen as an enabling technology for a new spectrum utilization paradigm called OSA or CSA. Irrespective of the kind of cognitive radio network, effective management of radio resource so that a fair share of the spectrum could be given to all cognitive members in the network is necessary. These issues would be discussed further in the CR network taxonomy and access mechanisms in the following sections.

16.2.2 Cognitive Radio Networks

A network is formed when individuals join together in order to facilitate sharing of resources and to perform intra communication. A network of cognitive radios can be considered as a self organizing system where it can understand the context it finds itself in and can respond to a given set of requirements. The best way to describe a Cognitive Radio Network (CRN) is *heterogeneous* as it co-exists with conventional wireless networks of various kinds. A CRN primarily consists of both the primary network composed of PUs and the cognitive devices also known as SUs. According to the network architecture, cognitive radio networks can be of three different types, namely, Ad hoc CR networks, infrastructure CR networks, and CR mesh networks [5], as described below and illustrated in Figure 16.2.

16.2.2.1 Ad-hoc CRN

Ad hoc CR networks do not have any infrastructure backbone. In this case CR users can communicate with other CR users on an ad hoc basis using both licensed and unlicensed spectrum. Nodes in Ad hoc networks forward data to other nodes dynamically based on the network connectivity. The network is formed on the fly, where nodes can join and leave the network at any time, thus it suffers from security drawbacks such as lack of trust and robust authentication.

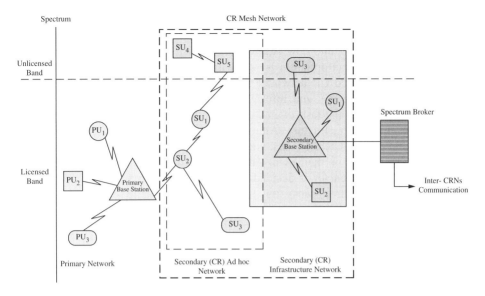

Figure 16.2   Cognitive Radio Networks.

16.2.2.2   Infrastructure CRN

The infrastructure CRN has cognitive radio base stations (also known as secondary base station) to enable single hop connectivity to users in the network, providing radio resource and security management and may also incorporate spectrum brokers whose functionality is to divide the radio resource amongst various clusters of CRNs for inter-CRN communication. Communication between members of different clusters is carried out via conventional networks. Usually different kinds of protocols and standards are supported by the base station which helps the user nodes to access communication networks.

16.2.2.3   Mesh CRN

Similar to Wireless Mesh Networks, CR mesh networks are an amalgamation of infrastructure and ad-hoc CRNs; it makes use of a mesh topology where each node acts as a gateway for other nodes and also disseminates its own data. Mesh networks operate using a backbone formed by connecting base stations and wireless access points which could also act as gateways providing increased connectivity into the wired infrastructure.

16.2.3   Cognitive Radio and OSI

A cognitive radio is made up of software and hardware components to facilitate the wide variety of configurations. Talking in terms of the Open Systems Interconnection

(OSI) layered architecture, the physical layer of CR devices is responsible for performing the job of spectrum sensing, the results of which are used for further analysis. It is worthwhile briefly looking into how Ad hoc and Infrastructure based CR network nodes perform once the sensed information is fed to the upper layers. Observations made by SUs in infrastructure based CR networks are provided to the central base station which subsequently makes decisions and informs cognitive devices for empty bands to use. In Ad hoc CR networks, the task of making decisions lies with the participating nodes due to absence of any infrastructure. Although SU have the cognitive capability in both infrastructure and Ad hoc CR networks, nodes in the latter case have the additional ability to reconfigure themselves on their own without instructions from a central entity. Once a frequency or set of frequencies is decided, the Medium Access Control (MAC) layer is tasked to provide and control access; this layer is also responsible for spectrum mobility, that is, to vacate the chosen white space by SU and shift to another one upon arrival of PU, resource allocation and spectrum sharing [6]. The network layer is assigned the job of maintaining the Quality of Service (QoS) and error control for the different services along with facilitating spectrum aware routing. The spectrum manager associates all three layers of the protocol stack and hence enables dynamic spectrum access (Figure 16.3).

The concept of cognitive radio, spectrum management, and types of cognitive radio networks have been defined, now the focus will be made to understand and describe how the spectrum observations are made by the physical layer, the process called spectrum sensing. How empty bands or the white spaces are actually sensed and fine tuned before being fed to the upper layers, and how the access control mechanisms finally assign and share the sensed resources and so on.

16.3 Spectrum Sensing in CRNs

As already described, a cognitive radio is a radio that understands the context in which it finds itself. To understand its context, a CR must be able to observe the outside world. A cognitive radio can either make autonomous local observations or it can receive

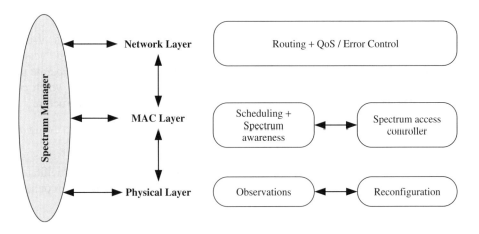

Figure 16.3 CR and OSI layers.

observations as input from other entities, that is, other CR devices. The former types of observations could be made by a cognitive radio using whatever resources (software plus hardware) it has available whereas in the latter case cognitive radios share and distribute information among themselves leading to the collective observations and giving them a global view. To gain a better understanding of the surrounding in which CRs operate, it is vital to have observations about the operation of other systems, that is, to seek knowledge of spectral occupancy. This process of determining how the spectrum is occupied is referred to as Spectrum Sensing. Referring to Figure 16.1, detection of white spaces is synonymous with the phenomenon of spectrum sensing. Spectrum sensing can be very challenging because the signal to be detected may have very low levels of power in the presence of high noise power. Spectrum sensing involves temporal and spatial spectrum sensing, that is, looking for occupied spectrum holes both in time and space.

16.3.1 *False Alarm and Missed Detection*

The main objective of spectrum sensing is to accurately detect the presence of a primary user, that is, signals transmitted by PU. A hypothesis test is usually used to make a decision as to whether a PU(s) exists or not [5]. The output of the test is analyzed to make the decision as illustrated in Figure 16.4.

The output of the test is either Null or Alternative hypothesis, and this determines the presence or absence of the PU. The hypothesis test may fail in several ways, for example a cognitive radio can result in a false alarm by making a positive decision about the presence of a primary user in its absence, and on the other hand the actual presence of the primary user could be overlooked to lead to the missed detection. Thus any method of observation or spectrum sensing used to detect the presence of a primary user should always aim to minimize the number of false alarms and missed detections because they would significantly affect the subsequent communication of the CR devices, a false alarm can result in a missed opportunity whereas missed detection could produce an increased number of collisions. Timely detection and freeing up of the empty white spaces is also of importance: sensing an empty space too late may result in a waste of the bandwidth and subsequently the transmission time available to the SU leading to inefficient usage of white space resulting into reduced throughput. Similarly freeing up of an empty space too late in the presence of the PU can result in increased interference. Efforts can be made to reduce false alarms and missed detections as explained in the following section.

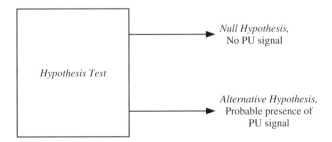

Figure 16.4   Hypothesis test for detection of primary signal.

16.3.2   Spectrum Sensing Techniques

With reference to Figure 16.3, the physical layer of the CR device captures received signal and decodes it using several signal processing techniques to analyze and understand its contents. This involves examining aspects of the received signal in order to make a firm decision about the presence or absence of a PU. The radio frequency (RF) front-end is at the disposal of the physical layer to provide a clean received signal before the signal processing techniques are employed, thus these very techniques are at the heart of the spectrum sensing process. There are three main signal processing techniques employed to detect the presence of a PU namely Energy Detection; Matched Filter; and Feature Detection. All these techniques are illustrated in Figure 16.5.

16.3.2.1   Energy Detection

As the name shows, for this technique the energy level of the received signal at a CR is compared with a threshold. Higher levels of energy compared to the threshold proves the presence of a primary transmitter. The simplicity of the energy detection mechanism makes it a very attractive technique to be employed in CR devices, but at the same time it has several drawbacks, for instance the choice of threshold level can be very difficult to select as it is impacted by varying noise and interference level.

16.3.2.2   Matched Filter

Unlike energy detection, matched filtering requires advanced knowledge of the primary users' signal. A CR would require the details of the primary signal of interest to be stored

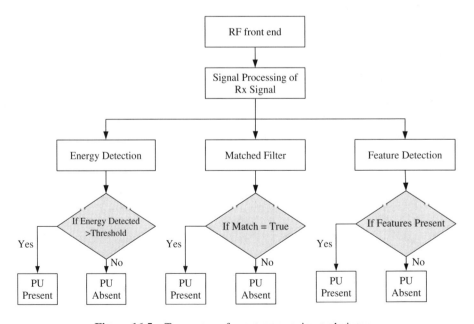

Figure 16.5   Taxonomy of spectrum sensing techniques.

in memory and compare these details with the received signal to avoid any false alarm or missed detection. Compared to energy detection, matched filtering can quickly reach a decision about presence or absence of PU transmitter with a higher confidence, so long as advanced knowledge of the PU signal is known a priori.

16.3.2.3 Feature Detection

Knowledge and storage of the PUs signal parameters can require substantial resources when using matched filter detection. An alternative approach involves the use of a feature detector. This technique is similar to matched filter detection, but instead of storing the entire PU signal, some signal features are used to aid detection. One of the well known feature detectors is the cyclostationary feature detector [7]. Cyclostationary feature detection works very well for low powered signals in the presence of noise or interference, because of the fact that noise unlike the true signal does not exhibit cyclic behavior. The cyclic behavior is introduced in the true signal by a process of modulation at the time of transmission. There are some drawbacks of the cyclostationary feature detector: it is computationally demanding and can require a long sensing time.

16.3.3 Types of Spectrum Sensing

In a typical cognitive radio network (CRN), every CR device is required to perform spectrum sensing and collect local observations from its surrounds (also known as standalone spectrum sensing). It is important to note that the observation range of a CR node may be much smaller compared to its transmission range, that is, its signal can be detected by nodes further away than any node it is able to detect, thus though individual CR nodes can make use of the above mentioned spectrum sensing techniques for primary signal detection, there can be a case of the signal-to-noise ratio (SNR), called the SNR wall, beyond which none of these techniques would be able to interpret any useful information. It is a well known fact that the power of the transmitted signal decreases with distance from the transmitter and that radio signals suffer from fading on their journey through the wireless channel. A cognitive radio device that is near to the primary user transmitter can miss its detection due to the signal fading. This is the cause for the hidden node problem in CRNs, where the false alarm could result in collisions and interference to the primary user. Both SNR wall and hidden node problems can be addressed by using cooperative spectrum sensing in CRNs [8].

For cooperative spectrum sensing, a group of CR nodes share sensing information about their local observations in order to get a better overview of spectrum utilization/occupancy. Thus CR nodes no longer work on their own but in the form of a cognitive radio network, where the network is formed of different SUs on the hunt to find the spectrum holes. Referring to Figure 16.2, nodes in both CR infrastructure and CR ad hoc networks may perform cooperative spectrum sensing. The base station in the former case receives the sensed information from all the CR nodes before a decision is made. The base station then delegates CR nodes to make use of an empty space. In the latter case the information is exchanged among peers by making use of a control frame called the Free Channel List (FCL), in a shared channel known as the Common Control Channel (CCC). Cooperative spectrum sensing can help to solve the hidden node problem because the nodes are spread over a certain suitable geographical region before the spectrum sensing information is exchanged and fused to enhance the confidence of a decision.

16.4 Spectrum Assignment–Medium Access Control in CRNs

Unused white spaces or spectrum holes can be assigned to CR users to enable them to pass information and execute transactions. The Medium Access Control (MAC) protocol helps CR users to access available spectrum holes without causing interference or collisions to primary users. This is known as spectrum sharing or spectrum assignment. Unlike classical MAC protocols, CR MAC protocols are closely tied to spectrum sensing functions to perform any reconfiguration. As far as classification of CR MAC protocols is concerned, they could be classified based on the nature of the channel access. Further classification is made with respect to how MAC protocols make use of the common control channel for exchange of FCL, which can be either global (Global Common Control Channel (GCCC)) or non-GCCC (Figure 16.6). An FCL contains the list of empty channels available in white spaces sensed by cooperatively communicating CR nodes, the availability of a CCC is very important in order to enable sharing of FCL among CR nodes.

16.4.1 Based on Channel Access

CR MAC protocols can either access the channel randomly in a de-centralized fashion (Random Access), or their access can be Time Slotted. The mixture of the two is also possible giving Hybrid channel access. All three types are briefly described as follows.

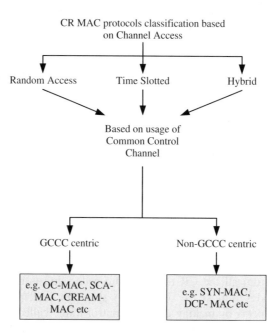

Figure 16.6 Taxonomy of CR MAC Protocols.

16.4.1.1   Random Access

Medium access control protocols in this class are generally based on the principle of carrier sense multiple access with collision avoidance (CSMA/CA) and do not require any time synchronization. The individual CR nodes contend for the medium and means of avoiding collision among the participants are employed. Examples of such MAC protocols are HC-MAC, and DOSS MAC and so on [9, 10].

16.4.1.2   Time Slotted Access

Time slotted MAC protocols divide the time into slots, it can happen for both the common control channel to perform transmission of FCL, or actual data transmission. Additionally there is a strict requirement for network-wide synchronization, for example C-MAC and so on [11]. Usually this class of protocols requires a centralized entity for assignment of time slots and to keep the network-wide time synchronization.

16.4.1.3   Hybrid Access

Hybrid access is a mixture of random and time slotted access. These protocols make use of the partially time slotted transmission for control signalling which generally occurs over synchronized time slots, followed by the random channel access for actual data transmission without requiring any time synchronization. There are some hybrid approaches which make use of identical pre-defined durations constituting a super frame for control and user data transmission, and access to the channels can then be either random or time slotted, for example, OS-MAC, SYN-MAC and so on [12, 13].

16.4.2   Based on Usage of Common Control Channel

All three CR MAC protocol types discussed above can either use a global common control channel (GCCC) or non-GCCC for exchange of control information, that is, FCL. So far the notion of a CCC has not been looked into in detail; it is worth looking into the CCC and understand its importance before going any further. A CCC is a free channel required by CR users to exchange FCL and to initialize communication among co-operating cognitive nodes. Before the pair of SUs can start sending and receiving actual data, they first have to coordinate their configuration using CCC. This includes deciding on the chosen white space(s). The pair of SUs exchange initial information in FCL, such as which white spaces to be used and how long will the communication last and so on. This information can also include exchange of Ready To Send (RTS) and Clear To Send (CTS) control frames in order to solve the hidden terminal problem and avoid collisions when using random access protocols. The selection criteria for the CCC can be static or dynamic. Under the static case, SUs can use the industrial, scientific, medicine (ISM) band [14] to exchange FCL, subsequently CCC is called global or GCCC. In the dynamic case, CCC can be one of the channels from the list of channels in sensed spectrum hole(s), in other words a channel from the FCL. In this case it is also called local CCC (or non-GCCC). Based on the static or dynamic assignment of the control channel, Cognitive Radio MAC protocols can be further classified as those which use Global CCC (GCCC) for example 2.4 GHz

ISM band and those which do not use GCCC usually called non-GCCC (Figure 16.6). Both GCCC and non-GCCC have their own advantages and disadvantages as described below.

16.4.2.1   Advantages of Using GCCC

- 24 - 7 Availability: Since GCCC make use of the ISM based, it is always available and can be used by CR nodes at any time (provided it is free of other ISM based wireless applications).
- License Fee: Due to the nature of the ISM band, CR nodes don't have to pay any licensing fee or require permission to use GCCC so long they operate within the bands restrictions on EIRP (Effective Isotropic Radiated Power).
- By communicating using GCCC, the pair of SUs can find the best channel based on a channel selection policy and agreed transmission parameters.

16.4.2.2   Drawbacks of GCCC (Advantages of Non-GCCC)

Some of the major drawbacks of using GCCC can be described as follows.

- No traffic differentiation, First Come First Served (FCFS) mechanism to access the GCCC.
- The higher the demand for GCCC, the higher will be the computational complexity and back off algorithm to access it. It also lowers the probability of availability of GCCC and it can have a serious effect on the QoS requirements of CR devices.
- The increased number of wireless applications has created a huge demand for the radio spectrum. In these circumstances, having a dedicated channel for exchange of the FCL and control frames is a waste of the precious resource.
- An adversary can impose a Denial of Service (DoS) attack on a dedicated GCCC by intentionally flooding it, thus it is a major security drawback [10].

As already described, CCC supports the spectrum related information exchange and transmission coordination among CR users. Although its main role is to help CR nodes exchange signaling and local observations relating to spectrum sensing, it can also help in neighbourhood discovery. Depending on whether the CCC shares the common data channel or uses a dedicated spectrum hole (channel in a white space), there are two main design approaches towards CCC that can further affect the operation and performance of CR MAC protocols: In-band-CCC and Out-of-band-CCC.

16.4.2.3   In-Band CCC

This approach is usually adopted by cognitive devices with one transceiver, that is, it does not require any additional transceivers because the licensed spectrum used for ongoing data transmission may also be used for the control messages, also known as in-band signalling. The MAC protocol controlling the channel access has a quiet period where spectrum sensing is performed before the actual data transmission takes place in the chosen white spaces [15]. Importantly, the extent of coverage of in-band CCC is lower

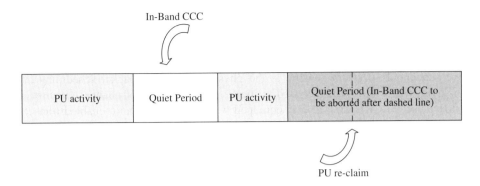

Figure 16.7   In-band CCC.

due to the fact that the channel used for sensing and sensed information exchange is the same as that of the data channel. There is also no spectrum switching cost associated with in-band CCC because the single transceiver in the CR devices do not need to frequently change the spectrum for control and data messages. In-band CCC does have drawbacks. One drawback is that the quiet period is significantly affected whenever a PU reclaims it, at this end the communication between two CR nodes would also be discontinued because of absence of the channel available for exchange of the control information. Similarly the quiet period may extend to long durations affecting the actual throughput of the PUs as illustrated in Figure 16.7. CR devices which employ a non-GCCC class of MAC protocols and equipped with single transceiver can benefit from in-band CCC to perform spectrum sensing.

16.4.2.4   Out-of-Band CCC

Out-of-Band CCC is the basis for both GCCC and non-GCCC class of CR MAC protocols where the spectrum sharing is either performed using a licensed channel or by employing the unlicensed ISM band, thus compared to in-band-CCC, the effect of PU reclaiming spectrum will have little effect on the performance of the CR nodes because they always have a higher probability of channel availability for the exchange of control information.

CR devices performing out of band CCC have more than one pair of transceivers, where one set of transceivers is continuously monitoring for the empty spaces and busy with exchanging control information. In this case since the data and the control signalling are communicated separately, thus the name out-of-band CCC. The absence of quiet periods result in more bandwidth available to both the PUs and SUs for actual data transmission leading to higher throughput.

16.4.3   CR Medium Access Control Protocols

A key challenge in Cognitive Radio Networks is to design an efficient MAC protocol that is capable of empowering the cognitive radio system to handle cross-layer parameter changes at the physical layer, eliminate collisions, save mobile energy and improve the network throughput by routing the packets to the destination with minimal delay. Since inception, there has been a number of MAC protocols for Cognitive Radio Networks that have been designed and developed. The Cognitive Radio-EnAbled Multi-channel MAC (CREAM-MAC) protocol is a GCCC based protocol which integrates the cooperative sequential spectrum sensing at the physical layer and packet scheduling at the MAC layer, where the former is used to improve the accuracy of channel sensing. Each secondary user in CREAM-MAC is equipped with a cognitive radio-enabled transceiver and multiple channel sensors. The proposed CREAM-MAC enables the secondary users to best utilize the unused spectrum while avoiding collisions among secondary users, and between the secondary and primary users. OC and SCA-MAC protocols help CR nodes use the spectrum efficiently and help to overcome the problem of hidden node terminal in multichannel CR networks. Initially nodes make use of GCCC and perform a three way hand shake to select a data channel from the exchanged FCL, before data transmission is acknowledged. Synchronization MAC (SYN-MAC), which belongs to the non-GCCC class of MAC protocols assumes every CR node to be equipped with two radios, where one of the radios is used for listening to the control signals (listening radio) and the other for both receiving and transmitting data (data radio). It chooses one of the channels available from the empty spaces to act as the common control channel for exchange of control information. Distributed coordination protocol (DCP-MAC) provides a common control channel selection process in a distributed way which is based on the appearance patterns of the PUs and their connectivity. Table 16.1 compares several CR MAC protocols in terms of different parameters.

Table 16.1 CR MAC protocols comparison

Protocol	Control Channel	Spectrum Access	No. of Transceivers	Type of Network
SYN-MAC [13]	Non-GCCC	Hybrid	Multiple	Ad hoc
OC-MAC [16]	GCCC	Not Addressed	Multiple	Ad hoc
SCA-MAC [17]	GCCC	Random Access	Not Mentioned	Ad hoc
C-MAC [11]	Non-GCCC	Time Slotted	Multiple	Ad hoc
A-MAC [18]	Non-GCCC	Random Access	Multiple	Ad hoc
DCP-MAC [19]	Non-GCCC	Random Access	Multiple	Ad hoc
HC-MAC [9]	GCCC	Random Access	Multiple	Ad hoc
CREAM-MAC [20]	GCCC	Hybrid	Multiple	Ad Hoc
CO-MAC [21]	Not Addressed	Random Access	Not Mentioned	Ad hoc
DOSS-MAC [10]	Non-GCCC	Random Access	Multiple	Ad hoc
DCM-MAC [22]	Not Addressed	Random Access	Multiple	Ad hoc
DC-MAC [23]	Not Addressed	Random Access	Not Mentioned	Ad hoc

To summarize, CR MAC protocols either make use of GCCC or non-GCCC for exchange of FCL before they can actually start any communication; accordingly drawbacks or advantages of each type of CCC apply to the corresponding CR MAC protocols thereby affecting their performance. The shortcomings of the GCCC class of CR MAC protocols can be avoided by employing non-GCCC mechanism which involves having a dynamic common control channel chosen from the list of channels, that is, FCL. One of the Dynamic CR MAC protocols which avoids shortcomings of GCCC and intelligently makes use of some of its advantages, called Dynamic Decentralized Hybrid MAC (DDH-MAC) has been developed and described in the following sections.

16.4.3.1 Dynamic Decentralized Hybrid MAC Protocol

The critical and most important aspect of cognitive radio devices is how to advertise the FCL among the participating nodes. Some protocols make use of GCCC for FCL exchange and suffer from all the disadvantages explained above [20, 24, 25]; others intelligently decide a local control channel from the available spectrum holes and advertise this to other nodes. The method of finding a local control channel is used by [11, 26], however lacks a clear methodology of finding the control channel within the white spaces. In particular authors of [10, 24] have made the assumption that the control channel is already found before the actual protocol starts its operation. It is important to note that no such assumption can be made in practice because finding a common control channel is the primary task of a CR MAC protocol and is the fundamental requirement in CR nodes before any subsequent communication can take place. This led to the design of a novel CR MAC protocol called DDH-MAC [27], which is a hybrid between the GCCC and non-GCCC class of MAC protocols. The DDH-MAC makes partial use of a GCCC to advertise the information about local control channel established from the white spaces available amongst cooperatively communicating cognitive nodes. It selects two white spaces called Primary Control Channel (PCCH) and Backup Control Channel (BCCH), both for the exchange of control information (i.e. FCL). PCCH and BCCH act as local control channels which are chosen from the list of available empty spaces in a cognitive node. The BCCH is only used once the PCCH is no longer available because of the PU re-claim, otherwise it can be employed as a normal white space for data exchange. The provision of local control channels such as PCCH & BCCH is to overcome the drawbacks of GCCC and provide benefits of the non-GCCC class of CR MAC protocols.

16.4.3.2 DDH-MAC Operation

Before FCL can be efficiently exchanged among the cognitive nodes, it is assumed that the nodes are aware of each others IDs, range and services. The DDH-MAC uses IEEE 802.11 standard as its benchmark, GCCC (ISM) in DDH-MAC is partially used to transmit a Beacon Management Frame (BF) which contains information about the advertisement of PCCH & BCCH (Figure 16.8). Importantly, all the header and Frame Check Sequence (FCS) fields and so on of the BF have the same size and contents as in IEEE 802.11 standard. The BF frame is launched in the GCCC which also includes the ID of the sender. Since it is a broadcast frame there is no need to include the destination ID. The purpose of the BF is to make the participating cognitive nodes aware of the chosen white

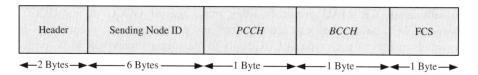

Header	Sending Node ID	PCCH	BCCH	FCS

←—2 Bytes—→←———— 6 Bytes ————→←——1 Byte ——→←——1 Byte——→←——1 Byte—→

Figure 16.8   Beacon management frame.

spaces acting as PCCH and BCCH. The complete operation of the DDH-MAC protocol can be described by using the flow chart shown in Figure 16.9.

1. *Scanning GCCC and launching beacon management frame:*

Upon initialization, cognitive nodes implementing DDH-MAC use one of the transceiver pairs to scan the GCCC for a BF. If the nodes fail to find a BF they then contend for the medium and the successful cognitive node is responsible for forming the beacon management frame and its launch in the GCCC. Any CR node which is implementing the DDH-MAC must have a minimum of three empty white spaces (channels) available. This threshold is set because two of them would be used for the PCCH/BCCH and the third one to start the initial communication. Since the CR devices are equipped with two sets of transceivers, the second pair would be continuously scanning to include other empty spaces in the FCL. The launch operation of the BF by a CR node is shown in Part ONE of the flowchart in Figure 16.9. In the other cases if the BF is successfully found by the CR node while scanning, it reads the BF to know about the PCCH/BCCH. Since PCCH and BCCH are empty spaces from the FCL maintained by CR nodes, there can be a case where some of CR nodes may not have information about these empty spaces which are represented by PCCH /BCCH. In this case they would update their FCL by adding information about the chosen local channels (PCCH/BCCH) to increase the number of white spaces maintained by them.

2. *Successful switch to PCCH for information exchange:*

Once the CR nodes have read about the PCCH/BCCH from the beacon frame and the FCL list is updated, they can start contending to send FCL in PCCH acting as a local control channel. The successful node then needs to check the PU re-claim on PCCH before it can finally exchange control information. If a PU re-claims the PCCH, rather than aborting the process, the BCCH is used to act as a local control channel. CR nodes can decide on the choice of the white space from the exchanged FCL. At this stage the DDH-MAC checks the continued availability of the chosen white space by applying the hypotheses test to find the presence of a PU. In case the chosen space is no longer available the operation is aborted and the selection of white space procedure is initiated again, otherwise CR nodes can go ahead to conclude their communication. If there is the need for any more transactions, the CR nodes again check the availability of the same chosen empty space for further communication to take place. It is believed that employment of BCCH and re-usage of the same white space for more transactions would help CR nodes to have better pre-transmission time (Pre-Tx time); importance of pre-transmission time is discussed in detail in the relevant section. DDH-MAC operation which involves contention, switching

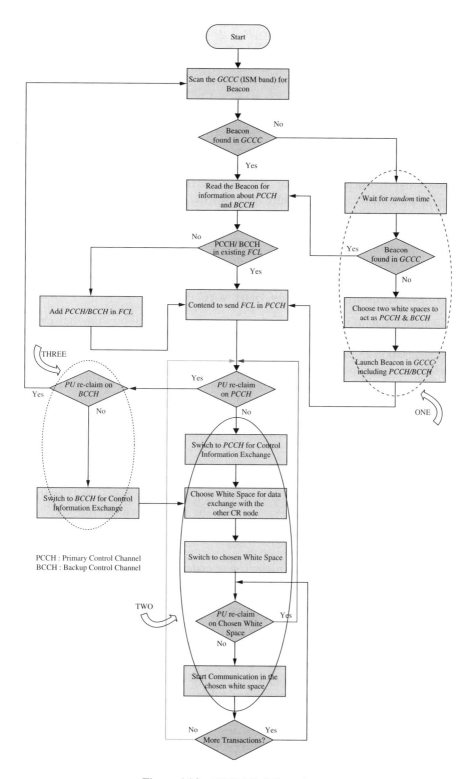

Figure 16.9 DDH-MAC flow chart.

to PCCH for FCL exchange and subsequent data communication after verification of chosen white space is depicted in part TWO of flowchart, Figure 16.9.

3. *PCCH re-claim by PU and switching to BCCH:*

Since CR nodes are parasite devices relying on the resources spared by the PUs, it is very likely that the PUs re-claim the chosen PCCH before CR nodes can use it for any control information exchange. In this case nodes in DDH-MAC simply switch to the backup control channel (BCCH) for seamless communication (see part THREE of flowchart, Figure 16.9). The BCCH is exactly the same as PCCH but it is only used in the scenario explained above. Dedicating a white space for the BCCH may give the impression of a loss of a white space, but it actually improves the overall network convergence time and it helps reduce the computational cost of the protocol, avoids the re-scanning of the GCCC and in turn helps CR nodes to conserve energy. In case both PCCH and BCCH are re-claimed, nodes revert back to the beginning of the operation.

16.4.3.3 Pre-Transmission Time for DDH-MAC Protocol

Since Cognitive Radios rely on the resources temporarily spared by PUs, they can spend lot of time searching for these resources before beginning any data transmission after control information exchange (FCL). Importantly, the PU reclaim of the resources from SUs can further interrupt their communication or completely halt it if there is no standby option available. In this case the communication protocols designed for CR networks should be flexible enough to adapt to these changes rapidly and seamlessly. The time spent for control information exchange before CR nodes can communicate is called the Pre-Transmission time and can undermine the performance of wireless communication protocols, especially those which serve applications with certain QoS requirements. Pre-Transmission time is one of the key performance indicators for CR network protocols in addition to other parameters like throughput, delay and so on.

With reference to ONE, TWO & THREE in Figure 16.9, there are several factors which contribute towards the DDH-MAC Pre-Transmission time, shown in Equation 16.1.

$$DDH - MAC_{pre-txtime} = T_{Scan} + T_{rand} + T_{Beacon} + T_{FCL} \qquad (16.1)$$

where T_{Scan} is the time taken by the participating nodes to scan the GCCC for any beacon frames. If there is no beacon found, T_{rand} is the time taken by contending CR nodes to avoid collision and multiple copies of the beacon frame before it is being launched in ISM band. T_{rand} also incorporates the contention time for CR nodes to send FCL in the PCCH. T_{Beacon} is the time taken either to read or launch the beacon frame and T_{FCL} is the time to exchange control information about the FCL in either the PCCH, or in the BCCH if there is a PU re-claim on the PCCH (Figure 16.10). After the pre-transmission time, cognitive radio nodes implementing DDH-MAC protocol switch to the chosen white space(s) and continue their communication. At this stage once again the PU re-claim of the chosen white space is verified. In the circumstances, where hidden terminals are to be avoided, the time for the control frames such as RTS and CTS will also form part of the pre-transmission time. With reference to Figure 16.10, the minimum and maximum

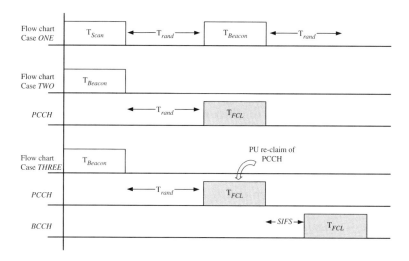

Figure 16.10   DDH-MAC pre transmission time.

values of the pre -transmission time for the DDH-MAC protocol are computed using Equations 16.2 and 16.3.

$$\text{Minimum Pre-Tx Time} = T_{Beacon} + T_{rand} + T_{FCL} \tag{16.2}$$

$$\text{Maximum Pre-Tx Time} = T_{Scan} + T_{Beacon} + (2 * T_{rand}) + SIFS + (2 * T_{FCL}) \tag{16.3}$$

The pre-transmission time has its maximum value at the time of initialization when no beacon frame is found in the GCCC. At this stage, the pre-transmission time includes a Short Inter Frame Space (SIFS) time value, beacon scanning and possibly one more random time value in case there is a PU re-claim on the PCCH. All wireless communication protocols, and especially for CR networks, should aim to have as little pre-transmission time as possible because it is an overhead before the actual data transmission can take place. Increased pre-transmission time can result the wastage of the precious bandwidth resource and it can subsequently incur considerable delays for the QoS sensitive applications.

16.5   Security in Cognitive Radio Networks

Cognitive radio devices can initiate communication using the resources spared by the PUs and these resources are detected through spectrum sensing. Sensed spectrum holes are formed into an FCL, and a common control channel is employed to exchange FCL between base station and SUs, in the case of infrastructure networks, or among individual SUs in the case of ad hoc cognitive radio networks. Cognitive radio networks like conventional wireless networks are also susceptible to several security threats and attacks which may include DoS, selfish misbehaviors, eavesdropping and so on. Additionally cognitive radio networks may also suffer from primary user emulation attacks and attacks on spectrum managers unless robust security mechanisms are in place [28].

Since the cognitive radio paradigm imposes human characteristics onto a radio device, a cognitive radio must exhibit good judgment to mitigate malicious manipulations of cognitive radios and cognitive networks into forced behaviors [29]. As already explained, cognitive radios face unique security problems not faced by conventional networks due to their very nature. It is important to distinguish primary users from secondary users to avoid any interference. It is also difficult for a secondary user to carry out accurate spectrum sensing on its own, thus cognitive radios perform co-operative spectrum sensing or is delegated empty spaces by base station (infrastructure networks). As the first line of security defense, the cognitive radio needs to be capable of judging whether the locally sensed data is real or falsified. At this stage, not only do the network messages require authentication but also the spectrum observations. Adversaries can exploit the communication taking place in the common control channel. One of the most common types of such attacks is the primary user emulation attacks, as illustrated in Figure 16.11. Here an adversary copies or emulates the characteristics of the primary transmitter or users' signal, thereby confusing the cognitive sensing devices which are then not able to interpret the original signal from the malicious one. As a result, sensing aggregation would potentially end up building a wrong list of empty spaces (FCL) thereby resulting in erroneous communication. In the case of spectrum emulation attacks, an adversary passes wrong information about the spectrum holes which are either not freely available or which do not exist at all. In these cases the SUs can end up having collisions with PUs resulting in data loss and unnecessary energy consumption.

Research shows that security in cognitive networks can be provided by identification of primary users, use of non-forgeable characteristics of the primary user signal, cognitive radio network admission control using IEEE 802.1x and identity management for user identification [30]. However, none of these techniques secure the communication between cognitive radio nodes in the common control channel. It is believed that cognitive radio networks have strict security requirements at two stages; during environment sensing [31]

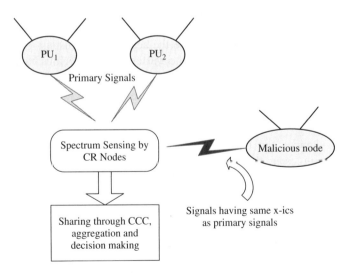

Figure 16.11 Primary user emulation attack in CRNs.

and during common control channel transactions. In distributed/co-operative communication, a secondary user collects local spectrum sensing results from neighbouring secondary users via a common control channel. Trustworthy distributed spectrum sensing ensures a correct decision is made from the fusion of the sensing results according to a fusion policy. Since the information resulting from co-operative communication is shared in the common control channel, this implies greater security needs for the specified channel. A robust common control channel security scheme is vital and can prevent the spread of falsified information which may result due to weak security during environment sensing [32]. Secure and authenticated transactions using the common control channel will ensure future security for the subsequent communication between the secondary users as illustrated in Figure 16.12. As the first line of security defense, the cognitive radio needs to be capable of judging whether the locally sensed data is real or falsified (Stage 1 in Figure 16.12 - environment sensing). At this stage, not only must the network messages be authenticated but also the observations of physical phenomena. Since the cognitive radio is utilizing not only its own observations as a basis for decision making, but also the observations from others, as a strong second line of defense, there is the obvious need to authenticate the shared observations (shared in a common control channel - Stage 2 in Figure 16.12). The cognitive radio needs assurance that messages are from who they claim they are from. After the authenticity of the source of collaborative cognitive radio network messages has been established, it is also important for cognitive radio nodes to judge whether the observations reported by other cognitive radio elements are real or falsified. This, combined with the ability to establish the authenticity of the source, is critical to prevent the propagation of attacker effects within the cognitive radio network.

16.5.1 Security in CRNs: CCC Security Framework

A co-operative communication cognitive radio MAC usually consists of a superframe, composed of common control channel (either GCCC or non-GCCC) and data sub channels. Distributed cognitive radio networks employ co-operative communication to exchange information about spectrum sensing in the common control channel. Information exchanged in the control channel must be secured to achieve authentication, confidentiality and integrity. It is strongly believed that the provision of security in the common control channel is vital and will further strengthen security against attacks such as DoS. In relation to information security, the control channel plays a key role in network availability and future communication among cognitive radio nodes. Since cognitive radio nodes exchange information about the sensed primary channels in the common control channel and further reserve a data-sub channel, the information about the data channel announcement, selection and reservation must be secured during all common control channel transactions. This will also provide protection from DoS attacks as malicious nodes/attackers will be unaware of the selected data channel. Additionally, the availability of numerous data-sub channels for selection and reservation adds a further level of security.

Any CR-MAC can be used to coordinate the exchange of information among cognitive radio nodes using the common control channel. Information security in the common control channel can be ensured by authenticating cognitive radio nodes that are engaged in the exchange of a channel table/list of free or available spectrum holes. Furthermore,

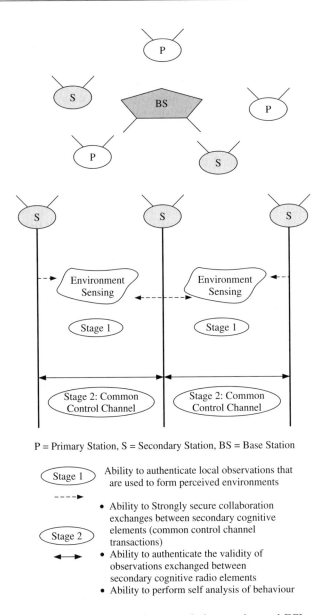

P = Primary Station, S = Secondary Station, BS = Base Station

| Stage 1 | Ability to authenticate local observations that are used to form perceived environments |

----▶

| Stage 2 | • Ability to Strongly secure collaboration exchanges between secondary cognitive elements (common control channel transactions)
• Ability to authenticate the validity of observations exchanged between secondary cognitive radio elements
• Ability to perform self analysis of behaviour |

◀──▶

Figure 16.12   CRNs security requirements during sensing and FCL exchange.

information exchanged in the common control channel between two co-operating cognitive radio nodes can be secured by employing cryptographic techniques such as encryption to provide confidentiality and message authentication codes or hash functions to provide integrity. Figure 16.13 describes such a common control channel security framework for cognitive radio networks [32]; the use of authentication and confidentiality in the common control channel ensures secure channel announcements, channel selection and reservation.

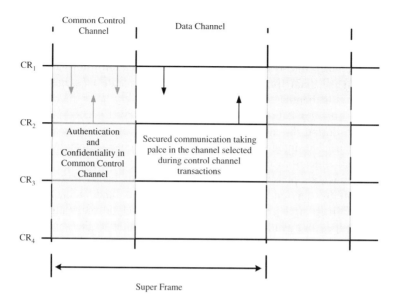

Figure 16.13   Common control channel security in the MAC superframe.

During channel negotiations in the common control channel, control frames, such as FCL, Channel Selection (CH-SEL) and Channel Reservation (CH-RES), are utilized [33]. To achieve further security, communication between the cognitive radio nodes in the data channel part of the super frame can be secured by employing encryption techniques.

Figure 16.14 illustrates the common control channel transactions and provides the core concept of the given security framework. The sender, CR_j is the host that transmits the MAC data frames and the receiver, CR_{j+1} is the host that receives MAC data frames.

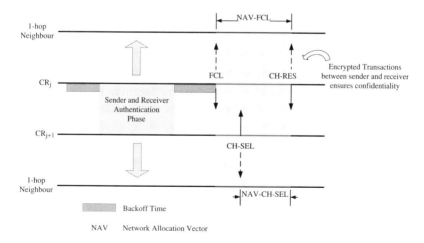

Figure 16.14   Common control channel security framework - distributed channel negotiation.

As a first line of defense, both sender and receiver must authenticate each other (any form of authentication scheme can be employed during this phase, such as ID authentication, secret sharing and so on). In order to save bandwidth and to reduce delay caused by additional overhead, the sender and receiver authentication information can either be piggybacked with the main frames, such as FCL, or separate messages can be used to achieve the desired authentication. To ensure confidentiality, after successful authentication, the two way communication between sender and receiver is encrypted using the keys transferred/shared during the authentication phase (encryption can be either private key or public key). If needed, integrity of the information exchanged can be implemented by using message authentication codes or hash functions.

After the authentication phase, identified fallow spectrum bands are mapped into logical channels and sent by the sender to the receiver after a random back-off time in the available channel list or free channel list (FCL) frame. Upon receiving the FCL frame, the receiver identifies the available data channels common to the sender and receiver. A data channel selection policy helps the receiver to select one data channel among all the common channels. The channel selection is announced by the receiver in a channel selection (CH-SEL) frame. After receiving the CH-SEL frame, the sender notifies its neighbours of the channel reservation via a channel reservation (CH-RES) frame. Both neighbour 1 and 2 refrain from transmitting by maintaining a Network Allocation Vector (NAV) specified in the FCL and CH-SEL frames during the channel negotiation process, NAV-FCL & NAV-CH-SEL respectively. From the perspective of the common control channel, a sequence of MAC control frames (FCL, CH-SEL, CH-RES) is exchanged for each channel negotiation process after a SIFS interval. This process is used by a sender and receiver to select a data channel for communication, avoiding any interference to the primary or main user.

In order to avoid possible interference among the neighbouring nodes, authentication information is overheard by the 1-hop neighbours of the sender and receiver nodes. 1-hop neighbours can also receive a copy of the keys during the channel broadcast/authentication phase, which are used to achieve confidentiality of the subsequent information exchanged between sender and receiver. Knowledge of the encryption keys allows 1-hop neighbours to learn of the reserved channel; this helps to avoid possible collisions as 1-hop neighbours know not to select the same (reserved) data channel for future communication; the FCL for 1-hop neighbours is accordingly updated. Since the cognitive radio nodes tend to operate using low transmission power levels [34], out of range 2-hop neighbours can receive the updated FCL from 1-hop neighbours in any future common control channel transactions as illustrated in Figure 16.15. The dynamic spectrum sensing policy can sense the availability of the particular reserved channel and add it back into the FCL once it is made free by the sender and receiver.

16.5.2   Security in CRNs: CCC Security Framework Steps

The steps required to establish security between two co-operatively communicating cognitive radio nodes are outlined below:

1. Sender: Sender takes over common control channel after a random back-off period.
2. Sender $\iff$ Receiver:

- Two way authentication between sender and receiver.
- Exchange of encryption keys.
- Communication is overheard by 1-hop neighbours.

3. Sender ⟹ Receiver:
   - Broadcast of FCL.
   - 1-hop neighbours overhear communications and set NAV-FCL to avoid collisions.

4. Receiver ⟹ Sender:
   - Application of channel selection policy to select common channel from FCL.
   - Broadcast of CH-SEL.
   - 1-hop neighbours overhear communications and set NAV-CH-SEL to avoid collisions.

5. Sender ⟹ Receiver:
   - Broadcast of CH-RES.

6. The common control channel is now vacant and made available to others.

Since the cooperatively communicating cognitive nodes make use of CCC for control information exchange, it is strongly believed that security of the common control channel is of utmost importance to ensure subsequent security among the communicating cognitive radio nodes. The framework presented above is aimed at providing the common control channel security. The key feature of the proposed framework is the

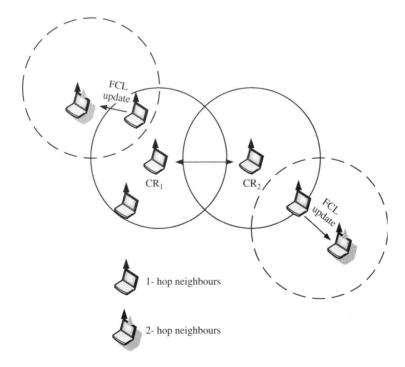

Figure 16.15   Common control channel security framework - transmission neighbours.

authentication requirement for any two cognitive radio nodes before any common control channel transactions take place. During the authentication phase, cognitive radio nodes securely exchange their public keys to achieve confidentiality both for the control channel and subsequently actual data transactions among participating CR nodes. Integrity of the information exchanged can be achieved by implementing hash functions or message authentication codes.

16.6 Applications of CRNs

Even though CRNs are in their infancy and there is little evidence of concrete applications, it is feasible to use them and they have applications in some of the following domains.

16.6.1 Commercial Applications

There is potential for CRNs to enhance current wireless systems and be an integral part in future wireless networks since they exploit unused spectrum for increased access. For example, once the spectrum becomes free when television channels switch from analogue to digital operation [35], an amount of white space will become available. Potentially, the digital switchover will result in a large amount of spectrum being released for a whole set of new services, thus CRNs could be used for provision of wireless broadband using these TV band frequencies. CRNs can also be used in short range communication systems which employ Ultra Wideband (UWB) technology. Since UWB transmits over a wide band and at low power, it can coexist with more powerful signals and not interfere in any significant manner allowing a UWB transmitter to transmit high data rates which can be used for short distances [36]. Since UWB signals can cause interference to low powered receivers such as GPS; equally a cognitive version of UWB could be used to overcome some of the issues related to it by knowing when, how and where to transmit by avoiding all sort of interferences.

16.6.2 Military Applications

Since the spectrum is a scarce resource, CRNs can ensure efficient use of the spectrum. It is a well known fact that the military is a major user of many different types of wireless systems which are always hungry for the spectrum [37, 38]. Examples may include weapon systems, sensors, navigation, location and so on. The systems employed might be heterogeneous having different data and communication needs and which are connected for inter-operability. CRNs can enable the independent configuration of different networks and manage co-existence. Since a cognitive radio device can take an input, and through reconfiguration translate it into another form to be acceptable by a different system, they can act as bridges between systems to provide a good means of inter-operability.

16.6.3 Public Safety Applications

Cognitive radios, once developed and deployed, will allow the disparate frequency bands used by public safety to be stitched together and available in a single end user device. Tunable spectrally adaptive radios and devices that can sense available transmit and receive frequencies, will provide stunning improvements in interoperability and channel capacity for systems in a region [39]. As with military applications, public safety can also incorporate requirements such as self-organization of networks, rapid deployment, inter-operability, varying bandwidth demands and shift of resources to the place where they are mostly needed. Thus, robust and reliable CRNs may be used to address these issues. Due to different regulatory environments, it is also believed that public safety would be seen as one of the favourable areas for usage of CRNs [40].

16.6.4   CRNs and LTE

16.6.4.1   Interference Management

Cognitive radios adapt dynamically to their RF environment, by observing the radio spectrum in real-time and making decisions based on reasoning. Interference management for Universal Mobile Telecommunications System (UMTS) LTE can be made possible by using cognitive base stations, where such base stations exploit their knowledge of the radio scene to intelligently allocate resources and to mitigate prohibitive co-channel interference [41].

16.6.4.2   Spectrum Sharing

Due to the fact that wider bandwidth is required for downlink peak throughput, spectrum scarcity is an inevitable issue in a LTE-Advanced system. A LTE-Advanced system can opportunistically benefit from the spectrum of digital video broadcasting (DVB) system by using spectrum sensing and sharing methods, but the tradeoff problem between complexity and detection probability always exists in spectrum sensing schemes, rendering most of these schemes not optimal enough. A cognitive based spectrum sharing scheme, including Auto-Correlation based Advanced Energy (ACAE) spectrum sensing method, can be employed for spectrum sharing between DVB and a LTE-Advanced system which provides improvements in detection probability leading to better spectrum fairness and efficiency [42].

16.6.4.3   Radio Resource Management

Cognitive functionalities can be used for enhanced Radio Resource Management (RRM) in LTE systems using orthogonal frequency division multiplexing (OFDM) access technology. OFDM features can be exploited, to help network segments properly adapt to the environment conditions by applying RRM algorithms for optimized sub-carriers' assignment, power allocation and adaptive modulation. At this stage, cognitive features are used to provide the system with knowledge that derives from past interactions with the environment; as a result, the system will be able to apply already known solutions in a timely manner to identify the problem that has already been addressed in the past, thus resulting in improved efficiency in terms of performance and network adaptation [43].

16.6.4.4   Capacity Augmentation

With rapid adoption of smart phones and LTE systems that allow end-users any-time, any-where Internet access, mobile data traffic is expected to increase exponentially. This has led to challenges like, how to scale the networks to achieve dramatic improvements in wireless access and system capacity while reducing the cost of deployment and network operations. Cognitive radio technologies can be employed to enable LTE systems to handle the data traffic growth, where DSA can be used for capacity augmentation in macro-cells, ultra-broadband small and femto cells using spectrum white spaces. The applications of CR can bring greater benefits to infrastructure networks by achieving balance between complexity, practicality, performance gains and market potential [44].

16.7   Summary

Cognitive radio networks are becoming an important part of the wireless networking landscape due to the ever increasing demand of spectrum resources. Cognitive radios are aware of their spectral environment and intelligently adapt their performance to the user needs. A cognitive radio is a software defined radio with a cognitive engine which responds to the changing environment by configuring the radio for different combinations of signal waveform, protocols, operating frequencies and so on. It enables flexible, efficient, reliable spectrum usage by adapting the radio's operating characteristics to real-time conditions. Co-operatively communicating cognitive radio nodes make use of the common control channel part of the CR MAC to perform free channel announcements, channel selection and reservation before any actual data transmission takes place.

This chapter started by looking into the need for cognitive radio networks before investigations were made to know different types of CRNs. Spectrum sensing was then explored in terms of its basics, techniques and different types. Spectrum sharing through medium access control protocols was also explained; together with a taxonomy of MAC protocols based on type of channel access and usage of common control channel. Drawbacks of global CCC (GCCC) in CR MAC protocols were highlighted before a new dynamic decentralized MAC protocol was introduced and explained which overcomes these drawbacks and incorporates the advantages of the local CCC (non-GCCC). This chapter also looked into the security requirements of CRNs, where some of the security threats specific to CRNs were explained. The importance of CCC security was introduced before a novel framework to achieve security of the CRNs was presented. Finally the chapter concluded by looking into applications of cognitive radio networks.

Acknowledgements

The author would like to express his sincere thanks to Dr. Gregory Epiphaniou and Dr. Mehmet Aydin for their help in producing the latex file for the chapter.

References

1. I. Mitola, J. and J. Maguire, G.Q., "Cognitive radio: making software radios more personal," *Personal Communications, IEEE*, vol. 6, no. 4, pp. 13–18, Aug. 1999.
2. E. L. Doyle, *Essentials of Cognitive Radio*. Cambridge University Press, 2009.
3. M. Cave, E. L. Doyle, and W. Webb, *Essentials of Modern Spectrum Management*. Cambridge University Press, 2007.
4. Y.-C. Liang, K.-C. Chen, G. Li, and P. Mahonen, "Cognitive radio networking and communications: An overview," *Vehicular Technology, IEEE Transactions on*, vol. 60, no. 7, pp. 3386–3407, Sept. 2011.

5. I. F. Akyildiz, W.-Y. Lee, and K. Chowdhury, "Crahns: Cognitive radio ad hoc networks," *Ad Hoc Networks*, vol. 7, no. 5, pp. 810–836, 2009. [Online]. Available: http://www.sciencedirect.com/science/article/pii/S157087050900002X.

6. A. De Domenico, E. Calvanese Strinati, and M. Di Benedetto, "A survey on mac strategies for cognitive radio networks," *Communications Surveys Tutorials, IEEE*, vol. PP, no. 99, pp. 1–24, 2010.

7. D. Cabric, S. Mishra, and R. Brodersen, "Implementation issues in spectrum sensing for cognitive radios," in *Signals, Systems and Computers, 2004. Conference Record of the Thirty-Eighth Asilomar Conference on*, vol. 1, Nov. 2004, pp. 772–776 Vol. 1.

8. B. Wang and K. Liu, "Advances in cognitive radio networks: A survey," *Selected Topics in Signal Processing, IEEE Journal of*, vol. 5, no. 1, pp. 5–23, Feb. 2011.

9. J. Jia, Q. Zhang, and X. Shen, "Hc-mac: A hardware-constrained cognitive mac for efficient spectrum management," *Selected Areas in Communications, IEEE Journal on*, vol. 26, no. 1, pp. 106–117, Jan. 2008.

10. L. Ma, X. Han, and C.-C. Shen, "Dynamic open spectrum sharing mac protocol for wireless ad hoc networks," in *New Frontiers in Dynamic Spectrum Access Networks, 2005. DySPAN 2005. 2005 First IEEE International Symposium on*, Nov. 2005, pp. 203–213.

11. M. Sha, G. Xing, G. Zhou, S. Liu, and X. Wang, "C-mac: Model-driven concurrent medium access control for wireless sensor networks," in *INFOCOM 2009, IEEE*, April 2009, pp. 1845–1853.

12. B. Hamdaoui and K. Shin, "Os-mac: An efficient mac protocol for spectrum-agile wireless networks," *Mobile Computing, IEEE Transactions on*, vol. 7, no. 8, pp. 915–930, Aug. 2008.

13. L. Pan and H. Wu, "Performance evaluation of the syn-mac protocol in multihop wireless networks," in *Computer Communications and Networks, 2007. ICCCN 2007. Proceedings of 16th International Conference on*, Aug. 2007, pp. 265–271.

14. J.-A. Park, S.-K. Park, D.-H. Kim, P.-D. Cho, and K.-R. Cho, "Experiments on radio interference between wireless lan and other radio devices on a 2.4ghz ism band," in *Vehicular Technology Conference, 2003. VTC 2003-Spring. The 57th IEEE Semiannual*, vol. 3, April 2003, pp. 1798–1801 vol. 3.

15. K. Chowdhury and I. Akyildiz, "Cognitive wireless mesh networks with dynamic spectrum access," *Selected Areas in Communications, IEEE Journal on*, vol. 26, no. 1, pp. 168–181, Jan. 2008.

16. S.-Y. Hung, Y.-C. Cheng, E.-K. Wu, and G.-H. Chen, "An opportunistic cognitive mac protocol for coexistence with wlan," in *Communications, 2008. ICC '08. IEEE International Conference on*, May 2008, pp. 4059–4063.

17. A. Chia-Chun Hsu, D. Weit, and C.-C. Kuo, "A cognitive mac protocol using statistical channel allocation for wireless ad-hoc networks," in *Wireless Communications and Networking Conference, 2007.WCNC 2007. IEEE*, March 2007, pp. 105–110.

18. G. Joshi, S. W. Kim, and B.-S. Kim, "An efficient mac protocol for improving the network throughput for cognitive radio networks," in *Next Generation Mobile Applications, Services and Technologies, 2009. NGMAST '09. Third International Conference on*, Sept. 2009, pp. 271–275.

19. M.-R. Kim and S.-J. Yoo, "Distributed coordination protocol for common control channel selection in multichannel ad-hoc cognitive radio networks," in *Wireless and Mobile Computing, Networking and Communications, 2009. WIMOB 2009. IEEE International Conference on*, Oct. 2009, pp. 227–232.

20. X. Zhang and H. Su, "Cream-mac: Cognitive radio-enabled multi-channel mac protocol over dynamic spectrum access networks," *Selected Topics in Signal Processing, IEEE Journal of*, vol. 5, no. 1, pp. 110–123, Feb. 2011.

21. H. Salameh, M. Krunz, and O. Younis, "Mac protocol for opportunistic cognitive radio networks with soft guarantees," *Mobile Computing, IEEE Transactions on*, vol. 8, no. 10, pp. 1339–1352, Oct. 2009.

22. Q. Zhao, L. Tong, A. Swami, and Y. Chen, "Decentralized cognitive mac for opportunistic spectrum access in ad hoc networks: A pomdp framework," *Selected Areas in Communications, IEEE Journal on*, vol. 25, no. 3, pp. 589–600, April 2007.

23. Y. Chen, Q. Zhao, and A. Swami, "Distributed cognitive mac for energy-constrained opportunistic spectrum access," in *Military Communications Conference, 2006. MILCOM 2006. IEEE*, Oct. 2006, pp. 1–7.

24. C. Cordeiro, M. Ghosh, D. Cavalcanti, and K. Challapali, "Spectrum sensing for dynamic spectrum access of tv bands," in *Cognitive Radio Oriented Wireless Networks and Communications, 2007. CrownCom 2007. 2nd International Conference on*, Aug. 2007, pp. 225–233.

25. A. Amanna and J. Reed, "Survey of cognitive radio architectures," in *IEEE SoutheastCon 2010 (SoutheastCon), Proceedings of the*, March 2010, pp. 292–297.

26. F. Wang, M. Krunz, and S. Cui, "Spectrum sharing in cognitive radio networks," in *INFOCOM 2008. The 27th Conference on Computer Communications. IEEE*, April 2008, pp. 1885–1893.

27. M. Shah, G. Safdar, and C. Maple, "Ddh-mac: A novel dynamic de-centralized hybrid mac protocol for cognitive radio networks," in *Roedunet International Conference (RoEduNet), 2011 10th*, June 2011, pp. 1–6.

28. G. Baldini, T. Sturman, A. Biswas, R. Leschhorn, G. Godor, and M. Street, "Security aspects in software defined radio and cognitive radio networks: A survey and a way ahead," *Communications Surveys Tutorials, IEEE*, vol. PP, no. 99, pp. 1–25, 2011.

29. J. Burbank, "Security in cognitive radio networks: The required evolution in approaches to wireless network security," in *Cognitive Radio Oriented Wireless Networks and Communications, 2008. CrownCom 2008. 3rd International Conference on*, May 2008, pp. 1–7.

30. R. Chen and J.-M. Park, "Ensuring trustworthy spectrum sensing in cognitive radio networks," in *Networking Technologies for Software Defined Radio Networks, 2006. SDR '06.1st IEEE Workshop on*, Sept. 2006, pp. 110–119.

31. R. Chen, J.-M. Park, Y. Hou, and J. Reed, "Toward secure distributed spectrum sensing in cognitive radio networks," *Communications Magazine, IEEE*, vol. 46, no. 4, pp. 50–55, April 2008.

32. G. Safdar and M. O'Neill, "Common control channel security framework for cognitive radio networks," in *Vehicular Technology Conference, 2009. VTC Spring 2009. IEEE 69th*, April 2009, pp. 1–5.

33. K. Bian and J.-M. Park, *MAC-Layer Misbehaviors in Multi-Hop Cognitive Radio Networks*, 2006, pp. 3–10.

34. T. X. Brown and A. Sethi, "Potential cognitive radio denial-of-service vulnerailities and protection countermeasures: A multi-dimensional analysis and assessment," in *Cognitive Radio Oriented Wireless Networks and Communications, 2007. CrownCom 2007. 2nd International Conference on*, Aug. 2007, pp. 456–464.

35. D. Setiawan, D. Gunawan, and D. Sirat, "Interference analysis of guard band and geographical separation between dvb-t and e-utra in digital dividend uhf band," in *Instrumentation, Communications, Information Technology, and Biomedical Engineering (ICICI-BME), 2009 International Conference on*, Nov. 2009, pp. 1–6.

36. A. Molisch, P. Orlik, Z. Sahinoglu, and J. Zhang, "Uwb-based sensor networks and the ieee 802.15.4a standard-a tutorial," in *Communications and Networking in China, 2006. ChinaCom '06. First International Conference on*, Oct. 2006, pp. 1–6.

37. A. Shukla, E. Burbidge, and I. Usman, "Cognitive radios-what are they and why are the military and civil users interested in them," in *Antennas and Propagation, 2007. EuCAP 2007. The Second European Conference on*, Nov. 2007, pp. 1–10.

38. A. Mody, M. Sherman, A. Trojan, K. Yau, J. Farkas, S. Sputz, T. McElwain, R. Bauer, J. Boksiner, and A. Fiuza, "On making the current military radios cognitive without hardware or firmware modifications," in *MILITARY COMMUNICATIONS CONFERENCE, 2010-MILCOM 2010*, 31 2010-nov. 3 2010, pp. 2327–2332.

39. N. Jesuale and B. Eydt, "A policy proposal to enable cognitive radio for public safety and industry in the land mobile radio bands," in *New Frontiers in Dynamic Spectrum Access Networks, 2007. DySPAN 2007. 2nd IEEE International Symposium on*, April 2007, pp. 66–77.

40. W. Lehr and N. Jesuale, "Public safety radios must pool spectrum," *Communications Magazine, IEEE*, vol. 47, no. 3, pp. 103–109, March 2009.

41. A. Attar, V. Krishnamurthy, and O. Gharehshiran, "Interference management using cognitive base-stations for umts lte," *Communications Magazine, IEEE*, vol. 49, no. 8, pp. 152–159, August 2011.

42. X. Zhao, Z. Guo, and Q. Guo, "A cognitive based spectrum sharing scheme for lte advanced systems," in *Ultra Modern Telecommunications and Control Systems and Workshops (ICUMT), 2010 International Congress on*, Oct. 2010, pp. 965–969.

43. A. Saatsakis, K. Tsagkaris, D. von Hugo, M. Siebert, M. Rosenberger, and P. Demestichas, "Cognitive radio resource management for improving the efficiency of lte network segments in the wireless b3g world," in *New Frontiers in Dynamic Spectrum Access Networks, 2008. DySPAN 2008. 3rd IEEE Symposium on*, Oct. 2008, pp. 1–5.

44. M. Buddhikot, "Cognitive radio, dsa and self-x: Towards next transformation in cellular networks extended abstract," in *New Frontiers in Dynamic Spectrum, 2010 IEEE Symposium on*, April 2010, pp. 1–5.

17

Antenna Design for Small Devices

Tim Brown

University of Surrey, UK

Antenna design for mobile communications has, for many years faced challenges due to the size of the mobile terminal often being considerably smaller than a single wavelength at the frequency of operation. Ideally, the antenna at the mobile terminal should be comparable to at least quarter of a wavelength in order to radiate and receive electromagnetic energy with the highest possible efficiency. At frequencies used in current and future cellular bands, this is not normally the case, particularly with mobile handsets and devices small enough to fit in a pocket. Therefore more compacted antennas have to be developed to fit within the physical constraints, which means that efficiency is compromised and also operating bandwidth is often reduced. When the antenna is in use and has user interaction impacting its electrical characteristics, the efficiency is further decreased. Over the years many antennas have been developed and methods have been researched in order to improve the ways in which these problems can be overcome.

For LTE and LTE Advanced, a number of further issues have to be considered with regards to the system [1, 2], which have an effect on the antenna design and are summarized as follows:

- Multiple bands with different centre frequencies will need to be accessed in order to increase the spectrum available that is required to address the anticipated user demand. This is known as "bandwidth aggregation", which will require the antenna to be tuned to all of the required bands, all of which have differing bandwidths.
- The use of multiple input multiple output (MIMO) will require up to eight antennas to be available at the base station or access point and up to four at the mobile terminal. Integrating four antennas onto a small device will bring further size reductions.
- Where small cells are required in intensively populated areas, they may be run by self organizing networks with the cell being so small it is known as a "femtocell". The design of a compact antenna on the femtocell (which must be a small, easy to install device) as well as the mobile terminal will therefore be required.

LTE-Advanced and Next Generation Wireless Networks: Channel Modelling and Propagation, First Edition.
Edited by Guillaume de la Roche, Andrés Alayón Glazunov and Ben Allen.

- Access points in small cellular areas, possibly serving a larger area than femtocells, may also be used as relay nodes to a main base station and such access points would still require small antennas in order to remain visually discrete.

This chapter is therefore constructed based on the requirements outlined above. To begin with, the chapter will introduce the reader to the fundamental theory behind antenna design and figures of merit used for both single antennas and multiple antennas at an access point or mobile terminal. Beyond this, it will then describe specific challenges with regards to tuning antennas to multiple frequency bands, installing multiple antennas on a small device and also show examples of such antennas used both at the access point and mobile terminal.

17.1 Antenna Fundamentals

The fundamental theory behind electromagnetism has been long defined using Maxwell's equations [3]. For the case of propagation of electromagnetic waves in free space, in simple terms they provide sufficient proof that for such propagation there is the presence of both electric and magnetic fields and that they are time and spatially variant. Two of the four equations derived by Maxwell state two important facts. The first is that the rate of change of magnetic field, $\mathbf{H}$, with respect to distance is proportional to the electric field, $\mathbf{E}$. This relationship also works with respect to the magnetic field in that the rate of change of electric field with respect to distance is proportional to the magnetic field. In order to understand the fundamentals of antenna theory, it is not necessary to understand the detail of Maxwell's equations. More importantly, the presence of the electric and magnetic fields radiated by an antenna are a direct result of current oscillating at a given frequency within the conducting elements of the antenna. The radiated electric and magnetic fields have a corresponding power density (i.e. a power per unit area), which is defined by what is known as a Poynting vector, $\mathbf{S}_{\text{Poynting}}$, that can be obtained from the following cross product:

$$\mathbf{S}_{\text{Poynting}} = \mathbf{E} \times \mathbf{H} \tag{17.1}$$

To simplify this equation, if there is only a vertical electric field, E_{V}, according to Maxwell's equations there must also be a horizontal magnetic field, H_{H}. The Poynting vector can be reduced down to a scalar quantity, which can now be defined as the power density, S_{d} due to these two fields:

$$S_{\text{d}} = \frac{|E_{\text{V}}||H_{\text{H}}|}{2} \tag{17.2}$$

This process also works in reverse for an antenna in receive mode. The received electric and magnetic fields (which have a corresponding power density), when they arrive at the antenna's conducting elements will induce an oscillating current into the antenna, which can consequently send a signal to the receiver.

The simplest antenna which can be considered is known as an *isotropic radiator*, which can be considered as a point source such that, if it is used as a transmit antenna, the source power, P_{s}, fed into the antenna will then radiate out of the antenna uniformly in a sphere.

At a distance, r, the sphere will have a surface area defined by $4\pi r^2$. Since the power is spread evenly over the surface area, the resultant power density of the isotropic radiator, S_{iso}, at any point distance, r, away from the source is therefore defined by:

$$S_{iso} = \frac{P_s}{4\pi r^2} \tag{17.3}$$

17.1.1   Directivity, Efficiency and Gain

An isotropic radiator is impossible to construct for a number of reasons, though for a mobile terminal such as a mobile handset or laptop computer, it is the intention to build an antenna that behaves as much like an isotropic radiator as possible. Such an antenna is known as an omni-directional antenna. At the mobile, as illustrated on the right hand side of Figure 17.1, the antenna does not radiate as much power density upwards or downwards, while it does radiate more power density to the left and the right in comparison to an isotropic radiator. The access point (such as that used in a femtocell base station or a wireless local area network) is attached to a wall and in this case it will need to radiate power density away from the wall and in the direction of where mobile terminals are going to be found. Therefore the access point antenna has a high directivity away from the wall as shown on the left hand side of Figure 17.1.

The directivity of an antenna is a measure of how much an antenna will radiate or receive power density in a single direction compared to what would be radiated or received by an isotropic antenna. The direction can be defined by two angles, θ and φ, which are identified in Figure 17.2 based on arbitrary three dimensional coordinates. For a fixed distance, r, the power density radiated or received is therefore a function of angles θ and φ so it can be denoted as $S_{ant}(\theta, \varphi)$. The directivity, $D(\theta, \varphi)$, is therefore:

$$D(\theta, \varphi) = \frac{S_{ant}(\theta, \varphi)}{S_{iso}} \tag{17.4}$$

The directivity is determined by how the antenna is constructed or its topology, such that if transmitting it will concentrate the power density to transmit in a certain direction or directions. It has been assumed up to now that all the source power input into the transmit antenna is then radiated. In reality the antenna will lose some of that energy as heat and will therefore have an efficiency of less than 100%. The efficiency of an antenna

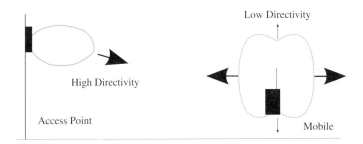

Figure 17.1   Illustration of the directivity on an access point and mobile terminal.

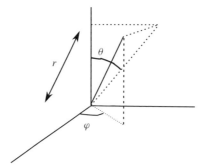

Figure 17.2   Three dimensional spherical coordinates.

is defined as the ratio of total power radiated by the antenna to the power input to the antenna. This efficiency, η, is simply multiplied by the directivity to give the gain of the antenna, thus the gain of the antenna can be defined as follows:

$$G(\theta, \varphi) = \eta D(\theta, \varphi) \qquad (17.5)$$

Antenna gain is more commonly used in defining how well the antenna radiates in a certain direction. It is normally the case that it is quantified in the decibel scale, such that $10\log_{10}G(\theta, \varphi)$ will return the gain in dBi (i.e. power density relative to an isotropic antenna). The gain of an antenna can be characterized in an anechoic chamber by comparing the power received by the antenna under test with a reference antenna where the gain is known, from which the measurement can be calibrated to resolve the dBi gain at any given angle. The gain is often analyzed versus φ with θ fixed at $90°$, which is known as the azimuth pattern. Another measurement commonly taken is to fix φ at $0°$ or $90°$ and measure gain versus θ, which is known as the elevation pattern.

17.1.2   Impedance and Reflection Coefficient

Not only does an antenna need to be efficient, it must also have an input impedance that is comparable to the impedance of the transmission line or power source that it is connected to. The antenna and transmission line are modelled as illustrated in Figure 17.3, where it is shown that there is a forward voltage going towards the antenna, whilst there is also a reflected voltage going back down to the transmission line due to a mismatch between the impedances. If the impedance of the antenna is significantly higher or lower than the transmission line, it will cause nearly all of the power to be reflected back, thus the antenna is not capable of transmitting or receiving.

To determine how much power the antenna will reflect back, or similarly fail to receive enough power into the transmission line in receive mode, the reflection coefficient can be calculated, which will be defined here as a parameter Γ. Its relation to the antenna impedance, Z_{Ant} and the source impedance Z_0 is defined as follows:

$$\Gamma = \frac{Z_{\text{Ant}} - Z_0}{Z_{\text{Ant}} + Z_0} \qquad (17.6)$$

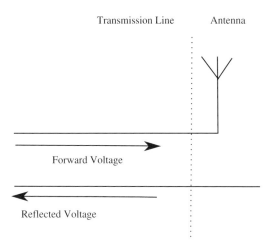

Figure 17.3 Reflections due to impedance mismatch.

This value can also be expressed in decibels and is known as return loss which is equivalent to $-20\log_{10}|\Gamma|$. In general it is required that the antenna has a return loss of more than 10dB within the bandwidth of operation and in this case just 10% of the power or less will get reflected back to the source. It should be noted that an efficient antenna may not have a good impedance match, while at the same time an antenna with a good impedance match may not be efficient. It is often assumed that an antenna with a good impedance mismatch is efficient, though in many instances this is not the case.

17.2 Figures of Merit and their Impact on the Propagation Channel

The impact of the chosen antenna on the propagation channel is important to identify in terms of how much it will improve the link or make it worse. Antennas close together will have low isolation, due to mutual coupling and therefore not behave as expected, which can reduce their efficiency. At the same time, the antenna gain patterns need to match the direction in which the incoming fields are arriving. This is measured by the mean effective gain and related parameters, which are discussed in this section.

17.2.1 Coupling and S-Parameters

In the previous section Γ was covered as a parameter to define the reflection coefficient of a single antenna connected to a feeder. This is also known as a single scattering parameter, or S-parameter and is termed s_{11}. If the antenna is in the proximity of one or more antennas in a device, then there will be at least four scattering parameters present, resulting in mutual coupling. For the simplest case, two antennas will be considered, as illustrated in Figure 17.4 where they have four scattering parameters:

1. The reflection coefficient of antenna 1 is defined as s_{11}.
2. The reflection coefficient of antenna 2 is defined as s_{22}.

3. The cross coupling from antenna 2 to antenna 1 is defined as s_{12}.
4. The cross coupling from antenna 1 to antenna 2 is defined as s_{21}.

In most cases, the coupling between the two antennas, which is defined in the simplest terms as the ratio of the voltage input to one antenna and then output at the other, is reciprocal. Therefore $s_{12} = s_{21}$, though if at least one of the antennas is loaded with ferrite material, this reciprocity rule is not true because the ferrite will cause the antenna to have different receive to transmit properties. This is due to different currents that will be induced in the ferrite in either case. As will be seen later, ferrite materials are sometimes used to help with reducing the size of antennas on mobile terminals. The scattering parameters are inserted into a matrix, **S**, defined as follows:

$$\mathbf{S} = \begin{pmatrix} s_{11} & s_{21} \\ s_{12} & s_{22} \end{pmatrix} \tag{17.7}$$

A common term used to define multiple antennas is the isolation, which is a measure of the cross coupling. The isolation can be defined in decibels as $-20\log10|s_{12}|$ or $-20\log10|s_{21}|$. A set of multiple antennas can be considered to have high isolation if the value is above 20dB, whilst isolation less than 10dB is considered to be low and most likely detrimental to the antenna's efficiency.

One important factor to consider when evaluating the S-parameters of multiple antennas is that the value of s_{11} will not only be dependent on antenna 1, but it will also be dependent on the cross coupling to antenna 2. Therefore, if the distance between the antennas were to change, not only would s_{21} change but also s_{11}. Similarly, s_{22} and s_{12} would also change. If three antennas are employed, the **S** matrix will be a 3×3 matrix, and for four antennas a 4×4 matrix. Thus, an $N \times N$ matrix can be created for N antennas and their interdependence determined by their proximity. It must be noted that when a single parameter, s_{nn} or s_{mn} is evaluated, it is assumed that all the other ports, which are not being evaluated at the time, will be terminated with a source impedance Z_0.

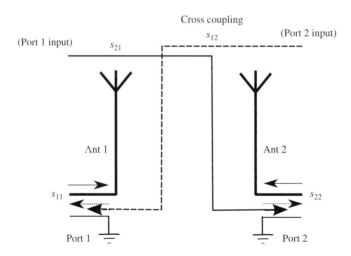

Figure 17.4   Illustration of two port S-parameters for antennas in close proximity.

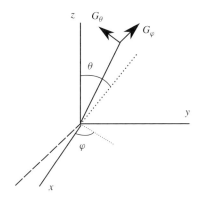

Figure 17.5   Illustration the antenna gain with orthogonal polarization.

17.2.2   *Polarization*

As has been established from Maxwell's equations, an antenna radiating a power density will always be radiating an E-field as well as an H-field. The antenna will radiate a quantity of vertical E-field parallel with the vertical axis in Figure 17.2. It will also radiate another E-field in the direction parallel with an axis orthogonal to the vertical one. In both cases there will be an orthogonal H-field but for purposes of analyzing the polarization of an antenna, only E-field is considered. If these two E-fields are compared, then the resulting gain can be determined using the same theory as presented in previous sections, only there will now be two gain terms used, that is, the gain for the E-field parallel to the vertical axis, G_θ and the gain for the E-field parallel to the orthogonal axis, G_φ. These two gains are illustrated in Figure 17.5. Whatever the value of θ or φ, G_θ is always parallel with the vertical z axis, whilst G_φ is parallel with the corresponding axis in the horizontal plane at angle φ as shown with a thick dotted line in Figure 17.5, which is not parallel with the x-axis or the y-axis in this case. When φ is $90°$, then the horizontal component is parallel with the x-axis, while at $0°$ it is parallel with the y-axis.

The values of G_θ and G_φ will change depending on the angles θ and φ and hence given the following notations, $G_\theta(\theta, \varphi)$ and $G_\varphi(\theta, \varphi)$, to show their dependency on the two angles. An antenna with a high G_θ and a low G_φ for all angles can be considered as a vertically polarized antenna, while if an antenna with high G_φ and low G_θ for all angles can be considered a horizontally polarized antenna. In practice, many antennas do not meet such criteria due to constraints of the device they are being used for. Antennas are often designed with the intention to have a vertical polarization, and the resulting horizontal polarization the antenna generates may be high or low. In this case, an antenna is defined has having a high or low cross polarization. A further term used to define antennas is the polarization purity, where a high polarization purity will mean there is a low cross polarization (i.e. the majority of the E-field is contained in a single polarization).

It is important to make a distinction between the antenna polarization and that of the environment. This is best explained by considering an isotropic receive antenna in an urban environment. If this antenna is vertically polarized (i.e. $G_\theta(\theta, \varphi) = 1$ and $G_\varphi(\theta, \varphi) = 0$), it will receive all vertical E-fields from random directions, all of which will be combined when they arrive at the isotropic antenna to produce a vertically polarized receive power, P_V. Now if this antenna is replaced with a horizontally polarized isotropic antenna (i.e. $G_\theta(\theta, \varphi) = 0$ and $G_\varphi(\theta, \varphi) = 1$), then it will result in a horizontally polarized power being received, P_H. If the ratio of the vertical to horizontal power was taken, then the Cross Polar Ratio (XPR), in dB can be defined thus:

$$\text{XPR(dB)} = 10\log_{10}\left(\frac{P_V}{P_H}\right) \tag{17.8}$$

XPR is difficult to measure in practice and several measurements as well as simulations have been carried out to determine statistically what different XPR values occur in a range of environments. Typically it will be in the range from 0 dB up to 20 dB if it is assumed the transmit source is vertically polarized. The higher values of XPR will tend to occur in rural environments, whilst in urban environments the XPR will be typically less than 10 dB.

17.2.3 Mean Effective Gain

The gain of the antenna for the two polarizations can be measured in an anechoic chamber (where free space conditions are maintained) or simulated using software-based modelling tools. Furthermore the efficiency of the antenna can be computed through taking several gain measurements over a range of angles, both in θ and φ, and knowing the transmit power fed into the antenna. These parameters are useful as an indication of whether the antenna has been built well, though, because a realistic environment does not have free space conditions, they do not give a clear indication as to how the antenna performs in such a case. Therefore the Mean Effective Gain (MEG) is used as an estimate to determine how well the antenna will perform on average in a real environment. This can be used to ultimately determine, on average, by how much the power from the transmitter to the receiver will be attenuated when the path loss of the radio environment is considered.

The simplest definition of MEG is therefore a ratio of how much power, P_{Ant}, the antenna will receive in a radio environment compared to what a dual polarized isotropic antenna would receive in the same environment. This can be defined in dBi as follows:

$$\text{MEG} = 10\log_{10}\left(\frac{P_{Ant}}{P_V + P_H}\right) \tag{17.9}$$

The MEG is difficult to measure in practice for the same reasons as it is difficult to measure XPR. Therefore, MEG is often evaluated using the following equation [4]:

$$\text{MEG} = \int_{-\pi}^{\pi}\int_0^{\pi}\left(\frac{\text{XPR}}{1 + \text{XPR}}G_\theta(\theta, \varphi)p_\theta(\theta, \varphi) + \frac{1}{1 + \text{XPR}}G_\varphi(\theta, \varphi)p_\varphi(\theta, \varphi)\right)\sin\theta d\theta d\varphi \tag{17.10}$$

If the antenna is considered to be in receive mode, the function $p_\theta(\theta, \varphi)$ is a power density function, which determines the proportion of vertically polarized total power that arrives at angles θ and φ. The function must therefore satisfy the following criteria:

$$\int_{-\pi}^{\pi}\int_0^{\pi} p_\theta(\theta, \varphi)\sin\theta d\theta d\varphi = 1 \tag{17.11}$$

Similarly, the function $p_\varphi(\theta, \varphi)$ is also a power density function, which determines the proportion of vertically polarized total power that arrives at angles θ and φ and the function must therefore satisfy the following criteria:

$$\int_{-\pi}^{\pi} \int_{0}^{\pi} p_\varphi(\theta, \varphi)\sin\theta\, d\theta\, d\varphi = 1 \qquad (17.12)$$

The terms $p_\theta(\theta, \varphi)$ and $p_\varphi(\theta, \varphi)$ are known as the Angle of Arrival (AOA). A further point to note is that MEG is also valid when the antenna is in transmit mode (where the AOA would be normally termed the Angle-of-Departure (AOD)), though this is harder to visualize and it is not necessary to evaluate it this way.

The MEG has a high value if there is high gain and in the same polarization in the direction that the majority of the power density is arriving at the antenna. The MEG would be low if the antenna gain has a mainly different polarization to that of the arriving power density, or if the antenna has high gain in the direction where there is little power density. It may be the case that power density is arriving from many angles at the mobile and so the antenna will need to be omni-directional in order to have a high MEG. It is often the case that when a mobile antenna has user interaction, its omni-directionality is reduced and thus its MEG is also reduced. MEG can also vary by as much as 10dBi depending on how the user is positioned and how they are handling the antenna. Therefore when designing antennas it is important to evaluate MEG and to determine by what degree it will vary.

Modelling the AOA in order to compute suitably reliable values of MEG is challenging, but required by the antenna designer to make useful comparisons between different antenna designs. A number of approaches have been considered, including the following:

1. To make a ray trace model [5], whereby a finite number of transmitted rays are modelled to propagate through a given radio environment from which they will reflect off surfaces, diffract off tall buildings and also refract through materials such as glass windows. Each ray has to be modelled individually and as the radio environment becomes more complex, the more complex the model becomes. A further point is that ray tracing also needs to consider the case where a vertically polarized source is transmitted and then it will result in vertical and horizontal polarized E-fields arriving at the receiver due to the XPR of the environment. Each of the rays can be analyzed separately to evaluate the AOA at a fixed point since the rays arrive at different angles.

2. To measure the AOA directly. This can be achieved by using either a single antenna with high directivity, which points in different directions to measure the power density at each angle, or several directive antennas could be constructed in a sphere so as to measure the AOAs simultaneously [6]. The antennas will have a degree of directivity which will limit the resolution to which they can measure the AOA. It is particularly difficult to use this method at lower frequencies where it is challenging to build compact enough antennas with high directivity.

3. Measurements can be taken with multiple antennas at the receiver, which are usually spaced in rows and columns by about half a wavelengh. This enables techniques such as the MUltiple Signal Identification and Classification (MUSIC) [7] or Space-Alternating Generalized Expectation Maximization (SAGE) [8] algorithms to be applied in post processing to extract the AOA information [9].

All three methods have their advantages and disadvantages. Accuracy of the AOA results in all three cases will be compromised either by the level of detail the ray trace model will use, the directivity of the antennas to measure AOA directly, or the limitations of the algorithms in terms of the assumptions behind which they are formed. It is often the case that comparisons are made between the three methods from which propagation channel models have adopted certain AOA models that not only make a good match to the radio environment but also because they are easy to model.

One model that has been used in the theoretical case is that produced by Taga [10], whereby measurements in an urban non line of sight environment have assumed a uniform AOA and a Gaussian angle of arrival in elevation. This has been supported by measurements in several urban locations and provides a simple means by which to implement functions $p_\theta(\theta, \varphi)$ and $p_\varphi(\theta, \varphi)$ for the MEG equation in software. It is also possible for the designer to vary the mean elevation AOA (which is typically 10° to 30° above the horizontal) together with the standard deviation (which varies from 20° to 40°).

Though it simplifies the evaluation of MEG, it is not by any means reasonable to assume there will always be a uniform AOA in any given environment. Several measurements including a mobile in a street canyon where most of the power is propagated down the street, a mobile indoor scenario where most power is arriving through the window, and the case with a line of sight are all examples of scenarios where there is not a uniform angle of arrival in azimuth. Modelling the azimuth AOA mathematically is difficult and a Gaussian model is not an appropriate mathematical approach. In many cases, use of real measurement data is applied to the MEG equation, from which evaluations of the MEG can be made [11]. A further point to note in this instance is that with a non uniform AOA, the azimuth rotation of the handset and user will mean there is a varying MEG, which should also be accounted for.

17.2.4   Channel Requirements for MIMO

As will have been established in Chapter 7 and preceding chapters, a Multiple Input Multiple Output (MIMO) channel (where in the case of Long Term Evolution (LTE) Advanced up to four antennas will be deployed at the mobile terminal) requires a rich scattering environment, which means the received Signal to Noise Ratio (SNR) at each of the four mobile terminal branches at any one time instant will be variable. Due to the nature of the multipath fading in a scattering rich environment, the SNR in each antenna will be constantly varying as the mobile moves, though the probability that two or more antennas fall into a very low SNR (which is known as a deep fade) is very rare in a case where there is high diversity richness. The four antennas therefore have a low correlation. In order to fully deploy diversity, it is also necessary that the average SNR on all four branches is the same. Therefore with regards to antenna design it is important that both a low correlation and a low branch power ratio is maintained, which will be covered in the following two sections.

17.2.5   Branch Power Ratio

The Branch Power Ratio (BPR) is simply the ratio of the average power received by two separate antennas at any one time. This can either be measured directly or it can be evaluated using the MEG of the two antennas. In this case, if antennas 1 and 2 have mean effective gains, MEG_1 and MEG_2 respectively, then the BPR is defined in dB as follows:

$$BPR = 10\log_{10}\left(\frac{\max\left(MEG_1, MEG_2\right)}{\min\left(MEG_1, MEG_2\right)}\right) \tag{17.13}$$

In the case of a mobile terminal with four antennas, it is easily possible to take any two of the antennas and evaluate their BPR. Therefore, in the case of the mobile terminal with four antennas, there are six unique BPR combinations. A high branch power ratio will directly correspond to the ratio of the SNR at the stronger antenna to the SNR at the weaker antenna. If the BPR is as high as 10 dB then the power in the weaker antenna is nearly suppressed and it therefore makes no difference to the diversity richness of the MIMO link, whether the weaker antenna is present or not. Ideally the BPR between all antennas should be 0 dB, though in practice this will not be the case. It is acceptable that the BPR can increase as high as 3 dB before the diversity is degraded [12].

17.2.6 Correlation

The correlation between two antennas is highly influenced by how the antennas are positioned on a mobile terminal as well as how they interact with each other and the user. When the antennas are in close proximity to each other, their field patterns will significantly change and this difference in the antenna patterns and also their polarization in some cases, will change the correlation and this can be analyzed in a similar way to the MEG if the AOA is known. First of all the antenna patterns need to be appropriately defined. If antenna 1 has gain values in dBi for two polarizations, $G_{1\theta}(\theta, \varphi)$ and $G_{1\varphi}(\theta, \varphi)$, then two field patterns can be defined as follows in linear form:

$$A_{1\theta}(\theta, \varphi) = 10^{\frac{G_{1\theta}(\theta,\varphi)}{20}} \tag{17.14}$$

$$A_{1\varphi}(\theta, \varphi) = 10^{\frac{G_{1\varphi}(\theta,\varphi)}{20}} \tag{17.15}$$

and the same can be defined for antenna 2:

$$A_{2\theta}(\theta, \varphi) = 10^{\frac{G_{2\theta}(\theta,\varphi)}{20}} \tag{17.16}$$

$$A_{2\varphi}(\theta, \varphi) = 10^{\frac{G_{2\varphi}(\theta,\varphi)}{20}} \tag{17.17}$$

then the correlation can be derived also based on AOA data for an antenna design as follows [4]:

$$\rho_{12} = \frac{\int_{-\pi}^{\pi}\int_{0}^{\pi}\left(XPR A_{1\theta}(\theta, \varphi)A_{2\theta}^*(\theta, \varphi)p_\theta(\theta, \varphi) + A_{1\varphi}(\theta, \varphi)A_{2\varphi}^*(\theta, \varphi)p_\varphi(\theta, \varphi)\right)\sin\theta d\theta d\varphi}{\sqrt{\sigma_1^2\sigma_2^2}}$$

$$\tag{17.18}$$

where

$$\sigma_n^2 = \int_{-\pi}^{\pi} \int_0^{\pi} \left(\text{XPR} |A_{n\theta}(\theta, \varphi)|^2 p_\theta(\theta, \varphi) + |A_{n\varphi}(\theta, \varphi)|^2 p_\varphi(\theta, \varphi) \right) \sin\theta d\theta d\varphi \quad (17.19)$$

The value ρ_{12} is known as the complex correlation. Another correlation term used is the envelope correlation, ρ_e, which in general can be related to the complex correlation by the relation $\rho_e = |\rho_{12}|^2$, though a more detailed analysis of this can be found in [12]. A complex correlation magnitude below 0.7 is suitable to achieve diversity, while for multiplexing deployed in MIMO as discussed in previous chapters, lower correlations will achieve higher MIMO capacity.

17.2.7 Multiplexing Efficiency

Another recently formed metric known as multiplexing efficiency, η_{mux} defined as follows:

$$\eta_{\text{mux}} = \frac{\text{SNR}_0}{\text{SNR}_\text{T}} \quad (17.20)$$

where SNR_0 is the required SNR to achieve a required capacity using a best case scenario giving fully de-correlated antenna branches. Such as an array antenna with more than half wavelength spacing giving good de-correlation. On the denominator, there is SNR_T, which is the required SNR to achieve the same capacity using a real mobile terminal array antenna. Inevitably the SNR required is going to be higher than that of the best case scenario, which will be influenced by the correlation between the antenna elements but also due to their efficiency, which is likely to be significantly less than that of an ideal array.

17.3   Challenges in Mobile Terminal Antenna Design

Having established the figures of merit used in antenna design, there are a number of challenges relating to mobile terminal antenna design that need to be considered where multiple antennas are implemented onto a small device. The following non exhaustive list summarizes some of the greatest challenges in designing such antennas:

- The antenna requires a matched impedance to the source for all frequency bands and must maintain high efficiency. As the antenna gets smaller compared to a wavelength, which is the case at lower frequencies, then while it may be possible to use techniques to achieve a good impedance match, the efficiency will be compromised due to the small size of the antenna. With smaller size it is also a further challenge to maintain the bandwidth at which the impedance is matched. Due to the limitations, this means that the lowest return loss can fall as low as 6 dB, rather than maintaining the preferred minimum of 10 dB. In such a case, at least 25 percent of the power transmitted or received by the antenna is lost.
- The feeder from the radio needs to be easily accessible. This may be restricted by other items on the mobile device occupying space such as the battery, digital signal processors and the radio transceiver. It is often the case that the antenna design will be

considered after the mobile device has been designed, from which a limited quantity of space will be made available to the antenna.

- Often antennas designed for small mobile terminals require the presence of a suitably large enough ground plane, usually the ground on the printed circuit board that is used. Inevitably the small size of the terminal will restrict this ground plane, which becomes less effective when held by a user's hand. The size of the ground plane can be particularly problematic in offsetting the resonant frequencies where there is a good impedance mismatch. This leads to the need to adapt or "tweak" a specific mobile antenna design so it resonates when in situ.

- The antenna must be designed in such a way that it is suitable for mass production. This means that it needs to be quickly and easily manufacturable, either by printing straight onto the printed circuit board of the mobile device, or by implementing an antenna which can be readily connected to a mobile device in the production line. All these factors influence the cost of the antenna.

- The correlation between two antennas is normally reduced at the base station or access point by means of spatial separation. At the mobile, there may be too little space available for there to be the option of low correlation. It is often the case that the antennas will be able to achieve different patterns through having directivity in differing directions, but also possibly by having separate polarizations.

- If positioned at opposite ends of a mobile terminal handset for example, the antennas will most likely have a large difference between their MEG values, especially if one of the antennas is covered by the user's hand. The mobile handset is particularly restricted in this regard, in that the antennas will have to be positioned into a small area of the small mobile terminal so that the chance of their BPR being high is minimized.

- The close proximity of the antennas, restricted by the need to have a good BPR, is then counteracted by the problem that the antennas coming into close proximity will have a low isolation. As mentioned before, this degrades the efficiency of the antennas when they are in close proximity to each other.

- Finally, many of the above problems will face further difficulty where the interaction of the user is considered. Such problems have had adverse effects on commercially available mobile handsets when handled in certain ways [13]. A mobile terminal may be used in different ways, for example a mobile handset may be used in "talk position" for a telephone conversation, whilst it may be used in "video position" where the user is using the handset to send a text message or use the internet. The different user handling of the mobile terminal in these circumstances will mean several variables, including efficiency, MEG, correlation and isolation will vary considerably, and as such the mobile terminal will require conformance testing in several scenarios.

17.4 Multiple-Antenna Minaturization Techniques

The antenna designs described in this section show solutions that fit into the appropriate space requirements on a mobile terminal whilst minimizing impact from user interaction. Consideration to the cost in terms of implementation is also considered since a number of printed antenna technologies are described. These employ dielectric and ferrite materials to assist in reducing the size to within the required limits. While these techniques have the capability of reducing the size of each individual antenna, they do little to assist in reducing the isolation between two terminal antennas, which will compromise the antennas' efficiency as well as other factors. The use of a neutralization line is also introduced as a useful example of how isolation between antennas can be reduced while still being possible to implement them using the same printed antenna methods, thus maintaining low cost.

17.4.1 Folded Antennas

For any antenna to be as omni-directional as possible, it should consist of a quarter wavelength long wire, vertically orientated and coming out of an infinitely large ground plane. This is known as a quarter wavelength monopole. In practice the ground plane on a mobile terminal is not infinitely sized, it is small and the best solution that can be achieved is to use a visible monopole antenna external to the handset. For the frequency bands used in LTE Advanced, this monopole length ranges from 28mm through to 107mm. Ideally, the antenna should have no user interaction in its proximity and the ground plane would be larger than the mobile terminal itself. In reality, particularly for mobile handset devices, this is not possible given the size of the handset and the closeness to the user. Consequently, the traditional "whip" antenna used in the earliest mobile handsets is now obsolete. In nearly all cases, the antenna is expected to be internal to the modern mobile handset to have acceptable aesthetics.

These factors bring challenges to the design of the antenna on a mobile handset which is therefore confined to an area at the top of the mobile handset, and with a volume in the order of $30 \times 15 \times 8$ mm. This is a very restrictive size, particularly where up to four antennas are possibly required to be implemented. There are two main ways in which the antenna can first of all be reduced in size from the ideal quarter wavelength. The first is to fold the antenna into a meander shape as illustrated in Figure 17.6. When this takes place, the antenna will have a physical height considerably less than the original quarter wavelength. In order to do the meandering technique, it may also be the case that more wire length is required. Although the physical length of the wire may be longer and the maximum physical dimension of the antenna considerably shorter than half a wavelength, the antenna can be modelled by a complex combination of several inductors and capacitors known as "parasitic elements". The combination of these parasitics will cause the antenna to have a defined electrical length, which is equivalent to quarter of a wavelength. The antenna will therefore have achieved a suitable return loss just like the quarter wavelength monopole, though at the same time it will have reduced its efficiency due to the reduced size. The key to a good design in this regard is that the antenna has only a small fraction of its energy lost.

Another means by which the folded antenna can be further reduced is to print the meandered structure onto a dielectric material also illustrated in Figure 17.6. The purpose of a dielectric material when placed between two metallic surfaces is to increase the capacitance per unit length between those two surfaces. Therefore if the capacitance increases, the parasitics in the folded antenna structure increase such that a smaller sized antenna will have an electrical length equivalent to a quarter wavelength. Thus the antenna is further minaturized while still allowing a good return loss. The disadvantage to using dielectric materials will mean that the bandwidth over which there is a suitable return loss is reduced and the efficiency will be compromised again. Thankfully, the dielectric

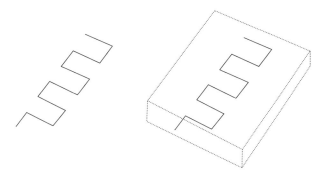

Figure 17.6 Illustration of meandered and dielectric loaded meandered monopole.

materials often used are not lossy at the frequencies of interest so the impact on efficiency is small.

Designing a folded antenna on a dielectric material is complex and often requires repeated trial and error with simulations and then building prototypes. When there are multiple antennas on a small handset, the interaction of the antennas will also influence the suitability of the design of the folded antenna. Another means by which antennas are folded is to insert a short circuit at an appropriate point along the antenna. This causes the antenna to resonate with the ground plane and it will largely assist with radiating electromagnetic fields because a high proportion of current is then present on the ground plane. The radiated electromagnetic fields are directly proportional to the current density on the mobile terminal. Such an antenna is known as an inverted-F antenna or a Planar Inverted-F Antenna (PIFA). The reason for this is because the source will start from the ground plane, then the antenna will normally be positioned parallel to the ground plane for a given length with a shorting line at some point near the source. Thus it looks like an inverted letter F touching the ground plane. An example of an inverted-F antenna can be found in [14], where there are two sets of antennas either side of the handset, though there are in fact three antennas inter-twined within each set, with different electrical length such that they will resonate over three frequency bands. Shorting points are also used and the branch lines are printed onto a low cost dielectric material, Fibreglass Epoxy Resin (FR-4).

17.4.2 Ferrite Antennas

The previous sub-section considered using dielectric materials, which assist in reducing the size through increasing the parasitic capacitance. It is also possible to use ferrite materials to increase the parasitic inductance since these materials have a higher inductance per unit length compared to that of free space. The increase in inductance creates a longer electrical length compared to that of the physical length of the antenna, thus assisting in minaturization. An example antenna with a similar inverted-F structure to that in Figure 17.7 uses a ferrite material between the ground and the top of the antenna as illustrated in Figure 17.7. The ferrite material used in this instance is shown to improve isolation between the two antennas, though its weakness is that there is a magnetic loss

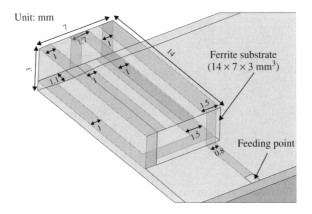

Figure 17.7   Illustration of a ferrite loaded antenna. © 2011 IEEE. Reprinted, with permission, from [15].

within the ferrite, which results in losing over 6 dBi of gain. Another disadvantage of ferrite materials is the cost of inserting them into the antenna structure as an extra substrate.

17.4.3   Neutralization Line

The method illustrated in Figure 17.8 uses the method of the neutralization line. The two PIFAs use a short conducting line between the two antennas, which is strategically positioned to interact with the shorting lines from which the current in one antenna will be prevented from leaking into the other. This is achieved because the neutralization line will have a suitable width to enable it to resonate at the desired frequency through which it is able to increase the isolation within the frequency bands by an order of over 10 dB. Thus the efficiency of the antennas are significantly improved. It is also possible to implement

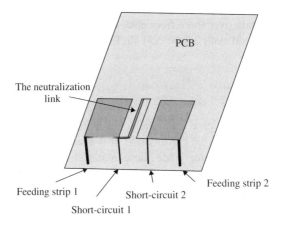

Figure 17.8   Illustration of a neutralization line. © 2011 IEEE. Reprinted, with permission, from [16].

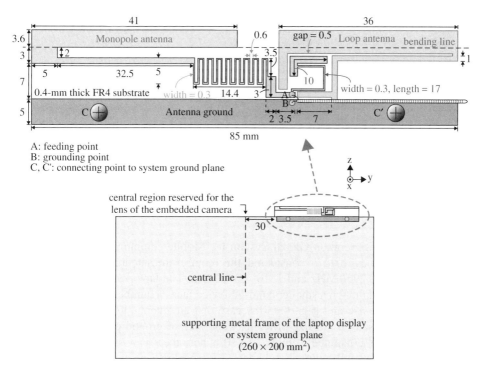

Figure 17.9 Illustration of laptop antenna design. Reproduced with permission of John Wiley & Sons, Inc. from [17].

neutralization lines in many other forms, either with conductors connecting the antennas or by placing the conducting strip in between the antennas but disconnected. In all cases, the neutralization line has simple implementation and maintains low cost of the antenna.

17.4.4 Laptop Antennas

In the case where a laptop computer is used as the mobile terminal, there is some improvement gained because the total size of the terminal is significantly larger and is therefore able to provide a larger ground plane. The implementation of antennas in a laptop at the top of the monitor means they are placed suitably far away from the user and they have space to be positioned further apart from each other, particularly at the higher frequency bands. It is still necessary to apply meandering and folding techniques as illustrated in Figure 17.9. The antenna to the left has a longer electrical length using meandered feed lines and a folded monopole for the lower frequency bands, while the antenna to the right uses folded and parasitic elements to resonate at the upper frequency bands.

17.5 Multiple Antennas with Multiple Bands

A further challenge for the antennas designed for LTE Advanced is their requirement not only to resonate at those frequency bands, but also the mobile device may be required to

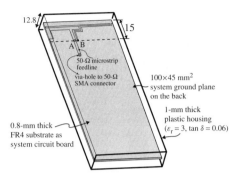

Figure 17.10   Illustration of printed multiband antenna. Reproduced with permission of John Wiley & Sons, Inc. from [18].

operate with second generation Global System for Mobile communication (GSM) bands. If all the required bands are to be considered, this requires the antenna to operate within the two main bands, 698–960 MHz and 1710–2690 MHz. The laptop antenna in Figure 17.9 has such capabilities, though a smaller handset device faces a number of further challenges due to the restricted size and it is often the case that an antenna to suit these bands will have to be tuned based on the size of handset and size of ground plane it is attached to. One such example is illustrated in Figure 17.10, where there is a longer antenna element along the top and a smaller antenna element underneath to resonate at the two bands. The proximity of these two antenna branches means that they will inevitably have cross coupling between them, so while one antenna is in use within its frequency band, the other will act as a parasitic element to that antenna to assist it in resonating within the desired band. A further example of such an antenna is illustrated in Figure 17.11 where

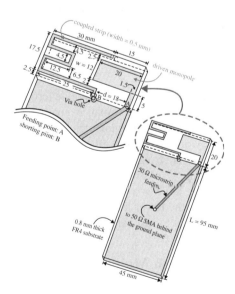

Figure 17.11   Illustration of printed multiband antenna using parasitics. Reproduced with permission of John Wiley & Sons, Inc. from [19].

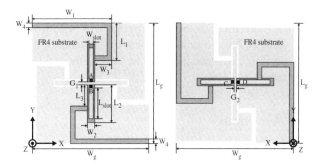

Figure 17.12 Illustration of printed multiband antenna with polarization. Reproduced with permission of John Wiley & Sons, Inc. from [20].

the proximity is not as close though the printed design is still simple and low cost FR-4 substrate is used. A final example is shown in Figure 17.12, where the two antennas, each one multi-band, have opposite polarization. This is a useful method to illustrate how two antennas can improve their isolation by using different polarizations while also reducing the correlation between them. This configuration assumes the radio environment is highly de-polarized and thus has a low XPR so that there are different polarizations available for the antenna to utilize.

17.6 Multiple Users and Antenna Effects

A final consideration relating to mobile terminals is their vulnerability to interference from other mobile users at the same frequency. An experiment to this effect was carried out and detailed in [21] whereby channel measurements of one base station to one mobile were measured simultaneously with a channel from another base station to a second mobile. The other important aspect of this measurement was that the interference from each base station to the other mobile was also measured simultaneously. Both base stations and both mobiles had four antennas each, therefore four separate 4×4 MIMO channel links were being measured simultaneously so that the interference from one mobile to the other could be evaluated. The measurement frequency in this instance was 5GHz, which is not in the LTE frequency bands though the principles concluded from this work would still apply.

The following conclusions can be drawn from the measurements:

- The BPR on a single terminal will vary according to how it is handled. This in turn will alter the Signal to Interference Ratio (SIR) of the mobile terminal by as much as 10 dB. The better the branch power ratio, the better the improvement on the signal to interference ratio.
- The BPR on the interfering terminal will also influence the interference it has on the other mobile terminal. A mobile terminal with a better BPR will have a better SIR.
- If the branch power ratios on both mobile terminals are similar, it provides more degrees of freedom to allow interference between the two links to be mitigated because more of the multiple antennas on the mobile terminal can be utilized.

Therefore the design of the antennas on a single mobile terminal are not only crucial to the user in terms of the radio link the mobile terminal will need to maintain with a base station, but also that where there are multiple users in the same or an adjacent frequency band, the antenna design on each terminal will be crucial to enable the mobile network to manage multiple users.

17.7 Small Cell Antennas

The majority of this chapter has focused on the design of compact antennas at the mobile terminal, where there are size constraints that bring challenges to the antenna design. At the base station it has been assumed that there are no such constraints and that there are means to build suitably large and efficient antennas to communicate with the mobile terminals. This may be true for macrocell and microcell base stations, though there is an increasing move towards small cells whereby the mobile network can meet its capacity demands from users. This is achieved by separating them into smaller cellular areas, particularly in busy locations such as railway stations and airports where the demand for mobile network usage is high at peak times. In particular, the growth of the femtocell [22], which is being deployed not only in the home but in public areas where there is a high density of mobile users. The antenna design is also critical in these cases whereby the femtocell will have to be small in size, low profile (i.e. it is not easily noticeable when deployed) and capable of working efficiently when deployed in an ad-hoc fashion, for example near to a shop sign that may block the radio propagation. Such antennas are required to be compact, suitable for propagating energy in different directions to users and also capable of transmitting and receiving two polarizations because a mobile user will be in close proximity and in a line of sight with a random polarization.

17.8 Summary

This chapter has covered the key figures of merit for antennas used on small devices for communication systems that will be suited to the LTE Advanced standard. In particular it has shown why antennas must maintain a good return loss and high efficiency as they are reduced in size, while at the same time they must maintain good isolation with other antennas or users they are in close proximity with. The antennas are also required to be low cost and positioned within a confined area of a mobile terminal such that the impact of users on the branch power ratio is minimized and thus the mobile's own figures of merit are maintained and the network is able to better control multiple user interference from neighboring cells, which will improve the network's quality of service. The space constraints and the requirements to operate in multiple bands mean that efficiency and return loss are compromised in some cases, which requires more effort by the mobile network to either transmit higher power from a large base station or create smaller cellular areas that will give better coverage to weak mobile terminals.

References

1. A. Ghosh, R. Ratasuk, and B. Mondal, "Lte-advanced: next-generation wireless broadband technology," *IEEE Wireless Communications*, vol. 17, no. 3, pp. 10–22, June 2010.

2. S. Sesia, I. Toufik, and M. B. et al., "Lte-the umts long term evolution: From theory to practice," Wiley, February 2009.

3. J. D. Kraus and D. Fleisch, "Electromagnetics with applications," McGraw Hill, 1999.

4. R. G. Vaughan and J. B. Andersen, "Channels, propagation and antennas for mobile communications," *IEE Press*, 2002.

5. G. E. Corazza, V. Degli-Esposti, and M. Frullone, "A characterization of indoor space and frequency diversity by ray-tracing modeling," *IEEE Journal on Selected Areas in Communications*, vol. 14, no. 14, pp. 411–419, August 1996.

6. K. Kalliola, H. Laitinen, K. Sulonen, L. Vuokko, and P. Vainikainen, "Directional radio channel measurements at mobile station in different radio environments at 2.15ghz," *Proceedings of the 4th European Personal Mobile Communications Conference*, February 2001.

7. Y. L. C. de Jong and M. H. A. J. Herben, "High-resolution angle-of-arrival measurement of the mobile radio channel," *IEEE Transactions on Antennas and Propagation*, vol. 47, no. 11, pp. 1677–1686, November 1999.

8. B. H. Fleury, M. Tschudin, R. Heddergott, D. Dahlhaus, and K. I. Pedersen, "Channel parameter estimation in mobile radio environments using the sage algorithm," *IEEE Journal on This paper appears in: Selected Areas in Communications*, vol. 17, no. 3, pp. 434–450, August 1999.

9. B. Allen and M. Ghavami, "Adaptive array systems: Fundamentals and applications," Wiley, 2005.

10. T. Taga, "Analysis for mean effective gain of mobile antennas in land mobile radio environments," *IEEE Transactions on Vehicular Technology*, vol. 39, no. 2, pp. 117–131, August 1990.

11. M. B. Knudsen and G. F. Pedersen, "Spherical outdoor to indoor power spectrum model at the mobile terminal," *IEEE Journal on Selected Areas in Communications*, vol. 20, no. 6, pp. 1156–1169, November 2002.

12. T. W. C. Brown, S. R. Saunders, and B. G. Evans, "Analysis of mobile terminal diversity antennas," *IEE Proceedings, Microwaves, Antennas and Propagation*, vol. 152, pp. 1–6, February 2005.

13. T. W. C. Brown and U. M. Ekpe, "When is clarke's approximation valid?" *IEEE Antennas and Propagation Magazine*, vol. 52, no. 3, pp. 171–181, June 2010.

14. S. Sande and E. S. et al., "Taking your iphone 4 to the max," Springer, 2010.

15. B.-G. Shin, M.-J. Park, Y.-S. Chung, B. Kim, H. Wi, B. Lee, C. W. Jung, W. Hong, and J. Lee, "Diversity and mimo antenna for multi-band mobile handset applications," *IEEE International Symposium on Antennas and Propagation*, pp. 419–421, July 2011.

16. J. Lee, Y.-K. Hong, S. Bae, G. Abo, W.-M. Seong, and G.-H. Kim, "Minature long-term evolution(lte) mimo ferrite antenna," *IEEE International Symposium on Antennas and Propagation*, pp. 603–608, July 2011.

17. A. Chebihi, C. Luxey, A. Diallo, P. L. Thuc, and R. Staraj, "A novel isolation technique for closely spaced pifas for umts mobile phones," *IEEE Antennas and Wireless Propagation Letters*, vol. 7, pp. 665–668, July 2008.

18. T.-W. Kang and K.-L. Wong, "Internal printed loop/monopole combo antenna for lte/gsm/umts operation in the laptop computer," *Wiley Microwave and Optical Technology Letters*, vol. 52, no. 7, p. 1673–1678, April 2010.

19. K.-L. Wong, M.-F. Tu, T.-Y. Wu, and W.-Y. Li, "Small-size coupled-fed printed pifa for internal eight-band lte/gsm/umts mobile phone antenna," *Wiley Microwave and Optical Technology Letters*, vol. 52, no. 9, p. 2123–2128, September 2010.

20. W.-Y. Li and K.-L. Wong, "Internal printed loop-type mobile phone antenna for penta-band operation," *Wiley Microwave and Optical Technology Letters*, vol. 49, no. 10, p. 2595–2599, October 2010.

21. M. Han and J. Choi, "Multiband mimo antenna using orthogonally polarized dipole elements for mobile communications," *Wiley Microwave and Optical Technology Letters*, vol. 53, no. 9, p. 2043–2048, September 2011.

22. T. W. C. Brown, P. C. F. Eggers, and G. F. Pedersen, "Analysis of user impact on interference between two 4 × 4 mimo links at 5ghz," *COST2100 Workshop: Multiple Antenna Systems on Small Terminals (Small and Smart)*, May 2009.

23. S. Saunders, S. Carlaw, A. Guilstina, R. R. Bhat, V. S. Rao, and R. Siegberg, "Femtocells: Opportunities and challenges for business and technology," Wiley, 2009.

18

Statistical Characterization of Antennas in BANs

Carla Oliveira, Michal Mackowiak and Luis M. Correia

IST/IT - Technical University of Lisbon, Portugal

18.1 Motivation

Body Area Networks (BANs) are at the heart of the next generation of wireless and mobile systems, linking personalization and convergence, through a network of sensors, either wearable or implanted into the human body. The integration of these sensors into compact devices, together with wireless gateways, ubiquitous communications and penetration of wireless technologies, brings enormous potential mobility solutions [1]. BANs have a plentiful range of potential applications, like healthcare and patient monitoring, sports monitoring, security/military/space usage, and business and multimedia entertainment, among others. The range of BAN applications emphasizes its interdisciplinary feature.

In healthcare, BANs can be applied to typical monitoring of vital parameters purposes, providing real time readings (i.e. electrocardiogram, electroencephalography, respiratory rate, and temperature of body), which not only make the monitoring more comfortable for the patient, but also save medical personnel's time. BANs may allow the detection of early signs of disease, and the monitoring of transient or infrequent events. Patient and elderly people monitoring in home environments is another attractive area.

In sports, BANs may be used to monitor fitness-related activities, including several sensors for measuring different physiological parameters, like heart rate, energy consumption, fat percentage (bio-resistance meter), body water content, or galvanic skin response. These sensors can measure and display on-time information and/or follow-up reports to a control entity (e.g. professional well-being and caring personnel). Sport applications demand for high capacity systems to deliver real time information.

Security and military applications comprise smart suits for fire fighters, soldiers and support personnel in battlefields. Smart clothes use special sensors to detect bullet wounds or to monitor the body's vital signals during combat conditions [2]. Space applications

LTE-Advanced and Next Generation Wireless Networks: Channel Modelling and Propagation, First Edition.
Edited by Guillaume de la Roche, Andrés Alayón Glazunov and Ben Allen.
© 2013 John Wiley & Sons, Ltd. Published 2013 by John Wiley & Sons, Ltd.

include biosensors for monitoring the physiological parameters of astronauts during space flights (e.g. ECG and temperature biotelemeters, and sensor pills), in order to understand the impact of space flights on living systems [3].

In the business environment, BAN devices can be used in numerous ways, such as touch-based authentication services using the human body as a transmission channel (e.g. data deliver on handshake). Several applications are possible, like electronic payment, e-business card, auto-lock, or login systems. User identification/authentication, associated with biometrics, plays a key role in here.

In entertainment, wireless applications are multimedia oriented, examples being personal video or wearable audio.

The technology segment of BANs is still at an early stage, with different standards actually competing to fulfil the requirements of potential applications, such as the IEEE 802.11 [4] or 802.15 families (e.g. Bluetooth [5], UWB [6], Zigbee [7]). Although some challenges do rely upon general wireless sensor network technologies, the peculiar characteristics of BANs demand for specific solutions to limit both radiated and consumed power, electromagnetic fields absorbed by the user, or the susceptibility of interferences [8]. In the meantime, a specific standard for BANs is being developed, that is, the IEEE 802.15.6 [9].

A key challenge in BANs is to account for the influence of the human body on the radio channel, the former being an integral part of the latter, and their separation being unfeasible at the moment. Moreover, a careful characterization of the channel is required, since on-body links exhibit great dissimilarities with the traditional wireless communication systems. The differences are tightly related to the performance of an antenna near body tissues, given an extensive range of conditions: the influence of human tissues, the dynamics of the human body, the short distances of propagation, the link geometry variability, as well as the arbitrary orientation of antennas, among others. Because of body movements (even when standing or sitting), a wearable antenna is constantly changing its direction of maximum radiation, which leads to significant changes in the radio link performance. Those variations are especially severe when antennas are mounted on the upper limbs (arms and hands), and become extreme while playing sports or similar activities. This on-body "random" channel suggests a statistical approach to the problem, which has not been considered by former studies. Moreover, in off-body radio channels, the body has no significant influence on the propagation, but rather just on the antenna itself. The importance of a statistical model for antennas near interfering objects, in analogy and in combination with statistical channel modelling, as a means to achieve a full channel model including antennas, is introduced in [10].

This statistical methodology is adopted throughout this chapter, the antenna radiation pattern no longer being considered to be deterministic, but, rather, a random component of the link budget, when developing radio channel models for BANs. Such a statistical result can be easily included in standards and radio channel simulators.

18.2 Scenarios

The biological system of a body is an irregularly shaped dielectric medium, and the amount of scattered energy depends on the electromagnetic properties of the body and its geometry, as well as on the frequency and polarization of the incident wave. Three frequencies are taken as examples for possible bands in BANs: 0.915 GHz in the cellular range, and 2.45 and 5.8 GHz in the Industrial, Scientific and Medical (ISM) one. The electric parameters (i.e. conductivity and relative permittivity) for the body tissues are obtained from the 4-Cole-Cole model [11].

In simulations, the body is usually defined as a geometrical shape with homogeneous filling. Different parts of the body can be modelled, with dimensions based on an average person, as given in [12], and cylindrical or elliptical shapes can be considered.

There are numerous techniques for modelling the interaction of electromagnetic fields with physical objects and the environment, for example, Ray-Tracing [13], Uniform Geometrical Theory of Diffraction (UTD) [14], and full wave methods [15]. The last one provides the most accurate results of antenna's parameters, such as input impedance, radiation pattern, gain, and coupling between elements. One technique that uses a full-wave model is the Finite Integration Technique (FIT), which is applied in Computer Simulation Technology (CST) Microwave Studio, [16].

CST can use realistic heterogeneous body models, consisting of large datasets obtained from Magnetic Resonance Imaging, Computer Tomography, and anatomical images. An example of heterogeneous body models is given in [17], known as the Virtual Family, which comprises voxel models of a female and male adult and two children. A practical segmentation of a body can be introduced, together with orientation, as shown in Figure 18.1.

In order to identify the location of the antenna on the body, a label related to a particular body segment and orientation is used. Also, in order to measure shifts of the antenna orientation, a reference coordinate system is taken, in which the reference direction is the front of the body, and azimuth φ and elevation ψ (or co-elevation θ) angles are calculated as presented in Figure 18.1, taking the usual spherical coordinate system. A BAN topology composed of wireless nodes distributed over the body, in locations that can correspond to a realistic usage scenario, is also shown in Figure 18.1.

A patch antenna is used in the 2.45 GHz band [18]. This antenna can be easily integrated in clothes or equipment, due to its small dimensions and flat configuration. When modelling the isolated patch antenna in CST, the following parameters have been

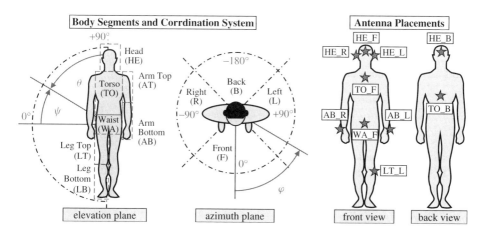

Figure 18.1 Body coordinate system and possible antenna placement.

calculated: gain, $G_a = 6.58\,\mathrm{dBi}$; total radiation efficiency, $\eta_a = 80.2\,\%$; resonance frequency, $f_r = 2.44\,\mathrm{GHz}$; input return loss, $s_{11} = -12.35\,\mathrm{dB}$.

18.3 Concepts

In a practical BAN scenario, the distance of the antenna to the body, h, can vary randomly in time, a Probability Density Function (PDF), $p(h)$, being used to describe its different realizations. In this case, the average value $\overline{E(\theta, \varphi)}$ for the electric field magnitude in the far-field region, $E(\theta, \varphi, h)$, and its standard deviation $\sigma_E(\theta, \varphi)$ can be obtained from:

$$\overline{E(\theta, \varphi)} = \int_0^\infty |E(\theta, \varphi, h)|\, p(h)dh \tag{18.1}$$

$$\sigma_E(\theta, \varphi) = \int_0^\infty \sqrt{\left(|E(\theta, \varphi, h)| - \overline{E(\theta, \varphi)}\right)^2}\, p(h)dh \tag{18.2}$$

The statistical parameters of electric field variations can be used to model the statistical radiation pattern of an antenna:

$$E(\theta, \varphi, h) = \overline{E(\theta, \varphi)} + u(p(h)) \cdot \sigma_E(\theta, \varphi) \tag{18.3}$$

where $u(p(h))$ is a coefficient depending on the distance probability $p(h)$.

In what follows, the electric field is normalized relative to the value of 1 V/m obtained from an isotropic radiator, in the far-field region.

By performing simulations or measurements of the antenna located at different distances from the body h_k ($k = 1, \ldots, N_h$), one obtains a set of radiation patterns $G(\theta_n, \varphi_m, h_k)$ (θ_n and φ_m being discrete values of the spherical coordinates). The distance sample h_k depends on the probability $p(h)$, and on the total number of distance samples N_h. Hence, from the set of radiation patterns, it is possible to calculate the average:

$$G_\mu(\theta_n, \varphi_m) = \frac{1}{N_h} \sum_{k=1}^{N_h} G(\theta_n, \varphi_m, h_k) \tag{18.4}$$

as well as the standard deviation:

$$\sigma_G(\theta_n, \varphi_m) = \sqrt{\frac{1}{N_h} \sum_{k=1}^{N_h} \left[G(\theta_n, \varphi_m, h_k) - G_\mu(\theta_n, \varphi_m)\right]^2} \tag{18.5}$$

To calculate the radiation pattern on a moving body, first, one has to obtain $G(\theta, \varphi)$, which is the average pattern for the desired placement on the static body. Then, a set of patterns is derived by introducing the shifts caused by the body movements. Using the dynamic model described in [19], the patterns can be generated by

$$G_k(\theta, \varphi) = G(\theta + \delta\theta_k, \varphi + \delta\varphi_k) \tag{18.6}$$

where $\delta\theta_k$ and $\delta\varphi_k$ are shifts of the normal to the body surface, in the elevation and azimuth planes, respectively. The index k denotes the current time frame of the body movement, $k = 1, \ldots, N_T$.

This modelling approach still does not reproduce all the phenomena that are caused by the presence of the moving body. For example, when the movement of the antenna is analyzed (e.g. located on the chest), some radiation directions can be additionally obstructed by the presence of other body segments (e.g. hands). It is worthwhile mentioning that obtaining realistic dynamic human models (e.g. voxel models) required in full-wave modelling approaches is, in general, very difficult, therefore, our geometrical approach is an excellent alternative to full-wave ones.

From the set of previously calculated radiation patterns for all time frames N_T of body movement, it is possible to calculate the average radiation pattern:

$$G_\mu(\theta, \varphi) = \frac{1}{N_T} \sum_{k=1}^{N_T} G_k(\theta, \varphi) \tag{18.7}$$

and the standard deviation:

$$\sigma_G(\theta, \varphi) = \sqrt{\frac{1}{N_T} \sum_{k=1}^{N_T} \left[G_k(\theta, \varphi) - G_\mu(\theta, \varphi) \right]^2} \tag{18.8}$$

In order to evaluate the difference between an arbitrary radiation pattern, G, and a reference one, G_{ref}, (e.g. the radiation pattern of the isolated antenna, that is, in free-space), the relative difference, taken for a particular direction, is calculated:

$$\Delta_{G_{nm}} = \frac{\left| G(\theta_n, \varphi_m) - G_{ref}(\theta_n, \varphi_m) \right|}{G_{ref}(\theta_n, \varphi_m)} \tag{18.9}$$

From (18.9), two parameters are defined, that is, the Pattern Average Difference (PAD):

$$\overline{\Delta_G} = \frac{1}{N_\varphi N_\theta} \sum_{n=1}^{N_\varphi} \sum_{m=1}^{N_\theta} \Delta_{G_{nm}} \tag{18.10}$$

and the Pattern Weighted Difference (PWD):

$$\overline{\Delta_{G_w}} = \frac{\displaystyle\sum_{n=1}^{N_\varphi} \sum_{m=1}^{N_\theta} \Delta_{G_{nm}} G_{ref}(\theta_n, \varphi_m)}{\displaystyle\sum_{n=1}^{N_\varphi} \sum_{m=1}^{N_\theta} G_{ref}(\theta_n, \varphi_m)} \tag{18.11}$$

where:

- N_θ: is number of discrete co-elevation angle samples,
- N_φ: is number of discrete azimuth angle samples.

The difference between these two parameters is that $\overline{\Delta_G}$ treats changes of the radiation pattern in all directions equally, whereas $\overline{\Delta_{G_w}}$ weights these changes by the radiation pattern of the reference antenna.

Some statistical distributions that best fit the selected data are used in this chapter, namely the Exponential [20], the Beta [21], the Kumaraswamy [22] and the Truncated Normal [23] Distributions. The goodness of fit is tested using the χ^2 or the R^2 tests [20]. For the χ^2 test, a value lower than the threshold for a 95 percent confidence interval ($\chi^2_{th95\%}$) means that the fitting is successful, otherwise it is not appropriate. Regarding the R^2, the optimum fitting is obtained when it is equal to 1.

The PDF of the Beta Distribution [21], for a random variable x distributed in the interval [0,1], is:

$$p_B(x; \alpha_B, \beta_B) = \frac{x^{\alpha_B - 1}(1 - x)^{\beta_B - 1}}{B_{\alpha_B, \beta_B}} \tag{18.12}$$

while the PDF of the Kumaraswamy Distribution [22] is:

$$p_K(x; \alpha_K, \beta_K) = \frac{\alpha_K \beta_K x^{\alpha_K - 1}(1 - x^{\alpha_K})^{\beta_K - 1}}{K_{\alpha_K, \beta_K}} \tag{18.13}$$

where:

- α_B, β_B, α_K and β_K are shape parameters of the Beta and Kumaraswamy Distributions, respectively,
- B_{α_B, β_B} and K_{α_K, β_K} are normalization constants for the Beta and Kumaraswamy Distributions, respectively, to ensure that the total probability integrates to unity.

The Truncated Normal Distribution corresponds to a normally distributed random variable, bounded in an interval [a,b], with mean μ and variance σ^2, being given by:

$$p_{TN}(x; \mu, \sigma, a, b) = \frac{\frac{1}{\sigma}\phi\left(\frac{x-\mu}{\sigma}\right)}{\Phi\left(\frac{b-\mu}{\sigma}\right) - \Phi\left(\frac{a-\mu}{\sigma}\right)} \tag{18.14}$$

where:

- $\phi()$ is the PDF of the standard Normal Distribution,
- $\Phi()$ is the Cumulative Distribution Function (CDF) of the standard Normal Distribution,
- The mean μ_{TN} and variance σ^2_{TN} of the Truncated Normal Distribution are given in [23].

18.4 Body Coupling: Theoretical Models

A first general understanding of the behaviour of a simple radiation source near the body can be obtained through very simple propagation models, considering different polarization or source orientations.

18.4.1 Elementary Source Over a Circular Cylinder

Considering the communication scenario between two on-body sensors, for example, the transmitting antenna is placed on the arm, while the receiving one is on the torso, the

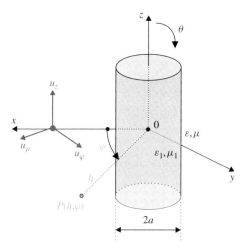

Figure 18.2   Geometry for an elementary source near a dielectric circular cylinder.

radiated signal may be approximated by the wave generated by an elementary electric source (i.e. a Hertz dipole) located near a dielectric circular cylinder. This problem requires a non-trivial solution involving extensive integral computations, however, a much simpler method to obtain the radiation pattern in such conditions has been proposed in [24], and can be used for this scenario. The author uses the principle of reciprocity in conjunction with the scattering of a plane wave from a dielectric circular cylinder at oblique incidence to determine the radiation pattern for the elementary electric source for three orthogonal directions u_ρ, u_φ and u_z in a cylindrical coordinate system, Figure 18.2. The solutions for the different orientations of the elementary source are given in [24].

For BAN communications, one can consider a further simplification, by confining the problem to the horizontal plane ($\theta = 90°$). One assumes that the transmitter is located in the $x - y$ plane, positioned at P(h, 0, 0), external to the cylinder ($h > a$). The observation point is positioned at P(h, φ, 0).

Table 18.1 summarizes the considered characteristics for the body, modelled as a homogeneous cylinder. In order to extract the statistics of the user's influence on the elementary source performance, the source was placed at different distances to the body surface according to Table 18.1. These distances range from the antenna attached to the skin to

Table 18.1   Distances used in the elementary source model and properties of the body

	Distance to the body			Body properties	
	h [cm]	h/λ		a [cm]	Tissue
		0.915 GHz	2.45 GHz		
Torso	[0, 20]	[0, 0.60]	[0, 1.63]	16	Fat
Leg	[0, 10]	[0, 0.30]	[0, 0.82]	11	Muscle
Arm	[0, 5]	[0, 0.15]	[0, 0.41]	5	Muscle

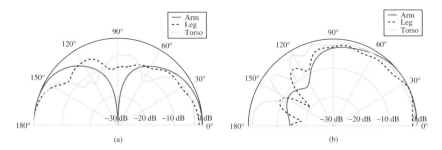

Figure 18.3 E-field patterns for different source orientations (2.45 GHz). © 2010 IEEE. Reprinted, with permission, from [25].

it being attached to the clothes, with the maximum distance being constrained by the flexibility of the clothes (e.g. if the antenna is a button in a jacket). Within each distance interval, equidistant points are considered, separated 0.05 cm from each other.

Two different sets of radiation patterns were obtained, at different distances from the body, one for each frequency, 0.915 and 2.45 GHz, on the $x - y$ plane of Figure 18.2. An overview of the main outcome and different statistics are given in what follows. The analysis focuses on the comparison of patterns for different orientations of the source, and at different body-source distances, modelling of field distributions, and average field fluctuations.

Figure 18.3 shows the average radiation patterns corresponding to the orientations u_φ and u_z, obtained at 2.45 GHz. Each pattern is normalized by its maximum. Note that only half of the radiation patterns is displayed, as they are symmetric with respect to the x axis.

Results are not shown for the u_ρ polarization, as it is observed that the cylinder has a minor impact on the radiation pattern (the source does not radiate on its own direction). For the u_φ and u_z polarizations, the influence of the body is clearly observed, particularly with an increase of directivity in some directions, and with the "absorption" of backward radiation by the body. This influence evidently depends on body dimensions, tissue's composition, distance to the source, and frequency. For the u_φ orientation, an increase of directivity is shown, while for the u_z one, the predominant effect is the attenuation of backward radiation. This phenomenon is particularly strong at 2.45 GHz.

Figure 18.4 shows the radiation patterns obtained for the u_z orientations, for different values of h, for the leg scenario. One can observe the presence of nulls in the backward direction, resulting from reflections on the cylinder. These reflections can be constructive or destructive, depending on the distance of the source to the body.

Table 18.2 shows the average and the maximum differences between the field obtained at distance h and the reference distance $h_0 = 0.05$ cm. The averaging is performed for all the considered angles.

For the u_ρ orientation, no major differences in the shape and gain of the patterns between 0.915 and 2.45 GHz are found. However, some fluctuations in the patterns are found at 2.45 GHz, probably resulting from numerical inaccuracy. For the u_φ orientation, at 0.915 GHz, the forward lobe presents a lower gain; at 2.45 GHz, a more directive pattern is observed. For the u_z orientation, higher gains are found for increased distances to the body. On average, larger variations with distance are found for 2.45 GHz.

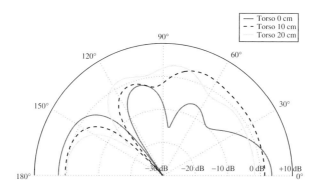

Figure 18.4 E-field u_z patterns for torso for different h (0.915 GHz). © 2010 IEEE. Reprinted, with permission, from [25].

Table 18.2 Statistical overview for different h for the leg: u_z orientation

h	0.915 GHz		2.45 GHz	
[cm]	$\overline{\Delta E^\rho}$ [dB]	$\overline{\Delta E^\rho_{max}}$ [dB]	$\overline{\Delta E^\rho}$ [dB]	$\overline{\Delta E^\rho_{max}}$ [dB]
1	2.5	7.0	5.0	20.3
2	7.2	12.6	7.3	24.8
5	11.7	20.0	8.8	25.7
7	12.8	22.0	8.4	25.8
10	13.8	22.9	9.3	26.9

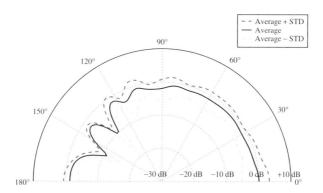

Figure 18.5 Statistical study for torso: E-field u_z patterns (2.45 GHz). © 2010 IEEE. Reprinted, with permission, from [25].

A comprehensive statistical study on the variations of radiation patterns was carried out for all scenarios described in Table 18.1. The average patterns and corresponding standard deviations (STD) are one of the main outcomes of this study. Figure 18.5 shows the results for a vertically polarized source near the torso, for distances up to 20 cm. Note that those values have been calculated in linear units and then expressed in logarithmic scale. Moreover, for each scenario, the radiation patterns are normalized to their maximum.

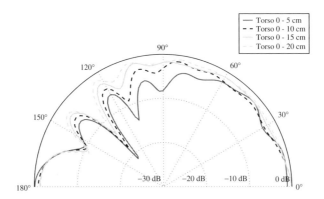

Figure 18.6 Average E-field u_z patterns for torso (2.45 GHz.).

The average patterns for the u_z orientation show a clear suppression of backward radiation, especially at 2.45 GHz, suggesting absorption by the body tissues. The fluctuations of the average radiation pattern depend on the angle of observation (φ). The maximum fluctuation is obtained for $\varphi = 180°$, that is, the backward direction.

For the leg and the torso, the large observation distances were divided in smaller intervals, the average pattern being then computed for each new distance interval. Figure 18.6 displays the obtained results. Larger variations occur at shorter distances, for example, one gets a 25 dB variation at 5 cm and 14 dB at 20 cm. A general trend is that the shape of the radiation patterns remains almost constant for the different average intervals; the more apparent modification is observed in gain values. Similar conclusions were gathered for 0.915 and 2.45 GHz, with small fluctuations at the higher frequency.

As a complement to the statistics given above, the PDFs of the envelope of the obtained E-field were studied. A simple visual inspection shows an exponential decay in most of the histograms, therefore, an exponential fitting function with mean $\overline{E^E}$ was plotted together with the results from the test. Figure 18.7 shows the PDFs for given values of φ.

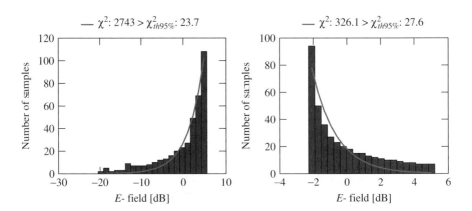

Figure 18.7 E-field patterns for different source orientations (2.45 GHz).

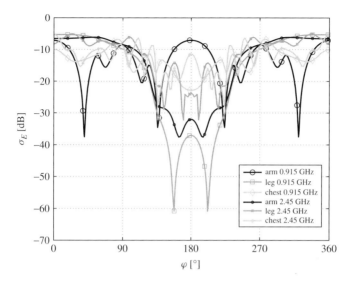

Figure 18.8 Fluctuations of σ_E: u_z orientation. © 2010 IEEE. Reprinted, with permission, from [25].

Large fluctuations of the field are found, for example, 11.6 dB and 2.3 dB at $\varphi = 0°$ and $\varphi = 180°$, respectively. In most of the cases, large fluctuations are found at $\varphi = 0°$, and small dispersion of values at $\varphi = 180°$ for the cases where there is suppression of backward radiation. From the PDFs analysis, it is concluded that the fitting functions are appropriated in the directions of strong lobes, with small fluctuations from the average (small values of standard deviation). But, whenever there are secondary lobes, or a higher dispersion of values, fitting is inefficient. Additionally, high values of $\overline{E^E}$ are found when there is no agreement with the fitting function.

Another outcome of this statistical analysis is the fluctuation of the standard deviation, depending on the angle of observation; an example is presented in Figure 18.8, for the u_z orientation. It is observed that σ_E depends on the angle of observation. The general trend is that σ_E decreases from $\varphi = 0°$ to $\varphi = 180°$, except for the 0.9 GHz arm scenario; this case experiences the highest variation of σ_E, with a maximum of -7 dB at $\varphi = 180°$. The torso is the scenario less affected by the presence of the cylinder. The variations of σ_E do not show a clear correlation of the results with the frequency, although higher differences are experienced at 2.45 GHz.

18.4.2 Elementary Source Over an Elliptical Cylinder

Another simple geometrical approach can be considered, where the body is modelled as an elliptical shaped medium, whose dimensions (i.e. a_m minor and a_M major axes) depend on the corresponding part of the body, Figure 18.9. The source is again an elementary dipole, parallel to the cylinder axis, hence, providing an omnidirectional pattern in the transversal plane to the cylinder.

In this case, the angle of incidence φ_i is different from the radiation angle φ and depends on the distance between the antenna and the body. The angle φ_i for given φ, h, a_m and

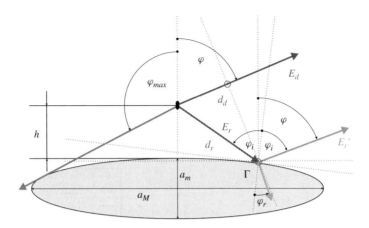

Figure 18.9 Simple elliptical model for the estimation of the radiation pattern of an antenna located in the vicinity of the body. © 2010 IEEE. Reprinted, with permission, from [26].

Table 18.3 Scenarios for an isolated antenna over an ellipse

Scenarios	Freq. [GHz]	Tissue	Dimensions (a_M, a_m) [cm]
Reference	2.45	2/3 Muscle	Torso (40, 10)
Other	0.915	Fat	Head (15, 15)
	5.8	Muscle	Arm (5, 5)
		Wet Skin	

a_M, can be found from geometrical considerations. For φ larger than the shadowing angle $\varphi_{\max}$, the antenna does not radiate directly in this direction, and the electric field is zero. Otherwise, the total field in the far-field is the result of the interference between the direct and reflected components, E_d and E'_r, respectively.

Calculations were performed for the scenarios described in Table 18.3. The scenario defined for 2.4 GHz was chosen as reference. Two distributions of the distance of the antenna to the body were considered: Uniform and "locally" Rayleigh. The Uniform Distribution analyzes the possible range of changes of the radiation pattern within the distance interval. The Rayleigh Distribution describes a realistic scenario in BANs, where the distribution parameters are defined by the features of the antenna location on the clothe (i.e. distance ranges between 0 and a large value, being concentrated around a given central value). The values corresponding to the Uniform Distribution were taken from the interval [0, 10] cm, while for Rayleigh Distribution one assumed an average equal to 2 cm.

Table 18.4 shows the average and the standard deviation of the E-field, which are denoted $\overline{E_U}$ and σ_{E_U} for the Uniform Distribution and $\overline{E_R}$ and σ_{E_R} for the Rayleigh Distribution, respectively. The minimum and maximum values are also given for both

Table 18.4 Statistical parameters for an elementary source over an ellipse for the reference scenario

$|E|, \sigma_{|E|}$[V/m]		φ						
		0°	15°	30°	45°	60°	75°	90°
pol.	$\overline{|E_U|}$	1.24	1.21	1.12	1.03	1.10	1.12	0.99
$\parallel$	$\sigma_{|E_U|}$	0.44	0.45	0.44	0.44	0.39	0.33	0.19
	$\overline{|E_R|}$	1.44	1.40	1.32	1.19	0.99	0.80	0.87
	$\sigma_{|E_R|}$	0.30	0.31	0.33	0.36	0.33	0.17	0.13
	$|E_{min}|$	0.28	0.40	0.30	0.36	0.42	0.65	0.77
	$|E_{max}|$	1.72	1.71	1.68	1.65	1.57	1.51	1.57
pol.	$\overline{|E_U|}$	1.24	1.21	1.15	1.08	1.19	1.25	0.92
$\perp$	$\sigma_{|E_U|}$	0.44	0.46	0.47	0.51	0.52	0.58	0.57
	$\overline{|E_R|}$	1.44	1.41	1.36	1.27	1.09	0.76	0.68
	$\sigma_{|E_R|}$	0.30	0.31	0.35	0.41	0.44	0.39	0.47
	$|E_{min}|$	0.28	0.39	0.26	0.25	0.19	0.10	0.01
	$|E_{max}|$	1.72	1.73	1.72	1.78	1.80	1.85	1.87

distributions. All these parameters are evaluated for the parallel and the perpendicular polarizations at selected angles of radiation.

For $\varphi = 0°$, all results for both parallel and perpendicular polarizations are the same. The dynamic range of the radiation pattern is very large (i.e. approximately from 0 to 1.9), depending on the angle of incidence. The average radiation pattern and standard deviation depend on the distance distribution, as well as on polarization. σ_{E_R} is always less than σ_{E_U} due to the fact that, in the Uniform Distribution, the antenna is changing position with the same probability, whereas in the Rayleigh Distribution distance changes less. For the parallel and the perpendicular polarizations, the maxima σ_{E_U} are 0.45 and 0.6, respectively, whereas the maxima σ_{E_R} are 0.35 and 0.5, respectively.

For the Uniform Distribution, the maximum standard deviation is almost 0.5 for muscle and 0.25 for fat, as shown in Figure 18.10. For φ less than Brewster's angle for parallel polarization, σ_{E_U} and σ_{E_R} for fat are two times less than for muscle. The minimum standard deviation for the parallel polarization is obtained always for radiation angles equal to the Brewster's angle, which depends on the electrical properties of body tissues. For fat and Rayleigh Distribution, the minimum value of the standard deviation is equal to 0.03 for $\varphi = 75°$.

The maximum radiation angle strongly depends on the dimensions of the body part, and also on the maximum distance from the antenna to the body, Figure 18.11.

Comparing the impact of the location of the antenna over torso, head and arm, the antenna is radiating up to 110°, 130° and 155°, respectively. For all positions on the body, the average radiation pattern, for angles up to the maximum radiation, changes from -3 dB to $+3$ dB. The maximum radiation is observed for 0°, a local minimum is observed for 90°, and sharp transition to 0 for the maximum radiation angle exists.

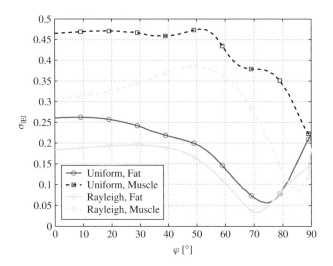

Figure 18.10 Comparison of standard deviation of E-field for parallel polarization, for fat and muscle tissues, for 2.45 GHz, and for the torso position. © 2010 IEEE. Reprinted, with permission, from [26].

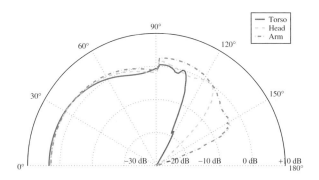

Figure 18.11 Comparison of average radiation pattern of E-field for Rayleigh Distribution, for various positions on the body (the radiation patterns are symmetric). © 2010 IEEE. Reprinted, with permission, from [26].

18.5 Body Coupling: Full Wave Simulations

18.5.1 Radiation Pattern Statistics for a Static Body

The CST full wave simulator is used in this study. The patch antenna has been modelled in free space and near to a female voxel model from Virtual Family [see Chapter 2], which was sliced into four parts to speed up simulations. The following on-body locations and antenna orientations were analyzed: HE_L, on the left side of the head; TO_F, on the front side of the chest; AB_L, on the left side of the arm's bottom part; LT_L, on the left side of the leg's upper part. The antenna was placed at different distances from the body

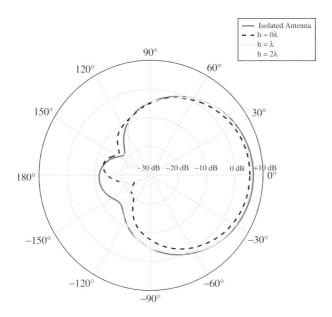

Figure 18.12 Antenna radiation pattern for various distances from the head. © 2011 IEEE. Reprinted, with permission, from [19].

surface according to two distance distributions: Uniform in the range $[0, 2\lambda]$ (i.e. $[0, 24.5]$ cm) and Rayleigh with a maximum probable distance of 2 cm.

Simulation results are composed of 3D radiation patterns, however, in here all radiation patterns are presented in the azimuth plane only, due to the lack of space. Nevertheless, the patterns differences previously defined were calculated for both planes.

The radiation patterns for the antenna located near the head for several distances, as well as the one for an isolated antenna, are presented on Figure 18.12.

The largest variations are observed in the direction of the head and to the sides. For the antenna attached to the head (i.e. $h = 0$), the antenna gain in the forward direction is the lowest, and also for all other distances it is less than the one for the isolated antenna.

In order to better investigate the change in the radiation pattern when the antenna is moving closer to the body, $\overline{\Delta_{G_w}}$ was calculated relative to the isolated antenna for all analyzed scenarios for the Uniform Distribution, Figure 18.13. For all body parts, the change of the radiation pattern is the strongest when the antenna is located at a null distance, because the coupling between the body and antenna is the highest, resulting in a significant reduction of antenna radiation efficiency, which affects the overall shape of the radiation pattern. In general, moving the antenna away from the body results in a smaller value for $\overline{\Delta_{G_w}}$; however, this is not a monotonous trend, especially in the case of the location on the chest.

The average radiation pattern G_μ, together with the standard deviation σ_G, are shown in Table 18.5, for selected angles, for all body parts.

The average radiation pattern in the forward direction (i.e. $\varphi = 0°$) varies slightly among body regions in $[6.1, 6.6]$ dBi. In the backward direction (i.e. $\varphi = 180°$), the

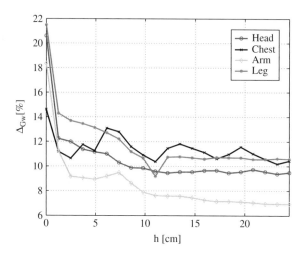

Figure 18.13 PWD for all analyzed body regions for a distance range of $[0, 2\lambda]$ at 2.45 GHz. © 2011 IEEE. Reprinted, with permission, from [19].

Table 18.5 G_μ and σ_G for selected φ and for $\theta = 90°$

	Distance distribution	φ	Head	Chest	Arm	Leg
		0°	6.2	6.5	6.2	6.1
	Uniform	90°	−4.9	−5.5	−4.8	−5.0
G_μ[dBi]		180°	−17.7	−22.8	−17.7	−21.4
		0°	6.3	6.6	6.4	6.2
	Rayleigh	90°	−5.6	−6.6	−6.1	−5.7
		180°	−19.2	−21.8	−16.6	−21.3
		0°	−6.1	−6.7	−4.6	−4.9
	Uniform	90°	−13.9	−13.6	−12.5	−13.1
σ_G[dB]		180°	−21.1	−26.1	−15.7	−24.3
		0°	−5.7	−7.7	−5.1	−4.2
	Rayleigh	90°	−16.5	−18.4	−13.4	−15.1
		180°	−22.0	−27.4	−15.3	−23.8

lowest radiation is obtained for chest and leg, due to the large attenuation of the body tissues. For all body regions, the highest σ_G is observed for the forward direction, and for both distance distributions for arm and leg, it is higher by 2 dB compared to other body regions. In the backward direction, independently of the distance distribution and of the body region, σ_G is very low, because the average radiation pattern takes low values.

The averages $\overline{\Delta_G}$ and $\overline{\Delta_{G_w}}$ are computed relative to the isolated antenna, for the analyzed positions of the antenna and both distributions, being presented in Table 18.6. For both distributions and all the considered scenarios, $\overline{\Delta_G}$ is higher than $\overline{\Delta_{G_w}}$. In the case of the Uniform Distribution, when the antenna is located at the maximum distance of 2λ, at $f = 2.45$ GHz (i.e. around 24.5 cm), $\overline{\Delta_{G_w}}$ and $\overline{\Delta_G}$ are lower compared to the

Table 18.6 PAD and PWD compared to the isolated antenna

	Distance distribution	Head	Chest	Arm	Leg
$\overline{\Delta_{G[\%]}}$	Uniform	12.1	22.2	9.7	16.0
	Rayleigh	18.1	24.3	18.7	22.9
$\overline{\Delta_{G_w[\%]}}$	Uniform	4.8	9.8	4.1	6.3
	Rayleigh	6.9	11.0	7.4	9.3

Rayleigh one, because the influence of the body is weaker when the antenna is moved away from the body.

18.5.2 Radiation Pattern Statistics for a Dynamic Body

Two scenarios are analyzed, Walk and Run, which are the most typical human day-to-day activities. Examples are given for fast and slow "movement of the body". Due to the periodic nature of the body dynamics (i.e. Run and Walk), the motion capture description contains only 30 looped time frames, with a duration of $1/30$ s each. These frames correspond to the duration of one walking or running period, starting with the right foot and the left hand forward.

Based on the statistical description of the antenna orientation, and the radiation pattern when the antenna is located on the arm on the static body (i.e. taken from the numerical simulations in CST), the statistical parameters of the radiation pattern (i.e. average G_μ, average and standard deviation $G_{\mu+\sigma}$, the minimum G_{min}, and maximum G_{max}) were obtained.

Figure 18.14 shows that the standard deviation of the radiation pattern is large, which is a result of the variation of the antenna orientation when it is placed on the arm for the Run scenario. As expected, on average, the gain has a maximum in the direction corresponding to the average shift in the normal to the body surface for the particular antenna placement. Table 18.7 presents the statistics in the direction of maximum radiation (i.e. $\overline{\delta\theta}$ and $\overline{\delta\varphi}$).

In the Walk scenario, the gain in the direction of maximum radiation is similar to the static one, ranging in $[6.1, 6.6]$ dBi. In general, the standard deviation in this direction is below -15 dB, which is more than 20 dB below the average value. In the Run scenario, the maximum gain for the arm placement is 4 dBi, which is more than 2 dB lower than in the similar Walk scenario. The standard deviation for the arm placement is very high, being 16 dB higher than in the case of the head, which is the most stable location for the antenna. Table 18.8 presents the statistics for selected directions of radiation.

On average, in the body forward direction (i.e. $\psi = 0°$ and $\varphi = 0°$), the highest gain is obtained when the antenna is placed on the chest. In the side direction (i.e. $\psi = 0°$ and $\varphi = 90°$) the highest gain is for the head and leg. For the arm placement, the average radiation is the highest in the backward direction (i.e. $\psi = 0°$ and $\varphi = 180°$).

When the antenna is placed on the head, the dynamic scenario almost does not change the radiation pattern. This placement results in the lowest standard deviation in the front and side directions. The standard deviation for both dynamic scenarios for the head and leg is very low, below -14 dB and -10 dB, respectively.

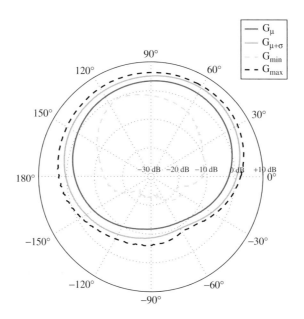

Figure 18.14 Statistics of the radiation pattern of the patch antenna placed on the arm for the Run scenario. © 2011 IEEE. Reprinted, with permission, from [19].

Table 18.7 G_μ and σ_G for the direction of maximum radiation $(\overline{\delta\theta}, \overline{\delta\varphi})$

	Movement	Head	Chest	Arm	Leg
G_μ[dBi]	Static	6.3	6.6	6.4	6.2
	Walk	6.3	6.6	6.3	6.1
	Run	6.3	6.3	4.0	6.1
σ_G[dB]	Walk	−18.2	−21.3	−14.7	−15.8
	Run	−16.4	−7.7	−0.4	−13.3

The average radiation patterns for the running body and all antenna placements, for the $\psi = 0°$ plane, are presented in Figure 18.15. When the antenna is located on the chest, the highest gain is in the front, while when it is located on the left side of the head, arm and leg, the radiation is in the left side direction, as expected. On average, when the antenna is located on the head or leg, the gain is higher than on the arm only for $\varphi = [45°, 150°]$. This means that the huge variation of the antenna orientation, in the case of the arm, results in a more uniform radiation pattern.

The average and standard deviation of the radiation pattern, in both planes (in the antenna coordinates, where $\varphi = 0°$ and $\psi = 0°$ corresponds to the direction of maximum radiation), for the Run scenario and arm placement is compared with the static one in Figure 18.16.

When the body is moving, the average radiation pattern is smoothed (i.e. gets more uniform). It is worth noticing that the standard deviation is the highest on the extremes

Table 18.8 G_μ and σ_G for selected φ and for $\psi = 0°$

	Movement	φ	Head	Chest	Arm	Leg
G_μ[dBi]		0°	−6.3	6.5	−5.2	−4.2
	Walk	90°	6.3	−6.7	4.4	6.0
		180°	−5.5	−20.6	−4.6	−7.3
		0°	−5.5	6.0	−2.4	−6.3
	Run	90°	6.3	−6.3	3.3	6.1
		180°	−6.1	−20.9	−2.3	−5.3
σ_G[dB]		0°	−18.5	−16.6	−13.7	−11.1
	Walk	90°	−18.6	−14.0	−5.0	−9.8
		180°	−17.8	−28.9	−13.7	−16.6
		0°	−14.1	−5.8	−2.8	−12.9
	Run	90°	−15.5	−7.9	0.2	−10.6
		180°	−15.9	−25.5	−2.1	−13.0

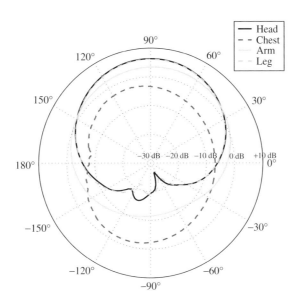

Figure 18.15 Comparison of average radiation patterns for the Run scenario. © 2011 IEEE. Reprinted, with permission, from [19].

of the radiation angles, and not in the direction of maximum radiation. The reason is that the antenna is oriented in this direction for a short time (i.e. the radiation is maximum), and then moves away from it (i.e. the radiation is low).

18.6 Body Coupling: Practical Experiments

A measurement campaign was carried out in an anechoic chamber. Two simple patch antennas were placed at 2 cm from the bodies of three human testers (a female and two

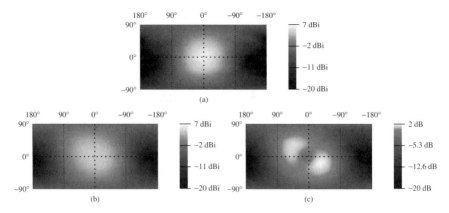

Figure 18.16 Antenna located on the arm. (a) static body, (b) average pattern for Run, (c) standard deviation for Run. © 2011 IEEE. Reprinted, with permission, from [19].

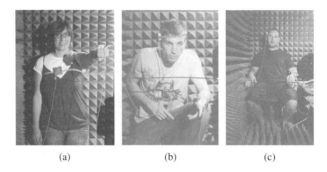

Figure 18.17 Body postures. (a) standing, (b) military, (c) sitting. Reproduced from [27] by permission of ICST.

males), radiating in one half of the hemisphere, with a 6.6 dBi gain and around 90° half power beam width. The transmitter (TX) patch was placed on the arm during all measurements, while the receiver (RX) patch was placed on the arm, the chest or the leg. Different on-body links were emulated corresponding to the standing, sitting and military postures, shown in Figure 18.17.

The transmitting power was set to 0 dBm, within the [2.4, 2.5] GHz band, with a 250 kHz step. Both TX and RX antennas were connected to the two ports of a vector network analyzer (Agilent E8361A), used to measure the s-matrix, with a data acquisition rate of 100 samples per measurement (an acceptable rate for statistical studies).

For each scenario, a matrix of connections was formed to identify whether a link is Quasi-Line-of-Sight (QLOS) or Non-LOS (NLOS). The QLOS condition is applied whenever the antennas do not have any obstacle between them, but takes into account that they are not aligned with each other in the maximum gain direction.

The analysis of the measurements is based on the extracted s parameters, namely on the input reflection loss, s_{11}, and on the transmission, s_{12}. The analysis of s_{11} is important for the link budget, as it quantifies the losses due to reflection or mismatch losses caused by the presence of the body. The s_{12} parameter quantifies the received power.

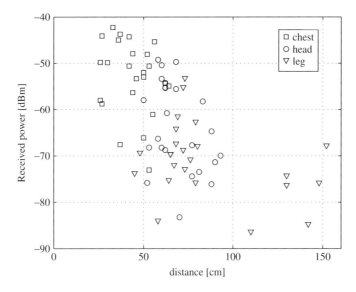

Figure 18.18 Overview of received power results (all postures). Reproduced from [27] by permission of ICST.

On average, the antenna performance on the body is 0.6 dB lower than in free space, which is reflected in a decrease of less than 1 percent on the antenna reflection efficiency. Additionally, there is a small detuning from free space resonance, which is less than 5 MHz, the practical implication being the increase of the standard deviation of the measured gain. At 2.45 GHz, s_{11} is lower than −8 dB, meaning that almost 85 percent of the input power is transmitted, which is in agreement with the specifications for short distance communications.

The military and sitting postures present a slightly worse antenna performance (around 0.5 dB) compared to the standing one. The standing posture presents a higher standard deviation for s_{11}, compared to the others, being especially sensitive near the resonant frequency.

Received power results in Figure 18.18 show that, on average, the chest-to-arm link results in a higher value than those obtained for the head-to-arm and leg-to-arm ones for all three postures, namely 18, 7 and 11 dB higher, for the military, standing and sitting postures, respectively. A plausible explanation could be that the low power levels are associated to with linking with visibility rather than distance between TX and RX. In addition, the obtained power levels were rather different (i.e. tens of dB of difference) even for similar lengths between TX and RX.

For the military posture, the dynamic range of both the head-to-arm and chest-to-arm links is as high as 15 dB. Flexing the knees produced an additional link variability of 7 dB for the chest-to-arm link. For the standing position, the dynamic range is even larger, for example, 33 dB for the leg-to-arm link. For the chest-to-arm link, it is also observed that the movement of the arm (up and down, and from the front to the back) results in a standard deviation of 8 dB, however, the largest variation, 34 dB, was obtained for the sitting posture and the head-to-arm link. Alternating the arm position between bend and straight down resulted in a standard deviation of 7 dB for the chest-to-arm link.

The large range of variations results from a mixture of different factors, such as the dielectric properties of the body and its posture. The amount of energy absorbed by the body depends on the position of the antenna and on the posture. Two individuals with exactly the same posture, having antennas in the same position, are likely to produce different results, mainly because of different body composition. However, the differences in received power can also be the result of hand rotation, hence, a channel can change from QLOS to NLOS very quickly, which results in increased path loss. An additional contributing factor is the polarization mismatch between TX and RX antennas.

An analysis of the CDF of the received power, corresponding to the different body postures, shows that the received power for the standing posture is 15 dB higher than the one at the military and the sitting postures with 50 percent probability.

The set of data formed with all the samples from all body postures was tested against the Normal Distribution using the χ^2 test. Results show that the hypothesis cannot be rejected at the 5 percent level in 98 percent of the cases. The estimated mean is −62 dBm, with a standard deviation of 11 dB.

Considering the matrix of connections, the CDF for NLOS and QLOS links is presented in Figure 18.19. These distributions were also tested against the Normal Distribution hypothesis using the χ^2 test. The hypothesis cannot be rejected at the 5 percent significance level, in 96 percent and 95 percent of the cases for the NLOS and QLOS links, respectively. The estimated mean and standard deviation for the NLOS links are −68 dBm and 6 dB, respectively; for the QLOS links, the corresponding results are −51 dBm and 6 dB, respectively. The average received power in QLOS is 15 dB higher than in NLOS, which can be decisive for the connectivity.

Figure 18.20 shows the received power as a function of the distance between the TX and the RX antennas, and also the range of variations within 3 cm long intervals, for the

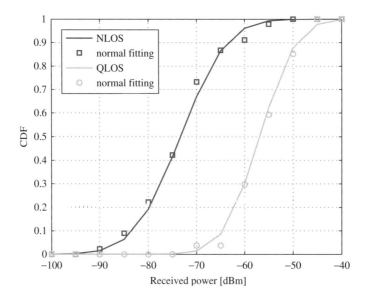

Figure 18.19 Overview of received power results (NLOS/QLOS). Reproduced from [27] by permission of ICST.

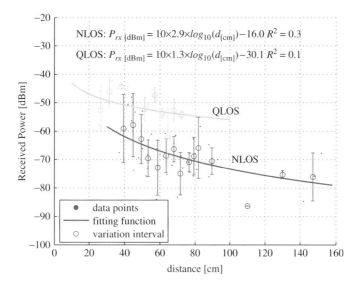

Figure 18.20 Overview of received power results (distance). Reproduced from [27] by permission of ICST.

QLOS and NLOS links. The standard deviation of the measured data is high for both links, being 8 dB for NLOS, which is 3 dB higher than in QLOS. The large variations of the received power are a direct result of the stochastic nature of the channel, which gives additional support to our modelling approach.

The usual average power decay model with distance was applied to the data, but a general poor correlation with this model was naturally observed (small R^2 values), due to the large variation of values.

A rather unexpected value for the path loss exponent equal 1.3 was obtained for the QLOS link, which is usually obtained in local guided propagation, like indoor corridor and outdoor street canyon environments. This is an interesting result that requires further investigation, as it may support the idea of a local guided propagation effect together with local (body) multipath and antenna coupling effects for short distance links, additional to potential effects of the radiation patterns.

Further studies of the scenario-based approach should be extended to comprise other relevant body postures, as well as other body dynamics and environments. In addition, the classification of the links should comprise several body segments, such as the front, the back and the two sides. These results can be used at the system design stage, to include, for example, more accurate fading margins.

18.7   Correlation Analysis for BANs

18.7.1   On-Body Communications

When researching on possible spatial diversity and multiplexing techniques for BANs, the spatial correlation among the multiple on-body links should be considered, as it is an important measure of the independence between links. In MIMO, the presence of orthogonal links is an important condition to get channel capacity gains. This section presents results for spatial correlation for on-body communications at 2.45 GHz.

This correlation study identifies and characterizes different classes of on-body links, giving an overview on the best positioned antennas to be selected for a cooperative MIMO scheme, where the available antennas cooperate together to form a virtual array. This is a simplified approach, as other issues must be considered, as power imbalance, for instance.

In our simulations, the CST software is used together with the 26-year-old female voxel model, no additional scatterers being considered other than the body. The BAN topology presented in Figure 18.1 is followed, except for the leg node.

In order to obtain results closer to reality, several distances of the antenna to the body are considered, instead of only one, as has been the case for other studies. The distance ranges from zero (antenna attached to the skin) to a maximum distance constrained by the flexibility of clothes (antenna attached to the clothes). A total number of 20 equidistant points was used.

The correlation between the links associated to antennas k and l, $\rho_{c_{k,l}}$, was computed via the received signals s_k and s_l. For the calculation of $\rho_{c_{k,l}}$, the signals are initially aligned (shifted) in time, through the cross-correlation function [20], so that the maximum value is obtained.

Each correlation pair can be classified according to its relative position to the excitation antenna, and to the orientation between RXs: SYM (nodes are symmetric to the TX), LOS (nodes in LOS), QLOS (nodes can be shaded by the body, being in LOS or NLOS) and NLOS (no LOS between nodes). For instance, if the TX is on the belt, the arm antennas belong to the SYM class, but, if the TX is on the right ear, then the arm antennas pair is classified as QLOS.

The correlation study was done by setting one of the antennas as the TX, and then computing the correlations between the 8 RXs, resulting in 28 correlation pairs. The study was conducted for the nine on-body antenna locations (Figure 18.1).

The correlation coefficients were obtained by computing the correlation matrix **R** for all possible links, considering the average correlation $\overline{\rho_c}$ and the corresponding standard deviation σ_ρ.

An overview of the global statistics for each class is presented in Table 18.9. The SYM nodes show the highest mean correlation values (0.9) and the smallest standard deviation of the variation with the distance to the body ($\sigma_\rho < 0.15$), which is intuitive, due to the symmetry of the body. The average correlation slightly decreases following the sequence LOS – QLOS – NLOS links, with higher variations for the LOS/QLOS classes.

The trend of the variation of correlation with distance, illustrated in Figure 18.21, shows that as antennas are moved away from the body, signals become more correlated; the standard deviation decreases with the antenna to body separation.

Table 18.9 Correlation statistics

SYM		LOS		QLOS		NLOS	
$\overline{\rho_c}$	σ_ρ	$\overline{\rho_c}$	σ_ρ	$\overline{\rho_c}$	σ_ρ	$\overline{\rho_c}$	σ_ρ
0.92	0.14	0.76	0.19	0.74	0.19	0.72	0.15

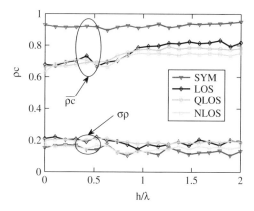

Figure 18.21 Trend of correlation statistics with distance.

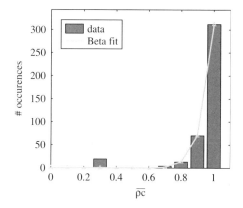

Figure 18.22 Global histogram of correlation with distance: SYM links.

A detailed study on the evolution of correlation with distance was also performed via the characterization of statistical distributions, that is, using the corresponding histograms. Figure 18.22 shows the results for the SYM scenario, with the Beta and Kumaraswamy Distributions fitting very well to this class ($R^2 = 1.0$). For LOS classes, these distributions have also a good fitting, with R^2 values around 0.9. For QLOS and NLOS links, better results are obtained for a Truncated Normal Distribution, although a bimodal approach could also be explored.

Figure 18.23 shows the average correlations when the TX antenna is placed on the belt (location WA_F in Figure 18.1). The highest correlation is obtained for SYM (with RX antennas placed on ears, HE_R/HE_L, or arms, AB_R/AB_L), while the lowest is observed for LOS (corresponding to the antennas in the back, TO_B/HE_B). The case for the latter seems counter-intuitive, since one would expect a high correlation in this situation, but it comes from the fact that these LOS links are on the back of the body, where signals from the TX are strongly attenuated.

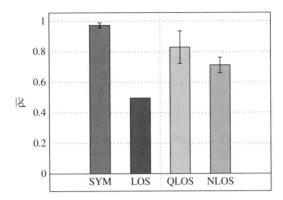

Figure 18.23 Correlation statistics for TX at the belt.

Based on the mean correlations and standard deviations for the considered on-body scenarios, one can conclude that the arms and the back are the best candidates for placing the antennas for MIMO BANs (correlation values around 0.7).

18.7.2 Off-Body Communications

This section studies the correlation between signals received at different antennas on the body, with a plane wave excitation coming from different directions, each wave representing a single multipath component. Note that in a multipath environment, the multiple copies of the signal can arrive from arbitrary directions, however, in here, one analyzes each component individually, in order to better understand the phenomena.

Due to the symmetry of the body, the correlation is studied for waves arriving from half-space of the azimuth plane (i.e. $[0°, 180°]$) with a resolution of $5°$. Various separations between the antenna and the body are considered, and the statistical variation of the correlation is presented.

This study can also provide information on the optimal location of the desired number of antennas on the body, based on the average signal cross-correlation between them.

For off-body communications, the signal is transmitted by an external antenna, propagates in the environment, and is captured by wearable antennas. The signals received at the various antennas on the body are necessarily different, since they are influenced by the electrical properties of different parts of the body. In order to evaluate this difference, the signals s_k and s_l received at different antennas are cross-correlated, as in the previous section.

The average cross-correlation and the standard deviation between the kth and l^{th} pair of antennas for a specific azimuth φ_m are:

$$\overline{\rho^{k,l}(\varphi_m)} = \frac{1}{N_k N_l} \sum_{k=1}^{N_k} \sum_{l=1}^{N_l} \rho^{k,l}(h_k, h_l, \varphi_m) \tag{18.15}$$

$$\sigma_\rho^{k,l}(\varphi_m) = \sqrt{\frac{1}{N_k N_l} \sum_{k=1}^{N_k} \sum_{l=1}^{N_l} (\rho^{k,l}(h_k, h_l, \varphi_m) - \overline{\rho^{k,l}(\varphi_m)})^2} \tag{18.16}$$

where:

- N_k and N_l: are the number of distance samples of kth and l^{th} antennas, respectively.
- h_k and h_l: are the distances from the body of kth and l^{th} antennas, respectively,
- φ_m: is the azimuth of the incoming wave.

As the cross-correlation depends on the mutual location of the antennas (i.e. Front, Back, Left and Right), the statistical analysis is performed for the following classes:

- Co-Directed (CD): both antennas are located on the same side.
- Cross-Directed (XD): one antenna is located on the back or in the front, whereas the other one is placed on the left or right.
- Opposite-Directed (OD): one antenna is located on the back and other one is placed in the front, or one antenna is located on the left and other one is placed on the right.

CST was run for 11 samples of the distance, which were calculated according to the Uniform Distribution in $[0, 2\lambda]$ (i.e. $[0, 24.5]$ cm). With an angle resolution of $5°$ over the half sphere (i.e. 36 angles) a total $11 \times 36 = 396$ cases were simulated. Each simulation took about 15 minutes using an Intel Xeon X5600 4xQuad-Core workstation.

Firstly, the average and the standard deviation of cross-correlation have been calculated for all possible combinations of pairs of antennas. In Table 18.10, the statistics of the cross-correlation and the allocation to the class are presented. The proposed classification of the mutual orientation of antennas is in agreement with the values obtained for the average correlation: for any pair of antennas coming from the OD class, the average cross-correlation is less than for any other coming from the XD one (the same is observed for XD and CD classes). Due to similar conditions of the links, the CD pairs of antennas are characterized by high values of cross-correlation, but, being quite lower than 1 as a result of coupling and interactions between the antenna and the body.

Table 18.10 The matrix with the definition of the classes, and simulated values of the average and standard deviation of cross-correlation

		\multicolumn{9}{Average correlation}							
		TO_F	WA_F	HE_F	HE_B	HE_L	HE_R	AB_L	AB_R	TO_B
Standard deviation	TO_F		0.97	0.95	**0.65**	0.77	0.81	0.78	0.79	**0.66**
	WA_F	0.03		0.95	**0.65**	0.77	0.81	0.79	0.79	**0.65**
	HE_F	0.07	0.06		**0.69**	0.77	0.81	0.78	0.79	**0.69**
	HE_B	**0.09**	**0.10**	**0.06**		0.76	0.78	0.77	0.80	0.94
	HE_L	0.15	0.15	0.15	0.14		**0.70**	0.96	**0.69**	0.77
	HE_R	0.12	0.12	0.12	0.13	**0.08**		**0.69**	0.95	0.77
	AB_L	0.14	0.14	0.14	0.14	0.06	**0.08**		**0.69**	0.77
	AB_R	0.12	0.12	0.12	0.12	0.10	0.08	**0.09**		0.77
	TO_B	**0.10**	**0.11**	**0.08**	0.16	0.14	0.13	0.14	0.13	

CD co-directed   XD cross-directed   **OD** oposit-directed

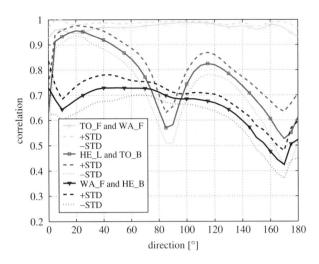

Figure 18.24 The average and standard deviation of cross-correlation for a selected pairs of antennas (i.e. TO_F & WA_F, HE_L & TO_B and WA_F & HE_B).

As expected, the cross-correlation is the lowest for antennas located on different sides of the body (i.e. OD), equal to 0.67, and the highest for antennas on the same side (i.e. CD), equal to 0.95. The largest variations in average cross-correlation are observed for the XD class, which is also characterized by the highest global (i.e. performed over all cases) average standard deviation, being equal to 0.13.

The statistics of cross-correlation for a selected pair of antennas (i.e. TO_F & WA_F, HE_L & TO_B and WA_F & HE_B), which are representative from all classes, depending on the azimuth of the wave, are presented in Figure 18.24.

The average cross-correlation for pair TO_F & WA_F is very high, above 0.93, with a very small standard deviation, especially in the side direction (i.e. 90°). In general, for the pairs of antennas located on the same side of the body, the average cross-correlation does not depend on the direction of the wave. In the case of the HE_L & TO_B pair, the cross-correlation depends on the direction of the wave and varies a lot. The lowest values of cross-correlation are obtained for the side and back directions, when one antenna is in LOS conditions and the other is shadowed. For the pair WA_F & HE_B, the lowest average cross-correlation is obtained for the 170° direction, and there is also a local minimum at 10°. In the case of 170°, the antenna HE_B is in LOS, whereas the other WA_F is shadowed by the torso. The attenuation and distortion caused by the head is weaker than the one from the torso, which is the reason why cross-correlation for 10° is higher than for 170°.

18.8   Summary

This chapter presents a new characterization of radio channels in BANs, adopting a statistical concept to characterize the user's influence. The statistical modelling of antennas in BANs can be separated into antennas in the vicinity of the body (including body dynamics) and propagation environment. The presented analysis starts from basic theoretical considerations, then presents full wave simulations enhanced by a realistic model of body movement, and is complemented with measurements. Also, correlation aspects for on- and off-body communications are analyzed.

Preliminary calculations with basic theoretical models show that the body causes important changes in the antenna radiation pattern. This reinforces the pertinence of the proposed concept of an antenna statistical model describing the changes in its overall performance (e.g. radiation pattern, efficiency, and input impedance), which can then be included in standards and radio channel simulators. Full wave simulations show that the relative change of the average radiation pattern for an antenna located on the chest can reach 24 percent, relative to the isolated antenna. When the antenna is placed on a very dynamic body segment (e.g. arm placement in a running scenario), the maximum gain of average radiation pattern is reduced. Moreover, the radiation pattern shape is more uniform relative to the static scenario, and the standard deviation is very high.

On-body measurements show a dynamic range of 33 dB for path links with the same length, as a consequence of antenna misalignment, which can turn a QLOS link into an NLOS one, increasing path loss up to 15 dB. This effect must be mitigated and taken into account in link calculations for BANs, as connectivity should be assured.

The study of spatial correlation among on- and off-body links gives important information for the use of MIMO. Correlation between on-body links depends on the orientation between RX antennas, and also on their position relative to the TX one. Different classes can be defined and statistically characterized. It is found that a symmetrical placement of the RX antennas relative to the TX one is not favorable for MIMO due to high correlation (values round 0.9). The Beta and the Kumaraswamy probability Distributions are good fits to the SYM (symmetric) and LOS links, for NLOS links the Truncated Normal Distribution is the best fit, while for QLOS a bimodal probability distribution should be studied. In off-body communications, the pairs of antennas can be split into three classes: co-directed (CD), cross-directed (XD) and opposite-directed (OD). As expected, the correlation is the lowest for antennas located on different sides of the body (i.e. OD), equal to 0.67, and is the highest for the antennas on the same side (i.e. CD), equal to 0.95. The direction of the incoming wave has a more significant impact on the correlation than the distance between the antenna and the body.

Acknowledgements

This work was partially funded by Fundacao para a Ciencia e a Tecnologia under grant SFRH/BD/46378/2008.

References

1. D. Ladeira, L. M. Correia, and S. Sweet, "Strategic Applications Agenda," eMobility Working Group on Leading Edge Applications, Tech. Rep., Jan. 2010. [Online]. Available: http://www.emobility.eu.org/SAA/eMobility_SAA_v3.01.pdf
2. S. Park and S. Jayaraman, "Enhancing the Quality of Life Through Wearable Technology," *IEEE Engineering in Medicine and Biology Magazine*, vol. 22, no. 3, pp. 41–48, May 2003.
3. J. W. Hines, "Medical and surgical applications of space biosensor technology," *Acta Astronaut*, vol. 38, no. 4-8, pp. 261–267, Feb. 1996.
4. G. Hiertz, D. Denteneer, L. Stibor, Y. Zang, X. P. Costa, and B. Walke, "The IEEE 802.11 universe," *IEEE Communications Magazine*, vol. 48, no. 1, pp. 62–70, Jan. 2010.

5. (2010, June) Bluetooth Core Specification v4.0, Bluetooth SIG. [Online]. Available: http://www.blue tooth.com

6. E. Karapistoli, I. Gragopoulos, I. Tsetsinas, and F. N. Pavlidou, "An overview of the IEEE 802.15.4a standard," *IEEE Communications Magazine*, vol. 48, no. 1, pp. 47–53, Jan. 2010.

7. A. Wheeler, "Commercial applications of wireless sensor networks using Zigbee," *IEEE Communications Magazine*, vol. 45, no. 4, pp. 70–77, Apr. 2007.

8. (2011, Dec.) BANET Project. [Online]. Available: http://www.banet.fr

9. (2011, Jan.) IEEE 802.15.6. [Online]. Available: http://www.ieee802.org/15/pub/TG6.html

10. A. Sibille, "Statistical antenna modeling," in *29th URSI General Assembly*, Chicago, IL, USA, Aug. 2008.

11. C. Gabriel, "Compilation of dielectric properties of body tissues at rf and microwave frequencies," Brooks Air Force, AL/OE-TR-1996-0037, San Antonio, TX, USA, Tech. Rep., Jan. 1996.

12. M. Henneberg and S. Ulijaszek, "Body frame dimensions are related to obesity and fatness: lean trunk size, skinfolds, and body mass index," *American Journal of Human Biology*, vol. 22, no. 1, pp. 83–91, Apr. 2009.

13. A. S. Glassner, *An Introduction to Ray Tracing*. New York, NY, USA: Academic Press, 1989.

14. P. H. Pathak, "Uniform geometrical theory of diffraction," in *Proc. of Applications of Mathematics in Modern Optics*, San Diego, CA, USA, Aug. 1982.

15. M. Sadiku, *Numerical Techniques in Electromagnetics*. Prairie View, TX, USA: CRC Press, 2000.

16. (2010, Oct.) Computer simulation technology. [Online]. Available: http://www.cst.com

17. A. Christ *et al.*, "The virtual family-development of surface-based anatomical models of two adults and two children for dosimetric simulations," *Physics in Medicine and Biology*, vol. 55, no. 2, pp. 23–38, Jan. 2010.

18. C. R. Medeiros, A. M. Castela, J. R. Costa, and C. A. Fernandes, "Evaluation of modelling accuracy of reconfigurable patch antennas," in *Proc. of ConfTele-Conference on Telecommunications*, Peniche, Portugal, May 2007.

19. M. Mackowiak, C. Oliveira, C. G. Lopes, and L. M. Correia, "A Statistical Analysis of the Influence of the Human Body on the Radiation Pattern of Wearable Antennas", in *Proc. of PIMRC 2011 – 22nd IEEE International Symposium on Personal, Indoor and Mobile Radio Communications*, Toronto, Canada, Sep. 2011.

20. S. J. Orfanidis, *Optimum Signal Processing, An Introduction*. Englewood Cliffs, NJ, USA: Prentice-Hall, 1996.

21. M. Abramowitz and I. Stegun, *Handbook of Mathematical Functions: with Formulas, Graphs, and Mathematical Tables*. New York, NY, USA: Dover Publications, 1965.

22. P. Kumaraswamy, "A generalized probability density function for double-bounded random processes," *Journal of Hydrology*, vol. 46, no. 1-2, pp. 79–88, Mar. 1980.

23. C. Robert, "Simulation of truncated normal variables," *Statistics and Computing*, vol. 5, no. 2, pp. 121–125, June 1995.

24. R. Lytle, "Far-field patterns of point sources operated in the presence of dielectric circular cylinders," *IEEE Transactions on Antennas and Propagation*, vol. 19, no. 5, pp. 618–621, Sep. 1971.

25. C. Oliveira and L. M. Correia, "A Statistical Model to Characterize User Influence in Body Area Networks," in *Proc. of VTC'2010 Fall – 72nd IEEE Vehicular Technology Conference*, Ottawa, Ontario, Canada, Sep. 2010.

26. M. Mackowiak and L. M. Correia, "A Statistical Approach to Model Antenna Radiation Patterns in Off-Body Radio Channels," in *Proc. of PIMRC 2010 – 21st IEEE International Symposium on Personal, Indoor and Mobile Radio Communications*, Istanbul, Turkey, Sep. 2010.

27. C. Oliveira, M. Mackowiak, C. Lopes, and L. M. Correia, "Characterisation of On-Body Communications at 2.45 GHz," in *Proc of BodyNets - International Conference on Body Area Networks*, Beijing, China, Nov. 2011.

Index

3D models, 91, 302, 308

Accuracy of models, 312, 339, 343, 349, 352, 383
Anechoic chamber, 221
Angular dispersion, 41
Angular distribution, 87, 417
Angular spread, 426
Antenna array, 241
Antenna design, 473
Antenna gain, 36
Antenna modeling, 356
Antenna pattern, 119, 142, 301, 436, 502
AWGN, 394

Beamforming, 403
Bit error rate, 235, 379, 393, 405
Body Area Network, 495, 497, 513
Body Area Network channel, 247, 495, 511
Box-Muller algorithm, 393
Branch power ratio, 482
Building penetration, 79, 132

Calibration, 143, 300, 339, 341, 367, 368
Capacity, 11
Carrier aggregation, 15
Cell edge user, 9
Cell spectral efficiency, 8
Channel emulato, 389
Channel selectivity, 107

Channel sounder, 155, 193
Cholesky factorization, 400
Cluster, 53, 108, 220, 416
Cognitive radio, 443, 446, 461
Confined scenario, 230
Cooperative coding, 182
Coordinated multi-point, 21
COST2100 model, 200
COST231 model, 117, 131, 330
Cross polarization, 88

Delay dispersion, 40
Delay spread, 90
Deterministic models, 41, 51, 127, 271, 322, 332–3
Deygout method, 330
Diffraction, 37, 276, 329
Diffuse scattering, 278
Direction of arrival, 44
Direction of departure, 44
Directivity, 475
Dominant path, 140
Doppler spectrum, 49, 434
Doppler spread, 114, 167

Empirical models, 127, 323, 325
Erceg's model, 325

Fade depth, 228, 231
Fading, 83, 125, 238, 305, 389
Fading generator, 394, 422
FDTD, 138, 294, 299
Ferrite antennas, 487

LTE-Advanced and Next Generation Wireless Networks: Channel Modelling and Propagation, First Edition.
Edited by Guillaume de la Roche, Andrés Alayón Glazunov and Ben Allen.
© 2013 John Wiley & Sons, Ltd. Published 2013 by John Wiley & Sons, Ltd.

Field synthesis, 426
Finite difference methods, 293
FIR Filter, 401
First order statistics, 103
Folded antennas, 486
Free space, 36, 221, 275
Frequency dependency, 217
Frequency dispersion, 40
Frequency selective fading, 226
Friis equation, 69

Gaussian random variables, 393
Geographic data, 317, 320
Geometrical optics, 174, 273, 416, 418

Home eNodeB, 26
Hybrid models, 142, 286, 309

Impulse response, 44, 56, 219
IMT-Advanced, 5, 7, 350
IMT-Advanced Channel, 357, 422
Indoor channel, 67–8, 76, 222, 305
Indoor to Outdoor, 310
Indoor to outdoor channel, 123
Interference management, 469

Jakes' model, 395–6, 417
Joint dispersion, 55

Kirchhoff theory, 39
Kronecker model, 399

Laptop antennas, 489
Large scale parameters, 69, 117, 167, 216, 361
Link level simulation, 372
LTE, 3, 123
LTE-Advanced, 3

M2M, 28
Maxwell's equations, 272, 294, 295
Mean effective gain, 480
Measurements, 222, 311, 340, 412, 506
MIMO channel, 44, 57, 91, 216, 375, 389, 390, 401, 411

MIMO channels, 83
MIMO over-the-air testing, 411
MIMO technique, 19, 239
MIMO-OFDMA, 389
Mobility, 10
MR-FDPF, 297, 310
Multi floor, 86, 303
Multi slope model, 70
Multi-probe systems, 421
Multipath, 40, 60, 100, 191, 220, 223
Multiple bounce scattering, 59

Nakagami distribution, 233

OFDM, 180, 241
OTA testing, 412–14
Outdoor channel, 97, 108
Outdoor to indoor, 310, 339
Outdoor to indoor channel, 130

ParFlow, 295, 296
Path loss, 117, 125, 166, 303
Path loss models, 360
Peak data rate, 13
Perturbation theory, 39
Polarization, 41, 45, 119, 479, 485, 502
Polarization power balance, 434
Power delay profile, 108, 165, 435
PRNG, 392
Probability distribution, 232
Pseudo random noise generator, 392

Random walk model, 397
Range dependency, 217
Ray launching, 52, 282, 310, 334
Ray tracing, 52, 138, 271, 280, 283, 298, 333
Rayleigh fading, 105, 432
Reflection, 36, 275, 476
Relay, 23, 181
Reverberation chamber, 413, 429
Rician distribution, 361
Rician fading, 105, 432
Rician K-Factor, 398
RMS delay, 85, 167

Scatterer, 175
Scattering, 38, 110, 277
SCE model, 422
SCM model, 422
Second order statistics, 106, 111
Security, 463
Self ptimizing networks, 29
Semi empirical model, 136
Semi-empirial models, 326
Shadowing, 47
Single bounce scattering, 59
Small cell antennas, 492
Small scale parameters, 48, 83, 103, 363
Spectrum, 5
Spectrum assignment, 452
Spectrum management, 444
Spectrum sensing, 448
Stationarity, 168
Statistical distribution, 500

Stochastic channel, 46
System level simulation, 350, 372

Time varying channel, 42, 157, 160
Transmission, 36, 275

Ultra wideband channel, 46, 215, 217, 244, 287

Vegetation, 277
Vehicular antenna, 157
Vehicular channel, 153, 159, 180

Walfish-Ikegami, 117
Waveguiding, 39, 332
Wideband channel, 155, 215, 223, 314, 337
WINNER model, 67, 200, 422

Zheng model, 396